T0180401

Lecture Notes in Computer Science 13593

Oscar Camara · Esther Puyol-Antón · Chen Qin ·
Maxime Sermesant · Avan Suinesiaputra ·
Shuo Wang · Alistair Young (Eds.)

Statistical Atlases and Computational Models of the Heart

Regular and CMRxMotion Challenge Papers

13th International Workshop, STACOM 2022
Held in Conjunction with MICCAI 2022
Singapore, September 18, 2022
Revised Selected Papers

 Springer

Editors
Oscar Camara ⓘ
Pompeu Fabra University
Barcelona, Spain

Esther Puyol-Antón ⓘ
King's College London
London, UK

Chen Qin
Imperial College London
London, UK

Maxime Sermesant ⓘ
Inria
Sophia Antipolis, France

Avan Suinesiaputra ⓘ
King's College London
London, UK

Shuo Wang
Fudan University
Shanghai, China

Alistair Young ⓘ
King's College London
London, UK

ISSN 0302-9743 ISSN 1611-3349 (electronic)
Lecture Notes in Computer Science
ISBN 978-3-031-23442-2 ISBN 978-3-031-23443-9 (eBook)
https://doi.org/10.1007/978-3-031-23443-9

This Springer imprint is published by the registered company Springer Nature Switzerland AG
The registered company address is: Gewerbestrasse 11, 6330 Cham, Switzerland

Preface

Cardiac image analysis has advanced significantly in recent years. Progress made in machine learning, particularly deep learning networks, has been gradually integrated with cardiac atlases and computational modelling of the heart to improve our understanding of the mechanism of heart diseases, treatment evaluation, and intervention planning.

However, significant clinical translation of these methods is constrained by the lack of complete and rigorous technical and clinical validation, as well as benchmarking with common data. This can be achieved through collaboration across the full research scope of cardiac imaging and modelling communities, which has been provided by the Statistical Atlases and Computational Modelling of the Heart (STACOM) workshop.

The 13th edition of STACOM (https://stacom.github.io/stacom2022/), was held in conjunction with the 25th MICCAI conference in Singapore on 18 September 2022. It followed the twelve successful previous editions (https://stacom.github.io/): STACOM 2010 (Beijing, China), STACOM 2011 (Toronto, Canada), STACOM 2012 (Nice, France), STACOM 2013 (Nagoya, Japan), STACOM 2014 (Boston, USA), STACOM 2015 (Munich, Germany), STACOM 2016 (Athens, Greece), STACOM 2017 (Quebec City, Canada), STACOM 2018 (Granada, Spain), STACOM 2019 (Shenzhen, China), STACOM 2020 (Lima, Peru), and STACOM 2021 (Strasbourg, France). Throughout these thirteen years, the STACOM workshop has provided a forum to discuss the latest developments in various areas of computational cardiac imaging, cardiac modelling, application of artificial intelligence and machine learning to cardiac image analysis, electro-mechanical modelling of the heart, and novel methods in preclinical/clinical imaging for tissue characterization and image reconstruction, as well as statistical cardiac atlases.

The STACOM 2022 workshop attracted 34 paper submissions, which was an increase from 25 papers in 2021. All submissions were accepted for presentation at the workshop and publication in this Proceedings following a double blind peer review process in which each submission received at least two reviews. The objectivity of the review process was ensured that all papers were reviewed by external reviewers, i.e., none of the reviewers reviewed papers from their own institution. The acceptance decision was made based on the total score of reviewers. Papers published in this Proceedings have been revised based on the reviewers' comments and feedback from the conference.

Topics varied from common cardiac segmentation and modelling problems to more advanced generative modelling for ageing hearts, learning cardiac motion using biomechanical networks, physics-informed neural networks for left atrial appendage occlusion, biventricular mechanics for Tetralogy of Fallot, ventricular arrhythmia prediction by using graph convolutional networks, and deeper analysis of racial and sex biases from machine learning-based cardiac segmentation. The workshop awarded the best oral and the best poster presenters as follows:

1. The best oral presenter award went to Jadie Adams, Nawazish Khan, Alan Morris, and Shireen Elhabian for the paper entitled *"Spatiotemporal Cardiac Statistical Shape Modelling: A Data-Driven Approach"*.
2. The best poster presenter award went to Dieuwertje Alblas, Christoph Brune, Kak Khee Yeung, and Jelmer M. Wolterink for the paper entitled *"Going Off-Grid: Continuous Implicit Neural Representations for 3D Vascular Modeling"*.

Besides the regular contributing papers, STACOM 2022 also hosted two challenges. One of the challenges, the CMRxMotion challenge, is included in this proceedings. The CMRxMotion challenge was aimed at assessing the effects of respiratory motion on cardiac MRI (CMR) imaging quality and examining the robustness of segmentation models in the face of respiratory motion artefacts. A total of 40 healthy volunteers were recruited to have a CMR exam with four different respiratory conditions: a) adhere to the breath-hold instructions; b) halve the breath-hold period; c) breathe freely, and d) breathe intensively. Two tasks were defined: 1) to assess the image quality and 2) to develop a segmentation method that is robust to respiratory motion artefacts. The challenge attracted 14 submissions that are included in this proceedings. All papers were reviewed in blinded fashion by 1–2 reviewers. Papers were revised based on reviewer feedback and also feedback from the conference.

We would like to express our gratitude to the external reviewers from King's College London (UK), Inria (France), the University of Oxford (UK), Fudan University (China), Stanford University (USA), and Universitat Pompeu Fabra (Spain), who volunteered their time to meticulously review papers in this proceedings. Finally, we would like to thank all authors who participated in this workshop.

September 2022

Esther Puyol-Antón
Oscar Camara
Chen Qin
Maxime Sermesant
Avan Suinesiaputra
Shuo Wang
Alistair Young

Organization

Program Committee

Camara, Oscar	Universitat Pompeu Fabra, Spain
Puyol-Antón, Esther	King's College London, UK
Qin, Chen	Imperial College London, UK
Sermesant, Maxime	Inria, France
Suinesiaputra, Avan	King's College London, UK
Wang, Shuo	Fudan University, China
Young, Alistair	King's College London, UK

External Reviewers

Dawood, Tareen	King's College London, UK
Harrison, Josquin	Inria, France
Karoui, Amel	Inria, France
Kashtanova, Victoriya	Inria, France
Ly, Buntheng	Inria, France
Li, Lei	University of Oxford, UK
Machado, Ines	King's College London, UK
Puyol-Anton, Esther	King's College London, UK
Rodríguez Padilla, Jesus Jairo	Inria, France
Ugurlu, Devran	King's College London, UK
Wang, Kang	Fudan University, China
Wang, Haoran	Fudan University, China
Wang, Vicky	Stanford University, USA
Xu, Hao	King's College London, UK
Yang, Yingyu	Inria, France

Contents

Regular Papers

Generative Modelling of the Ageing Heart with Cross-Sectional Imaging and Clinical Data

Mengyun Qiao[1,2,3(✉)], Berke Doga Basaran[1,2], Huaqi Qiu[1], Shuo Wang[7,8], Yi Guo[6], Yuanyuan Wang[6], Paul M. Matthews[3,4], Daniel Rueckert[1,5], and Wenjia Bai[1,2,3]

[1] Department of Computing, Imperial College London, London, UK
mq21@ic.ac.uk
[2] Data Science Institute, Imperial College London, London, UK
[3] Department of Brain Sciences, Imperial College London, London, UK
[4] UK Dementia Research Institute, Imperial College London, London, UK
[5] Klinikum rechts der Isar, Technical University of Munich, Munich, Germany
[6] Department of Electronic Engineering, Fudan University, Shanghai, China
[7] Digital Medical Research Center, School of Basic Medical Sciences, Fudan University, Shanghai, China
[8] Shanghai Key Laboratory of MICCAI, Shanghai, China

Abstract. Cardiovascular disease, the leading cause of death globally, is an age-related disease. Understanding the morphological and functional changes of the heart during ageing is a key scientific question, the answer to which will help us define important risk factors of cardiovascular disease and monitor disease progression. In this work, we propose a novel conditional generative model to describe the changes of 3D anatomy of the heart during ageing. The proposed model is flexible and allows integration of multiple clinical factors (e.g. age, gender) into the generating process. We train the model on a large-scale cross-sectional dataset of cardiac anatomies and evaluate on both cross-sectional and longitudinal datasets. The model demonstrates excellent performance in predicting the longitudinal evolution of the ageing heart and modelling its data distribution. The codes are available at https://github.com/MengyunQ/AgeHeart.

Keywords: Heart ageing · Conditional generative model · Cardiac anatomy modelling

1 Introduction

Heart ageing is a predominant risk factor of cardiovascular diseases. Understanding how ageing affects the shape and function of the heart is a key scientific question that has received substantial attention [4–6]. Due to the high dimensionality of the 3D cardiac shape data, researchers and clinicians often describe the anatomical shape using global metrics such as the volumes or ejection fraction. However, these metrics cannot reflect detailed information of local

O. Camara et al. (Eds.): STACOM 2022, LNCS 13593, pp. 3–12, 2022.
https://doi.org/10.1007/978-3-031-23443-9_1

shape variations. Describing the high-dimensional spatio-temporal anatomy of the heart and its evolution during ageing is still a challenging problem.

In this work, we propose a novel conditional generative model for the ageing heart, which describes the variations of its 3D cardiac anatomy, as well as its associations with age. The model is trained on a large-scale cross-sectional dataset with both cardiac anatomies and non-imaging clinical information. Once trained, given a cardiac anatomy and a target age, the model can perform counterfactual inference and predict the anatomical appearance of the heart at the target age. By evaluating on both cross-sectional and longitudinal datasets, we demonstrate that the predicted anatomies are highly realistic and consistent with real data distribution. The model has the potential to be applied to downstream tasks for cardiac imaging research, such as for analysing of the ageing impact on the anatomy, synthesising shapes for biomechanical modelling and performing data augmentation. The codes are available at https://github.com/MengyunQ/AgeHeart.

1.1   Related Work

Numerous efforts have been devoted into conditional generative modelling and synthesis of ageing. In this work, we focus on heart ageing synthesis using conditional generative modelling techniques. Existing literature can be broadly classified into the following two categories:

Generative Modelling. The field of generative modelling has made tremendous progress recently, driven by deep learning methods such as variational autoencoders (VAEs) [10,16], generative adversarial networks (GANs) [12], cycle-consistent GAN (CycleGAN) [23]. Generative models have been widely used in medical imaging. For example, Wang et al. proposed a CycleGAN-based model for cross-domain image generation, which generates pseudo-CT for PET-MR attenuation correction [17]. Yurt et al. proposed a multi-stream GAN architecture for multi-contrast MRI synthesis [21]. Pawlowski et al. formulated a structural causal model with deep learning components for synthesising and counterfactual inference of MNIST and brain MR images [15].

Synthesis of Ageing. Most ageing synthesis works focused on images of human face while some works explored the synthesis of brain MR images. These works investigated different ways of incorporating age information into the generating process. One way is to concatenate age vector with image feature vector to learn a joint distribution of age and image appearance in face ageing [2,22] or brain ageing [20]. Another way is to use a pre-trained age regression network, which provides guidance in age-related latent code generation [1,19]. Some works introduced an age estimation loss accounting for age distribution [7,13]. In [11], high-order interactions between the given identity and target age were explored to learn personalized age features. Although these methods are not designed for cardiac imaging, they provide valuable insights for modelling the ageing heart.

1.2   Contributions

The contributions of this work are three-fold: 1) We investigate the challenging problem of heart ageing synthesis, where both the structural variation and functional variation (anatomies in different time frames) need to be modelled. To this end, we develop a novel model which consists of an anatomy encoder and a condition mapping network that disentangles age and spatial-temporal shape information in the generating process; 2) We utilise multi-modal information including both imaging data and non-imaging clinical data so that the generative model can account for the impact of multiple clinical factors on the ageing heart; 3) We train the generative model using a large-scale cross-sectional datasets and demonstrate its performance quantitatively on both cross-sectional and longitudinal datasets. To the best of our knowledge, this is the first work to investigate generative modelling for ageing heart synthesis.

2   Methods

2.1   Problem Formulation

Figure 1 illustrates the proposed generative model. At the inference stage (Fig. 1 right), given a source cardiac anatomy image I_s with its clinical information (source age: a_s, gender: g) and the target age a_t, the network synthesises the cardiac anatomy $I'_t = G(I_s, a_t, g)$ conditioned on the target age a_t. Thus, the distribution of the synthetic anatomy approximates the distribution of the real data at the target age while maintaining subject-specific structure in the ageing process. Our model utilises a heart anatomy generator G, which consists of an encoder E, an anatomy decoder D and a condition mapping network M. During the training stage, the generator G learns the evolution of the anatomy I_s from the source age a_s to the anatomy I_t at target age a_t and vice versa in a cyclic manner.

2.2   Conditional Generative Modelling

Clinical Condition Incorporation. We incorporate two major clinical conditions, age and gender, into the generative model. The age space A is represented using a $m \times 1$ categorical vector, where m denotes the number of age groups. For age group i, an age vector $a_i \in A$ is generated:

$$a_i = \mathrm{A_i} + \varepsilon, \varepsilon \sim \mathcal{N}(0, \ \sigma^2) \tag{1}$$

where $\mathrm{A_i}$ is a one-hot encoded vector that contains one at the ith-element and zeros elsewhere and ε is random noise sampled from a prior distribution. The age vector a_i is concatenated with a one-hot gender vector g to form the clinical condition c.

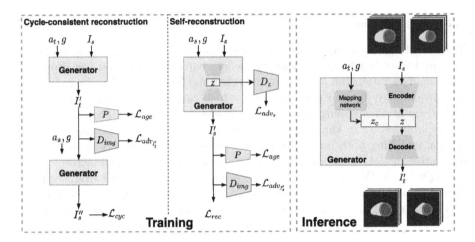

Fig. 1. The proposed generative model for the ageing heart. **Training:** The training scheme includes two parts: cycle-consistent reconstruction, which takes source image I_s at source age a_s as input, synthesise I_t' at target age a_t and then back to I_s'' at age a_s; self-reconstruction, which takes source image I_s as input and reconstruct I_s' at the same age. **Inference:** The Generator takes input image I_s at source age a_s, together with target age a_t and gender g, and generates the image I_t at target age a_t. Please refer to the text for detail.

Condition Mapping Network. Inspired by [8], we construct a condition mapping network $M(a, g)$ using a multi-layer perception (MLP). It embeds the input clinical condition c including age a and gender g to latent vector z_c in the conditional latent space. This latent representation integrates different clinical factors and enables exploration across the condition space.

Age Predictor. We construct an age predictor P to help the generative model focus on the age information in learning. P takes a cardiac anatomy as input and predicts its age. It is applied only at the training stage for both cycle-consistent reconstruction and self-reconstruction as shown in Fig. 1. We impose a distance loss between the predicted age $a_t' = P(I_t')$ and target age a_t, as well as between $a_s' = P(I_s')$ and a_s to guide the age prediction. The age predictor is implemented as a six-layer 3D convolutional network, followed by a fully connected layer to produce an age vector.

Anatomy Generator. The input to the generator G includes the source anatomy I_s and the condition code z_c generated from the condition mapping network $M(a, g)$. The generator follows an encoder-decoder structure, where the encoder E maps the input I_s into a subject-specific latent code z, the decoder concatenates z with the clinical condition code z_c and generates an output anatomy I_t'. The generation is described by,

$$I_t' = G(I_s, a_s, g) = D(E(I_s), M(a_s, g)) \tag{2}$$

We assume that the encoder $E(I_s)$ preserves the high-level subject-specific feature of the input anatomy I_s and the decoder D utilises this information as well as the clinical information to generate the anatomy I'_t. Adopting a cyclic design [23], we also generate an image $I''_s = G(I'_t, a_s, g)$ that maps the generated the image I'_t back to the source age a_s. This cycle-consistent generation is only applied in training process.

Discriminators. Two discriminator networks are imposed on the latent code z and the generated anatomy. D_z is designed to discriminate the latent code $z = E(I)$ by training z to be uniformly distributed. Simultaneously, E will be trained to compute z to fool D_z. Such an adversarial process forces the distribution of z to gradually approach the prior, which is the uniform distribution. Another discriminator D_{img} forces the generator to generate realistic cardiac anatomies.

2.3   Training Scheme

An overview of the training scheme is shown in Fig. 1. To generate realistic anatomies while modelling smooth continuous ageing, we use a multi-task loss function which combines cyclic reconstruction losses L_{rec} and L_{cyc}, adversarial losses $L_{adv_{I'_t}}$, $L_{adv_{I'_s}}$ for the anatomy discriminator D_{img}, an adversarial loss L_{adv_z} for the latent code discriminator D_z and an age loss L_{age} for the age predictor P. For heart ageing synthesis, the source anatomy I_s and the generated anatomy $I'_s = G(I_s, a_s, g)$ at the same age a_s are expected to be similar. A self-reconstruction loss between source image I_s and reconstruction image I'_s is applied to learn the identity generation. In addition, we employ the cycle consistency loss [23] between I_s and $I''_s = G(I'_t, a_s, g)$ for a consistent reconstruction from age a_s to age a_t and back to age a_s. L1 loss is used for reconstruction:

$$\mathcal{L}_{rec}(G) = \|I_s - I'_s\|_1, \mathcal{L}_{cyc}(G) = \|I_s - I''_s\|_1 \tag{3}$$

The generated images I'_t, I'_s are enforced to the target age space by minimizing the distance between the age predictor outputs $P(I'_t)$, $P(I'_s)$ and the age vectors a_t, a_s. A cross-entropy (CE) age loss is defined as,

$$\mathcal{L}_{age}(G) = \|a_t - P(I'_t)\|_{CE(a_t, P(I'_t))} + \|a_s - P(I'_s)\|_{CE} \tag{4}$$

An adversarial loss L_{adv_z} is used to impose an uniform distribution on the latent code $z = E(I_s)$:

$$\mathcal{L}_{adv_z}(E, D) = \mathbb{E}_{z^*}\left[\log D_z(z^*)\right] + \mathbb{E}_{I_s}\left[\log(1 - D_z(E(I_s)))\right] \tag{5}$$

where z^* denote random samples from a uniformed prior distribution.

In addition, two adversarial losses conditioned on the source and target ages of the real and synthetic anatomies are introduced, respectively:

$$\mathcal{L}_{adv_{I'_t}}(G, D) = \mathbb{E}_{I_s, a_s}\left[\log D_{img}(I_s, a_s)\right] + \mathbb{E}_{I'_t, a_t}\left[\log(1 - D_{img}(I'_t))\right] \tag{6}$$

$$\mathcal{L}_{adv_{I'_s}}(G, D) = \mathbb{E}_{I_s, a_s} [\log D_{img}(I_s, a_s)] + \mathbb{E}_{I_s, a_s} [\log(1 - D_{img}(I'_s))] \quad (7)$$

The adversarial losses presented in Eq. 6 and Eq. 7, minimizing the distance between the input and output images, forces the output anatomies to be close to the real ones.

Overall, the optimisation is formulated as an adversarial training process,

$$\min_{G,E} \max_{D_z, D_{img}} \lambda_0 \mathcal{L}_{rec}(G) + \lambda_1 \mathcal{L}_{cyc}(G) + \lambda_2 \mathcal{L}_{age}(G)$$
$$+ \lambda_3 \mathcal{L}_{adv_z}(E, D) + \lambda_4 \mathcal{L}_{adv_{I'_t}}(G, D) + \lambda_5 \mathcal{L}_{adv_{I'_s}}(G, D) \quad (8)$$

where the λ's are tunable hyperparameters weighting the loss terms.

3 Experiments

3.1 Datasets

Cross-Sectional Dataset. Short-axis cardiac images at the end-diastolic (ED) and end-systolic (ES) frames of 12,600 subjects from 44.6 to 82.3 years old, were obtained from the UK Biobank and split into training ($n = 11,340$) and test ($n = 1,260$) sets. The age is represented as seven categories ($m = 7$) with interval of five years: 44–50, 50–55, 55–60, 60–65, 65–70, 70–75 and 75–83. Most of datasets are from healthy volunteers and about 5%–6% have the cardiovascular diseases (CVD), which we would take into consideration in the future.

Longitudinal Dataset. A longitudinal dataset of 639 subjects from the UK Biobank is used, in which each subject undergoes imaging at two time points. The age ranges from 46.6 to 79.8 years old at the first imaging and 51.3 to 81.9 years at the re-imaging, with a median time gap of 3.2 years. The image resolution and size are the same as the cross-sectional dataset. All evaluations are performed on ED and ES frames of cardiac sequences.

Preprocessing. For both datasets, 3D cardiac anatomies at ED and ES frames are extracted from cardiac MR images using a publicly available segmentation network [3], then upsampled using a publicly available super-resolution model [18] followed by manual quality control. Subsequently, affine registration is performed to align all cardiac anatomies to the same orientation. The 3D cardiac anatomies are of an isotropic resolution of $1.8 \times 1.8 \times 2$ mm^3 and of size $128 \times 128 \times 64$ voxels.

3.2 Experimental Setup

Implementation Details. The encoder E consisted of five 3D convolutional layers and one flatten layer, outputting the latent code z. The decoder D consisted of one flatten layer and five 3D transposed convolution layer. The transposed convolution in encoder and decoder used a kernel size of $4 \times 4 \times 4$. All intermediate layers of each block use the ReLU activation function. The dimension of the latent variables z and z_c are both 32. The anatomy segmentations are transferred into one-hot map, and the output of $E(I)$ is restricted to $[-1,1]$ using the hyperbolic tangent activation function. For optimisation, the Adam optimizer [9] is used with learning rate of $2 \cdot 10^{-4}$ and weigh decay of $1 \cdot 10^{-5}$. We set $\sigma = 0.02$ in Eq. 1, $\lambda_0 = 1$, $\lambda_1 = 0.1$, $\lambda_2 = 0.01$, $\lambda_3 = 0.1$ and $\lambda_4 = 1$ in Eq. 8. The model was implemented using PyTorch [14]. At the inference stage, only the generator G is active, containing E, D and M as described in Eq. 2, while the other parts are not used. Our code will be made publicly available.

Baseline Methods. Two ageing synthesis methods, CAAE [22] and a modified version of Lifespan [13], are used as baselines. Since the original codes were developed for 2D face image synthesis, we re-implemented all the codes for 3D cardiac data synthesis. For Lifespan, we replaced the modulated convolution layers with 3D convolution layers to save GPU memory for 3D data.

3.3   Experiments and Results

Heart Ageing Synthesis. For each subject in the UK Biobank test set, we synthesise a series of anatomies for the same heart at age groups from 40 to 80 with interval of 5 years old. For example, in Fig. 2, the cardiac anatomy of a 50–55 years old female is taken as input and the anatomies of this heart at other ages are predicted using the proposed generative model. From the generated anatomies, we also derive clinical measures, including the left ventricular myocardial mass (LVM), LV end-diastolic volume (LVEDV), LV end-systolic volume (LVESV), RV end-diastolic volume (RVEDV) and RV end-systolic volume (RVESV). The bottom of Fig. 2 illustrates the trends of these clinical measures during heart ageing synthesis. Consistent with the literature [6], we observe a decreasing trend for LV or RV volumes. It demonstrates our model captures the relation between cardiac anatomical structure and age.

Distribution Similarity. Based the synthetic anatomies, we calculate the probability distribution $P_c(a)$ of each clinical measure c against age a and compare it to the probability distribution of the real data $Q_c(a)$. Here, c denotes one of the five clinical measures, e.g. LVM. The distance between $P_c(a)$ and $Q_c(a)$ is evaluated in terms of the Kullback-Leibler divergence (KL) and Wasserstein distance (WD). Table 1 compares the distribution similarities using different generative models. It shows the proposed method achieves a higher distribution similarity, compared to state-of-the-art ageing synthesis models.

Longitudinal Prediction. Using the repeated imaging scans from UK Biobank longitudinal dataset, we evaluate the predictive performance of the model. Given the anatomy at the first time point, the anatomy at the second time point is predicted and compared to the ground truth in terms of Dice metric, Hausdorff distance (HD) and average symmetric surface distance (ASSD), reported in Table 2. It shows that the proposed method achieves a good performance in prediction comparable to or better than other competing methods.

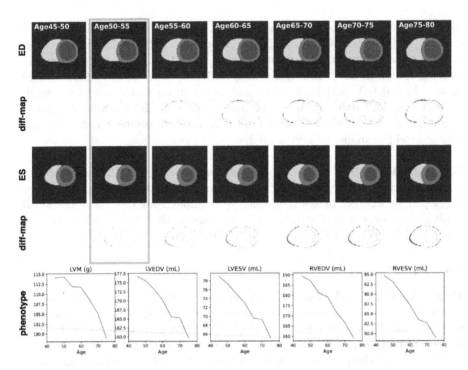

Fig. 2. An example of the synthetic ageing heart. The first and third rows show the cardiac anatomies at ED and ES frames, in which the blue rectangle denotes the original input anatomy of a 50–55 year old female and the other columns denote the synthetic anatomies at different ages. The second and fourth rows show the difference maps between an aged heart and the baseline anatomy at 45–50 year old. The fifth shows the predicted evolution of clinical measures including LVM, LVEDV, LVESV, RVEDV and RVESV during ageing. (Color figure online)

Table 1. The distribution similarity between synthetic and real data. The smaller the KL or WD distance, the higher the similarity.

	LVM		LVEDV		LVESV		RVEDV		RVESV	
	KL	WD	KL	WD	KL	WD	KL	WD	KL	WD
CAAE [22]	0.0266	19.8875	0.0355	**15.6661**	0.0737	9.8066	0.0343	**18.7905**	0.0467	11.9521
Lifespan [13]	0.0253	18.4733	0.0349	15.8506	0.0703	9.8091	0.0322	20.0034	0.0442	12.1956
Proposed	**0.0248**	**15.2829**	**0.0334**	15.7215	**0.0675**	**9.5658**	**0.0318**	19.1561	**0.0428**	**10.8607**

Table 2. The prediction performance on the UK Biobank longitudinal dataset. The higher Dice or lower HD and ASSD, the better the prediction.

	End-diastolic anatomy			End-systolic anatomy		
	Dice	HD	ASSD	Dice	HD	ASSD
CAAE [22]	0.727 (0.057)	30.431 (6.058)	2.777 (0.788)	0.769 (0.070)	15.904 (6.766)	2.528 (1.054)
Lifespan [13]	0.757 (0.064)	29.935 (5.988)	2.784 (0.806)	0.774 (0.072)	16.023 (6.527)	**2.490 (1.041)**
Proposed	**0.761 (0.066)**	**27.281 (7.436)**	**2.695 (0.835)**	**0.775 (0.073)**	**14.789 (7.224)**	2.524 (1.073)

4   Conclusion

To conclude, we propose a novel generative model for the ageing heart anatomy that allows preserving the identity of the heart while changing its characteristics across different age groups. The quantitative results on both cross-sectional and longitudinal datasets demonstrate the method achieves highly realistic synthesis and longitudinal prediction of cardiac anatomies, which are consistent with real data distributions.

Acknowledgements. This work was supported by EPSRC SmartHeart Grant (EP/P001009/1) and DeepGeM Grant (EP/W01842X/1). The research was conducted using the UK Biobank Resource under Application Number 18545. We wish to thank all UK Biobank participants and staff.

References

1. Alaluf, Y., Patashnik, O., Cohen-Or, D.: Only a matter of style: age transformation using a style-based regression model. ACM Trans. Graph. **40**(4), 1–12 (2021)
2. Antipov, G., Baccouche, M., Dugelay, J.L.: Face aging with conditional generative adversarial networks. In: IEEE International Conference on Image Processing, pp. 2089–2093 (2017)
3. Bai, W., Sinclair, M., Tarroni, G., Oktay, O., et al.: Automated cardiovascular magnetic resonance image analysis with fully convolutional networks. J. Cardiovasc. Magn. Reson. **20**(1), 65 (2018)
4. Bai, W., Suzuki, H., Huang, J., Francis, C., Wang, S., et al.: A population-based phenome-wide association study of cardiac and aortic structure and function. Nat. Med. **26**(10), 1654–1662 (2020)
5. Boon, R.A., Iekushi, K., Lechner, S., Seeger, T., Fischer, A., et al.: MicroRNA-34a regulates cardiac ageing and function. Nature **495**(7439), 107–110 (2013)
6. Eng, J., McClelland, R.L., Gomes, A.S., Hundley, W.G., Cheng, S., et al.: Adverse left ventricular remodeling and age assessed with cardiac MR imaging: the multi-ethnic study of atherosclerosis. Radiology **278**(3), 714–722 (2016)
7. Huang, Z., Chen, S., Zhang, J., Shan, H.: PFA-GAN: progressive face aging with generative adversarial network. IEEE Trans. Inf. Forensics Secur. **16**, 2031–2045 (2020)
8. Karras, T., Aittala, M., Hellsten, J., Laine, S., Lehtinen, J., Aila, T.: Training generative adversarial networks with limited data. Adv. Neural. Inf. Process. Syst. **33**, 12104–12114 (2020)
9. Kingma, D.P., Ba, J.: Adam: a method for stochastic optimization. In: International Conference for Learning Representations (2015)
10. Kingma, D.P., Welling, M.: Auto-encoding variational bayes. In: International Conference for Learning Representations (2013)
11. Makhmudkhujaev, F., Hong, S., Park, I.K.: Re-Aging GAN: toward personalized face age transformation. In: International Conference on Computer Vision, pp. 3908–3917 (2021)
12. Mirza, M., Osindero, S.: Conditional generative adversarial nets. arXiv preprint arXiv:1411.1784 (2014)

13. Or-El, R., Sengupta, S., Fried, O., Shechtman, E., Kemelmacher-Shlizerman, I.: Lifespan age transformation synthesis. In: Vedaldi, A., Bischof, H., Brox, T., Frahm, J.-M. (eds.) ECCV 2020. LNCS, vol. 12351, pp. 739–755. Springer, Cham (2020). https://doi.org/10.1007/978-3-030-58539-6_44
14. Paszke, A., Gross, S., Massa, F., Lerer, A., Bradbury, J., et al.: PyTorch: an imperative style, high-performance deep learning library. Adv. Neural. Inf. Process. Syst. **32**, 8026–8037 (2019)
15. Pawlowski, N., de Castro, D.C., Glocker, B.: Deep structural causal models for tractable counterfactual inference. Adv. Neural. Inf. Process. Syst. **33**, 857–869 (2020)
16. Sohn, K., Lee, H., Yan, X.: Learning structured output representation using deep conditional generative models. Adv. Neural. Inf. Process. Syst. **28**, 3483–3491 (2015)
17. Wang, C., Yang, G., Papanastasiou, G., Tsaftaris, S.A., Newby, D.E., et al.: DiCyc: GAN-based deformation invariant cross-domain information fusion for medical image synthesis. Inf. Fusion **67**, 147–160 (2021)
18. Wang, S., et al.: Joint motion correction and super resolution for cardiac segmentation via latent optimisation. In: de Bruijne, M., et al. (eds.) MICCAI 2021. LNCS, vol. 12903, pp. 14–24. Springer, Cham (2021). https://doi.org/10.1007/978-3-030-87199-4_2
19. Wang, Z., Tang, X., Luo, W., Gao, S.: Face aging with identity-preserved conditional generative adversarial networks. In: Proceedings of the IEEE Conference on Computer Vision and Pattern Recognition, pp. 7939–7947 (2018)
20. Xia, T., Chartsias, A., Tsaftaris, S.A., Initiative, A.D.N., et al.: Consistent brain ageing synthesis. In: International Conference on Medical Image Computing and Computer-Assisted Intervention, pp. 750–758 (2019)
21. Yurt, M., Dar, S.U., Erdem, A., Erdem, E., Oguz, K.K., Çukur, T.: mustGAN: multi-stream generative adversarial networks for MR image synthesis. Med. Image Anal. **70**, 101944 (2021)
22. Zhang, Z., Song, Y., Qi, H.: Age progression/regression by conditional adversarial autoencoder. In: Proceedings of the IEEE Conference on Computer Vision and Pattern Recognition, pp. 5810–5818 (2017)
23. Zhu, J.Y., Park, T., Isola, P., Efros, A.A.: Unpaired image-to-image translation using cycle-consistent adversarial networks. In: IEEE International Conference on Computer Vision (2017)

Learning Correspondences of Cardiac Motion from Images Using Biomechanics-Informed Modeling

Xiaoran Zhang[1]([✉]) [iD], Chenyu You[2], Shawn Ahn[1] [iD], Juntang Zhuang[1],
Lawrence Staib[1,2,3] [iD], and James Duncan[1,2,3] [iD]

[1] Biomedical Engineering, Yale University, New Haven, CT, USA
xiaoran.zhang@yale.edu
[2] Electrical Engineering, Yale University, New Haven, CT, USA
[3] Radiology and Biomedical Engineering, Yale School of Medicine,
New Haven, CT, USA

Abstract. Learning spatial-temporal correspondences in cardiac motion from images is important for understanding the underlying dynamics of cardiac anatomical structures. Many methods explicitly impose smoothness constraints such as the $\mathcal{L}_2$ norm on the displacement vector field (DVF), while usually ignoring biomechanical feasibility in the transformation. Other geometric constraints either regularize specific regions of interest such as imposing incompressibility on the myocardium or introduce additional steps such as training a separate network-based regularizer on physically simulated datasets. In this work, we propose an explicit biomechanics-informed prior as regularization on the predicted DVF in modeling a more generic biomechanically plausible transformation within all cardiac structures without introducing additional training complexity. We validate our methods on two publicly available datasets in the context of 2D MRI data and perform extensive experiments to illustrate the effectiveness and robustness of our proposed methods compared to other competing regularization schemes. Our proposed methods better preserve biomechanical properties by visual assessment and show advantages in segmentation performance using quantitative evaluation metrics. The code is publicly available at https://github.com/Voldemort108X/bioinformed_reg.

Keywords: Biomechanics-informed modeling · Cardiac motion · Magnetic resonance imaging

1 Introduction

A displacement vector field (DVF) estimated by medical image registration is crucial to infer the underlying spatial-temporal characteristics of anatomical structures. The DVF is especially important for cardiac motion, whose functional assessment is often analyzed using cine MRI or echocardiography [2,25 27]. Specifically, accurate prediction of the DVF across cardiac sequences enables

O. Camara et al. (Eds.): STACOM 2022, LNCS 13593, pp. 13–25, 2022.
https://doi.org/10.1007/978-3-031-23443-9_2

regional myocardial strain estimation, which contributes to the localization of myocardial infarction [9].

A wide range of deformable image registration methods have been studied for medical images to obtain reliable DVF measurement. Regularization is often imposed to ensure smoothness along with a dissimilarity measure. Such methods include B-splines [13], which use the bending energy of a thin-plate of metal, and optical flow [11], which uses total variation. Diffeomorphism is also a commonly used regularization assumption, which parameterizes the DVF as a set of velocity fields to ensure invertibility [3]. For cardiac motion estimation, incompressibility of the myocardium is also used to incorporate physiology-inspired prior into the registration process by enforcing, for example, a divergence-free constraint [7].

With the advent of deep learning, convolutional neural networks have been widely applied to unsupervised learning of deformable registration and segmentation [8,18–24]. Diffeomorphic constraints [6] and $\mathcal{L}_2$ norms on the displacement gradient [4] are usually used for regularization. However, biomechanical plausibility is usually not explicitly considered and thus the estimated displacement field might result in unrealistic motion. To encourage physically plausible transformation, pre-trained variational auto-encoders (VAEs) can be used as regularizers in the registration pipeline to enforce implicit constraints on the DVF [12,14,15]. However, such VAE regularizers are often trained separately on manually generated datasets which require physical simulations and thus can be time-consuming and complicated. Due to the intra- and inter-variability in clinical datasets, such network-based regularizers might also be biased when evaluating new datasets. In addition, explicit biomechanical constraints such as incompressibility are often considered only for specific regions of interest, such as the myocardium [1,12] for cardiac applications, while the preservation of geometric properties in other structures such as the right ventricle (RV) is not guaranteed.

In this work, we propose a biomechanics-informed regularization explicitly as a prior for DVF estimation to better preserve geometric properties without requiring myocardium segmentation. Using a separate network as a regularizer might introduce additional steps and thus increase complexity and computation. By contrast, our proposed methods show advantages in generalization when applying to new datasets. The consistency across different cardiac structures is also improved in our proposed methods due to the more generic assumption. We validate our methods on two public datasets and conduct extensive experiments to show its effectiveness and robustness over other competing regularization techniques.

2   Method

The proposed framework is illustrated in Fig. 1. We utilize an explicit biomechanical regularization on the estimated displacement vector field (DVF) from the registration network (RegNet), which takes pairs of moving and fixed images as inputs $(I_m, I_f) \in \mathbb{R}^{2 \times C \times H \times W}$ and outputs the estimated DVF, $\hat{u} \in \mathbb{R}^{2 \times H \times W}$. The following sections describe a detailed biomechanics-informed modeling (BIM) process (Sect. 2.1) and the derivation of the proposed loss function (Sect. 2.2).

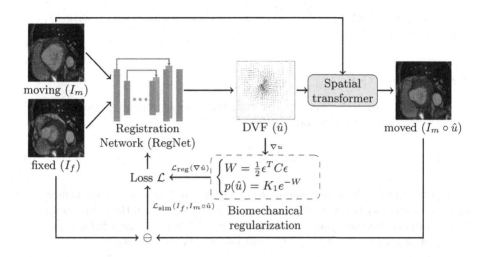

Fig. 1. Illustration of the proposed framework using biomechanics-informed modeling (BIM) for cardiac image registration.

2.1 Biomechanics-Informed Modeling

From the classical definition of the strain tensor $\epsilon \in \mathbb{R}^{3 \times H \times W}$ (defined in Eq. 1) in the infinitesimal linear elasticity model [16] for 2D displacement $\hat{u} = [\hat{u}_1, \hat{u}_2]^T$

$$\epsilon = \left[\frac{\partial \hat{u}_1}{\partial x_1}, \frac{\partial \hat{u}_2}{\partial x_2}, \frac{1}{2}\left(\frac{\partial \hat{u}_1}{\partial x_2} + \frac{\partial \hat{u}_2}{\partial x_1} \right) \right]^T, \tag{1}$$

we define the linear isotropic elastic strain energy density function for each pixel as follows:

$$W_{i,j} = \frac{1}{2} \epsilon_{i,j}^T C \epsilon_{i,j}, \tag{2}$$

where C is the stiffness matrix describing material properties of the deforming body

$$C^{-1} = \begin{bmatrix} \frac{1}{E_p} & \frac{-\nu_p}{E_p} & 0 \\ \frac{-\nu_p}{E_p} & \frac{1}{E_p} & 0 \\ 0 & 0 & \frac{2(1+\nu_p)}{E_p} \end{bmatrix}. \tag{3}$$

E_p is defined as stiffness and ν_p as Poisson ratio [10]. In this paper, we treat E_p as constant and ν_p as a hyperparameter. The prior probability density function (pdf) of the DVF can be written in Gibb's form:

$$p(\hat{u}) = k_1 e^{-W}. \tag{4}$$

The optimal DVF $\hat{u}^*$ can be obtained through maximum a posteriori (MAP) optimization:

$$\hat{u}^* = \arg\max_{\hat{u}} \left\{ p(\hat{u}|u) = \frac{p(u|\hat{u})p(\hat{u})}{p(u)} \right\} = \arg\min_{\hat{u}} \left\{ -\log p(u|\hat{u}) - \log p(\hat{u}) \right\}. \tag{5}$$

2.2   Loss Function

We assume the noise between the ground truth measurement u and the DVF estimate $\hat{u}$ is normally distributed $\mathcal{N}(0, \sigma^2)$

$$p(u|\hat{u}) = \frac{1}{\sqrt{2\pi}\sigma} e^{-\frac{||u-\hat{u}||_2^2}{2\sigma^2}}. \tag{6}$$

As the ground truth measurement u is usually hard to acquire, we utilize the image dissimilarity between the transformed image $I_m \circ \hat{u}$ and the reference image I_f to evaluate the difference of $\hat{u}$ with the ground truth motion field. From the prior pdf of DVF in Eq. 4, our proposed loss can be written as:

$$\mathcal{L}(I_m, I_f; \hat{u}) = \underbrace{\frac{1}{N}\sum_{i=1}^{N}||I_f^i - I_m^i \circ \hat{u}^i||_2^2}_{\mathcal{L}_{sim}} + \lambda \underbrace{\frac{1}{N}\sum_{i=1}^{N}||\epsilon^T C \epsilon||_2}_{\mathcal{L}_{reg}}, \tag{7}$$

where N is the number of samples. We apply $\mathcal{L}_2$ norm for the image after computing the strain energy density with scalar-valued entry for each pixel to penalize unrealistic motion. When segmentation masks are available for certain cardiac structures for an input image pair (I_m, I_f), we can further improve the approximated difference of $\hat{u}$ with ground truth previously computed using image dissimilarity by adding an auxiliary segmentation loss term

$$\mathcal{L}(I_m, I_f, s_m, s_f; \hat{u}) = \mathcal{L}(I_m, I_f; \hat{u}) + \gamma \underbrace{\frac{1}{N}\sum_{i=1}^{N}\sum_{j=1}^{K}\left(1 - \text{Dice}\left(s_m^{ij} \circ \hat{u}^i, s_f^{ij}\right)\right)}_{\mathcal{L}_{seg}}, \tag{8}$$

where (s_m^{ij}, s_f^{ij}) denotes the segmentation mask of cardiac structure j for image pair (I_m^i, I_f^i) with sample index i and $j = 1, 2, ..., K$.

3   Experiments

3.1   Dataset Details

We validate the effectiveness of our proposed methods on the publically-available ACDC 2017 dataset [5]. It contains 100 patients including the following five categories with even split: 1) healthy, 2) previous myocardial infarction, 3) dilated cardiomyopathy, 4) hypertrophic cardiomyopathy, and 5) abnormal right ventricle (RV). Segmentation labels of left ventricle (LV), RV, epicardium (Epi) boundaries for end-diastolic (ED) and end-systolic (ES) frames are given in the dataset. For this data, we use 60 patients for training, 20 for validation, and 20 for testing, after a random shuffle.

We also validate our method on another publicly available LV quantification dataset from 2019 [17]. It contains 56 subjects and includes 20 frames of the entire cardiac cycle. Segmentation labels of the endocardium (Endo) and epicardium boundaries are given for each frame in the cycle. In this paper, 34 patients are used for training, 11 for validation, and 11 for testing, after a random shuffle.

3.2   Experimental Setup

We set the ED frame as the moving image and the ES frame as the fixed image in the ACDC and LV quantification datasets in order to estimate cardiac contraction. Each slice in both datasets is cropped to 96×96 with respect to the myocardial centroid after min-max normalization. We compare our proposed framework with several state-of-the-art registration methods including 1) B-splines [13] implemented using SimpleElastix, 2) optical flow [11] implemented using scikit-image registration package, 3) a learning-based registration framework using VAE regularization (RegNet+VAE) [12], and 4) a learning-based registration framework using $\mathcal{L}_2$ regularization (RegNet+$\mathcal{L}_2$) [4]. The learning-based registration algorithms are implemented using Pytorch with Adam optimizer on 100 epochs. The learning-based models run 18–30 min on a single NVIDIA GTX 2080Ti GPU with 12 GB memory for training and validation. The loss curves in training and validation are shown in the Appendix (Fig. 5). We choose $\nu_p = 0.4$ (Poisson ratio), $\lambda = 0.05$ (weight for BIM loss $\mathcal{L}_{reg}$), and $\gamma = 0.01$ (weight for auxiliary segmentation loss $\mathcal{L}_{seg}$) in our proposed methods after optimizing hyperparameters. The hyperparameters of the competitor models are optimized as well.

3.3   Evaluation Metrics

To evaluate our proposed method, we compute the Dice coefficient (DC), Jaccard index (JD), Hausdorff distance (HD), and average symmetric surface distance (ASD) to evaluate segmentation conformance. We also compute the Jacobian determinant to evaluate the quality of the generated DVF.

4   Results

4.1   Visual and Quantitative Assessments

Visual assessment of our proposed methods compared to other approaches on the ACDC dataset is shown in Fig. 2. From the figure, we can see that B-splines [13] and optical flow [11] generate artifacts in the myocardium, especially in the highlighted blue patch shown in the third row. The RegNet+VAE [12] creates unrealistic motion in the right ventricle region, especially in the highlighted green patch shown in the last row, which might be due to penalization of the VAE regularizer mainly focusing on the myocardium only. Our proposed methods, in the last two columns, show improved motion compared to the RegNet+$\mathcal{L}_2$ with smoother and more biomechanically plausible motion that preserves the geometric features of cardiac structures. The visual results of our proposed methods compared with others on the LV quantification dataset is included in the Appendix (Fig. 4).

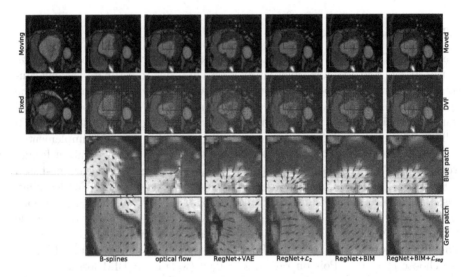

Fig. 2. Visual assessment of registration performance on the ACDC 2017 dataset [5].

A quantitative assessment of our proposed methods on the ACDC dataset is reported in Table 1. Our proposed methods outperform other methods in cardiac structures including the LV and epicardium across all evaluation metrics as well as the Jacobian determinant evaluation shown in Table 2. Our methods yield higher conformance in terms of DC and JD in RV and similar performance with B-splines and optical flow in terms of HD and ASD. A quantitative comparison of our proposed method with the others on the LV quantification dataset is included in the Appendix (Table 5).

We conduct paired t-tests for statistical significance analysis of our proposed method (RegNet+BIM+$\mathcal{L}_{\mathrm{seg}}$) with the others. Statistically significant improvement ($p < 0.05$) is observed for all other methods except for B-splines ($p = 0.09$), which might be due to the limited size of the testing set. The complete table is included in the Appendix (Table 4).

Table 1. Quantitative assessment of registration performance on the ACDC dataset [5].

	Method	Dice [%]↑	Jaccard [%]↑	HD [mm]↓	ASD [mm]↓
LV	Affine only	72.01	57.24	11.77	2.13
	B-splines [13]	89.96	82.62	5.37	0.48
	Optical flow [11]	87.07	77.87	7.65	0.74
	RegNet+VAE [12]	89.61	81.65	6.46	0.49
	RegNet+$\mathcal{L}_2$	89.60	81.68	8.00	0.54
	RegNet+BIM (ours)	90.32	82.70	5.51	0.37
	RegNet+BIM+$\mathcal{L}_{\text{seg}}$ (ours)	**92.77**	**86.76**	**4.39**	**0.19**
RV	Affine only	81.08	70.06	14.31	2.21
	B-splines [13]	83.59	74.33	**13.70**	2.43
	Optical flow [11]	84.96	75.94	**13.70**	**2.15**
	RegNet+VAE [12]	84.65	75.58	16.48	2.31
	RegNet+$\mathcal{L}_2$	84.80	75.73	14.75	2.25
	RegNet+BIM (ours)	85.07	76.16	14.55	2.22
	RegNet+BIM+$\mathcal{L}_{\text{seg}}$ (ours)	**85.93**	**77.54**	14.07	**2.15**
Epi	affine only	85.07	74.50	8.61	0.93
	B-splines [13]	92.33	86.12	5.97	0.33
	Optical flow [11]	92.02	85.64	6.09	0.39
	RegNet+VAE [12]	91.45	84.83	7.57	0.48
	RegNet+$\mathcal{L}_2$	89.92	82.22	7.77	0.59
	RegNet+BIM (ours)	91.53	84.75	6.20	0.40
	RegNet+BIM+$\mathcal{L}_{\text{seg}}$ (ours)	**92.57**	**86.48**	**5.54**	**0.32**

Table 2. Mean Jacobian determinant comparisons on the ACDC 2017 dataset [5].

Method	$|\det(J(\hat{u})) - 1|$
B-splines [13]	0.2489 ± 0.2735
Optical flow [11]	0.1283 ± 0.3432
RegNet+VAE [12]	0.0088 ± 0.0100
RegNet+$\mathcal{L}_2$	0.0038 ± 0.0051
RegNet+BIM (ours)	**0.0035 ± 0.0036**
RegNet+BIM+$\mathcal{L}_{\text{seg}}$ (ours)	0.0037 ± 0.0038

4.2 Effect of Hyperparameters

To investigate the impact of our proposed biomechanically-informed regularization and the auxiliary segmentation loss, we vary the weights for each loss term as $\lambda = \{0.005, 0.05, 0.5\}$ and $\gamma = \{0.001, 0.01, 0.1\}$ and report corresponding LV segmentation performance as shown in Table 3. We also evaluate the effect of Poisson ratio by varying $\nu_p = \{0.35, 0.4, 0.45\}$ in our proposed BIM term. From Table 3, we can see that the ν_p in the vicinity of 0.4 has little influence on the segmentation conformance. By increasing λ, we can see that the performance gradually improves from 0.005 to 0.05 and deteriorates quite significantly from 0.05 to 0.5. This performance loss is likely due to an underestimate of the DVF to avoid a large penalty when trained with a stronger regularization. By increasing γ, we observe a steady increase in segmentation performance. However, a stronger auxiliary segmentation regularization might yield unrealistic deformation estimates. The structural similarity index between the fixed and moved images for $\gamma = 0.01$ to 0.1 also drops from 0.70 to 0.65. We also perform an ablation study to illustrate the effectiveness for each term and the detailed statistics on both datasets are included in the Appendix (Table 6).

Table 3. Effect of hyperparameters in RegNet+BIM+$\mathcal{L}_{\text{seg}}$ on ACDC 2017 dataset [5]. ν_p, λ, and γ denote Poisson ratio, weight for BIM regularization, and weight for $\mathcal{L}_{\text{seg}}$, respectively. By varying each hyperparameter, we report left ventricle segmentation performance.

Parameter	ν_p			λ			γ		
Value	0.35	0.4	0.45	0.005	0.05	0.5	0.001	0.01	0.1
Dice [%]↑	92.74	92.77	92.66	91.51	92.77	81.03	91.14	92.77	93.27
Jaccard [%]↑	86.65	86.67	86.56	84.80	86.67	68.72	84.04	86.67	87.54
HD [mm]↓	4.21	4.39	4.51	7.57	4.39	8.25	5.33	4.39	4.17
ASD [mm]↓	0.19	0.19	0.22	0.35	0.19	1.04	0.33	0.19	0.14

5 Discussion

To analyze the potential registration performance among methods for people with different cardiac conditions, we show the Dice score of each method for patients under different pathologies from the ACDC dataset in the boxplot (Fig. 3). From the figure, we can see that our proposed methods yield higher segmentation conformance for dilated cardiomyopathy patients and healthy people. The performance of our proposed methods are similar to others for patients with previous myocardial infarction and hypertrophic cardiomyopathy. Optical flow seems to outperform our proposed method for patients who have abnormal right ventricles. Currently, we validate our proposed methods in the same 2D setting with other competing methods [11,12].

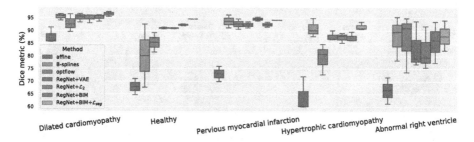

Fig. 3. Segmentation conformance in terms of the Dice metric for patients with different pathology from the ACDC dataset [5].

6   Conclusion

In this paper, we propose a data-driven approach for cardiac motion estimation with a novel biomechanics-informed modeling regularization loss. Our proposed methods generate more biomechanically plausible DVFs for all cardiac structures in the image without requiring myocardium segmentation or introducing additional steps compared to network-based regularizers. Our proposed methods outperform other state-of-the-art methods on both the ACDC 2017 dataset and the LV quantification 2019 dataset using quantitative evaluation metrics. Future work will validate current work on other modalities such as tagged MRI and show its effectiveness in more standardized 3D clinical datasets with additional metrics to assess local abnormalities.

Appendix

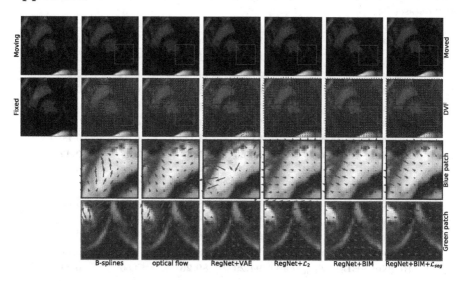

Fig. 4. Visual assessment of registration performance on the LV quantification 2019 dataset [17].

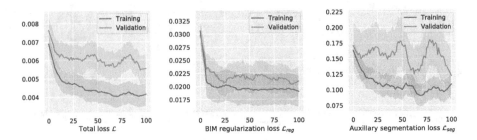

Fig. 5. Curves of total loss $\mathcal{L}$, BIM regularization loss $\mathcal{L}_{\text{reg}}$, and auxiliary segmentation loss $\mathcal{L}_{\text{seg}}$ in training and validation, respectively.

Table 4. Analysis of paired t-test for statistical significance between RegNet+BIM+$\mathcal{L}_{\text{seg}}$ (ours) and other methods on the ACDC 2017 dataset [5].

Method	Dice [%]↑	Jaccard [%]↑	HD [mm]↓	ASD [mm]↓
Affine only	$p < 0.05$	$p < 0.05$	$p < 0.05$	$p < 0.05$
B-splines [13]	$p = 0.09$	$p = 0.09$	$p = 0.09$	$p = 0.06$
Optical flow [11]	$p < 0.05$	$p < 0.05$	$p < 0.05$	$p < 0.05$
RegNet+VAE [12]	$p < 0.05$	$p < 0.05$	$p < 0.05$	$p < 0.05$
RegNet+$\mathcal{L}_2$	$p < 0.05$	$p < 0.05$	$p < 0.05$	$p < 0.05$

Table 5. Quantitative assessment of registration performance on the LV quantification 2019 dataset [5].

	Method	Dice [%]↑	Jaccard [%]↑	HD [mm]↓	ASD [mm]↓
Endo	Affine only	82.36	71.14	6.92	0.54
	B-splines [13]	84.54	74.30	**6.31**	0.47
	Optical flow [11]	83.65	73.01	6.52	0.48
	RegNet+VAE [12]	82.70	71.62	6.81	0.52
	RegNet+$\mathcal{L}_2$	82.49	71.33	6.91	0.53
	RegNet+BIM (ours)	84.24	73.87	6.63	0.48
	RegNet+BIM+$\mathcal{L}_{\text{seg}}$ (ours)	**86.65**	**77.48**	6.45	**0.42**
Epi	Affine only	93.72	88.26	4.62	0.13
	B-splines [13]	94.19	89.10	**4.27**	0.12
	Optical flow [11]	94.08	88.90	4.33	0.12
	RegNet+VAE [12]	93.74	88.29	4.64	0.13
	RegNet+$\mathcal{L}_2$	93.72	88.26	4.62	0.13
	RegNet+BIM (ours)	93.80	88.41	4.71	0.13
	RegNet+BIM+$\mathcal{L}_{\text{seg}}$ (ours)	**94.29**	**89.26**	4.50	**0.11**

Table 6. Ablation study of proposed methods in ACDC 2017 dataset [5] and LV quantification 2019 dataset [17].

	Method	Dice [%]↑	Jaccard [%]↑	HD [mm]↓	ASD [mm]↓
ACDC 2017 dataset [5]					
LV	RegNet	82.06	70.68	15.45	1.84
	RegNet+BIM (ours)	90.32	82.70	5.51	0.37
	RegNet+BIM+$\mathcal{L}_{seg}$ (ours)	**92.77**	**86.76**	**4.39**	**0.19**
RV	RegNet	84.29	75.14	17.15	2.40
	RegNet+BIM (ours)	85.07	76.16	14.55	2.22
	RegNet+BIM+$\mathcal{L}_{seg}$ (ours)	**85.93**	**77.54**	**14.07**	**2.15**
Epi	RegNet	86.67	77.06	10.60	0.83
	RegNet+BIM (ours)	91.53	84.75	6.20	0.40
	RegNet+BIM+$\mathcal{L}_{seg}$ (ours)	**92.57**	**86.48**	**5.54**	**0.32**
LV Quantification 2019 dataset [17]					
Endo	RegNet	84.03	73.58	6.76	0.48
	RegNet+BIM (ours)	84.24	73.87	6.63	0.48
	RegNet+BIM+$\mathcal{L}_{seg}$ (ours)	**86.65**	**77.48**	**6.45**	**0.42**
Epi	RegNet	93.71	88.25	4.70	0.13
	RegNet+BIM (ours)	93.80	88.41	4.71	0.13
	RegNet+BIM+$\mathcal{L}_{seg}$ (ours)	**94.29**	**89.26**	**4.50**	**0.11**

References

1. Ahn, S.S., Ta, K., Lu, A., Stendahl, J.C., Sinusas, A.J., Duncan, J.S.: Unsupervised motion tracking of left ventricle in echocardiography. In: Medical Imaging 2020: Ultrasonic Imaging and Tomography, vol. 11319, p. 113190Z. International Society for Optics and Photonics (2020)
2. Ahn, S.S., Ta, K., Thorn, S., Langdon, J., Sinusas, A.J., Duncan, J.S.: Multi-frame attention network for left ventricle segmentation in 3D echocardiography. In: de Bruijne, M., et al. (eds.) MICCAI 2021. LNCS, vol. 12901, pp. 348–357. Springer, Cham (2021). https://doi.org/10.1007/978-3-030-87193-2_33
3. Ashburner, J.: A fast diffeomorphic image registration algorithm. Neuroimage **38**(1), 95–113 (2007)
4. Balakrishnan, G., Zhao, A., Sabuncu, M.R., Guttag, J., Dalca, A.V.: VoxelMorph: a learning framework for deformable medical image registration. IEEE Trans. Med. Imaging **38**(8), 1788–1800 (2019)
5. Bernard, O., et al.: Deep learning techniques for automatic MRI cardiac multi-structures segmentation and diagnosis: is the problem solved? IEEE Trans. Med. Imaging **37**(11), 2514–2525 (2018)
6. Dalca, A.V., Balakrishnan, G., Guttag, J., Sabuncu, M.R.: Unsupervised learning for fast probabilistic diffeomorphic registration. In: Frangi, A.F., Schnabel, J.A., Davatzikos, C., Alberola-López, C., Fichtinger, G. (eds.) MICCAI 2018. LNCS, vol. 11070, pp. 729–738. Springer, Cham (2018). https://doi.org/10.1007/978-3-030-00928-1_82
7. Gao, Q., et al.: Optimization of 4D flow MRI velocity field in the aorta with divergence-free smoothing. Med. Biol. Eng. Comput. **59**(11), 2237–2252 (2021)

8. Li, L., et al.: MyoPS: a benchmark of myocardial pathology segmentation combining three-sequence cardiac magnetic resonance images. arXiv preprint arXiv:2201.03186 (2022)
9. Lu, A., et al.: Learning-based regularization for cardiac strain analysis via domain adaptation. IEEE Trans. Med. Imaging 40(9), 2233–2245 (2021)
10. Papademetris, X., Sinusas, A.J., Dione, D.P., Constable, R.T., Duncan, J.S.: Estimation of 3-D left ventricular deformation from medical images using biomechanical models. IEEE Trans. Med. Imaging 21(7), 786–800 (2002)
11. Pérez, J.S., Meinhardt-Llopis, E., Facciolo, G.: TV-L1 optical flow estimation. Image Process. Line 2013, 137–150 (2013)
12. Qin, C., Wang, S., Chen, C., Qiu, H., Bai, W., Rueckert, D.: Biomechanics-informed neural networks for myocardial motion tracking in MRI. In: Martel, A.L., et al. (eds.) MICCAI 2020. LNCS, vol. 12263, pp. 296–306. Springer, Cham (2020). https://doi.org/10.1007/978-3-030-59716-0_29
13. Rueckert, D., Sonoda, L.I., Hayes, C., Hill, D.L., Leach, M.O., Hawkes, D.J.: Non-rigid registration using free-form deformations: application to breast MR images. IEEE Trans. Med. Imaging 18(8), 712–721 (1999)
14. Sang, Y., Ruan, D.: Enhanced image registration with a network paradigm and incorporation of a deformation representation model. In: 2020 IEEE 17th International Symposium on Biomedical Imaging (ISBI), pp. 91–94. IEEE (2020)
15. Sang, Y., Ruan, D.: 4D-CBCT registration with a FBCT-derived plug-and-play feasibility regularizer. In: de Bruijne, M., et al. (eds.) MICCAI 2021. LNCS, vol. 12904, pp. 108–117. Springer, Cham (2021). https://doi.org/10.1007/978-3-030-87202-1_11
16. Spencer, A.J.M.: Continuum Mechanics. Courier Corporation (2004)
17. Xue, W., Brahm, G., Pandey, S., Leung, S., Li, S.: Full left ventricle quantification via deep multitask relationships learning. Med. Image Anal. 43, 54–65 (2018)
18. You, C., Dai, W., Staib, L., Duncan, J.S.: Bootstrapping semi-supervised medical image segmentation with anatomical-aware contrastive distillation. arXiv preprint arXiv:2206.02307 (2022)
19. You, C., et al.: Incremental learning meets transfer learning: application to multi-site prostate MRI segmentation. arXiv preprint arXiv:2206.01369 (2022)
20. You, C., Yang, J., Chapiro, J., Duncan, J.S.: Unsupervised wasserstein distance guided domain adaptation for 3D multi-domain liver segmentation. In: Cardoso, J., et al. (eds.) IMIMIC/MIL3ID/LABELS -2020. LNCS, vol. 12446, pp. 155–163. Springer, Cham (2020). https://doi.org/10.1007/978-3-030-61166-8_17
21. You, C., et al.: Class-aware generative adversarial transformers for medical image segmentation. arXiv preprint arXiv:2201.10737 (2022)
22. You, C., Zhao, R., Staib, L., Duncan, J.S.: Momentum contrastive voxel-wise representation learning for semi-supervised volumetric medical image segmentation. arXiv preprint arXiv:2105.07059 (2021)
23. You, C., Zhou, Y., Zhao, R., Staib, L., Duncan, J.S.: SimCVD: simple contrastive voxel-wise representation distillation for semi-supervised medical image segmentation. IEEE Trans. Med. Imaging 41(9), 2228–2237 (2022). https://doi.org/10.1109/TMI.2022.3161829
24. Zhang, X., et al.: Automatic spinal cord segmentation from axial-view MRI slices using CNN with grayscale regularized active contour propagation. Comput. Biol. Med. 132, 104345 (2021)
25. Zhang, X., Martin, D.G., Noga, M., Punithakumar, K.: Fully automated left atrial segmentation from MR image sequences using deep convolutional neural network

and unscented Kalman filter. In: 2018 IEEE International Conference on Bioinformatics and Biomedicine (BIBM), pp. 2316–2323. IEEE (2018)

26. Zhang, X., Noga, M., Martin, D.G., Punithakumar, K.: Fully automated left atrium segmentation from anatomical cine long-axis MRI sequences using deep convolutional neural network with unscented Kalman filter. Med. Image Anal. **68**, 101916 (2021)

27. Zhang, X., Noga, M., Punithakumar, K.: Fully automated deep learning based segmentation of normal, infarcted and edema regions from multiple cardiac MRI sequences. In: Zhuang, X., Li, L. (eds.) MyoPS 2020. LNCS, vol. 12554, pp. 82–91. Springer, Cham (2020). https://doi.org/10.1007/978-3-030-65651-5_8

Multi-modal Latent-Space Self-alignment for Super-Resolution Cardiac MR Segmentation

Yu Deng[1](✉), Yang Wen[2,3], Linglong Qian[4], Esther Puyol Anton[1], Hao Xu[1], Kuberan Pushparajah[1], Zina Ibrahim[4], Richard Dobson[4], and Alistair Young[1]

[1] School of Biomedical Engineering and Imaging Science, King's College London, London, UK
yu.deng@kcl.ac.uk
[2] Animal Imaging and Technology Core, Center for Biomedical Imaging, Ecole Polytechnique Fédérale de Lausanne, Lausanne, Switzerland
[3] Laboratory for Functional and Metabolic Imaging, Ecole Polytechnique Fédérale de Lausanne, Lausanne, Switzerland
[4] Department of Biostatistics and Health Informatics, Institute of Psychiatry, Psychology and Neuroscience, King's College London, London, UK

Abstract. 2D cardiac MR cine images provide data with a high signal-to-noise ratio for the segmentation and reconstruction of the heart. These images are frequently used in clinical practice and research. However, the segments have low resolution in the through-plane direction, and standard interpolation methods are unable to improve resolution and precision. We proposed an end-to-end pipeline for producing high-resolution segments from 2D MR images. This pipeline utilised a bilateral optical flow warping method to recover images in the through-plane direction, while a SegResNet automatically generated segments of the left and right ventricles. A multi-modal latent-space self-alignment network was implemented to guarantee that the segments maintain an anatomical prior derived from unpaired 3D high-resolution CT scans. On 3D MR angiograms, the trained pipeline produced high-resolution segments that preserve an anatomical prior derived from patients with various cardiovascular diseases.

Keywords: Super-resolution segmentation · Domain adaptation · Cardiac MR · CT angiogram · VAE

1 Introduction

To examine the hemodynamic function and anatomical structure of the heart, which are essential for deciding on medical or surgical intervention, it is necessary to examine the 3D cardiac structure and function with high resolution. This is possible using a 3D heart model reconstructed from segments drawn on cine images [9]. In a perfect world, 3D MR cine images would be desirable; however, there is a trade-off between high-resolution and low signal-to-noise ratio (SNR),

O. Camara et al. (Eds.): STACOM 2022, LNCS 13593, pp. 26–35, 2022.
https://doi.org/10.1007/978-3-031-23443-9_3

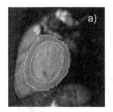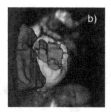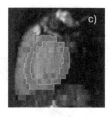

Fig. 1. Comparison of 3D models of the left and right ventricles (LV and RV). The LV cavity (Green), LV myocardium (Yellow), and RV cavity (Red) are best viewed in colour. a) 3D MR angiogram and high-resolution segments in the sagittal plane; b) 3D model from the high-resolution segments; c) 3D MR angiogram and low-resolution segments with the slice thickness of 8 mm; d) 3D model from the low-resolution segments. The surface of the 3D model is processed with a smoothness of 0.5 in 3D Slicer [6]. (Color figure online)

and the acquisition procedures for 3D MR images with acceptable SNR and segmentation are time-consuming and costly [2].

Standard clinical practice calls for the 3D reconstruction of the heart using 2D balanced steady-state free precession MR cine images stacked in their through-plane direction [1]. These stacked 2D MR images are chosen from multiple slice positions and maintain high in-plane resolution with good SNR and rapid acquisition [15]. Due to the relatively large slice thickness, however, the 2D MR images have a lower through-plane resolution than the 3D images (6–10 mm) [7]. This causes the segments on each slice to be separated from one another. Although interpolation is useful for reconstructing a 3D model, details in the anatomical shape of the heart are omitted, as shown in Fig. 1. This hinders the 3D model's pathological and anatomical analysis.

For precise reconstruction, high-resolution segmentation helps preserve fine-grained shape characteristics, such as wall thickness, chamber volumes, and landmarks. We seek to develop a compact pipeline for the segmentation of 2D MR cine images and to address a number of limitations of previous work in this area. Biffi *et al.* proposed a super-resolution technique for reconstructing a high-resolution heart from low-resolution segments [4]. However, registration of all short-axis (SAX) and long-axis (LAX) views via landmark-based registration is required. Delannoy *et al.* developed a reconstruction scheme that uses a generative adversarial network to generate high-resolution MR images and segments from low-resolution images [5]; however, high-resolution images and segments from manual segmentation are required for training. Wang *et al.* presented a latent optimisation method that super-resolves low-resolution segments and corrects inter-slice motion; however, high-resolution segments are necessary for training the network [17]. Oktay *et al.* argue that segmentation networks optimised by pixel-level losses (*e.g.*, cross-entropy and Dice) lack prior constraints from anatomical structures, and thus are susceptible to image corruption or artefacts [11]. For the high-resolution reconstruction task, they proposed a T-L network that required the latent representation of global shape features to

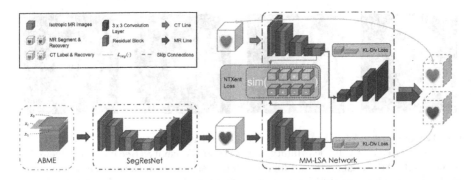

Fig. 2. A summary of our pipeline. In SegResNet and MM-LSA networks, the number of residual blocks in the convolutional layers of the encoder is 1, 2, 2, and 4, while the number of residual blocks in the convolutional layers of the decoder is 1. The skip connections between the decoder and encoder of an MM-LSA network are implemented, despite their absence from the figure. Throughout training, the size of the input image is $112 \times 112 \times 112$ and the size of the latent representation is 64×1024. Interpolation using pretrained asymmetric bilateral motion estimation (ABME) is performed independently and prior to training.

parameterize a segmentation network. Even though their method requires high-resolution segments, they emphasised the significance of the prior for segmentation and reconstruction of the heart.

As depicted in Fig. 2, we proposed in this study an end-to-end pipeline for super-resolution segmentation of the left and right ventricles (LV and RV). Considering the effect of slice thickness on 3D heart reconstruction, the 2D MR images were interpolated in the through-plane direction using a pretrained asymmetric bilateral motion estimation (ABME) network [13], and then Seg-ResNet [10] was used to generate initial segments from the interpolated 2D MR images. We propose a multi-modal latent-space self-alignment (MM-LSA) network to refine the initial segments using an anatomical prior of the heart translated from 3D high-resolution CT segments. The pipeline was then evaluated on 3D MR angiograms (MRAs) with high resolution 3D ground truth, revealing structurally consistent segment boundaries.

2   Methodology

2.1   Interpolation via ABME for 2D MR Images

To increase the though-plane resolution of stacked 2D MR image, the missing MR images between slice positions should be recovered. Those images are omitted since a series of slice positions divided by a slice thickness are planned prior to the scanning, as shown in Fig. 1. Every missing image $\hat{x}_i \in \mathbb{R}^{H \times W}$ is first synthesized from its succeeding image x_1 and its preceding image x_0 through a backward warping $\phi_B(\cdot)$ [18] as $\hat{x}_i = (1-i) \cdot \phi_B(\nu_{i \to 0}, x_0) + i \cdot \phi_B(\nu_{i \to 1}, x_1)$ with the symmetric bilateral motion fields $\nu_{i \to 0}$ and $\nu_{i \to 1}$. These two motion fields are assumed to be symmetric [12] so that $\nu_{i \to 1}$ is found by minimising the frame difference $\|\phi_{\mathrm{B}}(\nu_{i \to 0}, x_0) - \phi_{\mathrm{B}}(\nu_{i \to 1}, x_1)\| = \|\phi_{\mathrm{B}}(-i/(i-1)\nu_{i \to 1}, x_0) - \phi_{\mathrm{B}}(\nu_{i \to 1}, x_1)\|$. To overcome the disadvantage of the symmetric method when pixels obscure from x_0 to x_1, we apply the ABME [13] in this paper, which implements an asymmetric bilateral motion refinement for obtaining the motion fields and approximates $\hat{x}_i$.

The interpolation is conducted hierarchically so that all synthesized images between slice positions are from the previous result of interpolation, for which a detailed description can be found in [13]. The 2D MR images become high-resolved with the ABME network and the slice thickness decreased from 10 mm to 1.25 mm. ABME was trained on the Vimeo-90K dataset [19] that includes 89,800 video clips, without a fine-tuning on cardiac MR images.

2.2   Segmentation Network

The reconstruction of the heart relies on segments $\{Y_i \in \mathbb{R}^{H \times W \times D}\}_0^3$ of LV and RV cavities and LV myocardium from the 2D MR image $X \in \mathbb{R}^{H \times W \times D}$. In a supervised method, a prediction of the segments $\widehat{Y}_i$ is given by the SegResNet [10] as $\widehat{Y}_i = f(X; \theta)$. With each class of the segments $Y_i (i = 0, 1, 2, 3)$, the SegResNet's parameters θ are optimized that the error of predictions and segments is minimized as $\min_\theta \frac{1}{C} \sum_{i=0}^{C} \mathcal{L}_{\mathrm{seg}}(f(X; \theta), Y_i)$. In this paper, the $\mathcal{L}_{\mathrm{seg}}$ is a combination of Dice loss $\mathcal{L}_{\mathrm{Dice}}$ and Cross-entropy loss $\mathcal{L}_{\mathrm{CE}}$ as (Eq. 1) with a smooth coefficient $\epsilon = 1e - 16$.

$$\mathcal{L}_{\mathrm{seg}} = \mathcal{L}_{\mathrm{Dice}}\big(f(X; \theta), Y_i\big) + \mathcal{L}_{\mathrm{CE}}\big(f(X; \theta), Y_i\big)$$
$$= \frac{1}{4}\left(\sum_{i=1}^{3} \frac{2 \times f(X; \theta) \cap Y_i}{f(X; \theta) \cup Y_i + \epsilon} + \sum_{i=0}^{3} Y_i \cdot \log\big(f(X; \theta)\big)\right) \tag{1}$$

SegResNet is modified from U-Net [14] with additional residual blocks. A detailed description of this network could be found in [10]. However, training the SegResNet with interpolated 2D MR images is still ill-posed, because the segments are drawn on 2D MR images and then interpolated by the nearest neighborhood method to be as high-resolved as the images. This preserves arte-facts and misalignment between slices due to respiration motion, and thus the segments would disagree with the anatomical prior of the heart. We therefore considered translating the prior from high-resolution 3D CT segments.

2.3   Multi-modal Latent-Space Self-alignment Network

The anatomical prior implicitly encompasses the topology and morphology of the heart. It can be perceived as a latent representation z expected to follow a Gaussian distribution, from which we can recover the segments $\widehat{Y}$ through $p(\widehat{Y} \mid z) = \psi(z; \theta_{\mathrm{decoder}})$. This recovery is fulfilled by a decoder with parameters $\theta_{\mathrm{decoder}}$. To maximize the prior leveraged from 3D CT segments Y^{CT}, we adopt the VAE structure that avoids random generation. With an encoder parameterized by $\theta_{encoder}$, $p(z^{CT}) = \phi(Y^{CT}; \theta_{encoder})$ could be sampled as Eq. 2.

$$p(z^{CT}) = \int p(z^{CT} \mid Y^{CT})p(Y^{CT})dY^{CT}$$
$$= \mathcal{N}(\mu_{CT}, \sigma_{CT}) \cdot \int p(Y^{CT})dY^{CT} \tag{2}$$

where $p(z^{CT} \mid Y^{CT}) \sim \mathcal{N}(\mu_{CT}, \sigma_{CT})$. $\theta_{encoder}$ and $\theta_{decoder}$ (CT-line VAE) are optimized by minimizing the loss in (Eq. 1) between the Y^{CT} and recovered $\hat{Y}^{CT}$. The Gaussian distribution is defined for optimizing a KL loss $\mathcal{L}_{KL}$ where z^{CT} is reparameterized as $z^{CT} = \mu_{CT} + \xi \cdot \sigma_{CT}, \xi \sim \mathcal{N}(0,1)$. The total loss for the CT-line VAE is a combination of $\mathcal{L}_{seg}(\hat{Y}^{CT}, Y^{CT})$ and $\mathcal{L}_{KL}(\mu_{CT}, \sigma_{CT})$ in (Eq. 3).

$$\mathcal{L}_{CT} = \mathcal{L}_{seg}(\hat{Y}^{CT}, Y^{CT}) + \mathcal{L}_{KL}(\mu_{CT}, \sigma_{CT})$$
$$= \frac{1}{4}\left(\sum_{i=1}^{3} \frac{2 \times \hat{Y}_i^{CT} \cap Y_i^{CT}}{\hat{Y}_i^{CT} \cup Y_i^{CT} + \epsilon} + Y_i^{CT} \cdot \log(\hat{Y}_i^{CT})\right) \tag{3}$$
$$+ \frac{1}{2}\left(\mu_{CT}^2 + \sigma_{CT}^2 - \log(\sigma_{CT}^2) - 1\right)$$

After the above process, z^{CT} becomes an anchor for the following self-alignment method. The MR-line encoder maps the MR segments from SegResNet to another latent representation z^{MR}. Because the anatomical prior of the heart is domain-invariant, z^{MR} should align with z^{CT}. The distance between those two is minimized by a scalable metric learning loss, NTXent loss with cosine similarity $\text{sim}(z^{MR}, z^{CT})$ [16]. As shown in Fig. 2, the encoder of MR- and CT-line VAE have the same structure. The CT-line decoder with fixed parameters is shared to MR-line encoder for recovering MR segments $\hat{Y}^{MR}$, where the segmentation loss (Eq. 1) is adopted. The total loss for the MR-line encoder is formalized as a combination of $\mathcal{L}_{seg}(\hat{Y}^{MR}, Y^{MR})$, $\mathcal{L}_{NTXent}(z^{CT}, z^{MR})$ and $\mathcal{L}_{KL}(\mu_{MR}, \sigma_{MR})$ with tradeoff coefficients α and β and a temperature coefficient τ in (Eq. 4).

$$\mathcal{L}_{MR} = \alpha \cdot \mathcal{L}_{seg}(\hat{Y}^{MR}, Y^{MR}) + \beta \cdot \left(\mathcal{L}_{KL}(\mu_{MR}, \sigma_{MR}) + \mathcal{L}_{NTXent}(z^{CT}, z^{MR})\right)$$
$$= \alpha \cdot \mathcal{L}_{seg}(\hat{Y}^{MR}, Y^{MR}) + \beta \cdot \mathcal{L}_{KL}(\mu_{MR}, \sigma_{MR})$$
$$+ \beta \cdot \log\left(\frac{\exp\left(\text{sim}(z_i^{CT}, z_j^{MR})/\tau\right)}{\sum_{n=0}^{2N} \mathbb{I}_{n \neq i} \exp\left(\text{sim}(z_i^{CT}, z_j^{MR})/\tau\right)}\right) \tag{4}$$

2.4 Training Scheme and Implementation Details

The proposed pipeline was trained in two steps for 300 epochs. SegResNet and CT-line VAE were trained for the first 180 epochs using the Adam optimizer [8] with default coefficients and weight decay values of (0.99, 0.999) and 1e−3, when the losses $\mathcal{L}_{seg}$ and $\mathcal{L}_{CT}$ were minimised. In the subsequent 120 epochs, the loss $\mathcal{L}_{MR}$ was minimised by training the MM-LSA network with another Adam optimizer with the default parameters. On a single GPU (GeForce RTX3090, Nvidia, USA), all networks were trained with a batch size of eight. The cosine annealing learning rate scheduler was used to stabilise the training. In Eq. 4, the values of α, β, and τ were 1.0, 1.0, and 0.07, respectively.

Prior to training, 2D MR images were interpolated using the ABME network, while low-resolution manual segments were interpolated using the nearest neighbourhood method. High in-plane and through-plane resolutions are shared by the interpolated 2D MR images and segments. Interpolated 2D MR images were subjected to random scaling and shifting of the intensity, random contrast adjustments, and the addition of Gibbs and K-space spike noise during training. Data augmentations on CT segments are random rotation along the axis, random resizing, random swapping of labels indexes (ventriclular cavities and myocardiums were labelled with a number as 1, 2, or 3), and random slice shifting mimicking the respiration misalignment in 2D MR images. All images were cropped to include the manual segments of the heart while the image resolution was maintained. The method was implemented with Python 3.8.12, the PyTorch 1.11.0 framework, and the MONAI 0.8.1 repository.

3 Experiment

3.1 Datasets

Our pipeline was trained on low-resolution 2D MR images and 3D CT segments from the ACDC [3] and MM-WHS [20,21], respectively. During testing, 3D MRAs and high-resolution MM-WHS segments were utilised. Segments were labelled as belonging to one of three categories: the LV and RV cavities, and the LV myocardium.

ACDC Dataset. From the dataset, 100 2D MR images of patients were chosen, with 90 cases used for training and 10 for validation. The approximate size of the image was $224 \times 224 \times 13$. These patients are distributed evenly across five pathological categories: normal, myocardial infarction, dilated cardiomyopathy, hypertrophic cardiomyopathy, and abnormal right ventricle. The cine-MR images were acquired using two MRI scanners (1.5 T-Siemens Area, Siemens Medical Solutions, Germany and 3.0 T-Siemens Trio Tim, Siemens Medical Solutions, Germany) to acquire 2D MR sequences in the SAX view. The images have a slice thickness of 5 or 10 mm and an in-plane resolution of 1.34–1.68 mm. Images of end-diastole (ED) and end-systole (ES) were manually segmented.

MM-WHS Dataset. 20 CT segments and 14 3D MRAs were chosen from the dataset. After data augmentation, 100 CT cases were generated, of which 90 were used for training and 10 for validation. All 14 MRA cases were used exclusively for testing. The cardiovascular diseases of these patients included cardiac function insufficiency, hypertension (III), arrhythmia, tetralogy of Fallot (RV hypertrophy), LV dilated cardiomyopathy, and pulmonary artery stenosis.

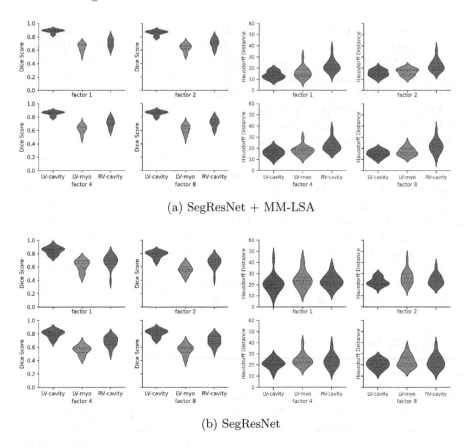

(a) SegResNet + MM-LSA

(b) SegResNet

Fig. 3. Influence of downsampling on pipeline-generated segments as measured by the Dice score and Hausdorff distance. 14 cases of the MM-WHS dataset are utilised to generate the results.

Two 64-slice CT scanners (Philips Medical Systems, Netherlands) were utilised to acquire axial 3D CT angiograms (CTAs) with an average slice thickness of 0.60 mm and an in-plane resolution of 0.44 mm. The 3D MRAs were acquired with a 1.5 T MRI scanner (Philips Healthcare, Netherlands) in axial view, with interpolated in-plane and through-plane resolution of 1 mm. Manual segmentation was performed on 3D CTAs at ED and 3D MRAs at ED and ES. Angiograms and manual segments with a size of $256 \times 256 \times 256$ pixels are rotated from axial to SAX view.

3.2   Segmentation Results

The pipeline was evaluated using 3D MRAs and high-resolution manual segments as the ground truth from the MM-WHS. Angiograms and segments were downsampled by 1, 2, 4, or 8 in the through-plane direction, and ABME was used

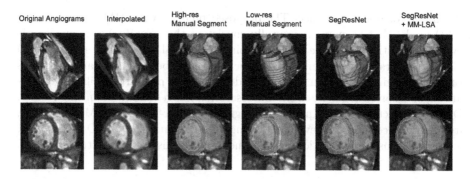

Fig. 4. Hard case comparison in terms of segmentation quality. The LV cavity (Green), LV myocardium (Yellow), and RV cavity (Red) are best viewed in colour (Red). Top row, left to right: 3D MR angiograms in the LAX view and 3D models reconstructed from the segments; bottom row, left to right: 3D MR angiograms in the SAX view and the LV and RV segments. To achieve the same resolution as the original angiograms, the interpolated angiograms were first downsampled by a factor of 8 and then interpolated via ABME. In segmentation, small unconnected components are eliminated. (Color figure online)

to recover resolution. To simulate respiration misalignment in 2D MR images, random slice shifting was performed on 3D MRAs, and the pipeline was expected to correct misalignment in generated segments. As shown in Fig. 3a, Dice score changes little when downsampling by larger factors, indicating that ABME helps preserve anatomical shapes in 3D MRAs for SegResNet. In the majority of cases, the Dice score of the LV cavity is greater than 0.8, and few outliers can be identified. The segmentation of the LV myocardium and RV cavity is difficult because angiograms with low contrast ratios have fuzzy boundaries. Even though the LV myocardium and RV cavity have relatively lower Dice scores, the high-resolution segments generated by the pipeline are accurate against the ground truth.

In addition, the pipeline is evaluated using 3D MRAs, whereas training is performed using 2D MR images. Different protocols affect the region of interest's intensity, contrast ratio, and SNR, making it difficult to verify the efficacy of the MM-LSA. To accomplish this, the generated segments are compared before and after MM-LSA processing. Comparing Fig. 3a and Fig. 3b, it is evident that MM-LSA processing resulted in segments with higher Dice scores and smaller variance. Due to the sensitivity of the Hausdorff distance to changes in segment boundaries, all segments exhibit large Hausdorff distances with some outliers. A conclusion can only be reached after visually inspecting generated segments. The 3D models depicted in Fig. 4 may aid in elucidating the distinction between generated and manual segments.

Interpolated angiograms preserve the anatomical shapes and details of the original 3D MRAs, including the left-right ventricle boundary. Respiration misalignment, on the other hand, was not corrected due to the fact that the ABME interpolates a new slice using only the previous adjacent slices, making the offset inevitable for every new slice. The proposed pipeline with MM-LSA generates a 3D model that is more structurally consistent than SegResNet alone, and is even more distinct than the 3D low-resolution manual model. The surface of 3D pipeline-generated models must be further refined to be as smooth and precise as those of 3D high-resolution manual models, but our pipeline's ability to generate super-resolved segments from low-resolution images shows promising results. Due to the fact that the pipeline is designed to super-resolve 2D MR segments, its performance should be evaluated using high-resolution 2D or 3D MR images, and corrections for respiration misalignment will be discussed in future research.

4   Conclusion

We proposed a pipeline for the segmentation of low-resolution 2D MR images at a high resolution. It is accomplished in three steps. Interpolation in the through-plane direction with ABME yields high-resolution 2D MR images, which are then processed by a SegResNet to yield intermediate segments. Using our Multi-modal Latent-space Self-alignment (MM-LSA) network, these segments are super-resolved with anatomical priors extracted from high-resolution 3D CT segments. The results of training the pipeline on 2D MR images and testing on 3D MRAs demonstrate the effectiveness of the heart's anatomical prior. Our method produces promising results for segmenting the left and right ventricle cavities and myocardium from low-resolution 2D MR images in terms of structural consistency and high-resolution.

Acknowledgement. This work was supported by the Kings-China Scholarship Council PhD Scholarship Program and the National Institutes of Health Grant (R01HL121754).

References

1. Alfakih, K., Reid, S., Jones, T., Sivananthan, M.: Assessment of ventricular function and mass by cardiac magnetic resonance imaging. Eur. Radiol. **14**(10), 1813–1822 (2004)
2. Barkhof, F., Pouwels, P.J., Wattjes, M.P.: The holy grail in diagnostic neuroradiology: 3T OR 3D? (2011)
3. Bernard, O., et al.: Deep learning techniques for automatic MRI cardiac multistructures segmentation and diagnosis: is the problem solved? IEEE Trans. Med. Imaging **37**(11), 2514–2525 (2018)
4. Biffi, C., et al.: 3D high-resolution cardiac segmentation reconstruction from 2D views using conditional variational autoencoders. In: 2019 IEEE 16th International Symposium on Biomedical Imaging (ISBI 2019), pp. 1643–1646. IEEE (2019)
5. Delannoy, Q., et al.: SegSRGAN: super-resolution and segmentation using generative adversarial networks-application to neonatal brain MRI. Comput. Biol. Med. **120**, 103755 (2020)
6. Fedorov, A., et al.: 3D slicer as an image computing platform for the quantitative imaging network. Magn. Reson. Imaging **30**(9), 1323–1341 (2012)

7. Greenspan, H., Oz, G., Kiryati, N., Peled, S.: MRI inter-slice reconstruction using super-resolution. Magn. Reson. Imaging **20**(5), 437–446 (2002)
8. Kingma, D.P., Ba, J.: Adam, a method for stochastic optimization. In: Proceedings of the 3rd International Conference on Learning Representations (ICLR), vol. 1412 (2015)
9. Mauger, C.A., et al.: Right-left ventricular shape variations in tetralogy of Fallot: associations with pulmonary regurgitation. J. Cardiovasc. Magn. Reson. **23**(1), 1–14 (2021)
10. Myronenko, A.: 3D MRI brain tumor segmentation using autoencoder regularization. In: Crimi, A., Bakas, S., Kuijf, H., Keyvan, F., Reyes, M., van Walsum, T. (eds.) BrainLes 2018. LNCS, vol. 11384, pp. 311–320. Springer, Cham (2019). https://doi.org/10.1007/978-3-030-11726-9_28
11. Oktay, O., et al.: Anatomically constrained neural networks (ACNNs): application to cardiac image enhancement and segmentation. IEEE Trans. Med. Imaging **37**(2), 384–395 (2017)
12. Park, J., Ko, K., Lee, C., Kim, C.-S.: BMBC: bilateral motion estimation with bilateral cost volume for video interpolation. In: Vedaldi, A., Bischof, H., Brox, T., Frahm, J.-M. (eds.) ECCV 2020. LNCS, vol. 12359, pp. 109–125. Springer, Cham (2020). https://doi.org/10.1007/978-3-030-58568-6_7
13. Park, J., Lee, C., Kim, C.S.: Asymmetric bilateral motion estimation for video frame interpolation. In: Proceedings of the IEEE/CVF International Conference on Computer Vision, pp. 14539–14548 (2021)
14. Ronneberger, O., Fischer, P., Brox, T.: U-net: convolutional networks for biomedical image segmentation. In: Navab, N., Hornegger, J., Wells, W.M., Frangi, A.F. (eds.) MICCAI 2015. LNCS, vol. 9351, pp. 234–241. Springer, Cham (2015). https://doi.org/10.1007/978-3-319-24574-4_28
15. Rousseau, F., Habas, P.A., Studholme, C.: A supervised patch-based approach for human brain labeling. IEEE Trans. Med. Imaging **30**(10), 1852–1862 (2011)
16. Sohn, K.: Improved deep metric learning with multi-class N-pair loss objective. Adv. Neural Inf. Process. Syst. **29** (2016)
17. Wang, S., et al.: Joint motion correction and super resolution for cardiac segmentation via latent optimisation. In: de Bruijne, M., et al. (eds.) MICCAI 2021. LNCS, vol. 12903, pp. 14–24. Springer, Cham (2021). https://doi.org/10.1007/978-3-030-87199-4_2
18. Wolberg, G.: Digital Image Warping, vol. 10662. IEEE Computer Society Press, Los Alamitos (1990)
19. Xue, T., Chen, B., Wu, J., Wei, D., Freeman, W.T.: Video enhancement with task-oriented flow. Int. J. Comput. Vision **127**(8), 1106–1125 (2019)
20. Zhuang, X.: Multivariate mixture model for myocardial segmentation combining multi-source images. IEEE Trans. Pattern Anal. Mach. Intell. **41**(12), 2933–2946 (2018)
21. Zhuang, X., Shen, J.: Multi-scale patch and multi-modality atlases for whole heart segmentation of MRI. Med. Image Anal. **31**, 77–87 (2016)

Towards Real-Time Optimization of Left Atrial Appendage Occlusion Device Placement Through Physics-Informed Neural Networks

Xabier Morales$^{(\boxtimes)}$, Carlos Albors, Jordi Mill, and Oscar Camara

PhySense, Department of Information and Communication Technologies,
Universitat Pompeu Fabra, Barcelona, Spain
xabier.morales@upf.edu

Abstract. The adoption of patient-specific computational fluid dynamics (CFD) simulations has been instrumental toward a better understanding of the mechanisms underlying thrombogenesis in the left atrial appendage. Such simulations can help optimize the placement of left atrial appendage occlusion (LAAO) devices in atrial fibrillation patients and avoid the generation of device-related thrombosis. However, integrating conventional solvers into clinical practice is cumbersome, as even the slightest change in model geometry involves computing the entire simulation from scratch. In contrast, neural networks can entirely circumvent this issue by transferring knowledge across models targeted at similar physical domains. Thus, in the present study, we introduced a neural network capable of predicting left atrial hemodynamics under different occlusion device configurations, relying solely on a single finite element simulation for training. To this end, we leveraged physics-informed neural networks (PINN), which embed the physical laws governing the domain of interest into the model, exhibiting far superior generalization capabilities than conventional data-driven models. Several device types and positions have been tested in two distinct left atrial geometries. By employing a single reference simulation per patient the network can predict the updated hemodynamics for a variety of device types and positions, orders of magnitude faster than with conventional CFD solvers.

Keywords: Physics-informed neural networks · Device related thrombosis · Computational fluid dynamics · Left atrial appendage occlusion

1 Introduction

Atrial fibrillation (AF) is the most common clinically relevant arrhythmia, characterized by irregular contraction of the left atrium leading to an increased risk of thrombosis [10]. In fact, non-valvular AF is responsible for 15–20% of all strokes, 99% of which can be ascribed to thrombi generated in the left atrial appendage (LAA) [7]. Although anticoagulant therapy is the current standard

© The Author(s), under exclusive license to Springer Nature Switzerland AG 2022
O. Camara et al. (Eds.): STACOM 2022, LNCS 13593, pp. 36–45, 2022.
https://doi.org/10.1007/978-3-031-23443-9_4

treatment, transcutaneous left atrial appendage occlusion (LAAO) has emerged as the preferred alternative for contraindicated patients [15]. However, beyond the risks associated with surgery, a sub-optimal occluder landing zone, the site where the device is anchored to the LAA wall, can lead to device-related thrombosis (DRT). Recent studies show that an uncovered pulmonary ridge (PR), the area connecting the LAA to the left superior pulmonary vein, following occluder implantation, may entail a 20-fold increase in DRT risk [9].

In this regard, computational fluid dynamics (CFD), has provided invaluable insight into the driver factors of DRT [13]. Accurate CFD simulations, however, are extremely time-consuming, which ultimately precludes their widespread use in clinical practice, including real-time user interaction with the simulations (e.g., for virtual and interactive testing of device settings [3]). As a result, neural networks are increasingly being introduced to complex dynamical systems, including the prediction of advanced cardiac hemodynamic parameters [8]. However, most such studies are based on fully-supervised data-driven models that require massive amounts of data to train. Physics-informed neural networks (PINN) simultaneously tackle these issues by embedding prior knowledge of the physical laws governing the dynamical system into the model [6].

Hence, in this study, we developed a fast surrogate of CFD simulations based on PINNs to quickly assess the hemodynamic changes derived from different LAAO device configurations. Unlike conventional finite element solvers, which must compute the simulation from scratch after any change in boundary conditions, our model can build on previous simulations by enforcing the Navier-Stokes equations in the target region. Most importantly, our model accurately predicts the complex time-resolved hemodynamics of any given window of the cardiac cycle, orders of magnitude faster than conventional numerical methods.

2   Methods

The present study first involved generating the ground-truth data by performing CFD simulations on the whole LA through finite-element solvers. Subsequently, the network was trained on a small rectangular segment of the LA divided into two domains, as shown in Fig. 1. In the proximity of the device (shown in blue) the model is free to generate the updated hemodynamics by enforcing the Navier-Stokes equations, while on the remainder (shown in grey) velocity data from a CFD simulation is employed as reference.

2.1   Computational Fluid Dynamics Simulations

Two AF patients were included in the study, provided by Hospital Haut-Lévêque (Bordeaux, France), after approval by the institutional Ethics Committee. The data consisted of pre-procedural high-quality computed tomography scans of patients undergoing LAAO implantation. A total of 8 simulations were completed, with the following variations: two device types per patient, including the Watchman FLX (Boston Scientific, Marlborough, Massachusetts, United States)

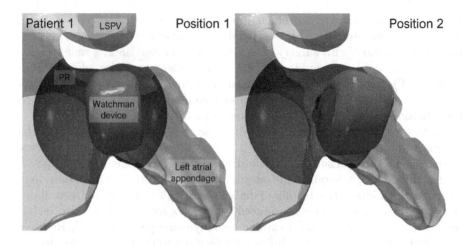

Fig. 1. Domain setup for the training of the physics-informed neural network. The model is left to predict the velocity field inside the blue sphere, enforcing the Navier-Stokes equations. Elsewhere, the network is fed the solution from a reference simulation. A virtually implanted Watchman device can be observed covering or exposing the pulmonary ridge (Position 1 and Position 2, respectively). PR: Pulmonary ridge, LSPV: Left superior pulmonary vein (Color figure online).

[2], and the Amplatzer Amulet (St.Jude Medical-Abbott, St. Paul, Minnesota, United States) [1]; and two different positions per device, one completely covering the PR (Position 1) and the other with the device deeper into the LAA, thus leaving the PR exposed (Position 2).

The virtual deployment of the LAAO device was simulated in Meshmixer (Autodesk Meshmixer v3.3.15)[1]. The device shape, scale, and location were chosen according to the guidelines (size and compression ratio) provided by the device manufacturers, based on the maximum and minimum diagonals of the LAA contour landing zone. The lobe, union, and disk parts of the Amplatzer Amulet device were treated separately to mimic real-life behavior. The final device setups underwent approval by an experienced clinician.

Simulations were computed on Ansys Fluent 19 R32 (ANSYS Inc,USA)[2]. The boundary conditions were set up according to the study by Mill et al. [14]. Passive LA motion induced by the longitudinal motion of the left ventricle, was simulated through a spring-based dynamic mesh. The fluid simulations included two full heartbeats, aiming to reach a steady state; during the training of the deep learning model, only the latter was considered as ground-truth.

2.2 Domain Setup

The domain setup for the training of the PINN is depicted in Fig. 1. To improve the sampling density without running into memory issues, we decided to narrow

---

[1] https://www.meshmixer.com/.
[2] https://www.ansys.com/products/fluids/ansys-fluent.

down the domain to a three-dimensional rectangular cut of the LA. This domain comprised all major elements influencing hemodynamics in the vicinity of the device, including the left superior pulmonary vein (LSPV), the LAA, the PR, and a piece of the mitral valve.

The training domain was then divided into two parts. On the one hand, the inference domain, is defined by a 20 mm sphere covering the PR and surface of the device (colored in blue in Fig. 1) in which the network is free to predict the new flow field. This simplification was performed under the assumption that the type and placement of the device only affects hemodynamics locally. Elsewhere the network was fed a sparse velocity field from the ground-truth simulation chosen as reference.

As we know that the solution from the finite-element solver must fully comply with the laws governing fluid dynamics, the Navier-Stokes equations were only enforced on a sphere just 2 mm larger than the inference domain. This overlapping was meant to avoid an abrupt discontinuity between the data-driven and unsupervised domains.

2.3   Physics-Informed Neural Network

Physics-informed neural networks (PINNs) seamlessly integrate information from both measurements and partial differential equations (PDEs) by embedding them as a soft-constrains into the loss function [11]. Although the current generation of PINNs cannot compete with classical finite-element solvers in terms of accuracy, they offer superior performance than any CFD solver in scenarios with scattered partial spatio-temporal data [6].

In our case, we are interested in training a neural network to approximate the velocity field, $u(x, t)$, as a function of space and time after a change of LAAO device configuration. This is achieved by minimizing the following weighted loss:

$$\mathcal{L} = w_1 \mathcal{L}_{PDE} + w_2 \mathcal{L}_{data} + w_3 \mathcal{L}_{BC}, \tag{1}$$

where $\mathcal{L}_{PDE}$ is the unsupervised PDE loss represented by the 3D Navier-Stokes equations for incompressible flow, $\mathcal{L}_{data}$ is the data loss from the reference flow field, $\mathcal{L}_{BC}$ is the boundary condition (BC) loss, and w_{1-3} are the weighting coefficients for the loss terms. If the Navier-Stokes equations are naively embedded in $\mathcal{L}_{PDE}$, the resulting loss values can easily lead to exploding gradients. Therefore, we chose to non-dimensionalize the equations through the following variable transformations:

$$x^* = \frac{x}{L}, \qquad u^* = \frac{u}{U}, \qquad t^* = \frac{t}{T}, \qquad p^* = \frac{p}{U^2, \rho} \tag{2}$$

$$\nabla u^* = 0, \qquad \frac{\partial u^*}{\partial t^*} + (u^* \cdot \nabla^*) u^* = -\nabla^* p^* + \frac{1}{Re} \nabla^{*2} u^*, \tag{3}$$

where u^*, p^* and t^* represent the non-dimensionalized velocity, pressure and time variables. The characteristic length, L, the characteristic velocity, U, and the characteristic time, T, were chosen so that the non-dimensionalized variables were normalized. The blood was assumed to be Newtonian with a viscosity $\mu = 0.0035$ Pa $\cdot$ s and density $\rho = 1060$ kg/m^3.

The model was implemented by leveraging Nvidia Modulus[3], a powerful Open-Source multi-physics framework for AI-driven physics problems. Mainly, we used the functionality enabling the definition of a domain geometry through triangular meshes, greatly simplifying the import of CFD data. Likewise, no-slip boundary conditions were set on the LA and device surfaces. As we considered LA motion, a separate no-slip constraint had to be defined for each time step. To avoid fluid leakage through the device, the no-slip loss was given a weight of 10. A batch size of 4000 uniformly sampled points was chosen for $\mathcal{L}_{BC}$ and $\mathcal{L}_{PDE}$, while only 1000 data points were sampled in each epoch to enforce $\mathcal{L}_{data}$. We selected a simple fully-connected neural network, as it was the least computationally demanding option. After a thorough grid search, we settled on an 8-layer-deep model, with a width of 256. Adam was chosen as the optimizer, using a learning rate of 0.001 with exponential decay. The model was trained on an NVIDIA Quadro RTX A6000 (48 GB of VRAM) for a total of 5000 epochs for each case.

Due to the memory constraints, we were limited to predicting five-time steps at a time. Three different time windows of the cardiac cycle were chosen, comprising end-diastole, diastasis, and end-diastole, in relation to the presence of low-velocity re-circulations near the occluder surface in the ground-truth simulations, based on observations in [12].

It should be noted that, unlike supervised models, PINN solvers require training for each instance making them more akin to classic numerical methods. To predict the hemodynamics for a given LAAO device configuration, a finite-element simulation needs to be chosen as $\mathcal{L}_{data}$ referred to as reference. Then, the geometry of the target device is used to define the new $\mathcal{L}_{BC}$. Finally, stochastic gradient descent is employed to minimize the weighted loss. Since, $\mathcal{L}_{PDE}$, is completely unsupervised, it allows the prediction of the hemodynamics adjusted to new geometry without the need for data inside the inference domain. Thus, the task assigned to the network is to reconstruct the flow field from partial data (reference simulation) while constrained to comply with the Navier-Stokes equations and no-slip boundary conditions. For each patient, we trained the network for all possible combinations of reference ground-truth simulations and device configurations, totaling 16 different runs per patient and time window.

3   Results

Each training run of the PINN solver took approximately 10 min (2 min per time-step). On the other hand, it took at least 24 h to complete every single ground-truth simulation for a total of 8 min per time step.

The accuracy results for all possible device configurations are given in Table 1, in terms of the mean absolute error (MAE) of the non-dimensionalized velocity

---

[3] https://developer.nvidia.com/modulus.

Table 1. Prediction accuracy results from the non-dimensionalized velocity in terms of the mean absolute error (MAE). The Self column corresponds to runs where the device setup of the simulation used as $\mathcal{L}_{data}$ is the same as the one predicted by the network. The remaining runs using a different device setup as $\mathcal{L}_{data}$ and $\mathcal{L}_{data}$ have been averaged in the mixed column to give a sense of the generalization error.

Timesteps	Patient 1								Patient 2							
	Self				Mixed				Self				Mixed			
	A1	A2	W1	W2	A1	A2	W1	W2	A1	A2	W1	W2	A1	A2	W1	W2
30–35	0.02	0.05	0.02	0.05	0.10	0.13	0.10	0.15	0.04	0.04	0.03	0.02	0.20	0.23	0.24	0.16
50–55	0.02	0.04	0.02	0.04	0.10	0.11	0.10	0.13	0.02	0.03	0.02	0.03	0.09	0.12	0.13	0.10
80–85	0.02	0.04	0.03	0.02	0.09	0.09	0.10	0.10	0.03	0.05	0.03	0.03	0.22	0.21	0.22	0.19

(normalized to be between 0 and 1). The runs in which the device setup in $\mathcal{L}_{data}$ and $\mathcal{L}_{BC}$ are the same have been separated from the rest in the Self column of Table 1. It can be observed that the cases in which the reference simulation comes from a distinct device (Mixed column) tend to have a worse MAE. It must be stated, however, that the MAE of the non-dimensionalized velocity between the ground truth CFD simulations of each patient ascended to 0.2 inside the training domain during the three analyzed time windows.

Flow streamlines (see Fig. 2), representing lines tangential to the instantaneous velocity direction, were created to qualitatively analyze the simulated blood flow patterns. A pair of predictions are shown for each of the analyzed cardiac cycle intervals. The reference CFD simulation for each time window pair is located in the top right corner.

Inspection of the streamlines seem to indicate that the predicted velocity magnitudes correlate better with CFD when the $\mathcal{L}_{data}$ velocity field originates from a simulation with the same device setup (corresponding to the first row of each time window pair). Nevertheless, the network appears to predict key characteristics of the hemodynamics associated with each occluder type and position, regardless of utilized reference flow field. For instance, the PINN model successfully captures a series of low-velocity re-circulations in all 3 devices showcasing an uncovered PR (Position 2), highlighted with a black arrow in Fig. 2. Conversely, the network predicts predominantly laminar flow in the surfaces of the devices properly covering the PR (Position 1). Besides, the network was even capable of predicting leakages in those devices in which the LAA was not completely sealed, as marked by the red arrow in the uncovered Amplatzer device (Position 2) in patient 1 during diastasis.

4   Discussion

Results in Fig. 2 that the PINN solver has effectively incorporated the Navier-Stokes equations into the model to generate complex time-resolved velocity fields. Using a single CFD simulation as a reference, the network successfully captured key characteristics of the hemodynamics on the surface of different LAAO devices. Low-velocity re-circulations, such as those predicted in devices in Position 2, have been identified as one of the main drivers of DRT [12,13].

Nonetheless, the results in Table 1 evidence better accuracy when the reference simulation and target device coincide. This is most likely attributable to the unexpected high discrepancy observed in the hemodynamics of the ground truth simulations. As the reference simulation differs so much from the real simulation at the boundaries of the inference domain, it becomes impossible for the network to reconstruct the target velocity field.

This can be clearly seen in Fig. 2 in the Position 2 Amplatzer example, where the network could not predict the vortices observed on the surface of the LAAO device with the uncovered PR (Position 2) as no such rotatory motion is observed on the reference simulation (device with covered PR, Position 1). Perhaps a less restrictive training configuration with a larger vacant domain could be beneficial giving the model more room to predict the updated hemodynamics.

While admittedly, given the accuracy edge provided by finite-element solvers the difference in computational time to predict individual time steps is not that high, CFD simulations need to be computed entirely from scratch with every single change on boundary conditions, while the network can efficiently take advantage of the previously generated solution. Ultimately, this yields a significant time and computational resource advantage to the PINN solver.

Unfortunately, the current training strategy is far from optimal owing to the current paradigm in PINNs. Since only the spatial and time coordinates are included as model inputs, given a new instance for inference, the network is incapable of understanding any new boundary conditions, as these are only included as guidance during stochastic gradient descent [11]. The inclusion of geometry parametrization could circumvent the need for several training runs [16]. In fact, major advancements in inverse designing could allow the simultaneous gradient-based design optimization of several domain parameters such as the position, size, and shape of the device given the right reward function [4].

Lastly, more weight-efficient models such as convolutional neural networks could have been employed [5]. However, no PINN model has yet been proposed to handle convolutions on unstructured data such as point clouds and meshes, which is often the preferred representation for anatomical data.

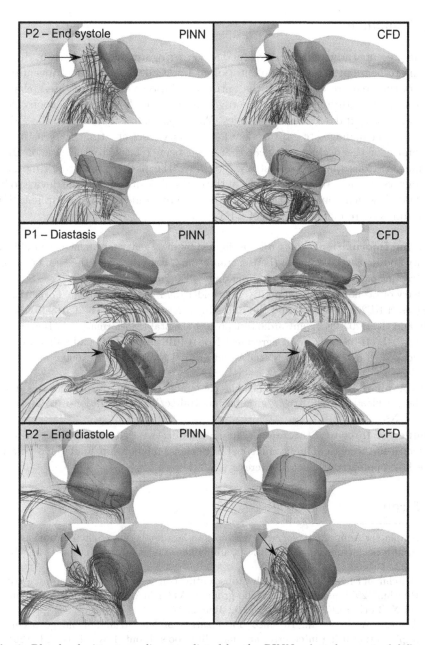

Fig. 2. Blood velocity streamlines predicted by the PINN solver for a set of different device configurations. For each pair of predictions the CFD simulation chosen as the reference velocity field to compute $\mathcal{L}_{data}$ is shown on the top right corner. The streamlines are colored based on velocity values on a scale from 0 to 0.2 m/s (blue to red). Black arrows indicate low velocity re-circulation areas. PINN: Physics-informed neural network, CFD: Computational fluid dynamics, P1: Patient 1, P2: Patient 2 (Color figure online).

5　Conclusion

In the present study, we have developed a fast surrogate of CFD solvers in the LA aiming to help optimize the implantation of LAAO devices. Unlike conventional solvers that require computing the entire simulation with the slightest change in boundary conditions, our model only requires a single reference velocity field to predict the updated hemodynamics. We have demonstrated that the PINN model solver manages to capture key low-velocity re-circulation patterns in the proximity of the occluder device, that have been related to the risk of DRT formation. Consequently, our study may pave the path toward real-time user interactive pre-operative assessment of the optimal positioning of the device to be implanted, which could drastically improve the outcome of LAAO implantation procedures by reducing the risk of DRT.

Funding Information. This work was funded by the Agency for Management of University and Research Grants of the Generalitat de Catalunya under the Grants for the Contracting of New Research Staff Programme - FI (2020 FI_B 00608) and the Spanish Ministry of Economy and Competitiveness under the Programme for the Formation of Doctors (PRE2018-084062), and the Retos Investigación project (RTI2018-101193-B-I00), and the H2020 EU SimCardioTest project (Digital transformation in Health and Care SC1-DTH-06-2020; grant agreement No. 101016496). Additionally, this research was supported by grants from NVIDIA and utilized NVIDIA RTX A6000.

Data Availability Statement. The files containing the code to reproduce this study can be downloaded from https://github.com/Xtaltec/RT-optimization-LAAO-placement-PINNs.

Conflict of Interest Statement. The authors declare that the research was conducted in the absence of any commercial or financial relationships that could be construed as a potential conflict of interest.

References

1. Amplatzer Amulet. https://www.cardiovascular.abbott/int/en/hcp/products/structural-heart/structural-interventions/amplatzer-amulet.html. Accessed 01 June 2022
2. Watchman FLX. https://www.bostonscientific.com/content/dam/bostonscientific/Rhythm%20Management/portfolio-group/WATCHMAN%20FLX/eu/Watchman_FLX_Brochure_EN.pdf. Accessed 02 June 2022
3. Aguado, A.M., et al.: In silico optimization of left atrial appendage occluder implantation using interactive and modeling tools. Front. Physiol. **10**, 237 (2019)
4. Allen, K.R., et al.: Physical design using differentiable learned simulators (2022). https://doi.org/10.48550/ARXIV.2202.00728, https://arxiv.org/abs/2202.00728
5. Bai, K., Wang, C., Desbrun, M., Liu, X.: Predicting high-resolution turbulence details in space and time. ACM Trans. Graph. **40**(6), 1–16 (2021)
6. Cai, S., Mao, Z., Wang, Z., Yin, M., Karniadakis, G.E.: Physics-informed neural networks (PINNs) for fluid mechanics: a review. Acta. Mech. Sin. **37**(12), 1727–1738 (2021)

7. Cresti, A., et al.: Prevalence of extra-appendage thrombosis in non-valvular atrial fibrillation and atrial flutter in patients undergoing cardioversion: a large transoesophageal echo study. EuroIntervention **15**(3), e225–e230 (2019)
8. Ferez, X.M., et al.: Deep learning framework for real-time estimation of in-silico thrombotic risk indices in the left atrial appendage. Front. Physiol. 12, 694545 (2021)
9. Freixa, X., et al.: Pulmonary ridge coverage and device-related thrombosis after left atrial appendage occlusion. EuroIntervention **16**(15), e1288–e1294 (2021)
10. Hirose, T., et al.: Left atrial function assessed by speckle tracking echocardiography as a predictor of new-onset non-valvular atrial fibrillation: results from a prospective study in 580 adults. Eur. Heart J.-Cardiovasc. Imaging **13**(3), 243–250 (2011)
11. Karniadakis, G.E., Kevrekidis, I.G., Lu, L., Perdikaris, P., Wang, S., Yang, L.: Physics-informed machine learning. Nat. Rev. Phys. **3**(6), 422–440 (2021)
12. Mill, J., et al.: Patient-specific flow simulation analysis to predict device-related thrombosis in left atrial appendage occluders. REC: Interv. Cardiol. **3**, 278–285 (2022) (English Edition)
13. Mill, J., et al.: Sensitivity analysis of in silico fluid simulations to predict thrombus formation after left atrial appendage occlusion. Mathematics **9**(18), 2304 (2021)
14. Mill, J., et al.: In-silico analysis of the influence of pulmonary vein configuration on left atrial haemodynamics and thrombus formation in a large cohort. In: Ennis, D.B., Perotti, L.E., Wang, V.Y. (eds.) FIMH 2021. LNCS, vol. 12738, pp. 605–616. Springer, Cham (2021). https://doi.org/10.1007/978-3-030-78710-3_58
15. Reddy, V.Y., et al.: PREVAIL and PROTECT AF Investigators: 5-year outcomes after left atrial appendage closure: from the PREVAIL and PROTECT AF trials. J. Am. Coll. Cardiol. **70**(24), 2964–2975 (2017)
16. Sun, L., Gao, H., Pan, S., Wang, J.X.: Surrogate modeling for fluid flows based on physics-constrained deep learning without simulation data. Comput. Methods Appl. Mech. Eng. **361**, 112732 (2020)

Haemodynamic Changes in the Fetal Circulation Post-connection to an Artificial Placenta: A Computational Modelling Study

Maria Inmaculada Villanueva[1]([✉]), Marc López[2], Sergio Sánchez[3],
Patricia Garcia-Cañadilla[3,4], Paula C. Randanne[3], Ameth Hawkins[3],
Elisenda Eixarch[3], Elisenda Bonet-Carne[3,7], Eduard Gratacós[3],
Fàtima Crispi[2,3], Bart Bijnens[3,5], and Gabriel Bernardino[6]

[1] Physense, DTIC, Universitat Pompeu Fabra, Barcelona, Spain
mariainmaculada.villanueva@upf.edu
[2] Universitat de Barcelona, Barcelona, Spain
[3] BCNatal - Barcelona Center for Maternal-Fetal and Neonatal Medicine (Hospital
Sant Joan de Déu and Hospital Clínic, August Pi i Sunyer Biomedical Research
Institute (IDIBAPS), University of Barcelona), Barcelona, Spain
[4] Cardiovascular Diseases and Child Development,
Institut de Recerca Sant Joan de Déu, Barcelona, Spain
[5] ICREA - Catalan Institution for Research and Advanced Studies, Barcelona, Spain
[6] Univ Lyon, Université Claude Bernard Lyon 1, INSA-Lyon, CNRS, Inserm,
CREATIS UMR 5220, U1294, 69621 Lyon, France
[7] Universitat Politècnica de Catalunya, BarcelonaTech, Barcelona, Spain

Abstract. Artificial placenta (AP) is a promising approach to improve
survival in extremely premature fetuses, removing them from the uterus
and replacing the placenta by an external oxygenator. In *in vivo* experiments with lambs, haemodynamic changes were observed in the umbilical
artery (UA) post-connection to the AP: decrease in flow pulsatility index
(PI), increase in mean pressure and heart rate, and flatter diastolic flow
with a secondary peak.

Computational models can help to understand the causes of observed
phenomena. Therefore, the clinical objective of this work was to investigate the causes of the aforementioned changes in UA velocity, while
the methodological objective was to implement a computational model
capable of reproducing the behaviour of an AP.

We used a closed lumped model of the whole fetal circulation, modified
to include the AP, and tested the haemodynamic changes in the UA after
altering heart rate and AP's resistance and compliance. We also added a
simple wave reflection model and studied its effect on pressure and flow
traces. We found that reducing AP's resistance increased mean flow, and
reducing compliance decreased velocity PI. When adding the reflection
model, the flatter diastolic flow observed in the UA was reproduced.

This study suggests that the observed haemodynamics changes in the
UA post-connection to the AP are due to a combination of decreased
compliance and resistance, as well as wave reflections from components
within the AP circuit.

O. Camara et al. (Eds.): STACOM 2022, LNCS 13593, pp. 46–55, 2022.
https://doi.org/10.1007/978-3-031-23443-9_5

Keywords: Artificial placenta · Fetal cardiology · Lumped cardiovascular model · Ultrasound · Wave reflection

1 Introduction

Preterm birth is a leading cause of neonatal death [1]. Individuals born extremely premature, before 28 weeks of gestation (GA) according to the World Health Organisation (WHO) [1], account for the majority of these deaths, given that they get oxygen and nutrients through the placenta, since their organs are still underdeveloped. Consequently, these individuals are not prepared for ex-uterine life [6,12].

Therefore, a promising approach to improve their survival is to connect them to an artificial placenta (AP) [3,8,11]. The fetus is surgically removed from the uterus and connected through the umbilical cord to a pumpless circuit, where the placenta is replaced by an external low resistance oxygenator. This allows the fetus to continue gestation in an artificial environment that provides oxygen and nutrients through the oxygenator.

In *in vivo* experiments with lambs, specific (Doppler-based) haemodynamic changes have been observed in the umbilical artery (UA) when the fetus is connected to an AP [10]. Specifically, diastolic velocity becomes constant with a small secondary peak, while pulsatility index (PI) decreases and mean flow, pressure and heart rate (HR) increase, which can be harmful and potentially lead to heart failure and ascites.

These changes may be explained by differences in the impedance between the biological and AP, the latter being less resistant and more rigid (less compliant). In addition, the natural placenta is optimised to maximise efficiency and avoid reflections [4]. It is composed of multiple ramifications and small subdivisions that act as wave-traps, in contrast to the filter-like oxygenator. Therefore, jointly with the cannula material properties being different from the natural vessels, the probability to have wave reflections (WR) is higher.

Computational models have been widely used to facilitate the study of the cardiovascular system, the mechanisms of haemodynamics and the relationships between its elements. To investigate the global distributions of blood flow and pressure under a range of physiological conditions and the interactions between the elements of the cardiovascular system, 0D models are generally used [16]. With this approach, a complete and consistent description is provided, and some fetal variables that are difficult to examine by non-invasive measurements can be investigated [14]. Previously published papers have modelled WR at the maternal-fetal interface [5,15], but the AP has been overlooked.

The objective of this work was to use computational models to investigate the potential causes of the changes in haemodynamic patterns observed in the UA when connecting a fetus to an AP, which we hypothesised to be a combination of the difference in placental impedance and the presence of WR. Our contributions are the implementation of WR at the oxygenator of an existing 0D lumped model of the cardiovascular system, and the simulation of the changes produced by the connection to the AP.

2   Methodology

2.1   *In vivo* experiments

110 days lamb fetuses (equivalent of a 23 weeks GA human fetus' pulmonary development [11]) were delivered by caesarean section, and connected to the AP circuit consisting of an approximately 40 cm long 1/4″ × 1/16″ soft polyvinyl chloride (PVC) tube (Sorin Group) and a hollow fiber oxygenator (Quadrox-ID Neonatal Oxygenator, Maquet), thanks to a polyurethane (PUR) cannula. Both UA were trimmed at a distance of approximately 6–8 cm. The oxygenator output was connected to the umbilical vein via another PVC tube and PUR cannula (Fig. 1A). Pressure and velocity in the intra-abdominal section of the UA immediately before and after the connection to the circuit were acquired via an intravascular catheter (Millar Mikro-Tip) and Doppler ultrasound Vivid iQX (General Electric). These devices were not synchronised and no electrocardiogram was recorded. The experimental procedure was approved by the local Animal Experimental Ethics Committee (code 11314).

2.2   Lumped Model

To study gross haemodynamic changes, we implemented a closed lumped model of the entire fetal circulation as described by Pennati *et al.* [14] in Matlab Simulink (Mathworks, vR2021b). Model parameters were scaled using the method proposed in [13] to simulate a 23 week GA fetus from the original 38 weeks model. The model consists of two distinct sections: the heart, with both ventricles and atria, which are modeled as time-varying elastances and the valves; and the blood circulation represented by 19 elements, combining arterial segments (RLC) and vascular bed compartments (RC), among which is the placenta. As shown in Fig. 1B, both biological and AP are modelled as a 2 element Windkessel, upstream connected to the UA resistor and downstream attached to the umbilical vein capacitor. The pressures and blood flows reported in this work were measured between the UA resistor and the placental capacitor.

In order to evaluate how the decrease in the placental impedance and the increase in HR influence the haemodynamics in the UA, placental resistance (R_{plac}) and compliance (C_{plac}) were decreased up to a 2-fold decrease, and HR was systematically increased from 130 to 220 beats per minute (bpm), while the remaining parameters were kept unchanged.

2.3   Wave Reflections

Pressure (P) and flow (Q) in the cardiovascular system propagate as waves [2], and therefore can be decomposed in a forward and backward wave, each moving with a wave speed c:

$$Q(x,t) = Q_f(x + ct) - Q_b(x - ct); \quad P(x,t) = P_f(x + ct) + P_b(x - ct) \quad (1)$$

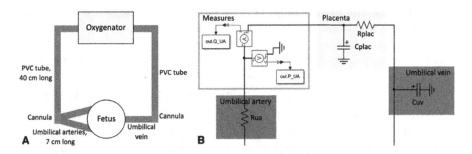

Fig. 1. (A) Scheme of the experimental setup and (B) equivalent lumped computational model of the placenta section, indicating where the UA blood flow and pressure are measured.

with x being the position and t the time. Note that pressure waves are added, while flow waves are subtracted since blood moving backwards counteracts the blood moving forwards. The wave speed c, also called pulse wave velocity (PWV), can be computed from the geometry and material characteristics of the vessel using Korteweg-Moens equation [7]:

$$c = \sqrt{\frac{E \cdot h}{2r\rho}} \qquad (2)$$

where E is the Young modulus, h the wall thickness, r the vessel radius and ρ the blood density. When flow/pressure forward waves encounter a change in PWV, for instance due to a change in the vessel properties, a backward wave is generated, with an intensity depending on the reflection coefficient Γ:

$$Q_b = \Gamma Q_f; \qquad P_b = \Gamma P_f \qquad (3)$$

The reflection coefficient Γ depends on the difference of the PWV between the two media (a and b):

$$\Gamma = \frac{c_a - c_b}{c_a + c_b} \qquad (4)$$

To assess the effect of WR in the UA, we added a simple reflection model to our lumped model, in which the UA flow (Q_{UA}) and pressure (P_{UA}) were considered to be a combination of original (Q_f, P_f), travelling forward, and reflected waves (Q_b, P_b), travelling backwards, satisfying the relationship of Eq. 3. The time delay (Δt), the phase shift between the forward and backward waves, was calculated from the PWV and the distance to the reflection origin.

Two possible WR origins were considered, the UA-PVC tube and the PVC tube-oxygenator interfaces. For the first case, Γ and Δt were estimated by assuming 7 cm of UA and computing the PWV in each material. For the second case, Δt was computed by assuming 7 cm of UA and 40 cm of PVC tube. However, the PWV in the oxygenator could not be calculated due to its complex geometry, so Γ could not be estimated. Therefore, the previous Γ was used and then

Table 1. Quantitative measurements of the velocity (v) and pressure (P) traces pre-
(first value) and post-connection to the AP (second value).

Indicator	Subject				Average
	#1	#2	#3	#4	
Heart rate (bpm)	131/205	134/210	141/245	124/147	133/202
FWHM P/v ratio	0.9/1.4	0.8/1.3	0.8/1.2	0.8/1.0	0.9/1.2
Pulsatility index	1.5/0.7	2.3/0.3	1.4/0.5	2.1/0.6	1.8/0.5
Mean pressure (mmHg)	65.4/91.7	38.7/37.2	32.5/97.9	49.6/61.5	46.5/72.1
Mean velocity (cm/s)	38.7/42.3	22.4/41.0	33.2/72.9	31.4/43.1	31.4/49.8

we did a parametric analysis changing its value to study the effect of different
reflection coefficients. The parameters needed to compute the PWV in the UA
were extracted from the literature [9,17], and those of the PVC from the man-
ufacturer's specifications, except for the Young modulus which was measured.

2.4   Signal Analysis of Pressure and Flow Traces

Several haemodynamic parameters were calculated to quantify both real and simulated blood flow velocity and pressure waveforms including: their mean value
pre- and post-connection to the AP, the pressure/flow full width at half maximum (FWHM) ratio, and the velocity PI. FWHM was defined as the difference
between the two instants of time when the curve is equal to half its maximum
value [18]. Also, PI was defined as the difference between the maximum systolic
velocity and the minimum diastolic velocity divided by the mean velocity.

3   Results

3.1   *In vivo* Pressure and Flow Signal Analysis

The collected Doppler velocity and catheter pressure traces of 4 individuals, measured at the UA, pre- and post-connection to the AP, are displayed in Fig. 2.
Note that data acquired were not time-synchronised due to experimental limitations, thus ventricular systole and diastole could not be indicated on the plots.
We can observe two main changes in the velocity profile: a decrease in PI and
a profile change in diastole, that before intervention presented an exponential
decay, and after intervention is transformed into a rapid decay followed by an
almost constant phase. In the pressure traces we can observe an increase in mean
pressure.

Table 1 shows the quantitative haemodynamic indicators for all 4 subjects
and the population average, before and after intervention. They show an increase
in heart rate, mean pressure and velocity and FWHM pressure/velocity ratio,
and a decrease in the UA PI after intervention.

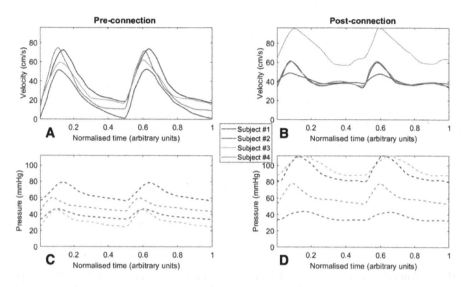

Fig. 2. UA Doppler velocity traces (A) pre- and (B) post-connection to the AP, and their corresponding (C) pre- and (D) post-connection to the AP pressure traces of 4 individuals. Ventricular systole and diastole could not be depicted due to the lack of synchronisation.

3.2 Lumped Model of the Artificial Placenta

We simulated the after-intervention haemodynamics in a lumped model of a 23 GA fetus by lowering the compliance and resistance of the placental component, as the oxygenator is stiffer and less resistant than the natural placenta, as well as increasing the HR. Results are displayed in Fig. 3, where each change is analysed separately. Ventricular systole and diastole are indicated on the graphs, although they may not be as accurate as in the heart. Lowering compliance decreased flow PI and increased pressure PI, lower resistances increased blood flow and decreased pressures, and higher heart rates resulted in lower PI. Afterwards, we manually selected the combination of HR = 200 bpm, C_{plac} = 0.0033 ml/mmHg and R_{plac} = 1.796 mmHg·s/ml as with these values we obtained the shape of the traces most similar to the experimental ones (Fig. 3D). However, the mean pressure was unrealistic and no significant change in the flow profile was observed.

3.3 Wave Reflections

We used the simple WR model to simulate possible WR at the UA-PVC interface or at the oxygenator level. Using the equations described in Sect. 2.3, we computed the theoretical PWV for the UA and PVC, obtaining 6.44 m/s and 39.02 m/s, respectively. For WR originated at the UA-PVC interface, Γ was estimated to be 0.72 and the time delay Δt to be 0.022 s, assuming 7 cm of UA. For

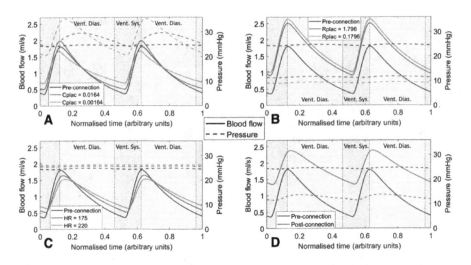

Fig. 3. Simulated UA blood flow (solid line) and pressure (dashed line) for a subject connected to the AP, with different values of (A) placental compliance, (B) placental resistance and (C) heart rate of the subject. In (D) the pre-connection and post-connection (HR = 200 bpm, C_{plac} = 0.0033 ml/mmHg and R_{plac} = 1.796 mmHg·s/ml) results are compared. Ventricular systole and diastole are depicted in green and yellow, respectively. (Color figure online)

WR originated at the PVC-oxygenator interface, the previous value of Γ was considered, and Δt was computed to be 0.042 s, assuming 7 cm of UA and 40 cm of PVC.

We added WR to the pressure and velocity traces obtained from the lumped model with AP to qualitatively compare the patterns. The resulting curves are shown in Fig. 4A and the quantitative measurements in Table 2. When WR model was added, we observed that UA pressure increased while UA flow decreased. We also observed that the diastolic UA flow had two differentiated phases as measured experimentally: a first phase of fast deceleration and a second phase of constant velocity, while without considering WR, diastolic flow decreased exponentially. Reflections generated at the oxygenator presented more pulsatility and a more flattened second phase of the flow with respect to WR at the UA-PVC interface.

We did a parametric analysis of Γ at the oxygenator level from 0.55 to 0.85 (Fig. 4B), maintaining Δt at 0.042 s. For higher Γ the pressure increased and the blood flow decreased, and the velocity decelerated faster.

4   Discussion

We implemented a 0D model of the fetal circulation at 23 weeks GA to study the haemodynamic changes in the UA produced by the connection of the fetus to an AP. While the pre-connection model flows were realistic, pressure traces

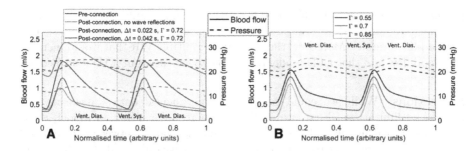

Fig. 4. Simulated UA blood flow (solid line) and pressure (dashed line) (A) pre-connection and post-connection to the AP, both with and without the WR model, and (B) considering WR with different values of reflection coefficient (Γ) at the oxygenator level. Ventricular systole and diastole are depicted in green and yellow, respectively. (Color figure online)

Table 2. Quantitative measurements of the velocity (v) and pressure (P) traces pre- and post-connection to the AP for the experimental data average and simulations with and without considering WR at different interfaces.

Indicator	Exp. data	No WR	UA-PVC	PVC-oxygenator
Heart rate (bpm)	133/202	130/200	130/200	130/200
FWHM P/v ratio	0.9/1.2	1.5/1.2	1.5/2.9	1.5/2.7
Pulsatility index	1.8/0.5	1.5/0.6	1.5/1.3	1.5/2.1
Mean pressure (mmHg)	46.5/72.1	24.6/12.8	24.6/22.0	24.6/22.0
Mean velocity (cm/s)	31.4/49.8	25.1/47.4	25.1/13.1	25.1/13.1

presented low PI compared to the real ones. Lowering compliance and resistance to model the AP resulted in an increase in UA blood flow and reduced PI, matching real data, but also a pressure drop that contradicted experimental observations. This was expected, as according to Ohm's law a lower resistance increases blood flow and reduces blood pressure. Moreover, simulated diastolic velocities presented an exponential decay rather than experimental data's flat profile. This could be explained by the presence of secondary waves that modulate the velocity.

We introduced a model of WR and assessed their impact on the blood flow and pressure patterns. By including the WR model, we were able to mimic the increase in FWHM pressure/velocity ratio, as part of the forward wave was counterbalanced by the backwards, thus increasing the width of the flow curve and flattening the velocity in diastole, two important patterns that were observed in experimental data. Also, higher Γ decreased the mean flow and increased the mean pressure, as was expected from Eq. 1. However, WR produced a mean pressure similar to pre-connection values, but with a significantly lower mean flow, which consequently increased the PI, despite the fact that the absolute pulsatility was lower. We hypothesise that in a real heart, there would be a pressure increase to elevate the blood flow, which would result in a preserved flow and much higher pressure, but these adaptation mechanisms are not present in our simplified model. Moreover, there could be multiple WR that would increase the blood flow and that a 0D model is not able to capture, but a 1D model could.

Limitations. While we could reproduce experimental data waveforms, our model had limitations. First, we used a non-personalised human model, while lambs were used in the real experiments, thus the match between the real and simulated traces was not to be expected. Secondly, we only modelled changes in the placental segment, and did not consider cardiac remodelling triggered by the removal from the natural environment, that would likely lead to increased blood flow. Thirdly, the WR model is purely phenomenological, and did not consider multiple reflections, for which a 1D model should be used. Finally, only data from 4 real subjects were provided, so no meaningful statistical test could be performed.

5   Conclusions

This study suggests that the observed changes in UA blood flow and pressure patterns, when connecting a fetus to an AP, are due to a combination of decreased placenta's compliance and resistance, as compared to a biological placenta, as well as the occurrence of WR from components within the AP circuit. In a future work, a 1D personalised model should be implemented in order to model the probable multiple WR present in the experiments.

Acknowledgements. The research leading to these results has received funding from "la Caixa" Banking Foundation. This work has been performed under FI 2022 grant number 00237, awarded by the Agency for Management of University and Research Grants (AGAUR), Generalitat de Catalunya.

References

1. Preterm birth. https://www.who.int/news-room/fact-sheets/detail/preterm-birth
2. Bramwell, J.C., Hill, A.V.: The velocity of pulse wave in man. Proc. Roy. Soc. Lond. Ser. B, Containing Papers Biol. Charact. **93**(652), 298–306 (1922). https://doi.org/10.1098/rspb.1922.0022, https://royalsocietypublishing.org/doi/10.1098/rspb.1922.0022
3. Bryner, B., et al.: An extracorporeal artificial placenta supports extremely premature lambs for 1 week. J. Pediat. Surg. **50**(1), 44–49 (2015). https://doi.org/10.1016/j.jpedsurg.2014.10.028
4. Cahill, L.S., et al.: Wave reflections in the umbilical artery measured by Doppler ultrasound as a novel predictor of placental pathology. EBioMedicine **67**, 103326 (2021). https://doi.org/10.1016/j.ebiom.2021.103326
5. Hill, A.A., Surat, D.R., Cobbold, R.S.C., Langille, B.L., Mo, L.Y.L., Adamson, S.L.: A wave transmission model of the umbilicoplacental circulation based on hemodynamic measurements in sheep. American Physiological Society (1995)

6. Howson, C.P., Kinney, M., McGougall, L., Lawn, J.: Born too soon : the global action report on preterm birth. Reproduct. Health, 9 (2013). https://doi.org/10.1186/1742-4755-10-S1-S1

7. Korteweg, D.J.: Ueber die Fortpflanzungsgeschwindigkeit des Schalles in elastischen Röhren. Annalen der Physik, 525–542 (1878). https://doi.org/10.1002/andp.18782411206

8. Miura, Y., et al.: Novel modification of an artificial placenta: pumpless arteriovenous extracorporeal life support in a premature lamb model. Pediat. Res. **72**(5), 490–494 (2012). https://doi.org/10.1038/pr.2012.108

9. Myers, L.J., Capper, W.L.: A transmission line model of the human foetal circulatory system. Med. Eng. Phys. **24**(4), 285–294 (2002). https://doi.org/10.1016/S1350-4533(02)00019-X

10. Ozawa, K., et al.: Fetal echocardiographic assessment of cardiovascular impact of prolonged support on EXTrauterine Environment for Neonatal Development (EXTEND) system. Ultrasound Obstet. Gynecol. **55**(4), 516–522 (2020). https://doi.org/10.1002/uog.20295

11. Partridge, E.A., et al.: An extra-uterine system to physiologically support the extreme premature lamb. Nat. Commun. **8** (2017). https://doi.org/10.1038/ncomms15112

12. Patel, R.M., et al.: Causes and timing of death in extremely premature infants from 2000 through 2011. New Engl. J. Med. **372**(4), 331–340 (2015). https://doi.org/10.1056/nejmoa1403489

13. Pennati, G., Fumero, R.: Scaling approach to study the changes through the gestation of human fetal cardiac and circulatory behaviors. Ann. Biomed. Eng. **28**(4), 442–452 (2000). https://doi.org/10.1114/1.282

14. Pennati, G., Bellotti, M., Fumero, R.: Mathematical modelling of the human foetal cardiovascular system based on Doppler ultrasound data. Med. Eng. Phys. **19**(4), 327–335 (1997). https://doi.org/10.1016/S1350-4533(97)84634-6

15. Saghian, R., et al.: Interpretation of wave reflections in the umbilical arterial segment of the feto-placental circulation: computational modeling of the feto-placental arterial tree. IEEE Trans. Biomed. Eng. **68**, 3647–3658 (2021). https://doi.org/10.1109/TBME.2021.3082064

16. Shi, Y., Lawford, P., Hose, R.: Review of zero-d and 1-d models of blood flow in the cardiovascular system. BioMed. Eng. Online **10** (2011). https://doi.org/10.1186/1475-925X-10-33

17. Van Den Wijngaard, J.P., Westerhof, B.E., Faber, D.J., Ramsay, M.M., Westerhof, N., Van Gemert, M.J.: Abnormal arterial flows by a distributed model of the fetal circulation. Am. J. Physiol. - Regulat. Integrat. Comparat. Physiol. **291**(5), 1222–1233 (2006). https://doi.org/10.1152/ajpregu.00212.2006

18. Weisstein, E.W.: Full Width at Half Maximum - from Wolfram MathWorld. https://mathworld.wolfram.com/FullWidthatHalfMaximum.html

Personalized Fast Electrophysiology Simulations to Evaluate Arrhythmogenicity of Ventricular Slow Conduction Channels

Dolors Serra[1], Paula Franco[2], Pau Romero[1], Ignacio García-Fernández[1], Miguel Lozano[1], David Soto[3], Diego Penela[3], Antonio Berruezo[3], Oscar Camara[2], and Rafael Sebastian[1(✉)]

[1] Computational Multiscale Simulation Lab (COMMLAB), Department of Computer Science, Universitat de Valencia, Valencia, Spain
rafael.sebastian@uv.es
[2] Physense, BCN Medtech, Department of Information and Communication Technologies, Universitat Pompeu Fabra, Barcelona, Spain
[3] Cardiology Department, Heart Institute, Teknon Medical Center, Barcelona, Spain

Abstract. After suffering a myocardial infarction, patient's tissue shows a complex substrate remodeling that combines dead and viable tissue in the scar region. Within such regions, slow conduction channels (SCC) might be present, being formed by viable tissue with altered electrical properties that can change the normal ventricle activation sequence, and sustain a ventricular tachycardia (VT) [8]. Computational models can help to stratify patients at risk, but they usually require large computational resources. In this study, we present a fast pipeline based on fully automatic modeling and simulation of patient's electrophysiology to assess the potential arrhythmogeneity of SCCs based on hundreds of simulated scenarios per patient. We apply our pipeline to four patients that have suffered a myocardial infarction, reproducing successfully predicting patient arrhythmogeneity in all cases with low computational times compatible with clinical routine (less than 4 h).

Keywords: Slow conduction channel · Cardiac electrophysiology simulation · Therapy planning · Ventricular tachycardia

1 Introduction

Ventricular arrhythmias might occur after myocardial infarction and remain a common cause of sudden cardiac death. In its chronic state, the infarction can show numerous paths of viable tissue that extend across the scar and form the so-called slow conduction channels (SCC). It has been reported that some SCC can be responsible for sustaining ventricular tachycardia (VT), and therefore it is key to identify them, since they can lead to lethal arrhythmias such as sudden cardiac death (SCD) [1]. In some technological advanced clinical centers,

O. Camara et al. (Eds.): STACOM 2022, LNCS 13593, pp. 56–64, 2022.
https://doi.org/10.1007/978-3-031-23443-9_6

SCC can be identified pre-operatively by contrast-enhanced magnetic resonance imaging (MRI), by performing an image-based reconstruction of the 3D paths. The volume of the scar and SCC tissue have also been proposed as biomarkers to determine the risk to arrhythmia in chronic myocardial infarction [14]. However, the anatomical characterization of the SCCs is not sufficient to determine if they can sustain and arrhythmia, or to determine the potential reentrant pathways that can occur in a given patient.

Computational approaches based on patient-specific electrophysiology simulations, i.e., digital twins [6], have emerged to provide complementary dynamical information that can help in the identification of potential arrhythmia pathways, and critical substrates that should be eliminated by catheter ablation [10,15]. In general, those technologies make use of biophysical models that can reproduce faithfully cardiac electrophysiology at cellular and tissue levels [2,11]. One of the major drawbacks of such models is that they are difficult to personalize to each patient, since simulations are computational demanding, which makes it impractical to run large parametric simulations sweeps of unknown parameters. Therefore, the risk of a patient to trigger or sustain an arrhythmia is evaluated from a few simulations with fixed configurations, or otherwise performed in a large computer facility with long computational times.

Translating digital twin technology into the clinics requires a different approach, in which patient-specific image-based information can be easily and efficiently integrated, and simulations are accelerated to be run in short computational times without the requirement of having a specific computer infrastructure [5,12].

In this paper, we present a fast modelling pipeline based on electrophysiological simulations modelled with cellular automata [13] that permits to go from raw clinical imaging data to an structured arrhythmia report, in clinical times, using a conventional desktop computer, considering hundreds of potential scenarios for each patient. The methodology is applied to patients that have suffered a myocardial infarction for which it is important to study if they show arrhythmias, and in such case to identify the critical paths that should be eliminated.

2   Material and Methods

2.1   Clinical Data

We have retrospectively included in this study four different patients from Teknon Clinical Center (Barcelona) that have suffered a myocardial infarction. Two of them showed VT after catheter pacing, and were candidates for catheter-based ablation of SCC. All patients were imaged with late gadolinium enhancement (LGE) MRI to analyze the scar configuration pre-operatively. Table 1 summarizes the clinical information related to each patient.

LGE-MRI scans were processed with the commercial software Adas3D (Galgo Medical, Barcelona, Spain, https://www.adas3d.com) to segment the endocardial and epicardial surfaces from both ventricles, together with the scar area, differentiating the border zone (BZ) from core zone (CZ) regions. Adas3D uses

Table 1. Clinical data for the patients in the study. VT: VT triggered from catheter; BZ: border zone; CZ: core zone; SCC: slow conduction channel.

Patient	LV mass(g)	BZ+CZ(g)	BZ(g)	CZ(g)	SCC(g)	LVEF	Age	Sex	VT
P1	197.7	47.6	42.2	5.4	21.1	28	84	M	Y
P2	105.9	18.6	13.8	4.8	3.6	57	63	M	Y
P3	79.8	29.7	25.7	4.0	14.9	47	78	F	N
P4	181.7	65.5	52.2	13.3	28.6	38	74	M	N

the maximum intensity methods to perform the segmentation. Figure 1 shows the resulting segmented models including the scar regions in the left ventricle (LV), where it is appreciable their complexity. From the 3D segmentation the weight (in grams) of the BZ and CZ were obtained, since they could be used as predictors for arrhythmia risk (see Table 1). In addition, Adas3D automatically provides the location of all the potential SCC's (Fig. 1 magenta tubes) that are defined as continuous areas of BZ tissue surrounded by CZ tissue that connect two areas of healthy tissue. All patients showed an abundant number of SCC that could sustain a VT if an ectopic focus is triggered. However, arrhythmogenity cannot be confirmed only based on imaging data due to the scar and SCC complexity

2.2   Digital Twin Construction

For the construction of the four digital twins we used a pre-processing tool from our software Arrhythmic3D [13], that automatically extracts a voxelized 3D biventricular mesh from the LGE-MRI, and adds all the necessary properties to run electrophysiology simulations. Those properties include the myocardial fiber orientation based on the Streeter model, the labeling of tissue (healthy, BZ or CZ), the labeling of heart regions (17 AHA segments), and the labeling of cell types (endocardial, mid-myocardial and epicardial).

2.3   Cardiac Electrophysiology Modeling

The software Arrhythmic3D was used to simulate cardiac electrophysiology in the digital twins [13]. The software is based on an Eikonal solver for the electrical diffusion in tissue coupled to an automaton that provides the cellular dynamics. At cellular level, the automaton encodes functions to reproduce the human ventricular ten Tusscher model [16], including action potential duration (APD) restitution properties, APD memory and safety-factor. All that information was obtained from a large biophysical simulation study, that included healthy and BZ tissue, and cells from endocardium, midmyocardium and epicardium. The event-based system does not imply strong restrictions on the time and spatial discretization, and can obtain results comparable to a biophysical model [13]. The system can run a cardiac simulation of 1200 ms in 4,8 s in a single CPU

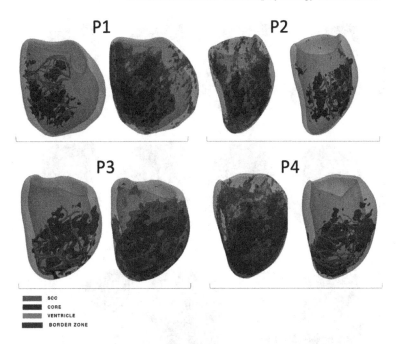

Fig. 1. Left ventricular anatomies extracted from the segmentation of magnetic resonance images of the four analyzed patients (P1-P4). The infarcted area is included and divided into CZ region (blue area) and BZ region (red area). For each model, it is also included the CZ region together with the SCC shown as magenta tubes. Segmentation was performed using the maximum intensity provided by ADAS3D software. (Color figure online)

core, which allows to run thousands of simulations in a short amount of time. Note that an equivalent biophysical simulation will take several days running in a single CPU core.

2.4   Batch Simulations

To assess the risk of arrhythmia in a digital twin, our approach is based on simulating a large number of scenarios, varying the desired input parameters within physiological ranges, so that all possible cases are taken into account. Next, we check whether any of the simulated scenarios produces a sustained VT (SVT), and which SCC is involved in the identified arrhythmia. The procedure is automatized, and performs simulations in which we emulate a pace-mapping procedure from a fixed number of locations that correspond to the 17 AHA regions at both endocardium and epicardium. A S1-S2 stimulation protocol is applied, where the coupling intervals are decreased to mimic the clinical procedure. In a S1-S2 protocol, first a train of stimuli (six in this study) with a constant basic cycle length (BCL) is given, followed by a single S2 stimulus with a shorter BCL.

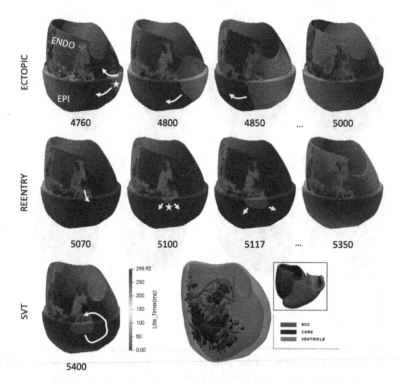

Fig. 2. Snapshots at different time points provided in ms from a VT simulation in patient P1 LV. Sustained VT (SVT) was induced by a pace-mapping protocol triggered from the RV mid-septal wall (yellow star mark). After 5000 ms the VT was sustained with an epicardial exit point at the LV anterior wall (white star mark). White arrows show the direction of the wavefront. The location of the CZ and SCC is included, as well as the orientation of the biventricular model. LV snapshots show the endocardial wall (upper half) and epicardial wall (lower half). ENDO: endocardium; EPI: epicardium; Life time refers to the time required for a cell to return to resting state in ms. (Color figure online)

The parameters that we varied in the protocol were: i) S2 frequency [305, 270] ms in intervals of 5 ms; ii) number of S2 stimulus (2, 3, 4); iii) location of the extra stimulus (34 locations); iv) AHA segments at endocardium and epicardium). The parameters that were fixed for all simulations were: i) frequency of S1 stimuli (600 ms); ii) the number of S1 stimuli (6) for stabilization; iii) weight of CV memory between consecutive stimuli (5%); iv) electrotonic effects between neighboring cells (85%).

Table 2. Simulation results. Time: Simulation Time.

Patient	#Configs	#Sim	#VT	Seg	S2 (ms)	Time	#Nodes	#Elems
P1	768	369	2	9	[275,280]	5 h 3 m	293293	253527
P2	816	521	2	9	[275,285]	3 h 43 m	221199	186973
P3	672	402	0	–	–	2 h 13 m	123793	104989
P4	816	402	0	–	–	3 h 50 m	268484	232441

3 Results

The digital twin was automatically built for each processed patient. Segmenting the biventricular model together with the scar region for each patient required around 10 min. Following, we built the computational models for simulations from the segmented models automatically, which took 90 s for each digital twin. Next, we started the simulation batch that included for each digital twin around 700–900 simulations (see Table 2, column 'Configs').

Each simulation batch for a digital patient took between 2–5 h, depending on the number of simulations. Not all simulations were successful, since for short coupling intervals they might still be in refractory period. This is specially critical when the extra stimuli is placed in a border zone region. In cases where the extra stimulus falls in a CZ region, the simulation fails, i.e. the activation cannot be triggered from the stimulation point. Hence for each digital twin the number of successful simulations are indicated in Table 2 (column 'Sims'). For patient P1 and P2, we obtained a sustained VT for two configurations (Table 2 column 'VT'). In both cases, the VT was triggered from the RV septal wall (AHA segment 9, LV Mid-ventricular inferoseptal), which was validated with the EPLab electro-anatomical map recordings with positive tachycardias stimulated from catheter. Note that differences in computation time also depend on the number of nodes and elements of the model (Table 2 column 'Time').

Figure 2 shows several snapshots of one of the positive VT simulations in patient P1. The ectopic focus was delivered at region 9 (see yellow star mark), with a S2 frequency of 270 ms. At 5070 ms of simulation the wavefront that was slowly traveling through the SCC's from LV septal endocardium to RV septal endocardium, finding the tissue repolarized and starting a new ventricular activation that was sustained for 25 cycles (until simulation end). In the region of the breakthrough, it could be observed a SCC (see Fig. 2 purple tubes in subfigure below). Note that, it was not the most complex region from the SCC point of view, which indicates that only with image-based information, it is difficult to predict which SCC could sustain a VT.

Figure 3 shows the results of an S2 activation for each of the models, in which streamlines following the wavefront propagation are depicted, and colored with the cell state from resting (blue) to activated (red). For P1 and P2, the tissue is reactivated at the septal region from the LV endocardium (transmural view), while the rest of the models show a complete and ordered activation of the LV.

We did not find a positive correlation between patients showing a VT and the scar mass, or the SCC mass, which suggests that the most important factor is the geometry of the SCC instead.

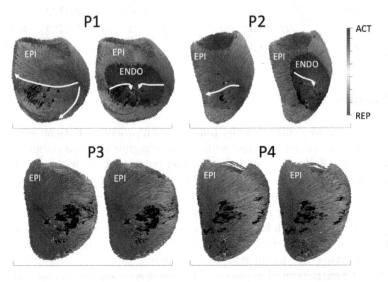

Fig. 3. Streamslines showing the direction of the propagation wavefront for a given S2 stimulus. Patients P1 and P2 show a reentry in the middle septum region (red colored area) propagating from LV to RV (not shown). Streamlines are color-coded with the state of the cell, from repolarized/resting (blue) to activated (red). (Color figure online)

4 Discussion and Conclusions

We have presented an automatic pipeline to study the potential arrhythmogene-ity of patient scars, based in simulating the electrophysiology on a large number of scenarios. In this study, the results obtained for four patients were positively confirmed by catheter-based electro-anatomical maps. The SCC supporting the arrhythmia could be visually identified by combining the imaging and simulation information, which could be very useful to determine the most critical paths that cross the scar tissue. Other studies have obtained similar results, using detailed biophysical models, based on a limited number of simulations [7]. One of the powerful benefits of our approach is the easiness to go from the clinical data to the final simulation report, and the number of scenarios that can be simu-lated (around 800 per patient) in clinical times (max. 5 h). Although, it was not required in this particular study, it could be important to assess the effect of varying CV in both healthy and BZ tissue, as well as the APD restitution curves within ranges observed in the population to obtain more information about risk to arrhythmia of each patient. In this study we based our results in standard val-ues used in the literature [10]. In the future, performing larger parameter sweeps will require a robust uncertainty quantification analysis. We neglected the effect of the Purkinje system, that could be included if EAM data were available [3,4]. Finally, it would be also important to be able to compare the patient's ECG during VT episodes with the simulated ECG [9], which currently is not available in our implementation yet.

References

1. Adabag, A.S., Therneau, T.M., Gersh, B.J., Weston, S.A., Roger, V.L.: Sudden death after myocardial infarction. JAMA **300**(17), 2022–2029 (2008). https://doi.org/10.1001/jama.2008.553
2. Arevalo, H.J., et al.: Arrhythmia risk stratification of patients after myocardial infarction using personalized heart models. Nat. Commun. 7, 11437 (2016). https://doi.org/10.1038/ncomms11437
3. Barber, F., et al.: Estimation of personalized minimal purkinje systems from human electro-anatomical maps. IEEE Trans. Med. Imaging **40**(8), 2182–2194 (2021)
4. Cárdenes, R., Sebastian, R., Soto-Iglesias, D., Berruezo, A., Camara, O.: Estimation of purkinje trees from electro-anatomical mapping of the left ventricle using minimal cost geodesics. Med. Image Anal. **24**(1), 52–62 (2015)
5. Chen, Z., et al.: Biophysical modeling predicts ventricular tachycardia inducibility and circuit morphology: a combined clinical validation and computer modeling approach. J. Cardiovasc. Electrophysiol. **27**(7), 851–60 (2016). https://doi.org/10.1111/jce.12991
6. Corral-Acero, J., et al.: The 'digital twin' to enable the vision of precision cardiology. Eur. Heart J. **41**(48), 4556–4564 (2020). https://doi.org/10.1093/eurheartj/ehaa159
7. Deng, D., Prakosa, A., Shade, J., Nikolov, P., Trayanova, N.A.: Characterizing conduction channels in postinfarction patients using a personalized virtual heart. Biophys. J. **117**(12), 2287–2294 (2019). https://doi.org/10.1016/j.bpj.2019.07.024
8. Fernández-Armenta, J., et al.: Three-dimensional architecture of scar and conducting channels based on high resolution ce-cmr: insights for ventricular tachycardia ablation. Circ. Arrhythm Electrophysiol. **6**(3), 528–37 (2013). https://doi.org/10.1161/CIRCEP.113.000264
9. Ferrer-Albero, A., et al.: Non-invasive localization of atrial ectopic beats by using simulated body surface p-wave integral maps. PloS one **12**(7), e0181263 (2017)
10. Lopez-Perez, A., Sebastian, R., Ferrero, J.M.: Three-dimensional cardiac computational modelling: methods, features and applications. Biomed. Eng. Online **14**, 35 (2015). https://doi.org/10.1186/s12938-015-0033-5
11. Lopez-Perez, A., et al.: Personalized cardiac computational models: From clinical data to simulation of infarct-related ventricular tachycardia. Front. Physiol. **10**, 580 (2019). https://doi.org/10.3389/fphys.2019.00580
12. Relan, J., et al.: Coupled personalization of cardiac electrophysiology models for prediction of ischaemic ventricular tachycardia. Interface Focus **1**(3), 396–407 (2011). https://doi.org/10.1098/rsfs.2010.0041
13. Serra, D., et al.: An automata-based cardiac electrophysiology simulator to assess arrhythmia inducibility. Mathematics **10**(8), 1293 (2022)
14. Soto-Iglesias, D., et al.: Cardiac magnetic resonance-guided ventricular tachycardia substrate ablation. JACC Clin. Electrophysiol. **6**(4), 436–447 (2020). https://doi.org/10.1016/j.jacep.2019.11.004

64   D. Serra et al.

15. Trayanova, N.A., Pashakhanloo, F., Wu, K.C., Halperin, H.R.: Imaging-based simulations for predicting sudden death and guiding ventricular tachycardia ablation. Circ. Arrhythm Electrophysiol. **10**(7) (2017). https://doi.org/10.1161/CIRCEP.117.004743
16. ten Tusscher, K.H.W.J., Noble, D., Noble, P.J., Panfilov, A.V.: A model for human ventricular tissue. Am. J. Physiol. Heart Circ. Physiol. **286**(4), H1573-89 (2004). https://doi.org/10.1152/ajpheart.00794.2003

Self-supervised Motion Descriptor for Cardiac Phase Detection in 4D CMR Based on Discrete Vector Field Estimations

Sven Koehler[1,2](✉) , Tarique Hussain[3], Hamza Hussain[1], Daniel Young[3], Samir Sarikouch[2,4,5], Thomas Pickardt[4], Gerald Greil[3], and Sandy Engelhardt[1,2]

[1] Department of Internal Medicine III, Group Artificial Intelligence in Cardiovascular Medicine, Heidelberg University Hospital, 69120 Heidelberg, Germany
sven.koehler@med.uni-heidelberg.de
[2] DZHK (German Centre for Cardiovascular Research), Heidelberg, Germany
[3] Department of Pediatrics, Division of Cardiology; Department of Radiology; Adv. Imaging Research Center, UT Southwestern Medical Center, Dallas, TX, USA
[4] German Competence Network for Congenital Heart Defects, Berlin, Germany
[5] Department of Cardiothoracic, Transplantation and Vascular Surgery, Hannover Medical School, Hannover, Germany

Abstract. Cardiac magnetic resonance (CMR) sequences visualise the cardiac function voxel-wise over time. Simultaneously, deep learning-based deformable image registration is able to estimate discrete vector fields which warp one time step of a CMR sequence to the following in a self-supervised manner. However, despite the rich source of information included in these 3D+t vector fields, a standardised interpretation is challenging and the clinical applications remain limited so far. In this work, we show how to efficiently use a deformable vector field to describe the underlying dynamic process of a cardiac cycle in form of a derived 1D motion descriptor. Additionally, based on the expected cardiovascular physiological properties of a contracting or relaxing ventricle, we define a set of rules that enables the identification of five cardiovascular phases including the end-systole (ES) and end-diastole (ED) without usage of labels. We evaluate the plausibility of the motion descriptor on two challenging multi-disease, -center, -scanner short-axis CMR datasets. First, by reporting quantitative measures such as the periodic frame difference for the extracted phases. Second, by comparing qualitatively the general pattern when we temporally resample and align the motion descriptors of all instances across both datasets. The average periodic frame difference for the ED, ES key phases of our approach is 0.80 ± 0.85, 0.69 ± 0.79 which is slightly better than the inter-observer variability (1.07 ± 0.86, 0.91 ± 1.6) and the supervised baseline method (1.18 ± 1.91, 1.21 ± 1.78). Code and labels are available on our GitHub repository. https://github.com/Cardio-AI/cmr-phase-detection.

Keywords: Self-supervised · Cardiac phase detection · Cardiac magnetic resonance · Deep learning

© The Author(s), under exclusive license to Springer Nature Switzerland AG 2022
O. Camara et al. (Eds.): STACOM 2022, LNCS 13593, pp. 65–78, 2022.
https://doi.org/10.1007/978-3-031-23443-9_7

1   Introduction

Analysis of cardiac function is tremendously important for diagnosis and monitoring of cardiac diseases. Therefore, modalities such as CMR are predominantly time-resolved for assessing global pump function and local contractility. Identification of key phases in such sequences are beneficial to determine the underlying myocardial mechanics with respect to contraction and relaxation. The end-diastolic (ED) and end-systolic (ES) phases are used to compare standardized clinical measures of the cardiac function, such as the left ventricular (LV) ejection fraction and peak systolic strain. An exact identification of these phases is crucial and has a major impact on the accuracy of strain measurements as shown by Mada et al. [14]. In addition, Zolgharni et al. [20] recognised a median disagreement of three frames between different observers in manually detecting the ED/ES frames. A better phase detection is feasible by evaluating the QRS-complex from electrocardiogram (ECG) signals. However, there is often no ECG-signal permanently stored.

2   Related Work

Several automatic ECG-free detection approaches have been already suggested. Most of them used echo [4–9,17] or CMR image sequences [13,19]. Early semi-automatic works [4,6,9] required either manual definition of landmarks, the ED frame or an initial contour. Gifani et al. [8] and Shalbaf et al. [17] solved the problem by using nonlinear dimensionality reduction techniques such as locally linear embedding and isomap. Evaluation was conducted on small echo cohorts of 8 and 32 patients. Recently, fully automatic deep learning based methods for CMR [13,19], echo [5,7] and fluoroscopy with contrast agent [3] evolved. All of these approaches are supervised, e.g., they require target labels. Three out of five predicted the ED and ES 'key frame' directly [3,5,13], in contrast to Xue et al. [19] and Fiorito et al. [7], who classified each frame of the sequence into either diastole or systole. Another obvious approach is to use a LV segmentation approach [12] and infer min/max volume from it as ES and ED.

Besides their very promising results, there are limitations. Similar in spirit as a segmentation-based approach, the method of Dezaki and Kong et al. [5,13] posed the task in a way to detect the maximum/minimum of a regressed LV volume curve. However, this method is based on a synthetic volume curve which is only at two points related to the underlying CMR. Furthermore, Kong et al. [13] evaluated the approach on a homogeneous in-house database with CMR sequences of unique lengths that starts with the ED phase and achieved an impressive average frame difference (aFD) of 0.38/0.44 (ED,ES). However, their loss definition assumes CMR sequences to start with the ED phase, which is not the case in our multi-centric datasets. The methodological extension by Dezaki et al. [5] was evaluated on a bigger cohort of echo sequences with varying length, collected from one clinic. Their supervised approach achieved an aFD of 0.71/1.92 (ED/ES), with the mitral valve as strong indicator for the ES phase, visible in their modality. Fiorito et al. [7] defined a binary classification task to detected two phases on echo sequences and achieved an aFD of 1.52, 1.48 (ED/ES). All fore-mentioned deep-learning approaches are supervised and need either labels of the left ventricle (LV)-blood pool [5,13] or cardiac phase labels [7]. In addition, they rely on the LV-volume change over time for cardiac phase estimation, which is in some aspects an assumption that does not always hold true: For example, in the isovolumetric contraction and relaxation phases around ED and ES the myocardium changes without affecting ventricular volumes. Therefore, our hypothesis is that phase detection based on myocardial deformation might be more accurate than using LV volume change as a marker.

In this work, we show how to efficiently reduce a 3D+t deformable vector field into a 1D motion descriptor representing the general dynamic pattern of a cardiac cycle in a self-supervised manner (cf. Fig. 1). We evaluate the plausibility of this motion descriptor and apply it to the task of cardiac phase detection. Here, we define a rule-set based on the expected cardiovascular physiological properties of a contracting/relaxing ventricle and extract the ED/ES phase at the turning point of deformation. To verify the plausibility of the motion descriptor in the intermediate frames, we further extend this rule-set to extract three additional physiologically relevant time points (cf. Sect. 4) and compare them with clinical labels from an experienced pediatric cardiologist on two heterogeneous cohorts.

3   Methods

This work is based on the assumption that a sequential deformable volume-to-volume registration field roughly represent the dynamics of the heart along a cardiac cycle. However, such 3D+t vector fields might become large, they are often spatially not aligned between patients and represent also deformation of the surrounding area which makes an automatic interpretation in a patient cohort challenging. Therefore, our rationale is to compress essential information into a 1D motion descriptor, which is decorrelated from the image grid (cf. Fig. 1).

3.1   Model Definition

Let a 3D+t CMR sequence be defined as x where each cardiac volume x_t is a 3D image at time point $t = \{1, ..., T\}$. We defined a deformable registration task (cf. Fig. 4.a. in the Appendix) as $\phi, \hat{M} = f_\theta(M, F)$ with M, F the moving and fixed volume pair, θ the learnable parameters of f, ϕ the resulting discrete vector field and $\hat{M}$ the moved volume after ϕ was applied with a spatial transformer layer to M. In our 4D case, for $M = x_t$ and $F = x_{t+1}$ we obtain $\phi_t, \hat{x}_{t+1} = f_\theta(x_t, x_{t+1})$. Additionally, due to the periodic behaviour, the last volume of the sequence x_T is warped to the first x_1. Therefore, ϕ_T is defined as $\phi_T, \hat{x}_1 = f_\theta(x_T, x_1)$.

We define one 3D focus point $C_n \in \mathbb{Z}^3$ that we use for each volume x_t (cf. Fig. 4b. in the Appendix), which can represent an anatomical landmark or a point that is calculated without prior knowledge (cf. Sect. 4). For each vector $\vec{v} \in \phi_t$, we calculate the angle $\alpha = cos(\vec{v}, \vec{w})$, where the vector $\vec{w}$ points from

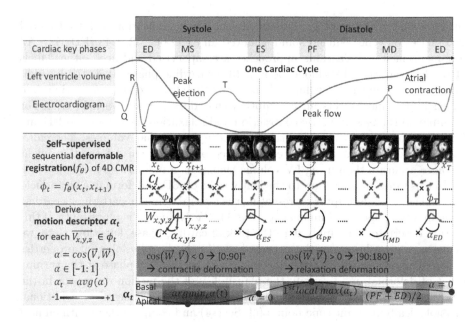

Fig. 1. Concept of the self-supervised 1D motion descriptor α_t, derived from cine short-axis CMR sequences. Upper third: clinical definition of the five labeled cardiac phases (cf. Sect. 3.2) aligned with a corresponding electrocardiogram (ECG) and the left ventricular (LV) volume curve. Middle third: 2D visualisation of the sequential 3D+t registration approach and the expected main deformation pattern ϕ_t (cf. Sect. 3.1). Lower third: derivation of the vector-wise cosine angle α depending on different focus points C (cf. Sect. 4) together with the main concept for systolic/diastolic differentiation and an example curve of α_t. First, slice-wise avg. of α, color coded with blue: movement towards C, red: movement away from C. Second, volume-wise avg. (α_t) as interpolated black curve together with a simplified version of the rule-set (cf. Eq. 4).

the corresponding x, y, z grid position to C_n. Therefore, $\vec{v}$ with $\alpha \in [-1, 0[$ will point towards the focus point C_n and with $\alpha \in]0, 1]$ away from it. This enables voxel-wise differentiation of a contractile deformation vector from a vector that describes relaxation, which is the main rationale we followed for phase extraction.

Additionally, we calculate the 70th quantile on the temporally averaged euclidean norm $|\vec{v}|$ of ϕ and use it as threshold to filter non-cardiac motion information and noise from the cardiac deformation (cf. row d. in Fig. 3).

Finally, for the extraction of the 1D curves over time (motion descriptor α_t and $|\vec{v}|_t$) we average α and $|\vec{v}|$ per 3D volume, apply a Gaussian filter ($\sigma = 2$) on α_t and min/max normalise both into a range of $[-1, 1]$ and $[0, 1]$, respectively. The scaling of α_t may introduce small shifts (<1 frames) to the zero-crossing time points t_{ED} and t_{ES} but also removes diffuse zero crossings for weak pathological relaxation phases.

3.2   Loss Function and Key Frame Extraction

The registration loss consists of an image similarity component $\mathcal{L}_{sim}$ and a regularizer $\mathcal{L}_{smooth}$ and is defined by Eq. 1. It turned out that the structural similarity index measure (SSIM) [18], which is based on a luminance, contrast and structure measurement, performs better than the mean squared error as $\mathcal{L}_{sim}$. Here, we average the 2D SSIM per 3D volume, the equal weighted general 2D form is given by Eq. 2 with μ_x, μ_y as the average and σ_x^2, σ_y^2 as the variance of a $N \times N$ region in x and y, which correspond to the two neighbouring time steps. σ_{xy} is given by the co-variance of x and y and ϵ_1, ϵ_2 are two variables included to avoid instability. For $\mathcal{L}_{smooth}$ (cf. Eq. 3) of ϕ we used the same diffusion regularizer of the spatial gradients as introduced by Balakrishnan et al. [1], with Ω representing each voxel in x_t and set the regularization parameter $\lambda = 0.001$.

$$\mathcal{L}(F, M, \phi) = \mathcal{L}_{sim}(F, M(\phi)) + \lambda \mathcal{L}_{smooth}(\phi). \tag{1}$$

$$SSIM(x, y) = \frac{(2\mu_x\mu_y + \epsilon_1)(2\sigma_{xy} + \epsilon_2)}{(\mu_x^2 + \mu_y^2 + \epsilon_1)(\sigma_x^2 + \sigma_y^2 + \epsilon_2)}. \tag{2}$$

$$\mathcal{L}_{smooth}(\phi) = \sum_{p \in \Omega} ||\nabla\phi(p)||^2. \tag{3}$$

We define a set of rules that derives five cardiac key time points from the compressed 1D signal α_t, including the ED, mid-systole (MS; maximum contraction resulting in a peak ejection between ED and ES), ES, peak flow (PF; peak early diastolic relaxation) and mid-diastole (MD; phase before atrial contraction at the on-set of the p-wave). As shown in Fig. 1, the succession of these labels is temporally correlated, however, we observed that the starting cardiac phase of the CMR sequences varies between acquisitions (cf. Sect. 3.4). In order to handle CMR sequences with varying starting phases we first detected the time-point with maximum contraction (MS) and applied the other rules sequentially to the cyclic sub-sequences. Please note that this rule-set (especially for the MD rule, where we could also rely on the last peak as indicator for the atrial contraction) is a work-in-progress-trade-off between accuracy and the generalisation capability towards cut-off or pathological sequences. For this work we preferred simple rules that are based on a physiological reasoning over complicated ones. Especially, the intermittent diastolic pattern of α_t differs between pathological patients and sometimes leads to multiple diffuse relaxation peaks, which are difficult to assign to the presumed peak flow or to the atrial contraction close to the MD phase.

$$
\begin{aligned}
MS &= t_m \text{ with } \alpha(t_m) \le \alpha(t) & &\text{with } t \in T \\
ES &= \min_t \alpha(t) = 0 & &\text{with } t \in [MS; PF] \\
PF &= \min_t \alpha'(t) = 0 \wedge \alpha''(t) < 0 & &\text{with } t \in [ES; MS] \qquad (4) \\
ED &= \max_t \alpha(t) = 0 & &\text{with } t \in [PF; MS] \\
MD &= (PF + ED)/2
\end{aligned}
$$

3.3 Deep Learning Framework

Our Deep Learning model consists of a 3D CNN-based deformable registration-module followed by the direction-module. For the sequential volume-to-volume deformable registration we use a slightly modified time distributed 3D U-Net as introduced by Ronneberger et al. [15] followed by a spatial transformer layer, such as by Balakrishnan et al. [1]. For further details we refer to our GitHub repository. Our final model expects a 4D volume as input with $b \times 40 \times 16 \times 64 \times 64$ as batchsize, time, spatial slices and x-/y dimension. Subsequently, the direction-module calculates voxel-wise α, $|\vec{v}|$ and per 4D volume the 1D curves α_t and $|\vec{v}|_t$ which are used in our rule-set. Please note: all parts are differentiable, learning is end-to-end and could be used in a supervised approach.

3.4 Datasets

For evaluation we used two 4D cine-SSFPs CMR SAX datasets. The annotations were made by an experienced paediatric cardiologist. Distributions and mean occurrence of the phases are shown in Table 2 and Fig. 6a in the Appendix.

First, the publicly available *Automatic Cardiac Diagnosis Challenge* (ACDC) dataset [2] (100 patients, 5 pathologies, 2 centers) was used. The mean$\pm SD$ number of frames is 26.98 ± 6.08, within a range of $[12, 35]$. Furthermore, not all 4D sequences capture an entire cardiac cycle (cf. Fig. 5 in the Appendix). A detailed description of the dataset can be found in [2] and in the Appendix. We corrected and extended the original cardiac phase labels (e.g.: the ED phase was uniformly labelled at frame 0 over the entire cohort, which is a rough approximation). Now, 75 sequences start close to the MS and 25 close to the ED phase. Our labels will be released on our GitHub repository. The inter-observer error between the original phase labels and our re-labelling is 0.99 ± 1.23, with a maximal distance of 6 (ED) and 10 (ES), respectively.

Second, a multi-centric dataset (study identifier: NCT00266188) [10,12,16] of 278 patients with Tetralogy of Fallot (TOF), which is a complex congenital heart disease, was used. The mean number of frames is 21.92 ± 4.02, within a range of $[12, 36]$. The sequence length of each cardiac cycle is 743 ± 152ms, within a range of $[370, 1200]$. 191 sequences start close to the MS and 84 close to the ED phase. The other three phases occurred once at the sequence start.

4 Evaluation and Experiments

We extend the previously used [5,7,8,13,17] average Frame Difference ($aFD = |p_i - \hat{p}_i|$) to account for the periodicity of the cardiac cycle and refer to it as

$$pFD(p_i, \hat{p}_i) = \min(|p_i - \hat{p}_i|, T - \max(p_i, \hat{p}_i) + \min(p_i, \hat{p}_i)) \qquad (5)$$

with $i \in [ED, MS, ES, PF, MD]$ and p_i, $\hat{p}_i$ the ith ground truth and predicted label. This is important for permuted sequences when the annotated phase p_i is

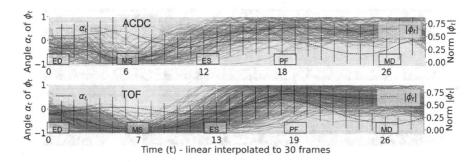

Fig. 2. Qualitative results for the motion descriptor α_t on both datasets. Both plots show the per-cohort average of α_t (blue/left axis) and $|\vec{v}|_t$ (black/right axis) together with the temporal aligned, resampled and averaged phase indices (x-axis), without any post-processing. We overlaid the per-instance curves of α_t but for the sake of clarity, we omitted the instances curves for $|\vec{v}|_t$. Please note: $\alpha_t < 0$ correspond to systolic and $\alpha_t > 0$ to diastolic frames. The peaks in $|\vec{v}|_t$ are close to the mid-systolic (MS) and peak flow (PF) phases. We aligned/resized the data to visualise the general properties of α_t and $|\vec{v}|_t$, while the original data for training or inference remained unaligned. (Color figure online)

labelled at $t = 1$ but, $\hat{p}_i$ predicts $t = T$ and vice versa. The pFD would be 1; in the original aFD formulation, the distance would be T.

Each experiment was carried out in a four-fold-cross validation manner. We resampled x with linear interpolation to a spacing of 2.5 mm³ and repeated x_t along t until we reached the network's input size of 40. Following that, we focus crop with different focus points C_n in 3D (cf. next paragraph), clipped outliers with a quantile of .999 and standardised per 4D. We did not apply any image-based augmentation as we noticed no over-fitting in $\mathcal{L}_{sim}$.

We compare our results with a supervised LV-volume-based approach on the same data and refer to it as *base*. Four U-net based segmentation models were trained on the public ACDC data (LV DICE: 0.91 ± 0.02). Next, we applied a connected component filter and identified the ED/ES frames based on the min/max LV volume. Later we applied one of them on the TOF dataset to provide a supervised baseline.

Quantitatively, we report the pFD (cf. Eq. 5) per dataset and cardiac key-point in the original temporal resolution in Table 1. Additionally, we investigate the sensitivity of different C_n on the pFD and compare the LV blood-pool center of mass C_{lv}, the mean septum landmark (center between the average anterior and inferior right ventricular insertion points (RVIP) [11]) C_{sept}, the CMR-volume center C_{vol} and the center of mass for a quantile-threshold mean squared error mask averaged along the temporal axis C_{mse}.

Finally, we qualitatively evaluate the general pattern of α_t and $|\vec{v}|_t$ on both datasets (cf. Fig. 2) and show different views for one random patient (cf. Fig. 3).

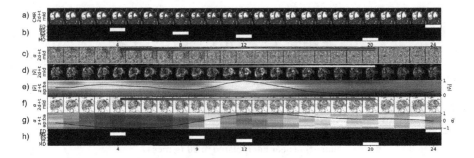

Fig. 3. Visualisation for one cardiac cycle of a random TOF patient. a) Mid-cavity CMR slice, b) GT phase labels, c) Mid-cavity view of the direction α without percentile masking (blue: towards focus point, red: away from), d) Mid-cavity view of $|\vec{v}|$ (masked by the 70th percentile of $|\vec{v}|$), e) Color coded slice-wise average of $|\vec{v}|$ (base2apex) in background. Volume-based average of $|\vec{v}|_t$ as curve with axis on the right, f) Mid-cavity view of α (same color-coding, masked by the 70th percentile of $|\vec{v}|$), g) Color coded slice-wise average of α (base2apex) in background. Volume-based average of α_t as curve with axis on the right, h) Predicted phase. (Color figure online)

5   Results

Based on the per instance and global average curves in Fig. 2, we provide a qualitative and more clinical interpretation of our results. The general systolic/diastolic pattern of α_t across all patients of both cohorts aligns with our physiological expectations and show a negative course (mainly contractile direction of ϕ) during the systole followed by multiple positive local maxima during the diastole. Furthermore, we would expect a change from contractile to relaxile deformation (ES) or vice versa (ED) where α_t has a zero pass. From our observations and confirmed by the pFD in Table 1, this is usually true.

In general, α_t shows a clear global minimum close to the MS-phase, that refers to the time-point where most of the masked voxels (percentile-based threshold of $|\vec{v}|$) represent the greatest contractile deformation direction. The relaxation part is often more diffuse but with mostly two peaks in α_t (the systolic negative course is greater than the diastolic positive course if we omit the re-scaling). The time point of minimal and maximal deformation should correspond to the peaks and valleys as shown in Fig. 2 for the cohort-based average of $|\vec{v}|_t$. On both datasets we have a local maxima of $|\vec{v}|_t$ close to the MS during systole and close to the PF or shortly after the MD phases during diastole. The global minimum of $|\vec{v}|_t$ is often around the MD phase (cf. Fig. 2 and Fig. 3). From a visual point of view, the threshold $|\vec{v}|_t$ mask is able to eliminate most of the non-cardiac information (cf. row f. in Fig. 3).

As quantitative measure we present the pFD (cf. Eq. 5) for both cohorts and for each phase in Table 1. In addition we report the error distribution per key time point in Fig. 6b in the Appendix. The cardiac contraction happens usually in one coherent contractile deformation, which results in a clear negative course of

Table 1. Sensitivity of the resulting pFD for two cohorts with respect to different focus points C_n (cf. Sect. 4) and comparison to supervised segmentation-based approach *base*. $^1 = C$ based on anatomical GT knowledge, $^2 = C$ based on more generic information (unsupervised).

Data	C_n	All	ED	MS	ES	PF	MD
ACDC	base	–	1.13 ± 1.82	–	$\mathbf{0.95 \pm 1.29}$	–	–
	C^1_{lv}	1.36 ± 1.37	1.13 ± 1.23	$\mathbf{0.97 \pm 0.95}$	1.05 ± 1.09	1.87 ± 1.98	1.77 ± 1.59
	C^1_{sept}	1.32 ± 1.21	1.09 ± 1.09	$\mathbf{0.97 \pm 0.83}$	0.96 ± 0.87	1.68 ± 1.63	1.91 ± 1.65
	C^2_{vol}	1.56 ± 1.86	1.37 ± 2.01	1.24 ± 1.40	1.19 ± 1.60	1.99 ± 2.14	2.01 ± 2.15
	C^2_{mse}	$\mathbf{1.29 \pm 1.25}$	$\mathbf{1.08 \pm 1.26}$	1.02 ± 0.94	0.97 ± 0.95	$\mathbf{1.66 \pm 1.56}$	$\mathbf{1.73 \pm 1.54}$
TOF	base	–	1.18 ± 1.91	–	1.21 ± 1.78	–	–
	C^1_{lv}	0.99 ± 0.91	0.81 ± 0.93	1.07 ± 0.79	0.72 ± 0.79	0.90 ± 0.82	$\mathbf{1.46 \pm 1.22}$
	C^1_{sept}	$\mathbf{0.95 \pm 0.89}$	0.82 ± 0.88	$\mathbf{0.87 \pm 0.72}$	0.70 ± 0.76	$\mathbf{0.78 \pm 0.83}$	1.58 ± 1.26
	C^2_{vol}	1.02 ± 0.97	0.86 ± 1.04	1.06 ± 0.83	0.76 ± 0.80	0.88 ± 0.90	1.56 ± 1.28
	C^2_{mse}	0.97 ± 0.91	$\mathbf{0.80 \pm 0.85}$	0.94 ± 0.76	$\mathbf{0.69 \pm 0.79}$	0.85 ± 0.86	1.57 ± 1.27

α_t, that makes phase extraction straight-forward, which is visible in lower pFD scores for the key-points (ED, MS, ES). This is in contrast to the relaxation of the heart, which does not follow such an homogeneous pattern. In fact, we observed multiple peaks that may result from basal to apical regions relaxing at different rate (cf. Fig. 2). The pFD for the diastolic phases (PF and MD) are slightly worse and represent the difficulties to assign these peaks to either the ventricle contraction during the peak flow (PF) or to the atrial contraction shortly after the mid-diastolic (MD) phase. Both experiments that are based on prior knowledge (C_{lv} and C_{sept}) provided similar results except for the PF and MD phases, where both performed once better. Our experiment C_{vol}, results, as expected, in the highest pFD and SD. Using the center of mass of the temporally sequential mean squared error of x_t and x_{t+1} as focus point C (cf. Sect. 4) closed the gap of the unsupervised approaches with similar or slightly better pFD while removing the need of prior knowledge or labels.

6   Discussion and Conclusion

In this work we compute a motion descriptor based on the mean direction and norm of a sequential deformable registration field ϕ_t in a self-supervised manner according to different focus points C_n, to derive the cardiac dynamics over time.

Furthermore, according to the expected properties of a vector field that mainly represents myocardial contraction and relaxation, we define a set of rules and extend the state-of-the-art by extracting not only two but five cardiovascular key-time frames on CMR sequences with any length and independent of the starting phase. To the best of our knowledge this has not been done before. We evaluate the reliability of the motion descriptor on two challenging multi-center datasets and compare our method to a supervised baseline. Even though the set of rules was defined empirically, we could quantitatively and qualitatively confirm that the self-supervised motion descriptor α_t is able to express the expected, underlying cardiovascular physiological motion properties. We will release our extended ACDC phase labels to enable future comparison.

The pFD (ED,ES) of the completely self-supervised experiment (TOF, C_{mse}: 0.80±0.85, 0.69±0.79) is slightly better than the recognised inter-observer error (ACDC: 1.07±0.86, 0.91±1.6) and significantly ($p < 0.001$) better than the supervised baseline (TOF, $base$: 1.18±1.91, 1.21±1.78).

Supervised methods may achieve promising/comparable results (cf. 1st row ACDC in Table 1: 1.13±1.82,0.95±1.29), nevertheless their performance often drops (ES: -0.26) when they are applied to unseen datasets, due to the inherent domain shift (cf. 2nd row TOF in Table 1). This is were self-supervised methods might unfold their strengths, since model re-training or adjustment does not rely on annotations and can be easily done on domain shifted data.

This work assumes that CMR sequences capture an entire cardiac cycle. Cut-off sequences may result in unphysiological peaks in $|\vec{v}|_t$ and hence slightly worse results for phase detection (cf. ACDC dataset in Table 1). However, we show how to benefit from these information to automatically detect cut-off sequences in a self-supervised way for quality control purposes (cf. Fig. 5 in the Appendix). These cut-off sequences refrained us from using the magnitude directly as keyframe detector, e.g. identifying the peak ejection/peak flow phases based on the largest overall vector magnitudes. In one experiment we weight α of each voxel by the corresponding magnitude $|\vec{v}|$, unfortunately this descriptor performed worse for cardiac phase detection. In future work, we will show the value of this descriptor for inter-patient comparison and cardiovascular pathology description.

Acknowledgments. This work was supported in parts by the Informatics for Life Project through the Klaus Tschira Foundation, by the Competence Network for Congenital Heart Defects (Federal Ministry of Education and Research/grant number 01GI0601) and the National Register for Congenital Heart Defects (Federal Ministry of Education and Research/grant number 01KX2140), by the German Centre for Cardiovascular Research (DZHK) and the SDS@hd service by the MWK Baden-Württemberg and the DFG through grant INST 35/1314-1 FUGG and INST 35/1503-1 FUGG.

A   Dataset Properties

The ACDC dataset was collected at two sites and covers adults with normal cardiac anatomy and four cardiac pathologies: systolic heart failure with infarction, dilated cardiomyopathy, hypertrophic cardiomyopathy and abnormal right ventricular volume. Each pathology is represented by 20 patients with pre-defined ES and ED phase. The CMR volumes have an average resolution of $220.12 \pm 34.04 \times 247.14 \pm 39.44 \times 9.51 \pm 2.40$ and a spacing of $1.51\pm0.19\times1.51\pm0.19\times9.34\pm1.67$ mm^3 (X/Y/Z). The resolutions and spacings (same ordering) are within in the following ranges: [154, 428], [154, 512], [6, 18], [0.70, 1.92],[0.70, 1.92],[5, 10]. Each 4D CMR volume has between 84 and 450 2D slices with a mean of 253 ± 72.

The second cohort includes patients over 8 years with a complex congenital heart defect called Tetralogy of Fallot (TOF). We used short-axis 4D cine series from 278 patients. They have an average resolution of $244.49 \pm 60.02 \times 252.37 \pm 48.82 \times 14.08 \pm 3.20$ and a spacing of $1.37 \pm 0.18 \times 1.37 \pm 0.18 \times 8.03 \pm 1.25$ mm^3 (X/Y/Z). The resolution and spacing (same ordering) are within the following min, max ranges: $[126, 512]$, $[156, 512]$, $[8, 28]$, $[0.66, 2.08]$,$[0.66, 2.08]$,$[6, 16]$. Each 4D CMR volume has between 112 and 700 2D slices with a mean of 313 ± 106 slices.

Table 2. First row: Mean$\pm SD$, second row: min and max occurrence of the ground truth cardiac key frame indexes per ACDC and TOF dataset. Especially the cardiac key frame ranges in the TOF dataset, which comes from 14 different sites, illustrates the possible variability/permutations in clinical data.

Data	ED	MS	ES	PF	MD
ACDC	20.42±12.26	4.24±1.42	9.91±2.67	15.08±3.36	22.92±15.14
	[1 : 35]	[3 : 12]	[6 : 23]	[9 : 30]	[13 : 34]
TOF	15.54±10.09	4.91±1.72	9.39±2.11	13.22±12.41	18.44±3.34
	[1 : 34]	[2 : 18]	[2 : 22]	[3 : 25]	[7 : 30]

B Visual Examples

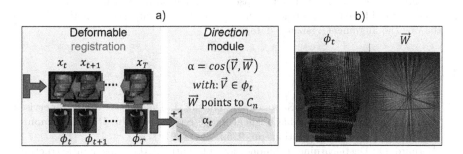

Fig. 4. a) The registration module outputs ϕ, which represents the deformation between neighbouring time steps x_t and x_{t+1}. Following, the direction module which calculates the mean deformation angle/motion descriptor α_t between each vector $\vec{v} \in \phi$ and $\vec{w}$ pointing to C_n. b) Example ϕ_t and the corresponding focus matrix with vectors $\vec{w}$ pointing to focus point C_n. Here, ϕ_t is masked by the left ventricle contours for visualisation purposes.

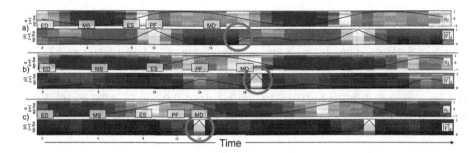

Fig. 5. Motion descriptor (α_t and $|\vec{v}|_t$) derived from three patients, the green marker highlights the cycle-end; information is repeated until 40 time steps are filled (input format into network). A high peak reflects long deformation vectors in $|\vec{v}|_t$ towards the following time step. In the highlighted case it is $|\vec{v}|_T$, which is the deformation from the last volume x_T to x_1. If $|\vec{v}|_t$ is an outlier with a high relative value compared to the other time steps it is a strong indicator for a cut-off CMR sequence. a) No cut-off. b) Moderate cut-off. c) Strong cut-off.

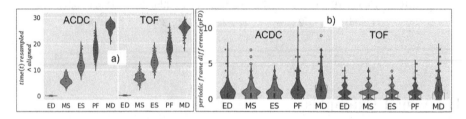

Fig. 6. a) The distribution of the GT label show a much easier problem if we align and resize both datasets, which is not the case for clinical data (cf. Table 1). b) Average periodic frame difference (pFD) for each phase of the self-supervised method (C_{mse}) on the raw (no alignment/resizing) datasets.

References

1. Balakrishnan, G., Zhao, A., Sabuncu, M., Guttag, J., Dalca, A.V.: An unsupervised learning model for deformable medical image registration. In: CVPR: Computer Vision and Pattern Recognition, pp. 9252–9260 (2018)
2. Bernard, O.: Deep learning techniques for automatic MRI cardiac multi-structures segmentation and diagnosis: is the problem solved? IEEE Trans. Med. Imaging **37**(11), 2514–2525 (2018). https://doi.org/10.1109/TMI.2018.2837502
3. Ciusdel, C., et al.: Deep neural networks for ecg-free cardiac phase and end-diastolic frame detection on coronary angiographies. Comput. Med. Imaging Graph. **84**, 101749 (2020)
4. Darvishi, S., Behnam, H., Pouladian, M., Samiei, N.: Measuring left ventricular volumes in two-dimensional echocardiography image sequence using level-set method for automatic detection of end-diastole and end-systole frames. Res. Cardiovasc. Med. **2**(1), 39 (2013)

5. Dezaki, F.T., et al.: Cardiac phase detection in echocardiograms with densely gated recurrent neural networks and global extrema loss. IEEE Trans. Med. Imaging **38**(8), 1821–1832 (2018)

6. Dominguez, C.R., et al.: Classification of segmental wall motion in echocardiography using quantified parametric images. In: Frangi, A.F., Radeva, P.I., Santos, A., Hernandez, M. (eds.) FIMH 2005. LNCS, vol. 3504, pp. 477–486. Springer, Heidelberg (2005). https://doi.org/10.1007/11494621_47

7. Fiorito, A.M., Østvik, A., Smistad, E., Leclerc, S., Bernard, O., Lovstakken, L.: Detection of cardiac events in echocardiography using 3d convolutional recurrent neural networks. In: 2018 IEEE International Ultrasonics Symposium (IUS), pp. 1–4. IEEE (2018)

8. Gifani, P., Behnam, H., Shalbaf, A., Sani, Z.A.: Automatic detection of end-diastole and end-systole from echocardiography images using manifold learning. Physiol. Meas. **31**(9), 1091 (2010)

9. Kachenoura, N., Delouche, A., Herment, A., Frouin, F., Diebold, B.: Automatic detection of end systole within a sequence of left ventricular echocardiographic images using autocorrelation and mitral valve motion detection. In: 2007 29th Annual International Conference of the IEEE Engineering in Medicine and Biology Society, pp. 4504–4507. IEEE (2007)

10. Koehler, S., et al.: Unsupervised domain adaptation from axial to short-axis multi-slice cardiac MR images by incorporating pretrained task networks. IEEE Trans. Med. Imaging **40**(10), 2939–2953 (2021). https://doi.org/10.1109/tmi.2021.3052972

11. Koehler, S., et al.: Comparison of evaluation metrics for landmark detection in CMR images. In: Bildverarbeitung für die Medizin 2022. I, pp. 198–203. Springer, Wiesbaden (2022). https://doi.org/10.1007/978-3-658-36932-3_43

12. Koehler, S., et al.: How well do U-Net-based segmentation trained on adult cardiac magnetic resonance imaging data generalize to rare congenital heart diseases for surgical planning? In: Medical Imaging 2020: Image-Guided Procedures, Robotic Interventions, and Modeling, vol. 11315, pp. 409–421. International Society for Optics and Photonics, SPIE (2020). https://doi.org/10.1117/12.2550651

13. Kong, B., Zhan, Y., Shin, M., Denny, T., Zhang, S.: Recognizing end-diastole and end-systole frames via deep temporal regression network. In: Ourselin, S., Joskowicz, L., Sabuncu, M.R., Unal, G., Wells, W. (eds.) MICCAI 2016. LNCS, vol. 9902, pp. 264–272. Springer, Cham (2016). https://doi.org/10.1007/978-3-319-46726-9_31

14. Mada, R.O., Lysyansky, P., Daraban, A.M., Duchenne, J., Voigt, J.U.: How to define end-diastole and end-systole?: impact of timing on strain measurements. JACC: Cardiovasc. Imaging **8**(2), 148–157 (2015). https://doi.org/10.1016/j.jcmg.2014.10.010

15. Ronneberger, O., Fischer, P., Brox, T.: U-Net: convolutional networks for biomedical image segmentation. In: Navab, N., Hornegger, J., Wells, W.M., Frangi, A.F. (eds.) MICCAI 2015. LNCS, vol. 9351, pp. 234–241. Springer, Cham (2015). https://doi.org/10.1007/978-3-319-24574-4_28

16. Sarikouch, S., et al.: Impact of gender and age on cardiovascular function late after repair of tetralogy of fallot. Circ.: Cardiovasc. Imaging **4**(6), 703–711 (2011). https://doi.org/10.1161/CIRCIMAGING.111.963637

17. Shalbaf, A., AlizadehSani, Z., Behnam, H.: Echocardiography without electrocardiogram using nonlinear dimensionality reduction methods. J. Med. Ultrasonics **42**(2), 137–149 (2015)

18. Wang, Z., Bovik, A., Sheikh, H., Simoncelli, E.: Image quality assessment: from error visibility to structural similarity. IEEE Trans. Image Process. **13**(4), 600–612 (2004). https://doi.org/10.1109/TIP.2003.819861
19. Xue, W., Brahm, G., Pandey, S., Leung, S., Li, S.: Full left ventricle quantification via deep multitask relationships learning. Med. Image Anal. **43**, 54–65 (2018). https://doi.org/10.1016/j.media.2017.09.005
20. Zolgharni, M.: Automatic detection of end-diastolic and end-systolic frames in 2d echocardiography. Echocardiography **34**(7), 956–967 (2017)

Going Off-Grid: Continuous Implicit Neural Representations for 3D Vascular Modeling

Dieuwertje Alblas[1(✉)], Christoph Brune[1], Kak Khee Yeung[2,3],
and Jelmer M. Wolterink[1]

[1] Department of Applied Mathematics, Technical Medical Centre,
University of Twente, Enschede, The Netherlands
d.alblas@utwente.nl
[2] Department of Surgery, Amsterdam UMC Location Vrije Universiteit Amsterdam,
Amsterdam, The Netherlands
[3] Amsterdam Cardiovascular Sciences, Microcirculation, Amsterdam,
The Netherlands

Abstract. Personalised 3D vascular models are valuable for diagnosis, prognosis and treatment planning in patients with cardiovascular disease. Traditionally, such models have been constructed with explicit representations such as meshes and voxel masks, or implicit representations such as radial basis functions or atomic (cylindrical) shapes. Here, we propose to represent surfaces by the zero level set of their signed distance function (SDF) in a differentiable implicit neural representation (INR). This allows us to model complex vascular structures with a representation that is implicit, continuous, light-weight, and easy to integrate with deep learning algorithms. We here demonstrate the potential of this approach with three practical examples. First, we obtain an accurate and watertight surface for an abdominal aortic aneurysm (AAA) from CT images and show robust fitting from as few as 200 points on the surface. Second, we simultaneously fit nested vessel walls in a single INR without intersections. Third, we show how 3D models of individual arteries can be smoothly blended into a single watertight surface. Our results show that INRs are a flexible representation with potential for minimally interactive annotation and manipulation of complex vascular structures.

Keywords: Implicit neural representation · Vascular model · Abdominal aortic aneurysm · Signed distance function · Level set

1 Introduction

Accurate and patient-specific models of vascular systems are valuable for diagnosis, prognosis and treatment planning in patients with cardiovascular disease. Personalised vascular models might be used for stent-graft sizing in patients with abdominal aortic aneurysms [27] or for computational fluid dynamics (CFD) [30],

O. Camara et al. (Eds.): STACOM 2022, LNCS 13593, pp. 79–90, 2022.
https://doi.org/10.1007/978-3-031-23443-9_8

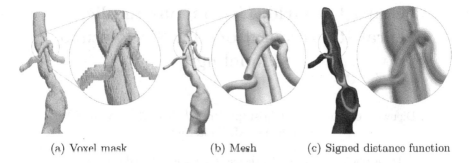

(a) Voxel mask   (b) Mesh   (c) Signed distance function

Fig. 1. Different representations of a 3D aortofemoral tree [31]. (a) and (b) are *explicit* representations; (a) has non-smooth boundaries, whereas boundaries of (b) are locally smooth. Both (a) and (b) are restricted to this resolution. (c) *implicitly* represents the surface with smooth boundaries, at any resolution.

or for doctor-patient communication and shared decision-making [29]. However, extracting these models from medical image data can be cumbersome. Commercial software and open-source software packages [2,3,14] traditionally rely on the construction of cylindrical models [24] in three steps. First, the lumen centerline is identified for each vessel. Then, local cross-sectional contours are determined and used to construct a watertight mesh model using (spline) interpolation. Finally, polygon mesh models of multiple vessels are blended to obtain a connected vascular tree. In this approach, tortuosity of the centerline can cause self-intersections of the orthogonal contours, resulting in surface folding of the final mesh model [9]. Moreover, smoothly connecting triangular meshes around bifurcations is challenging.

Deep learning has made great progress towards automatic model building from images [6,16]. However, popular convolutional neural network-based methods return 3D voxel masks. Because voxel masks merely discretize an underlying continuous shape, their quality heavily depends on the resolution of the image data, and they are not guaranteed to be contiguous. Hence, voxel masks typically require additional processing steps before use in, e.g., CFD. There is a need for a vascular model shape representation that is continuous, modular, and can be easily integrated with existing deep learning methods.

In this work, we adapt the work of Gropp et al. [10] to model vascular systems as combinations of level sets of signed distance functions represented in differentiable neural networks. Implicit representations and level sets have a substantial history in both segmentation [4,17] and 3D modeling [11,13] of vascular structures. Recently, there have been significant advances in signal representations using neural networks, i.e., implicit neural representations (INRs) [18,19,22,26]. Continuous implicit neural representations of the signed distance function can be easily transformed into any explicit representation, while conversion between explicit representations is non-trivial (Fig. 1). In this paper, we show how a continuous representation can be obtained from an explicit one. Moreover, an INR has a limited memory footprint that is independent of resolution, while the memory requirements of voxel masks and triangular meshes grow rapidly with an increase in resolution.

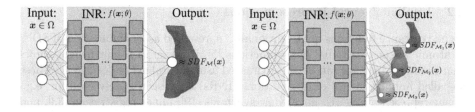

Fig. 2. Schematic overview of an implicit neural representation, parametrising a single SDF *(left)*, or three SDFs simultaneously *(right)*. Coordinate values can be queried and the network returns local SDF values.

We here demonstrate that INRs are a potentially valuable tool to bridge the gap between vascular modeling on the one hand and deep learning on the other hand. First, we show how INRs can be used to reconstruct an implicit surface from a sparse point cloud and form an alternative to conventional annotation procedures. We evaluate the efficiency and robustness of this approach. Secondly, we show that we can use a single INR to represent multiple surfaces, and demonstrate the effectiveness on nested shapes in an AAA case study. Finally, we demonstrate the added value of implicit shape representations in the smooth blending of separate structures in the reconstruction of an aortofemoral tree.

2   Method

We represent the surface of a vascular structure by the zero level set of its signed distance function (SDF), which we embed in a multilayer perceptron as an implicit neural representation.

2.1   Signed Distance Functions

A surface can be *implicitly* described by the zero level set of its signed distance function (SDF). In the case of a 2D surface $\mathcal{M}$ embedded in a 3D domain, $SDF_{\mathcal{M}}(\boldsymbol{x}) : \mathbb{R}^3 \to \mathbb{R}$ is defined as:

$$SDF_{\mathcal{M}}(\boldsymbol{x}) = \begin{cases} -d(\boldsymbol{x}, \mathcal{M}) & \boldsymbol{x} \text{ inside } \mathcal{M} \\ 0 & \boldsymbol{x} \text{ on } \mathcal{M} \\ d(\boldsymbol{x}, \mathcal{M}) & \boldsymbol{x} \text{ outside } \mathcal{M}, \end{cases} \quad (1)$$

$$\text{therefore: } \mathcal{M} = \{\boldsymbol{x} \in \mathbb{R}^3 | SDF_{\mathcal{M}}(\boldsymbol{x}) = 0\}. \quad (2)$$

In Eq. (1), $d(\boldsymbol{x}, \mathcal{M}) = \min_{\boldsymbol{y} \in \mathcal{M}} ||\boldsymbol{x} - \boldsymbol{y}||$, the minimal distance to the surface. An SDF satisfies the Eikonal equation, hence $||\nabla_{\boldsymbol{x}} SDF_{\mathcal{M}}(\boldsymbol{x})|| = 1, \forall \boldsymbol{x}$. As this function is continuous on $\mathbb{R}^3$, the surface $\mathcal{M}$ is represented independently of resolution.

2.2   Implicit Neural Representations

Signed distance functions are traditionally represented as a combination of SDFs of atomic shapes, or radial basis functions [5]. However, a recent insight shows that functions in space can be approximated using neural networks [22,26]. These so-called implicit neural representations (INRs) have been widely used for, e.g., image representations [7], novel view synthesis [19], image registration [33], and sparse-view CT reconstruction [28]. In this work, we use INRs to approximate the signed distance functions of one or multiple surfaces. This INR is a fully connected multilayer perceptron (MLP), $f(\boldsymbol{x};\theta)$, that takes $\boldsymbol{x} \subset \Omega := [\ 1,1]^3 \subset \mathbb{R}^3$ as input, and outputs $SDF_{\mathcal{M}}(\boldsymbol{x})$, as shown in Fig. 2 *(left)*. During optimization of the INR, desired properties of the SDF can be directly imposed on the network via the loss function. These properties include conditions on gradients, as they can be computed for the network through backpropagation [26]. An INR thus represents a function or signal, and not a mapping between two functions, as is often the case with convolutional neural networks. As an INR is trained on continuous coordinates, they allow for representing a function at any resolution.

2.3   Optimising an INR

We use an INR to represent the SDF of a 3D vascular model, under the condition that the model should be easily obtainable from a 3D medical image volume. We start from a small set of points on the vessel surface: $\mathcal{X} = \{\boldsymbol{x}_i\}_{i=1,\,...,\,N} \subset \Omega$. We aim to simultaneously reconstruct an SDF with $\mathcal{X}$ on the zero level set and embed it in an INR. This amounts to solving the Eikonal equation with pointwise boundary conditions; an ill-posed problem. Gropp et al. [10] tackle this ill-posedness by minimising the following loss function:

$$\ell(\theta) = \frac{1}{N} \sum_{1 \le i \le N} (|f(\boldsymbol{x}_i;\theta)|) + \lambda \mathbb{E}_x \left(\|\nabla_x f(\boldsymbol{x};\theta)\| - 1\right)^2. \qquad (3)$$

The first term enforces that all points from $\mathcal{X}$ lie on the zero level set of the SDF, and thus on the surface $\mathcal{M}$. The second term is a regularising Eikonal term, encouraging that the INR $f(\boldsymbol{x};\theta)$ satisfies the Eikonal equation, just like the SDF it represents. The key observation of Gropp et al. is that minimising Eq. (3) yields an SDF with a smooth and plausible zero isosurface. This approach allows for flexibility in the consistency of $\mathcal{X}$, in contrast to similar methods [20,23], that require local normals or points to lie within the same plane. In Sect. 3.1, we investigate the quality and robustness of the reconstruction under variation of the point cloud size and locations.

2.4   Fitting Multiple Functions in an INR

The walls of blood vessels consist of layers, and in patients with, e.g., atherosclerosis or dissections it is important to model these layers separately. For example,

in patients with abdominal aortic aneurysms (AAAs) the lumen and the surrounding solidified thrombus structure should be modeled separately [35]. While
INRs have mostly been used to represent single functions, they can be extended
to fit multiple functions on the same domain. Figure 2 *(right)* visualizes how by
adding additional output nodes to the neural network, multiple SDFs can be
fitted simultaneously, in this case for the lumen, the inner wall, and outer wall
of the AAA. This means that mutual properties of separate SDFs, such as for
nested shapes, could be learned by a single neural network. This INR is optimised by minimising Eq. (3) on each of its output channels. We will compare
this approach to multiple INRs representing structures separately in Sect. 3.2.

2.5 Constructive Solid Geometry

Signed distance functions make it very easy to determine the union, difference,
or intersection of multiple shapes, a cumbersome and challenging task in explicit
polygon mesh representations [12,34]. Let $SDF_{\mathcal{M}_1}(\boldsymbol{x})$ and $SDF_{\mathcal{M}_2}(\boldsymbol{x})$ be SDFs
for shapes $\mathcal{M}_1$ and $\mathcal{M}_2$ in Ω, respectively. Then their union is determined as:

$$\mathcal{M}_1 \cup \mathcal{M}_2 = \min(SDF_{\mathcal{M}_1}(\boldsymbol{x}), SDF_{\mathcal{M}_2}(\boldsymbol{x})) \tag{4}$$

To allow for smoother blending between interfaces of shapes, a smoothed min
function can be considered [15]:

$$
\begin{aligned}
(\mathcal{M}_1 \cup \mathcal{M}_2)_{\text{smooth}} &= \min(SDF_{\mathcal{M}_1}(\boldsymbol{x}), SDF_{\mathcal{M}_2}(\boldsymbol{x})) - \gamma_{\text{smooth}}, \\
\gamma_{\text{smooth}} &= \frac{1}{4k} \cdot \max(k - |SDF_{\mathcal{M}_1}(\boldsymbol{x}) - SDF_{\mathcal{M}_2}(\boldsymbol{x})|, 0)^2,
\end{aligned}
\tag{5}
$$

where k represents a smoothing parameter, i.e. a set region around the interfaces
of the two surfaces. We will use this strategy to join structures represented by
INRs in Sect. 3.3.

2.6 Data

For all experiments conducted, we used 3D vascular models from two publicly
available datasets. First, a dataset containing 19 3D geometries of abdominal
aortic aneurysms [32]. Each of these geometries consists of three nested structures: lumen, inner- and outer wall. Second, the vascular model repository [31],
containing 119 mesh models of varying regions of the vascular system, designed
for use in computational fluid dynamics.

  For our experiments on robustness and nested shapes, we selected three different cases from the AAA dataset, with varying shape complexity. Case 1 contains a bifurcation and two major dilations of the aorta; the lumen mesh contains
11,490 vertices and 67,944 edges. Case 2 is a single vessel, with constant dilation;
the mesh consists of 9,753 vertices and 57,387 edges. Similarly, Case 3 is a single
vessel with no bifurcations, however there is some additional curvature in the
center. The lumen mesh of Case 3 contains 9,816 vertices and 58,113 edges.

For the constructive geometry experiment, we randomly selected an aortofemoral tree from the vascular model repository. This geometry consisted of nine separate vessels, each of them given as a triangular mesh, as shown in Fig. 5. The number of vertices and edges for these vessels varied between 8,608 and 70,284, and 51,636 and 421,692, respectively.

Since the INR operates on the domain $\Omega = [-1, 1]^3$, the data first requires some preprocessing to match this domain. For the nested AAA geometries, we rescaled the union of the vertices of the lumen, inner and outer wall to Ω. For each of the three nested shapes, we randomly sampled a point cloud on their respective surfaces, that can be used for surface reconstruction using an INR. The original meshes are used to assess the quality of the surface reconstructions, using the average symmetric surface distance and Dice similarity coefficient.

For the meshes from the vascular model repository, we individually rescaled the mesh of each vessel to Ω. Note that the physical domain each INR operates on hence differs between vessels. For each rescaled mesh, we randomly sampled a point cloud of a size proportional to the surface of the mesh, that is used to train the INR.

3   Experiments and Results

We conducted three experiments: (1) assessing the robustness of the INR against variations in the point cloud, (2) testing the effectiveness of representing multiple shapes with a single INR and (3) demonstrating how our method can be used to reconstruct a vascular system from separate vessels and blend them naturally. For all experiments we used MLPs consisting of an input layer with three nodes, six fully connected hidden layers with 256 nodes and ReLU activations, and a final prediction layer consisting of one or three nodes, embedding a single shape or three nested shapes, respectively. Similar to [10], we used a skip connection, connecting the input to the third hidden layer. We base our implementation on PyTorch code[1] provided with [10].

3.1   Robustness

We demonstrate how INRs can be integrated in the existing pipeline for vascular analysis. Instead of acquiring a tubular mesh by manual contour annotations, we fit an INR representation of the SDF to a point cloud. We assess the robustness against variations in this point cloud. For this experiment, we used three cases of the AAA dataset [32] as described in Sect. 2.6. We sampled 20 different point clouds of predetermined sizes from the lumen mesh of each case. To train the INR reconstructing the geometries, we used an Adam optimizer with a learning rate of 0.0001, and set λ in the loss function (Eq. (3)) to 0.1, as suggested in [10]. We trained the network for 25,000 epochs on an NVIDIA Quadro RTX 6000 GPU, taking 10–15 minutes per network.

---

[1] https://github.com/amosgropp/IGR.

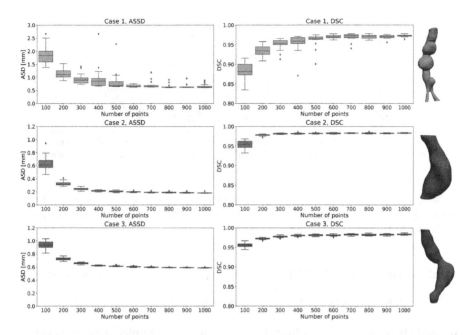

Fig. 3. The Dice similarity coefficient (DSC) and average symmetric surface distance (ASSD) of reconstructed vascular surface for varying number of reference points. Each SDF is reconstructed using 20 different point clouds of each size.

Figure 3 shows the quality of the surface reconstruction for each vascular structure, in terms of the Dice similarity coefficient (DSC) and average symmetric surface distance (ASSD). The DSC was computed using a binarized grid of the original mesh as a reference, the ASSD was computed based on the vertices of the original mesh. The median DSC for Case 2 and 3 is above 0.95 in all cases, even when only 100 points are used. For Case 1, which includes a bifurcation at the bottom, the median DSC at 100 points is slightly below 0.9. The median surface distances for Cases 2 and 3 are all sub-millimeter. The quality of the reconstruction of Case 1 lags compared to the other two, less complex, cases in terms of ASSD. For all three cases, we observe more variation in both ASSD and DSC when the reconstruction is based on fewer points. Medians of both metrics stabilize after 400 points for Case 2 and 3, and after 600 points for Case 1. Moreover, the standard deviations are small here, implying that the locations of the points will have a small impact on the final reconstructed shape.

3.2 Reconstructing Nested Shapes

In our second experiment, we use the same AAA geometries as in Sect. 3.1. However, we now use all three meshes of the vascular model: lumen, inner wall and outer wall of the aorta. Here, there is an anatomical prior that we want our model to capture. Namely, the lumen should be entirely nested inside the inner

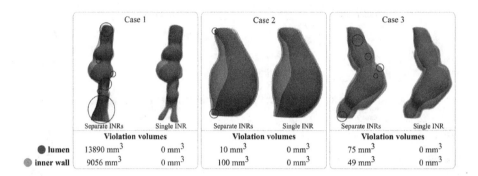

Fig. 4. Three nested AAA structures represented by a single INR, or separate INRs each representing a single shape. Using a single INR preserves the nested property, violations of this anatomical prior are marked with circles.

wall of the aorta, which is itself nested inside the outer wall. This implies:

$$SDF_{\text{outer wall}}(\boldsymbol{x}) \leq SDF_{\text{inner wall}}(\boldsymbol{x}) \leq SDF_{\text{lumen}}(\boldsymbol{x}), \forall \boldsymbol{x} \in \Omega. \qquad (6)$$

We used three AAA models from [32]. For each, we sampled a point cloud of 200 points for each of the three surfaces; a proper trade-off between point cloud sparsity and reconstruction quality, per the results presented in Sect. 3.1. We fit a joint INR to these point clouds, as well as three separate INRs, one for each of the three surfaces. Figure 4 shows the resulting surfaces for both approaches. In all three cases, we see unwanted extrusions of the inner structures when separate INRs are used. Case 1 displays this effect most clearly; the surfaces of the lumen and inner wall (red and green structures respectively) fail to capture the bifurcation and are inconsistent with the outer wall. On the other hand, representation of the shapes obtained using a single INR consistently were free of such extrusions. Moreover, the thickness of the vessel wall remains constant for this approach, in agreement with ground-truth data.

We quantified the unwanted extrusions in terms of violation volumes of the lumen and inner wall, also shown in Fig. 4. These represent the volumes of the regions where each nested structure violates Eq. (6). The violation volumes are defined as the volumes of the following sets:

$$VV_{\text{lumen}} = \{\boldsymbol{x} \in \Omega : SDF_{\text{lumen}}(\boldsymbol{x}) < 0 \wedge (SDF_{\text{inner}}(\boldsymbol{x}) > 0 \vee SDF_{\text{outer}}(\boldsymbol{x}) > 0)\}$$
$$VV_{\text{inner}} = \{\boldsymbol{x} \in \Omega : SDF_{\text{inner}}(\boldsymbol{x}) < 0 \wedge SDF_{\text{outer}}(\boldsymbol{x}) > 0\}.$$

The values reported in Fig. 4 are the volumes scaled back to physical sizes. We observe that using separate INRs for the reconstruction of Case 1, the most complex geometry, results in extremely high violation volumes compared to the other two cases. As we observed in our qualitative assessment, unwanted extrusions do consistently not appear when separate INRs are used.

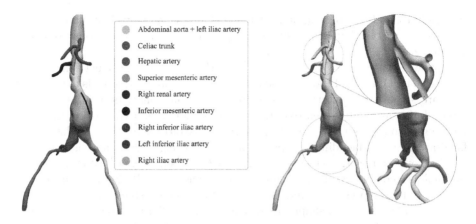

Fig. 5. Subdomains of separate INRs, indicated by different colours (left) and smooth blending of these INRs, with an inside-view at the top (right).

3.3 Constructive Geometry

In our third experiment, we demonstrate how INRs enable easy and smooth surface blending. This is critical for personalised vascular modeling, where many arteries of different calibers need to be blended into one single model for, e.g., CFD. Here, we use an aortofemoral tree from [31], that we preprocessed as described in Sect. 2.6. This sample consists of nine separate mesh structures: aorta, superior and inferior mesenteric arteries (SMA/IMA), renal arteries, hepatic artery, and common and internal iliac arteries, shown in Fig. 5 (left). We fit an INR to a point cloud of each of the vessels. We optimize these networks using an Adam optimizer for 50,000 epochs.

To blend the vessels, we generate a Cartesian grid encompassing the entire vascular tree. Hence, this domain contains the respective subdomains that all nine INRs were trained on. To evaluate the INRs for each structure on this grid, we map each grid point to the corresponding coordinate in each INR domain and query the SDF value by forwarding it through the network. Next, we rescale the SDF value from the INR domain to the real size. This results in a grid containing patches of SDFs, that we blend together using Eq. (5), with smoothing parameter $k = 0.1$. The result in Fig. 5 (right) shows a smooth transition between the different vascular structures, without any intrusions on the inside, making this structure suitable for CFD analysis [13].

4 Discussion and Conclusion

In this work, we have shown the value of implicit neural representations for 3D vascular modeling. We have demonstrated that INRs can be used to fit surfaces in abdominal aortic aneurysms using a small number of annotated points on the surface. Moreover, results show that fitting multiple nested surfaces in one INR is less likely to lead to violation of anatomical priors. Finally, we have used simple arithmetics to blend multiple surfaces represented in INRs into one complex vascular model.

Following the work of [10], we have shown that we can reconstruct continuous surfaces based on (small) point clouds, where their implicit prior leads to surfaces that are smooth and interpolate well in space. This contrasts with, e.g., slice interpolation as is often used to obtain dense voxel masks. INRs are *continuous*, allowing us to represent the vessel surface at any arbitrary resolution. This could facilitate seamless blending of separate arteries for, e.g., CFD analysis [13].

We have demonstrated how a single INR can simultaneously parametrise multiple nested surfaces in an abdominal aortic aneurysm. We showed that using a single INR to represent nested shapes benefits the consistency of the surfaces. The SDF values of nested shapes are similar, as they differ by a local offset. Using a single network to reconstruct them enables the network to learn these offsets by sharing weights and information from the other shapes during training. This nested property is, however, not a topological guarantee, but can be enforced by additional regularisation in the loss function.

Explicit representations such as polygonal meshes [2,3,14] and voxel masks are ubiquitous in vascular modeling. A direct comparison in this work between explicit representations and INRs is challenging, as we used triangular meshes as ground-truth geometries. However, we argue that instead of replacing explicit representations with INRs, they can cooperate. We demonstrated how INRs can be obtained from an explicit representation, and that smoothly combining them is straightforward. Afterwards, a mesh representation of the combined shape can be obtained at any desired resolution, obviating the need of cumbersome mesh operations [12].

One limitation of INRs is their lack of generalisation; a new INR needs to be optimised for each shape. This may be overcome by enriching the network input with a latent code, in which case INRs could represent a whole distribution of shapes [22]. Alternatively, INRs could be embedded in existing deep learning methods for (cardiovascular) image segmentation, e.g., by including local image features [7] or training hypernetworks [25]. Also, our work could lead to inter-active annotation tools. The combination of recently proposed rapid INR fitting approaches [21] and uncertainty quantification [8] could guide the annotator to areas that need additional annotations.

Besides their use in manual annotation procedures, INRs can be used in addition to automatic segmentation of vascular structures. For example, in [1] contour points of the lumen and outer wall of atherosclerotic carotid arteries are automatically acquired from MR images. These points can be used for optimizing the INR and acquiring a continuous surface representation of the nested lumen and outer wall of the carotid arteries.

In conclusion, INRs are a versatile surface representation that are easily acquired and integrated in existing frameworks for vascular modeling.

References

1. Alblas, D., Brune, C., Wolterink, J.M.: Deep learning-based cartotid artery vessel wall segmentation in black-blood MRI using anatomical priors. In: SPIE Medical Imaging (2022)
2. Antiga, L., Piccinelli, M., Botti, L., Ene-Iordache, B., Remuzzi, A., Steinman, D.A.: An image-based modeling framework for patient-specific computational hemodynamics. MBEC **46**(11), 1097–1112 (2008). https://doi.org/10.1007/s11517-008-0420-1
3. Arthurs, C.J., et al.: CRIMSON: an open-source software framework for cardiovascular integrated modelling and simulation. PLoS Comp. Biol. **17**(5), e1008881 (2021)
4. van Bemmel, C.M., Spreeuwers, L.J., Viergever, M.A., Niessen, W.J.: Level-set based carotid artery segmentation for stenosis grading. In: Dohi, T., Kikinis, R. (eds.) MICCAI 2002. LNCS, vol. 2489, pp. 36–43. Springer, Heidelberg (2002). https://doi.org/10.1007/3-540-45787-9_5
5. Carr, J.C., Fright, W.R., Beatson, R.K.: Surface interpolation with radial basis functions for medical imaging. IEEE Trans. Med. Imaging **16**(1), 96–107 (1997)
6. Chen, C., et al.: Deep learning for cardiac image segmentation: a review. Front. cardiovasc. med. **7**, 25 (2020)
7. Chen, Y., Liu, S., Wang, X.: Learning continuous image representation with local implicit image function. In: CVPR, pp. 8628–8638. IEEE/CVF (2021)
8. Gal, Y., Ghahramani, Z.: Dropout as a Bayesian approximation: representing model uncertainty in deep learning. In: ICML, pp. 1050–1059. PMLR (2016)
9. Gansca, I., Bronsvoort, W.F., Coman, G., Tambulea, L.: Self-intersection avoidance and integral properties of generalized cylinders. Comput. Aided Geom. Des. **19**(9), 695–707 (2002)
10. Gropp, A., Yariv, L., Haim, N., Atzmon, M., Lipman, Y.: Implicit geometric regularization for learning shapes. In: ICML, pp. 3789–3799 (2020)
11. Hong, Q., et al.: High-quality vascular modeling and modification with implicit extrusion surfaces for blood flow computations. Comput. Methods Programs Biomed. **196**, 105598 (2020)
12. Jiang, X., Peng, Q., Cheng, X., Dai, N., Cheng, C., Li, D.: Efficient Booleans algorithms for triangulated meshes of geometric modeling. Comput. Aided Des. Appl. **13**(4), 419–430 (2016)
13. Kretschmer, J., Godenschwager, C., Preim, B., Stamminger, M.: Interactive patient-specific vascular modeling with sweep surfaces. IEEE Trans. Vis. Comput. Graph. **19**(12), 2828–2837 (2013)
14. Lan, H., Updegrove, A., Wilson, N.M., Maher, G.D., Shadden, S.C., Marsden, A.L.: A re-engineered software interface and workflow for the open-source SimVascular cardiovascular modeling package. J. Biomech. Eng. **140**(2) (2018)
15. Li, Q.: Smooth piecewise polynomial blending operations for implicit shapes. In: Computer Graphics Forum, vol. 26, pp. 157–171. Wiley Online Library (2007)
16. Litjens, G., et al.: State-of-the-art deep learning in cardiovascular image analysis. JACC Cardiovasc. Imaging **12**(8 Part 1), 1549–1565 (2019)
17. Lorigo, L.M., et al.: Curves: curve evolution for vessel segmentation. Med. Image Anal. **5**(3), 195–206 (2001)
18. Martel, J.N., Lindell, D.B., Lin, C.Z., Chan, E.R., Monteiro, M., Wetzstein, G.: Acorn: adaptive coordinate networks for neural scene representation. ACM Trans. Graph. **40**(4), 1–13 (2021)

19. Mildenhall, B., Srinivasan, P.P., Tancik, M., Barron, J.T., Ramamoorthi, R., Ng, R.: NeRF: representing scenes as neural radiance fields for view synthesis. In: Vedaldi, A., Bischof, H., Brox, T., Frahm, J.-M. (eds.) ECCV 2020. LNCS, vol. 12346, pp. 405–421. Springer, Cham (2020). https://doi.org/10.1007/978-3-030-58452-8_24

20. Mistelbauer, G., Rössl, C., Bäumler, K., Preim, B., Fleischmann, D.: Implicit modeling of patient-specific aortic dissections with elliptic Fourier descriptors. In: Computer Graphics Forum, vol. 40, pp. 423–434. Wiley Online Library (2021)

21. Müller, T., Evans, A., Schied, C., Keller, A.: Instant neural graphics primitives with a multiresolution hash encoding (2022)

22. Park, J.J., Florence, P., Straub, J., Newcombe, R., Lovegrove, S.: DeepSDF: learning continuous signed distance functions for shape representation. In: CVPR, pp. 165–174. IEEE/CVF (2019)

23. Schumann, C., Oeltze, S., Bade, R., Preim, B., Peitgen, H.O.: Model-free surface visualization of vascular trees. In: EuroVis, pp. 283–290 (2007)

24. Shani, U., Ballard, D.H.: Splines as embeddings for generalized cylinders. Comput. Vis. Graph. Image Process. **27**(2), 129–156 (1984)

25. Sitzmann, V., Chan, E.R., Tucker, R., Snavely, N., Wetzstein, G.: MetaSDF: meta-learning signed distance functions. In: NeurIPS (2020)

26. Sitzmann, V., Martel, J.N., Bergman, A.W., Lindell, D.B., Wetzstein, G.: Implicit neural representations with periodic activation functions. In: NeurIPS (2020)

27. Sobocinski, J., et al.: The benefits of EVAR planning using a 3D workstation. Eur. J. Vasc. Endovasc. Surg. **46**(4), 418–423 (2013)

28. Sun, Y., Liu, J., Xie, M., Wohlberg, B., Kamilov, U.S.: Coil: coordinate-based internal learning for tomographic imaging. IEEE Trans. Comput. Imaging **7**, 1400–1412 (2021)

29. Swart, M., McCarthy, R.: Shared decision making for elective abdominal aortic aneurysm surgery. Clin. Med. **19**(6), 473 (2019)

30. Tran, K., Yang, W., Marsden, A., Lee, J.T.: Patient-specific computational flow modelling for assessing hemodynamic changes following fenestrated endovascular aneurysm repair. JVSL: Vasc. Sci. **2**, 53–69 (2021)

31. Wilson, N.M., Ortiz, A.K., Johnson, A.B.: The vascular model repository: a public resource of medical imaging data and blood flow simulation results. J. Med. Devices **7**(4) (2013)

32. Wittek, A., et al.: Image, geometry and finite element mesh datasets for analysis of relationship between abdominal aortic aneurysm symptoms and stress in walls of abdominal aortic aneurysm. Data Br. **30**, 105451 (2020)

33. Wolterink, J.M., Zwienenberg, J.C., Brune, C.: Implicit neural representations for deformable image registration. In: MIDL. PMLR (2022)

34. Wu, J., Ma, R., Ma, X., Jia, F., Hu, Q.: Curvature-dependent surface visualization of vascular structures. Comput. Med. Imaging Graph. **34**(8), 651–658 (2010)

35. Zhu, C., Leach, J.R., Wang, Y., Gasper, W., Saloner, D., Hope, M.D.: Intraluminal thrombus predicts rapid growth of abdominal aortic aneurysms. Radiology **294**(3), 707–713 (2020)

Comparison of Semi- and Un-Supervised Domain Adaptation Methods for Whole-Heart Segmentation

Marica Muffoletto[1]([✉]), Hao Xu[1], Hugo Barbaroux[1], Karl P. Kunze[1,2],
Radhouene Neji[1,2], René Botnar[1], Claudia Prieto[1], Daniel Rueckert[3,4],
and Alistair Young[1]

[1] School of Biomedical Engineering and Imaging Sciences, King's College London,
St Thomas' Hospital, 4th Floor Lambeth Wing,
Westminster Bridge Road, London SW1 7EH, UK
marica.muffoletto@kcl.ac.uk
[2] MR Research Collaborations, Siemens Healthcare Limited, Frimley, UK
[3] Department of Computing, Imperial College London,
Biomedical Image Analysis Group, London, UK
[4] Institute for Artificial Intelligence and Informatics in Medicine,
Klinikum Rechts der Isar, Technical University of Munich, Munich, Germany

Abstract. Quantification of heart geometry is important in the clinical diagnosis of cardiovascular diseases. Changes in geometry are indicative of remodelling processes as the heart tissue adapts to disease. Coronary Computed Tomography Angiography (CCTA) is considered a first line tool for patients at low or intermediate risk of coronary artery disease, while Coronary Magnetic Resonance Angiography (CMRA) is a promising alternative due to the absence of radiation-induced risks and high performance in the evaluation of cardiac geometry. Yet, the accuracy of an image-based diagnosis is susceptible to the quality of volume segmentations. Deep Learning (DL) techniques are gradually being adopted to perform such segmentations and substitute the tedious and manual work performed by physicians. However, practical applications of DL techniques on a large scale are still limited due to their poor adaptability across modalities and patients. Hence, the aim of this work was to develop a pipeline to perform automatic heart segmentation of multiple cardiac imaging scans, addressing the domain shift between MRs (target) and CTs (source). We trained two Domain Adaptation (DA) methods, using Generative Adversarial Networks (GANs) and Variational Auto-Encoders (VAEs), following different training routines, which we refer to as un- and semi- supervised approaches. We also trained a baseline supervised model following state-of-the-art choice of parameters and augmentation. The results showed that DA methods can be significantly boosted by the addition of a few supervised cases, increasing Dice and Hausdorff distance metrics across the main cardiac structures.

Keywords: Deep Learning · Automatic segmentation · Domain adaptation · Generative Adversarial Networks · Variational Auto-Encoders

O. Camara et al. (Eds.): STACOM 2022, LNCS 13593, pp. 91–100, 2022.
https://doi.org/10.1007/978-3-031-23443-9_9

1   Introduction

1.1   Problem

Cardiovascular diseases (CVDs) are a leading cause of death worldwide and their prevention and treatment hugely benefits from advances in imaging techniques. Cardiac segmentation is a fundamental prerequisite step which can influence the correct analysis of the acquired images, but this is often a challenging task and heavily susceptible to differences in image modality, scanner and acquisition protocol. An extensive literature already covers reliable and accurate methods for segmentation in cardiac imaging [3,5,20]. These are usually trained with manual domain-specific label maps segmented by expert radiologists [1,12], whose labelling takes long time and still suffers from inter- and intra- observer variability. Hence, the most important challenge remains the adaptation of these models from one modality to another without the need for vast amounts of labelled data in both domains. This problem arises from the fact that models trained on a fixed distribution are meant to perform well on data drawn from the same distribution, and not on a different target distribution [2], with zero to very few labelled cases. In this context, we refer to the expression "domain gap" as the difference in distribution between a large set of labelled source data and an unlabelled test set of target data [10,19]. In this paper, we aim to investigate how different Deep Learning architectures can tackle this problem, in un-supervised and semi-supervised settings.

1.2   Related Work

In Yan et al. [18], the effect of the domain gap has been demonstrated through an adversarial attack performed on CMR images by applying a perturbation which hardly affects human vision, but leads to failure of the UNet architecture. To bridge this problem, which significantly compromises the performance of state-of-the-art supervised approaches to segmentation task, we can adopt Deep Learning based Unsupervised Domain Adaptation (UDA) methods. These seek to overcome such problems by transferring the segmentation ability acquired in one domain to the other. Recently, UDA has been widely explored for classification and detection tasks [18,19], and Variational Auto Encoders (VAEs) or Generative Adversarial Networks (GANs) constitute core approaches. These are both generative models which employ reconstruction or adversarial losses to reduce the domain discrepancy at the image and feature level. Chen et al. [8] have discussed the relevance of these models in medical image computing for unsupervised learning problems and mentioned that a combination of VAE and GAN could produce sharp images that substantially decrease the reconstruction loss. In Skandarani et al. [15], a VAE-GAN model is trained to learn the latent distribution of cardiac shapes, and generate an arbitrary large number of cardiac maps, thus overcoming the problem of scarce ground-truth labels.

1.3   Contributions and Aim

The success of experiments on UDA from 3D CTs to MRs for the task of Whole-Heart Segmentation (WHS) has been very limited. Using the available 51 matching pairs of Cardiac Computed Tomography Angiographies (CCTAs) and Cardiac Magnetic Resonance Angiographies (CMRAs) from [4], we aim to draw a comprehensive comparison between UDA methods. Specifically, we want to test the performance of such techniques on the adaptation from an easier task (CT segmentation) to a harder one (MR segmentation). With this scope in mind, we design a study which can be applied to multi-domain data, in order to compare two state-of-the-art architectures: the GAN-based model SIFA [7] and the VAE architecture VARDA [16]. We modify both of these to accept some supervised cases in the target domain, allowing for a comparison with a fully supervised method trained on the same amount of data. The pipeline of this work is illustrated in Fig. 1: a) obtain accurate CCTA segmentations from method in [17]; b) perform a multi-modal registration using manual correction of landmarks followed by a non-linear registration method; c) identify and perform essential pre-processing steps for each method; d) training of the original unsupervised techniques, the new semi-supervised approach and a baseline method.

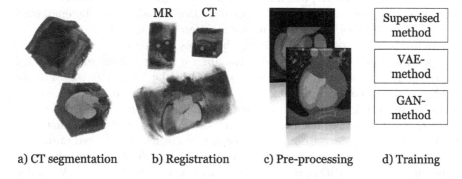

a) CT segmentation   b) Registration   c) Pre-processing   d) Training

Fig. 1. Overall pipeline. Step a) shows the CCTA volume and the overlayed segmentation obtained by method [17]. Step b) shows the unregistered volumes with matching landmarks, and an example of the succesful registration where the CTA segmentation is now aligned to the MRA. Finally step c)shows the images after registration, slicing and cropping, representing an example of the training data fed into the three trained methods - step d).

2   Method

2.1   Dataset

We used a dataset of 51 matching cases of CCTAs and CMRAs, including Coronary Artery Disease (CAD) patients and healthy subjects. The acquisition of the

3D whole-heart isotropic sub-millimeter resolution CMRAs was performed using a free-breathing 3D balanced steady state free precession (bSSFP) with image navigator (iNAV) and non-rigid motion compensated reconstruction, described in [4]. Six patients are excluded from the study, due to unresolved misalignment after the registration step (Fig. 1b). The CMRA images were reconstructed to $0.6\,mm^3$ isotropic resolution, while CCTA have constant $0.5\,mm$ resolution in the z-axis, and in-plane resolution of $0.2{\sim}0.4\,mm \times 0.2{\sim}0.4\,mm$.

2.2 Experiments

To gather the necessary experiments for our comparative analysis, we followed the pipeline in Fig. 1. Firstly, we used the method developed in [17] to obtain automatic segmentations of the 45 CCTA images, which we used as ground truth labels for all our models. These included the following structures: 1) Left Ventricle (LV), 2) Left Ventricle Myocardium (LVM), 3) Right Ventricle (RV), 4) Left Atrium (LA), 5) Right Atrium (RA), 6) Aorta (AO), 7) Pulmonary Artery (PA), 8) Left Pulmonary Vein (LPV), 9) Right Pulmonary Vein (RPV), 10) Left Ascending Aorta (LAA). We then performed a landmark-based registration using 3DSlicer [11], and manually selected the following landmarks: the intersection of the left and right main coronary arteries with the aorta, a bifurcation of the left anterior descending (LAD) and a bifurcation of the right coronary artery and the left ventricular apex. Finally, we further improved the alignment by employing the NiftyReg-f3d tool which uses the method described in [14].

As pre-processing steps, we reduced the labels to the first six structures, we cropped the center of the heart using the aligned segmentation and cut off 2% of the intensity histogram; finally we standardised the image intensity. For the GAN-based approach we rescaled each axial slice to the range [–1,1], as this seems to substantially improve training of GANs. Each slice is then fed to the networks in an unpaired fashion, hence there is no correspondence in anatomies between source and target domains, although we start from a paired dataset.

To analyse the performance of UDA approaches, we trained a supervised method as baseline. This was a 2D Dynamic Unet implemented with MONAI [9]. We added the affine, Gaussian noise, Gaussian blur, scale intensity and mirror augmentation techniques provided by MONAI, to emulate nnU-Net [13] training protocols.

For the GAN-based method, we chose the architecture introduced in Chen et al. [6,7], and referenced as SIFA (Synergistic Image and Feature Adaptation). An overview of this is shown in Fig. 2a. This method uses a generator to perform a source-to-target image transformation and a shared encoder, which takes as inputs the real target x^t and the fake one $x^{s\rightarrow t}$ and is connected to a decoder and a pixel-wise classifier. The former reconstructs both images into a fake and cycle source, while the latter performs the image segmentation task. The original architecture is optimised by a combination of adversarial losses, reconstruction losses and a source segmentation loss which are shown in Fig. 2a. We modify this by adding an alternative segmentation loss for the target domain.

For the VAE-based method, we used the architecture developed by Wu et al. [16]. Two VAEs are used for reconstructing source and target domains. The total loss is a combination of two reconstruction terms, the Kullback-Leibler (KL) Divergence term, and a discrepancy loss which is introduced as an explicit metric to directly reduce difference between the latent variables from the two domains. The classifier takes features from the encoder and predicts a segmentation from both the source and target images. While originally this is optimised solely by a source segmentation loss, we add a target segmentation loss if supervised cases are available. Both of these techniques are 2D, and hence, as illustrated in Fig. 2, they are applied to single slices derived from the 3D volumes. The additions made for training a semi-supervised approach are highlighted in red.

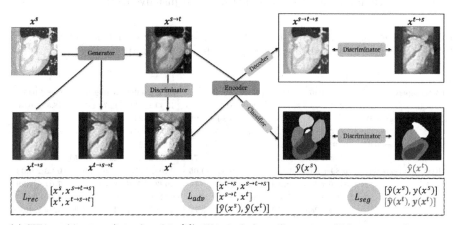

(a) SIFA architecture (introduced in [7]). This includes a Generator which operates the transformation from source domain to target domain, an Encoder which takes as input target images, either real or fake, and feeds either a Decoder or a Classifier. The former reconstructs them into source, while the latter predicts the label maps. Finally, Discriminators distinguish between fake and real source and target images and between predictions of the real and fake target images.

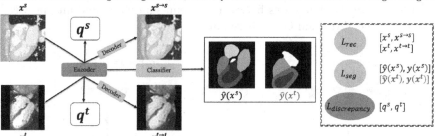

(b) VARDA architecture (introduced in [16]). Both the source and target images are fed into a shared Encoder, and reconstructed by a separate Decoder. The latent features are also fed into a Classifier which predicts label maps for both domains.

Fig. 2. Diagrams of Domain Adaptation methods included in the study. The dashed boxes contain the main losses ($L_{rec}, L_{adv}, L_{seg}, L_{discrepancy}$) used during training, with the new supervised loss highlighted in red. Each slice is labelled as corresponding to either an image (x) or a label map (y), and we use the (s,t) notation to describe the source and target domains respectively. Figure 2a refers to the GAN-based approach and Fig. 2b includes the VAE architecture. (Color figure online)

3   Results

We show the results obtained by training all the methods discussed in the previous section using a total of 40 cases in both source and target domains, including zero (SIFA unsupervised, VARDA unsupervised) or five labelled target cases (SIFA semi-supervised, VARDA semi-supervised, UNet). The 2D Dice/HD scores reported in Table 1 represent the outcome of the post-processed predictions obtained by each method on an unseen hold-out dataset of five cases. They show that semi-supervised approaches for both SIFA and VARDA increased the performance of the trained models, and that the UDA methods performed generally better than a supervised model trained on just five cases.

Table 1. Performance of trained methods on test set. Odd rows include average $\pm$ std 2D Dice for single labels, while even ones refer to the Hausdorff distance.

	LV	LVM	RV	LA	RA	AO
SIFA (unsupervised)	0.87 ± 0.07	0.69 ± 0.07	0.79 ± 0.02	$\mathbf{0.72\pm0.09}$	$\mathbf{0.86\pm0.04}$	0.78 ± 0.07
	12.65 ± 7.64	17.20 ± 9.99	54.15 ± 17.69	13.49 ± 4.44	12.74 ± 7.70	18.85 ± 6.91
SIFA (semi-supervised)	0.88 ± 0.10	$\mathbf{0.74\pm0.08}$	$\mathbf{0.84\pm0.05}$	0.59 ± 0.22	0.83 ± 0.03	0.75 ± 0.11
	$\mathbf{10.75\pm5.54}$	17.02 ± 6.87	18.05 ± 12.25	22.52 ± 9.65	20.98 ± 8.71	18.76 ± 6.75
VARDA (unsupervised)	0.79 ± 0.06	0.53 ± 0.09	0.82 ± 0.04	0.58 ± 0.22	0.76 ± 0.08	0.75 ± 0.07
	11.58 ± 4.17	25.93 ± 7.34	$\mathbf{13.12\pm7.63}$	15.04 ± 7.30	$\mathbf{12.30\pm5.51}$	10.30 ± 3.53
VARDA (semi-supervised)	$\mathbf{0.89\pm0.04}$	0.68 ± 0.13	$\mathbf{0.84\pm0.05}$	0.70 ± 0.21	0.80 ± 0.08	$\mathbf{0.83\pm0.03}$
	11.16 ± 3.34	22.65 ± 5.82	15.43 ± 5.71	$\mathbf{10.60\pm5.35}$	15.30 ± 9.28	$\mathbf{7.01\pm1.99}$
UNet	0.76 ± 0.07	0.55 ± 0.09	0.74 ± 0.08	0.51 ± 0.28	0.68 ± 0.10	0.71 ± 0.07
	39.97 ± 28.60	54.70 ± 13.48	17.53 ± 8.71	33.81 ± 17.72	32.45 ± 15.21	32.65 ± 10.01

This is confirmed by Fig. 3, where higher variation (LA/RA labels) and lower scores (LV, LVM, RA) are found in the UNet plot, compared to the others. We can also observe the differences between the semi-supervised and unsupervised approach in the VAE- and GAN- based architectures.

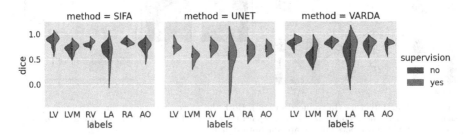

Fig. 3. Quantitative comparison of SIFA, UNet and VARDA on the test set. Each violin plot is split according to the presence of supervision during training. Since UNet is a supervised method, this plot is only one sided.

Qualitative results are reported in Fig. 4, 5. They show a comparison of the segmentation predicted by the methods under analysis on the same slice (Fig. 4) and the consistency across anatomical planes obtained by a semi-supervised approach (Fig. 5).

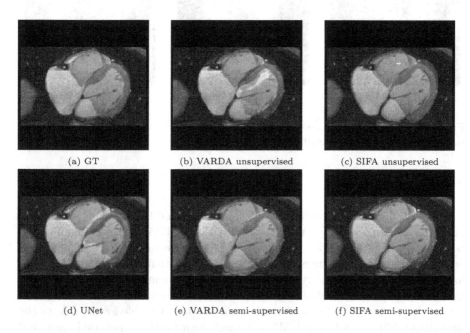

(a) GT   (b) VARDA unsupervised   (c) SIFA unsupervised

(d) UNet   (e) VARDA semi-supervised   (f) SIFA semi-supervised

Fig. 4. Qualitative comparison of SIFA, UNet and VARDA on a single slice. The top row shows the Ground Truth (a) and the results of the unsupervised methods, while the bottom row shows the prediction by the fully supervised method (d) and by the new semi-supervised methods adapted from the originally unsupervised SIFA and VARDA architectures.

In Fig. 4, we evaluate the quality of each method on a single mid-axial slice which features five of the six structures included in the study. The label maps shown in Fig. 4e and 4f (semi-supervised approaches) are closer to the ground truth segmentation provided in Fig. 4a than their unsupervised counterparts. The label map predicted by UNet is quite similar to the GT one, but misses some pixels in the LA and LV, and it fails to detect an accurate boundary between the LV myocardium and the LV cavity. The prediction by the SIFA semi-supervised approach (Fig. 4f) is able to overcome this obstacle. In Fig. 5, we show that, although the model was only trained on 2D axial slices, this prediction features 3D consistency, and performs equally well on axial, sagittal and coronal views.

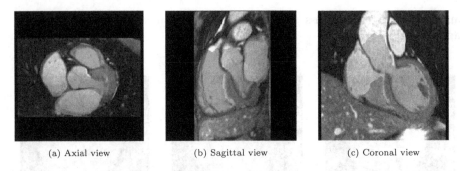

(a) Axial view   (b) Sagittal view   (c) Coronal view

Fig. 5. Prediction of single test case across anatomical planes. Results of SIFA semi-supervised approach (cf. Fig. 4f).

4   Conclusion

The results above lead us to conclude that the selected DA methods show some interesting differences: the VAE-based approach (VARDA) improves significantly with more supervision (cf Fig. 4b, 4e, Table 1), whereas the GAN-based approach is less affected by this change. There is a strong variation in the results obtained for each label. Figure 3 clearly shows that the LA segmentation is generally worse in the semi-supervised techniques than in the unsupervised ones. This might be due to the quality of the ground truth segmentation available for the LA, or the underlying differences in the CMRA dataset between the supervised cases and the test cases. Despite these flaws, we generally observe good quality WHS across the test set provided by the SIFA semi-supervised approach (cf. Table 1, Fig. 4f), which outperforms both the fully supervised method and the VAE-based one in the challenging tasks of myocardium and ventricles' boundary detection.

4.1   Future Work

In this analysis, due to the limited number of target manual segmentations, we keep the training and test sets fixed. This includes the supervised cases used to train both DA methods and the UNet, whose number is restricted to five. In the future, we plan to extend the validity of this study by obtaining ground truths for the entire CMRA dataset, hence operating a 5-fold cross validation and experimenting with an increasing amount of supervised cases. Based on the results achieved through cross-validation, we intend to select the best working method to improve the performance on small areas such as the LV myocardium, and to re-train it on unregistered cases. Hence, tackling the second limitation of this study: the manual detection of landmarks in the overall pipeline (Fig. 1b). Based on preliminary results, we deem the registration step as unnecessary, and we believe that with some extra processing steps, it will be possible to remove it and perform a UDA segmentation that does not rely on the alignment of the two domains. As a long-term plan, both the GAN- and VAE- based approach can be modified to directly work on 3D inputs to tackle 4D challenges such as the segmentation of 3D cardiac CINE MRI, investigating motion and geometry of the heart.

Acknowledgments. Research supported by ESPRC and Siemens Healthineers.

References

1. Abdeltawab, H., et al.: A deep learning-based approach for automatic segmentation and quantification of the left ventricle from cardiac cine MR images. Comput. Med. Imaging Grap. **81**, 101717 (2020). https://doi.org/10.1016/j.compmedimag.2020.101717
2. Ben-David, S., Blitzer, J., Crammer, K., Kulesza, A., Pereira, F., Vaughan, J.W.: A theory of learning from different domains. Mach. Learn. **79**(1–2), 151–175 (2010). https://doi.org/10.1007/s10994-009-5152-4
3. Bernard, O., et al.: Deep learning techniques for automatic MRI cardiac multi-structures segmentation and diagnosis: is the problem solved? IEEE Trans. Med. Imaging **37**(11), 2514–2525 (2018)
4. Bustin, A., et al.: 3D whole-heart isotropic sub-millimeter resolution coronary magnetic resonance angiography with non-rigid motion-compensated PROST. J. Cardiovasc. Magn. Reson. **22**(1), 24 (2020). https://doi.org/10.1186/s12968-020-00611-5
5. Chen, C., et al.: Deep learning for cardiac image segmentation: a review. Front. Cardiovasc. Med. **7**, 25 (2020). https://doi.org/10.3389/fcvm.2020.00025
6. Chen, C., Dou, Q., Chen, H., Qin, J., Heng, P.: Synergistic image and feature adaptation: towards cross-modality domain adaptation for medical image segmentation. In: AAAI (2019)
7. Chen, C., Dou, Q., Chen, H., Qin, J., Heng, P.A.: Unsupervised bidirectional cross-modality adaptation via deeply synergistic image and feature alignment for medical image segmentation (2020)
8. Chen, X., Pawlowski, N., Rajchl, M., Glocker, B., Konukoglu, E.: Deep generative models in the real-world: an open challenge from medical imaging. ArXiv abs/1806.05452 (2018)
9. Consortium, M.: Monai: Medical open network for AI (2022). https://doi.org/10.5281/zenodo.6639453. If you use this software, please cite it using these metadata
10. Dou, Q., Ouyang, C., Chen, C., Chen, H., Heng, P.: Unsupervised cross-modality domain adaptation of convnets for biomedical image segmentations with adversarial loss. In: IJCAI (2018)
11. Fedorov, A., et al.: 3D slicer as an image computing platform for the quantitative imaging network. Magn. Reson. Imaging **30**(9), 1323–1341 (2012)
12. Habijan, M., Leventic, H., Galic, I., Babin, D.: Whole heart segmentation from CT images using 3D U-Net architecture. In: 2019 International Conference on Systems, Signals, and Image Processing, 121–126 (2019). https://doi.org/10.1109/IWSSIP.2019.8787253
13. Isensee, F., Jaeger, P.F., Kohl, S.A.A., Petersen, J., Maier-Hein, K.H.: nnU-Net: a self-configuring method for deep learning-based biomedical image segmentation. Nat. Methods **18**(2), 203–211 (2021). https://doi.org/10.1038/s41592-020-01008-z
14. Rueckert, D., Sonoda, L., Hayes, C., Hill, D., Leach, M., Hawkes, D.: Nonrigid registration using free-form deformations: application to breast MR images. IEEE Trans. Med. Imaging **18**(8), 712–721 (1999). https://doi.org/10.1109/42.796284

15. Skandarani, Y., Painchaud, N., Jodoin, P.M., Lalande, A.: On the effectiveness of GAN generated cardiac MRIs for segmentation. ArXiv abs/2005.09026 (2020)
16. Wu, F., Zhuang, X.: Unsupervised domain adaptation with variational approximation for cardiac segmentation. CoRR abs/2106.08752 (2021), https://arxiv.org/abs/2106.08752
17. Xu, H., Niederer, S.A., Williams, S.E., Newby, D.E., Williams, M.C., Young, A.A.: Whole heart anatomical refinement from CCTA using extrapolation and parcellation. In: Ennis, D.B., Perotti, L.E., Wang, V.Y. (eds.) FIMH 2021. LNCS, vol. 12738, pp. 63–70. Springer, Cham (2021). https://doi.org/10.1007/978-3-030-78710-3_7
18. Yan, W., et al.: The domain shift problem of medical image segmentation and vendor-adaptation by Unet-GAN. In: Shen, D., et al. (eds.) MICCAI 2019. LNCS, vol. 11765, pp. 623–631. Springer, Cham (2019). https://doi.org/10.1007/978-3-030-32245-8_69
19. Zhang, T., Yang, J., Zheng, C., Lin, G., Cai, J., Kot, A.C.: Task-in-all domain adaptation for semantic segmentation. In: 2019 IEEE International Conference on Visual Communications and Image Processing, VCIP (2019). https://doi.org/10.1109/VCIP47243.2019.8965736
20. Zhuang, X., et al.: Evaluation of algorithms for multi-modality whole heart segmentation: an open-access grand challenge. Med. Image Anal. **58**, 101537 (2019). https://doi.org/10.1016/j.media.2019.101537

Automated Quality Controlled Analysis of 2D Phase Contrast Cardiovascular Magnetic Resonance Imaging

Emily Chan[1](✉), Ciaran O'Hanlon[1], Carlota Asegurado Marquez[1],
Marwenie Petalcorin[1], Jorge Mariscal-Harana[1], Haotian Gu[7],
Raymond J. Kim[6], Robert M. Judd[6], Phil Chowienczyk[1,7],
Julia A. Schnabel[1,2,3], Reza Razavi[1,4], Andrew P. King[1], Bram Ruijsink[1,4,5],
and Esther Puyol-Antón[1]

[1] School of Biomedical Engineering and Imaging Sciences,
King's College London, London, UK
emily.m.chan@kcl.ac.uk
[2] Technical University Munich, Munich, Germany
[3] Helmholtz Center Munich, Munich, Germany
[4] Guy's and St Thomas' NHS Foundation Trust, London, UK
[5] Department of Cardiology, Heart and Lung Division, University Medical Center
Utrecht, Utrecht, The Netherlands
[6] Division of Cardiology, Department of Medicine, Duke University,
Durham, NC, USA
[7] British Heart Foundation Centre, King's College London, London, UK

Abstract. Flow analysis carried out using phase contrast cardiac magnetic resonance imaging (PC-CMR) enables the quantification of important parameters that are used in the assessment of cardiovascular function. An essential part of this analysis is the identification of the correct CMR views and quality control (QC) to detect artefacts that could affect the flow quantification. We propose a novel deep learning based framework for the fully-automated analysis of flow from full CMR scans that first carries out these view selection and QC steps using two sequential convolutional neural networks, followed by automatic aorta and pulmonary artery segmentation to enable the quantification of key flow parameters. Accuracy values of 0.998 and 0.828 were obtained for view classification and QC, respectively. For segmentation, Dice scores were >0.964 and the Bland-Altman plots indicated excellent agreement between manual and automatic peak flow values. In addition, we tested our pipeline on an external validation data set, with results indicating good robustness of the pipeline. This work was carried out using multivendor clinical data consisting of 699 cases, indicating the potential for the use of this pipeline in a clinical setting.

Keywords: Cardiac magnetic resonance · Deep learning · Quality control · Cardiac function · View-selection · Multi-vendor

B. Ruijsink and E. Puyol-Anton—Joint last authors.

O. Camara et al. (Eds.): STACOM 2022, LNCS 13593, pp. 101–111, 2022.
https://doi.org/10.1007/978-3-031-23443-9_10

1 Introduction

Cardiac Magnetic Resonance (CMR) is one of the most comprehensive modalities for assessment of cardiovascular function, enabling the quantification of cardiac volumes, myocardial motion and tissue characteristics. Aside from these properties, CMR allows for the measurement of blood flow in the major vessels. These measurements can be useful for quantifying a range of aspects of cardiovascular function, including cardiac output, the shunting of blood through intra-cardiac connections, the severity of valvular lesions (including regurgitation), and vascular properties. While 2D flow can potentially be measured in any blood vessel using CMR, pulmonary and aortic flow measurements are the most important measures used in clinical assessments [10]. After acquisition of 2D Phase Contrast (PC) images, flow measures are normally obtained semi-automatically; a physician manually delineates the target blood vessel and a semi-automatic (pixel threshold based) method propagates this segmentation over the full cardiac cycle. After this, manual adjustment of the target segmentation is performed and, importantly, the physician interrogates the images and obtained flow data for quality. This quality control (QC) includes checking for errors in imaging, such as fold-over and aliasing of the blood flow signal, and checking the appearance of the flow curve for irregular (unphysiological) behaviour [14].

Several works have proposed the semi-automatic or automatic quantification of flow using PC-CMR images. For example, Bidhult et al. [2] incorporated shape constraints to aid in active contour based semi-automatic segmentation of the ascending aorta and pulmonary artery. Similarly, Bratt et al. [3] utilised a modified U-Net to develop a method for fully-automatic aortic flow quantification. However, these solutions ignore highly important steps in the automation of flow assessment: image classification, to select the correct images from a full CMR scan automatically, and the aforementioned quality control, which is an integral part of CMR (flow) analysis.

In recent years, machine learning-enabled QC of CMR images has received increasing interest, in particular for cine images. Tarroni et al. [19] developed a decision forest-based solution for heart coverage estimation, inter-slice motion detection and image contrast estimation in the cardiac region, while Ruijsink et al. [16] proposed an end-to-end pipeline with automated QC techniques to analyse cine CMR sequences. Other works focused on detecting missing slices [23,24] and motion artefacts [11], while some addressed QC of automatic image segmentation [1,4,15,21]. Additionally, Vergani et al. [20] recently proposed the use of convolutional neural networks for the automatic quality controlled selection of cine images for analysis. To the best of our knowledge, the use of deep learning (DL) for view selection and QC prior to flow quantification has not been explored with PC-CMR images. Moreover, existing solutions for automated flow quantification have generally been trained on single-centre/highly selective data sets, making their generalisation to external data sets challenging.

In this paper, we propose a novel framework that takes into account these important aspects to provide a fully-automated pipeline from image-to-report. We develop a DL-based image classification algorithm to detect 2D PC flow imaging, and classify/select the most important target sequences for analysis (aortic flow and pulmonary flow); a QC algorithm to detect image quality related errors; and a robust segmentation pipeline for 2D pulmonary and aortic flow. Our pipeline calculates key parameters of interest in flow analysis: peak, net, total forward and total backward flow. Furthermore, we utilise an external validation set to assess the generalisability of our framework.

2   Methods

The framework we present is composed of four steps: 1) view selection aimed at identifying conventional PC-CMR views, 2) QC to detect image quality related artefacts, 3) automated segmentation of PC images, and 4) parameter extraction (see Fig. 1 for a summary). These steps are described below:

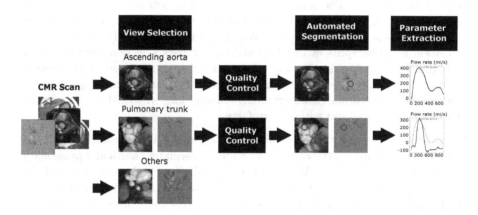

Fig. 1. Overview of the proposed AI-based framework for automatic flow quantification from a full CMR scan. The first block, view selection, aims to identify conventional phase contrast views; the second block aims to perform quality control to detect image quality related artefacts; the third block aims to perform automatic segmentation of the quality controlled selected images; and the fourth block aims to extract clinical parameters using the resulting segmentations.

1. View Selection: The first part of our framework aims to identify the standard 2D PC views used for the analysis of cardiac function and haemodynamics from a CMR acquisition. The manually classified data were divided as follows: 80% was used for training, 10% was used for validation, and 10% for testing, with the splitting carried out in a stratified manner with respect to the different views. The training, validation and test data cohorts had a mutually exclusive subject pool, i.e. acquisitions from the same subject could only be used in one of the three cohorts. Prior to training, all images were cropped to a standard size of 192 × 192 pixels. In order to classify between the ascending aorta, pulmonary trunk and other views, we implemented seven different architectures that have previously proven robust for classification: AlexNet [8], Densenet-121 [6], ConvNeXt Small [9], MobileNet v2 [17], RegNetX_3.2GF [13], ResNet-18 [5], VGG11 [18]. Training was carried out with batch size 16 and learning rate 0.001 with stochastic gradient descent for 200 epochs with cross entropy·loss to classify images into the three classes described. For the training data, data augmentation was performed on-the-fly using random translations (±30 pixels), rotations (±90°), flips (50% probability) and scaling (up to 20%) for each mini-batch of images before feeding them to the network. The probability of augmentation for each of the parameters was 50%.

2. Quality Control: QC of image acquisition is of particular importance when measuring flow because errors in the quantification of net flow can arise from a range of different image acquisition or processing quality issues, including tortuosity of the aorta, planning of the acquisition plane, arrhythmia and breathing artefacts. These quality issues can lead to errors in flow analysis and therefore need to be detected. The same seven classification networks as used for view selection were implemented as binary classifiers to detect such artefacts, utilising manual labels provided by a cardiologist as ground truth. The same training, validation and test splits of the data utilised for view selection were also used here. Similarly, the same pre-processing and data augmentation scheme were carried out for this step in the framework, with the same network parameters used for training, although binary cross entropy loss was used to classify the images as those with and without artefacts.

3. Automated Segmentation: We used the nnU-Net framework [7] to segment all frames through the cardiac cycle for the ascending aorta and pulmonary artery classes, respectively. We used the 2D nnU-Net implementation and the models were trained for 1000 epochs. Training was performed with a batch size of 32 using stochastic gradient descent with Nesterov momentum (μ=0.99) and an initial learning rate of 0.01. The loss function was the sum of the cross entropy and Dice losses. Data augmentation was performed on the fly and included rotations, scaling, Gaussian noise, Gaussian blur, brightness, contrast, simulation of low resolution, gamma correction and mirroring. An ensemble of five different models trained on subsets of the training data was used to achieve the final predicted segmentations on the test set. All deep learning models used for this work were trained on an NVIDIA Quadro RTX 6000 GPU.

4. Flow Curve Estimation and Parameter Estimation: pixel values were converted to velocity using the established formula [22] as follows:

$$\text{Velocity} = \frac{10\pi R}{\text{VENC}} * P * M$$

where P and M represent the raw pixel values from the phase and magnitude maps, VENC is an adjustable scanner parameter representing the maximum measurable flow velocity extracted and R is a reconstruction scaling factor specified from the PC-CMR DICOM files.

Flow was calculated from the automatic segmentation map of a given PC scan as:

$$\text{Net Flow} = \sum_{n=0}^{N} \sum_{i=0}^{I} S_{n,i} V_{n,i} a \Delta_t$$

where n is the frame index, N the number of temporal frames in the scan, I the number of pixels in each frame, S the binary segmentation map, V the velocity map calculated using the previous equation, a is the pixel area (in cm^2), and Δ_t the time interval between frames.

From the flow curves, we quantify the peak, total forward, total backward and net flow.

3   Materials

This is a retrospective multivendor study conducted on a large set of clinical 2D PC-CMR data from three centres that includes the full spectrum of cardiac disease phenotypes. Details of the databases are provided below:

1. UK Biobank (UKBB): This database contains a mixture of 80 patients and healthy volunteers. CMR imaging was performed using a 1.5 T Siemens MAGNE-TOM Aera (see [12] for further details of the image acquisition protocol).

2. Guy's and St Thomas' NHS Foundation Trust (GSTFT): This database contains 517 patients acquired under routine clinical CMR practice with a range of different cadiovascular diseases (CVDs). CMR imaging was performed using different CMR scanners (Philips Achieva 1.5T/3.0 T, Philips Ingenia 1.5 T and Siemens Aera 1.5 T).

3. Duke University Hospital (Duke): This database contains 102 patients acquired in routine clinical CMR practice with a range of different CVDs. CMR imaging was performed using different CMR scanners (Siemens Avanto 1.5 T, Siemens Sola 1.5 T, Siemens Verio 3.0 T and Siemens Vida 3.0 T). This database is used as an external validation data set.

For the three clinical databases, all data were manually classified and annotated though the full cardiac cycle by multiple cardiologists, and an additional review of all segmentations was performed to ensure the annotations were of high quality. Table 1 summarises the cases used for view classification, segmentation and QC.

4   Experiments and Results

We present the results in terms of the internal data set and the external data set. The internal data set comprises the UKBB and GSTFT data sets, which were merged and then randomly split into the training/validation/test splits described in Sect. 2. External data set is used to refer to the Anon Centre 2 data set, the entirety of which was used for external validation.

Table 1. Number of images used in this study for view selection, automated segmentation and quality control (QC). Ao: ascending aorta, PA: pulmonary artery.

Database	View selection			QC		Segmentation	
	Ao	PA	Other	No Artefact	Artefact	Ao	PA
UKBB	80	0	0	64	16	79	0
GSTFT	284	215	18	440	61	246	185
Duke	54	48	0	88	12	41	37

1. View Selection: We present the results for the view selection obtained using ResNet-18, which was the best performing network of the seven archictures tried. The performance of the network was evaluated in terms of precision, recall and the F1-score for each of the three classes, as well as the accuracy achieved for all classes together. Table 2 summarises these results for the internal data set and the external data set. High performance is shown on the internal data set, with an accuracy of 0.998, while the accuracy of 0.833 on the external data set demonstrates good generalisability for the view selection task.

Table 2. View classification results, in terms of precision, recall and F1 score for each class, as well as the global accuracy.

	Internal data set			External data set		
	Precision	Recall	F1-score	Precision	Recall	F1-score
Ascending Aorta	1.000	0.999	1.000	0.727	1.000	0.842
Pulmonary Artery	0.998	1.000	1.000	1.000	0.700	0.824
Other	0.999	1.000	0.998	–	–	–
Accuracy	0.998			0.833		

2. Quality Control: As with the view classification network, the performance of the quality control network was measured using the precision, recall, F1-score and accuracy. For QC, Densenet-121 achieved the best performance; the results obtained are reported in Table 3. The results demonstrate good performance on the internal data set, with an accuracy of 0.828. A reduction in performance was noted for the external data set, with an obtained accuracy of 0.650. This is likely due to a shift in the distribution of the data compared to the internal data caused by the difference in protocols used to acquire the data sets.

Table 3. Quality control performance, in terms of the precision, recall, F1-score and accuracy.

	Precision	Recall	F1-score	Accuracy
Internal data set	0.864	0.828	0.825	0.828
External data set	0.773	0.650	0.693	0.650

3. Segmentation: The segmentation network for each view was evaluated in terms of the average Dice coefficient between the manual and automatic segmentations, and the results for the internal and external validation sets can be seen in Table 4. Dice scores for the internal data set were excellent, with good values also obtained for the external data set, where values of greater than 0.964 and greater than 0.801 were achieved, respectively.

Table 4. Segmentation results in terms of the mean Dice coefficient for the two segmentation networks (ascending aorta and pulmonary artery) for the full cardiac cycle.

	Internal data set	External data set
Ascending Aorta	0.964	0.873
Pulmonary Artery	0.970	0.801

4. Parameter Estimation: The peak, net, total forward and total backward flow values in the aorta and pulmonary artery estimated using the automatic segmentations were compared to the manually obtained values using Pearson's correlation and Bland-Altman plots. Table 5 shows the correlation coefficients; for all views, correlations are strong, with coefficients >0.914 ($p<0.001$). Figure 2 shows the generated Bland-Altman plots; there is good agreement between automated and manual analysis, without a significant bias for any of the views and narrow limits of agreement.

Table 5. Pearson correlation coefficient values obtained when comparing the manual and automatic values for peak, net, total forward and total backward flow ($p<0.001$).

	Internal data set				External data set			
	Peak	Net	Forward	Backward	Peak	Net	Forward	Backward
Ascending Aorta	0.998	0.980	0.977	0.998	0.992	0.956	0.914	0.974
Pulmonary Artery	0.994	0.996	0.998	0.999	0.983	0.933	0.941	0.961

5   Discussion

In this work, we have proposed a novel end-to-end pipeline enabling automatic flow analysis from full CMR scans, which encompasses four main steps: view selection to detect conventional 2D PC-CMR scans; quality control to identify images with artefacts; automatic segmentation of the ascending aorta and pulmonary artery views; and the estimation of key flow parameters. To the best of our knowledge, this is the first automatic flow analysis framework to include the automatic quality-controlled selection of conventional 2D PC-CMR views, which mimics crucial steps ordinarily carried out by physicians. We utilised a

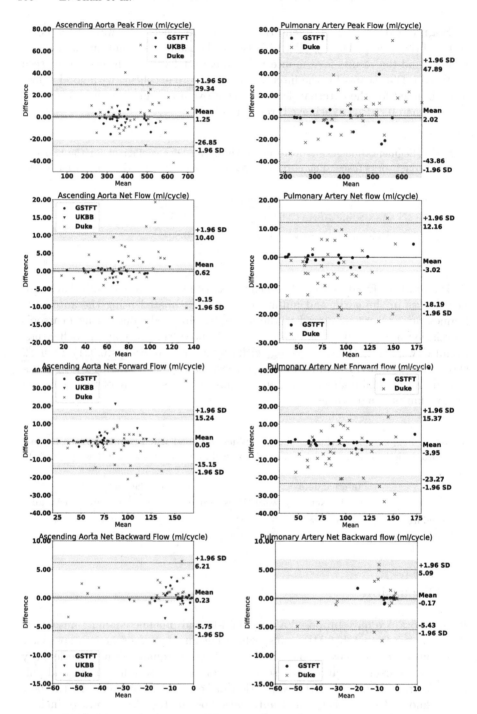

Fig. 2. Bland-Altman plots for peak flow (first row), net flow (second row), forward flow (third row) and backward flow(final row), for the ascending aorta (left column) and the pulmonary artery (right column).

large scale database from multiple centres, vendors and field strengths, with one data set being used for external validation.

Our results show a good performance at each step in the pipeline, with accuracy values of 0.998 and 0.828 for view classification and QC, respectively, and Dice scores >0.964, indicating that our pipeline achieves close to human-level performance and is comparable to other automatic flow quantification frameworks [2,3]. The obtained flow curves can be used to calculate further metrics of interest in cardiovascular medicine, including ejection times and first-phase ejection fraction (EF1). Additionally, we obtained good results on the external validation set, indicating that our pipeline has good generalisability. This is noteworthy, as the external data set included images that were obtained with a different imaging strategy (higher image-acceleration factors and breath-held imaging), leading to significant differences in image characteristics.

Future work includes the improvement of the QC by better accounting for the difference in protocols across data sets acquired in different centres. Furthermore, in this work, we excluded views other than the ascending aorta and pulmonary artery from analysis and also trained on relatively few cases comprising the 'other' class for view selection. In the future, we aim to extend our pipeline to further views, such as descending aorta, branch pulmonary artery and pulmonary vein flows, and increase the number of cases in the 'other' class to strengthen the prediction. In addition, our experiments only considered the correctly-classified cases when moving onto the next task. It would therefore be useful to conduct an analaysis comparing the results achieved when also using incorrectly classified samples in the downstream tasks. Finally, the output automatic segmentations are not checked for quality before being used for the computation of the flow parameters. Future work will therefore also include the addition of an extra quality control step after the segmentation of the vessels to ensure the quality of the parameter extraction.

Acknowledgements. This work was supported by the UKRI London Medical Imaging & Artificial Intelligence Centre for Value Based Healthcare, and the Wellcome EPSRC Centre for Medical Engineering at the School of Biomedical Engineering and Imaging Sciences, King's College London (WT 203148/Z/16/Z). The authors also acknowledge financial support from the National Institute for Health Research (NIHR) Cardiovascular MedTech Co-operative award to the Guy's and St Thomas' NHS Foundation Trust and the Department of Health National Institute for Health Research (NIHR) comprehensive Biomedical Research Centre award to Guy's & St Thomas' NHS Foundation Trust in partnership with King's College London.

References

1. Albà, X., Lekadir, K., Pereañez, M., Medrano-Gracia, P., Young, A.A., Frangi, A.F.: Automatic initialization and quality control of large-scale cardiac MRI segmentations. Med. Image Anal. **43**, 129–141 (2018)
2. Bidhult, S., et al.: A new vessel segmentation algorithm for robust blood flow quantification from two-dimensional phase-contrast magnetic resonance images. Clin. Physiol. Funct. Imaging **39**(5), 327–338 (2019)

3. Bratt, A., et al.: Machine learning derived segmentation of phase velocity encoded cardiovascular magnetic resonance for fully automated aortic flow quantification. J. Cardiovas. Mag. Reson. **21**(1) (2019)
4. Hann, E., et al.: Quality control-driven image segmentation towards reliable automatic image analysis in large-scale cardiovascular magnetic resonance aortic cine imaging. In: Shen, D., et al. (eds.) MICCAI 2019. LNCS, vol. 11765, pp. 750–758. Springer, Cham (2019). https://doi.org/10.1007/978-3-030-32245-8_83
5. He, K., Zhang, X., Ren, S., Sun, J.: Deep residual learning for image recognition (2015). https://arxiv.org/abs/1512.03385
6. Huang, G., Liu, Z., Van Der Maaten, L., Weinberger, K.Q.: Densely connected convolutional networks. In: Proceedings of the IEEE Conference on Computer Vision and Pattern Recognition, pp. 4700–4708 (2017)
7. Isensee, F., et al.: nnU-Net: a self-configuring method for deep learning-based biomedical image segmentation. Nat. Methods **18**(2), 203–211 (2021)
8. Krizhevsky, A., Sutskever, I., Hinton, G.E.: ImageNet classification with deep convolutional neural networks. In: Advances in Neural Information Processing Systems, pp. 1097–1105 (2012)
9. Liu, Z., Mao, H., Wu, C.Y., Feichtenhofer, C., Darrell, T., Xie, S.: A convnet for the 2020s (2022). https://arxiv.org/abs/2201.03545
10. Nayak, K.S., et al.: Cardiovascular magnetic resonance phase contrast imaging. J. Cardiovasc. Magn. Reson. **17**(1), 1–26 (2015)
11. Oksuz, I., et al.: Automatic CNN-based detection of cardiac MR motion artefacts using k-space data augmentation and curriculum learning. Med. Image Anal. **55**, 136–147 (2019)
12. Petersen, S.E., et al.: Uk biobank's cardiovascular magnetic resonance protocol. J. Cardiovasc. Magn. Reson. **18**(1), 8 (2015)
13. Radosavovic, I., Kosaraju, R.P., Girshick, R., He, K., Dollár, P.: Designing network design spaces (2020). https://arxiv.org/abs/2003.13678
14. Rebergen, S.A., van der Wall, E.E., Doornbos, J., de Roos, A.: Magnetic resonance measurement of velocity and flow: Technique, validation, and cardiovascular applications. Am. Heart J. **126**(6), 1439–1456 (1993)
15. Robinson, R., et al.: Automated quality control in image segmentation: Application to the UK Biobank cardiovascular magnetic resonance imaging study. J. Cardiovas. Mag. Res. 21(1) (2019)
16. Ruijsink, B., et al.: Fully automated, quality-controlled cardiac analysis from cmr: validation and large-scale application to characterize cardiac function. Cardiovas. Imaging **13**(3), 684–695 (2020)
17. Sandler, M., Howard, A., Zhu, M., Zhmoginov, A., Chen, L.C.: Mobilenetv 2: Inverted residuals and linear bottlenecks (2018). https://arxiv.org/abs/1801.04381
18. Simonyan, K., Zisserman, A.: Very deep convolutional networks for large-scale image recognition. In: 3rd International Conference on Learning Representations, ICLR 2015 - Conference Track Proceedings, pp. 1–14 (2015)
19. Tarroni, G., et al.: Learning-based quality control for cardiac MR images. IEEE Trans. Med. Imaging **38**(5), 1127–1138 (2019)
20. Vergani, V., Razavi, R., Puyol-Antón, E., Ruijsink, B.: Deep learning for classification and selection of cine cmr images to achieve fully automated quality-controlled cmr analysis from scanner to report. Front. Cardiovas. Med. **8**(742640) (2021)
21. Wang, S., et al.: Deep generative model-based quality control for cardiac mri segmentation. In: Martel, A.L., et al. (eds.) MICCAI 2020. LNCS, vol. 12264, pp. 88–97. Springer, Cham (2020). https://doi.org/10.1007/978-3-030-59719-1_9

22. Watanabe, T., et al.: Accuracy of the flow velocity and three-directional velocity profile measured with three-dimensional cine phase-contrast mr imaging: verification on scanners from different manufacturers. Magn. Reson. Med. Sci. **18**(4), 265–271 (2019)
23. Zhang, L., et al.: Automated quality assessment of cardiac mr images using convolutional neural networks. In: Tsaftaris, S.A., Gooya, A., Frangi, A.F., Prince, J.L. (eds.) SASHIMI 2016. LNCS, vol. 9968, pp. 138–145. Springer, Cham (2016). https://doi.org/10.1007/978-3-319-46630-9_14
24. Zhang, L., Gooya, A., Frangi, A.F.: Semi-supervised assessment of incomplete LV coverage in cardiac MRI using generative adversarial nets. In: Tsaftaris, S.A., Gooya, A., Frangi, A.F., Prince, J.L. (eds.) SASHIMI 2017. LNCS, vol. 10557, pp. 61–68. Springer, Cham (2017). https://doi.org/10.1007/978-3-319-68127-6_7

An Atlas-Based Analysis of Biventricular Mechanics in Tetralogy of Fallot

Sachin Govil[1]([⊠]), Sanjeet Hegde[2], James C. Perry[2], Jeffrey H. Omens[1], and Andrew D. McCulloch[1]

[1] Department of Bioengineering, University of California San Diego, San Diego, USA
sagovil@eng.ucsd.edu
[2] Division of Cardiology, Rady Children's Hospital San Diego, San Diego, USA

Abstract. The current study proposes an efficient strategy for exploiting the statistical power of cardiac atlases to investigate whether clinically significant variations in ventricular shape are sufficient to explain corresponding differences in ventricular wall motion directly, or if they are indirect markers of altered myocardial mechanical properties. This study was conducted in a cohort of patients with repaired tetralogy of Fallot (rTOF) that face long-term right ventricular (RV) and/or left ventricular (LV) dysfunction as a consequence of adverse remodeling. Features of biventricular end-diastolic (ED) shape associated with RV apical dilation, LV dilation, RV basal bulging, and LV conicity correlated with components of systolic wall motion (SWM) that contribute most to differences in global systolic function. A finite element analysis of systolic biventricular mechanics was employed to assess the effect of perturbations in these ED shape modes on corresponding components of SWM. Perturbations to ED shape modes and myocardial contractility explained observed variation in SWM to varying degrees. In some cases, shape markers were partial determinants of systolic function and, in other cases, they were indirect markers for altered myocardial mechanical properties. Patients with rTOF may benefit from an atlas-based analysis of biventricular mechanics to improve prognosis and gain mechanistic insight into underlying myocardial pathophysiology.

Keywords: Congenital heart disease · Biomechanics · Statistical atlases

1 Introduction

While shape, wall motion, and hemodynamics are important indicators of regional ventricular wall mechanics, they form an incomplete picture. Regional stresses and strains are more direct and sensitive indicators of regional myocardial mechanics because they are related *via* intrinsic myocardial material properties, such as active myofiber contractility and passive stiffness [1–3]. Several groups have demonstrated the feasibility of patient-specific biomechanical modeling [4, 5], though this depends on accurate measures of myocardial anatomy, hemodynamics, activation patterns, and wall motion to identify unknown mechanical properties, and still requires material assumptions that

© The Author(s), under exclusive license to Springer Nature Switzerland AG 2022
O. Camara et al. (Eds.): STACOM 2022, LNCS 13593, pp. 112–122, 2022.
https://doi.org/10.1007/978-3-031-23443-9_11

cannot be measured *in vivo*. Patient-specific models of biomechanics are also highly nonlinear, and consequently, their solutions can significantly amplify measurement uncertainties.

The use of statistical shape modeling to characterize major features of ventricular shape and wall motion at the population-level has been insightful for better understanding many cardiac pathologies including several congenital heart diseases. By identifying major modes of variation in biventricular shape and wall motion in various patient cohorts, many groups have observed statistically significant relationships between ventricular shape modes and ventricular function or clinical outcomes [6–10]. These analyses, however, do not establish causal relationships between shape and function or outcomes. When statistical shape models are integrated with cardiac biomechanics models, however, mechanistic relationships between observable features of cardiac shape, function, and intrinsic myocardial mechanical properties can be tested further, which may be important for understanding and predicting ventricular remodeling. This is particularly relevant in patients with repaired tetralogy of Fallot (rTOF) that face long-term, adverse right ventricular (RV) and/or left ventricular (LV) remodeling as a consequence of surgical repair [11–14].

The current study proposes an efficient strategy for exploiting the statistical power of cardiac atlases to investigate the extent to which clinically significant variations in ventricular shape explain corresponding differences in ventricular function directly or whether they are indirect markers of altered myocardial mechanical properties. Patient-specific relationships between end-diastolic shape and systolic wall motion were assessed for consistency with properties representative of the cohort mean and, in the case of deviations, were explored for their ability to explain differences in clinical outcomes. Specifically, in a rTOF patient cohort, we tested the hypothesis that shape variations can be direct determinants of variations in systolic function, but may also be indirect markers of altered myocardial mechanical properties that could help stratify clinical outcomes in rTOF.

2   Methods

2.1   Study Population and Geometry Fitting

A previously identified cohort of 84 rTOF patients was employed in this study [15]. Cardiovascular magnetic resonance (CMR) images and associated clinical data for this cohort were retrospectively collected from the Cardiac Atlas Project database (https://www.cardiacatlas.org) [16]. Deidentified datasets were contributed from two clinical centers (Rady Children's Hospital, San Diego, CA, US and The Center for Advanced Magnetic Resonance Imaging, Auckland, NZ) with approval from local institutional review boards *via* waiver of informed consent (UCSD IRB 201138 and HDEC 16/STH/248, respectively). All patients underwent standard-of-care CMR examination, and CMR images were acquired using 1.5T MRI scanners, including Siemens Avanto (Siemens Medical Systems) and GE Discovery (GE Healthcare Systems), as described previously [15]. Three-dimensional, patient-specific geometric models were customized to a biventricular subdivision surface template mesh using manually drawn contours at end-diastole (ED) and end-systole (ES) in Segment (Medviso) along with manually annotated anatomical landmarks, as described previously [15].

2.2 Atlas-Based Analysis of Systolic Wall Motion

A statistical atlas of systolic wall motion (SWM) was generated from the patient-specific geometric models at ED and ES. Patient-specific surface points at ED and ES were first aligned to population mean ED surface points by a rigid registration using a Procrustes alignment. A vector field of SWM was then computed between ED and ES for each patient. Principal component analysis (PCA) was then used to evaluate the distribution of SWM across the rTOF cohort.

2.3 Sensitivity of Biventricular Function to Systolic Wall Motion

The effects of the first ten SWM modes on LV and RV function were analyzed by varying individual modes and computing corresponding LV and RV ejection fractions (EFs). For each SWM mode, wall motion displacements were calculated for Z-scores between −2 and 2 in steps of 0.5 and were added to the mean ED shape of the rTOF cohort to yield an ES shape. The mean ED model volumes and computed ES model volumes were used to calculate EFs for each score along each mode. The sensitivity of changes in LV and RV EF to changes along an individual SWM mode was then computed using linear regression.

2.4 Association of End-Diastolic Shape with Systolic Wall Motion

Associations between ED shape modes and SWM modes were assessed *via* univariate regression analysis. ED shape modes were taken from an ED shape atlas previously derived from the same cohort of rTOF patients [15]. ED shape modes that were previously found to be significantly associated with differences in prognosis were correlated with the five SWM modes that had the greatest effect on combined LV/RV function. ED shape modes that had significant correlations with systolic wall motion modes were perturbed in a finite element (FE) mechanics framework to assess the degree to which shape markers are direct determinants of systolic function.

2.5 Finite Element Analysis of Biventricular Biomechanics

Biomechanics simulations were performed using biventricular, cubic-Hermite FE meshes generated from the ED shape atlas. A compressible, nonlinearly elastic, transversely isotropic constitutive model was used to define the passive material properties of the myocardium in the fiber direction (a, a_f, b, b_f) [17, 18]. Mesh fiber directions were assigned using a rule-based approximation, where fibers were defined relative to the circumferential direction in the LV and RV epicardium as $-60°$ and $-25°$, respectively, and in the LV and RV endocardium as $60°$ and $0°$, respectively [19]. In order to define a physiologically realistic transmural gradient, fiber directions were linearly interpolated from epicardium to endocardium. Active contraction of the myocardium was governed by a transversely isotropic active tension model, where the transverse direction active force was specified to be 70% of the fiber direction active force. An unloaded, stress-free reference geometry for each simulation was approximated using a previously described iterative, inflation-deflation algorithm [5], with convergence criteria defined as a change in computed ED volumes between successive iterations of less than 2%. Pressure boundary conditions were applied normal to the endocardial surfaces using average LV and RV ED and ES catheterization pressures from rTOF patients with available clinical data. The average LV and RV EDPs were 1.05 kPa and 0.95 kPa, respectively, and the average LV and RV ESPs were 11 kPa and 5.15 kPa, respectively. Additionally, nodal boundary conditions were implemented to constrain epicardial longitudinal and circumferential displacement at the valve planes.

Mean passive material properties were approximated by estimating the reference geometry for the mean ED shape. Material constants, b and b_f, were taken from previously reported values from human subjects with heart failure [5], while material constants, a and a_f, were estimated by altering these parameters until the FE-computed ED model volumes at mean ED pressure matched the atlas mean ED model volumes. The material anisotropy ratio of a to a_f was maintained so that only a single parameter was estimated. Mean active material properties were approximated by linearly increasing the active tension developed in the myocardium until the FE-computed ES model volumes at mean ES pressure matched the atlas mean ES model volumes. These mean passive and active material properties were then used in additional FE simulations with perturbations of ED geometry corresponding to Z-scores of $\pm$ 2 for tested ED shape modes. The computed ED and ES mesh geometries for each shape perturbation were sampled on the endocardial and epicardial surfaces in order to compute wall motion displacements, which were then projected onto the SWM atlas to compute Z-scores for corresponding SWM modes of interest. SWM mode Z-scores from FE-computed models were compared with those predicted from linear regression models fit to patient data. All FE analysis was performed in *Continuity 6* (https://www.continuity.ucsd.edu), a problem-solving environment for multiscale biomechanics and electrophysiology.

2.6 Statistical Analysis

Statistical analysis was carried out using the SciPy Python library (https://www.scipy.org). Statistical associations in the regression analysis are denoted by p-values with a significance level of 0.05. A Bonferroni correction was used to adjust the significance level to correct for multiple comparisons.

3   Results

3.1   Atlas of Systolic Wall Motion and Associations with End-Diastolic Shape

An atlas of SWM was constructed from a cohort of 84 rTOF patients. The first ten SWM modes explained approximately 70% of the variation in the study population (Fig. 1A), and their effects on LV and RV function were analyzed (Fig. 1B). ED shape modes that were previously found to be significantly associated with differences in prognosis

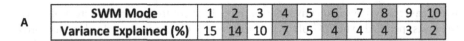

SWM Mode	1	2	3	4	5	6	7	8	9	10
Variance Explained (%)	15	14	10	7	5	4	4	4	3	2

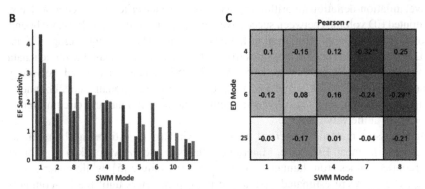

Fig. 1. A) Variance explained (%) per SWM mode. B) Magnitude of the sensitivity of LV EF (green), RV EF (blue), and combined LV/RV EF (grey) to variations in SWM modes (pp/SD). SWM modes were ranked from greatest to least effect on combined LV/RV EF. C) Correlations between clinically significant ED shape modes and SWM modes that had the greatest effect on combined LV/RV EF. Symbols indicate statistical significance (**$p < 0.01$). (Color figure online)

in rTOF correlated with SWM modes that had the greatest effect on combined LV/RV function (Fig. 1C).

ED4, which is a specific marker of opposing RV apical dilation and LV dilation, was significantly correlated with SWM7, which appears to be a specific marker of displacement around the RV base, RV apex, and the entire LV free wall (Fig. 2A). ED6, which is a specific marker of RV basal bulging and LV conicity, was significantly correlated with SWM8, which appears to be a specific marker of displacement around the posterior RV base, LV base, and LV apex (Fig. 2B). These ED shape modes were perturbed in a FE model to assess the degree to which shape differences are direct determinants of observed changes in SWM.

3.2 Finite Element Analysis of Biventricular Biomechanics

Parameterization of the material properties of the mean FE-computed model resulted in an average absolute error compared with the mean atlas model of 1.06 mm and 1.44 mm at ED and ES, respectively, both of which are within voxel resolution of the original CMR images used to make the patient-specific geometric models (0.6 – 1.75 mm). The estimated mean passive material properties are shown in Table 1, where the approximated stiffness of the RV was twice that of the LV. The mean peak active tension developed in the myocardium was estimated to be 164 kPa in both the LV and RV. A summary of global volume and functional measurements for the mean FE-computed model and atlas model are shown in Table 2. Overall, the FE-computed model measurements demonstrate good agreement with the atlas model measurements. Lastly, the resulting SWM Z-scores

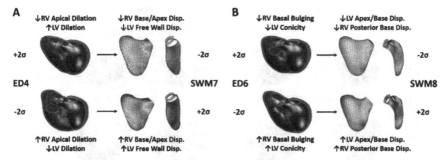

Fig. 2. Summary morphological characteristics for associations between ED shape and SWM, (A) ED4 ∝ SWM7 and (B) ED6 ∝ SWM8. For ED shape modes, the LV endocardial surface, RV endocardial surface, and epicardial surface are shown in green, blue, and maroon, respectively, and the mitral, tricuspid, aortic, and pulmonary valves are shown in cyan, pink, yellow, and green, respectively. For SWM modes, the LV and RV free walls are shown and are colored based on the systolic displacement relative to the mean, inward (blue) and outward (red). (Color figure online)

Table 1. Estimated mean passive material properties.

Region	a (kPa)	a_f (kPa)	b (–)	b_f (–)
LV	1.197	0.8925	9.726	15.779
RV	2.394	1.785	9.726	15.779
Base	11.97	8.925	9.726	15.779

Table 2. Comparison of mean FE-computed model and atlas model volumes and EFs.

Measure	Atlas Model	FE-Computed Model
LV EDV (mL)	78.3	77.9
LV ESV (mL)	38.8	39.7
RV EDV (mL)	153.4	153.3
RV ESV (mL)	89.2	87.9
LV EF (%)	50.5	49.0
RV EF (%)	41.9	42.7

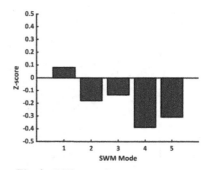

Fig. 3. Difference between mean FE-computed model (blue bar) and atlas model (black line) SWM Z-scores. (Color figure online)

for the mean FE-computed model were compared to those for the mean atlas model, which are zero by definition, for the first five SWM modes (Fig. 3). Overall, the Z-score differences between the FE-computed model and atlas model are within an acceptable range.

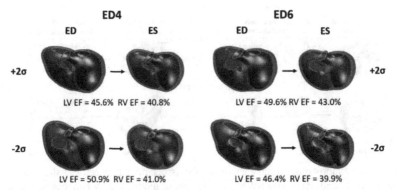

Fig. 4. FE-computed ED shape perturbations with corresponding deformed ES shapes. Computed LV and RV EFs for each shape perturbation are also shown. For all models, the LV endocardial surface, RV endocardial surface, and epicardial surface are shown in green, blue, and maroon, respectively, and the mitral, tricuspid, aortic, and pulmonary valves are shown in cyan, pink, yellow, and green, respectively. (Color figure online)

FE simulations with perturbed ED shapes were executed with mean material properties, and their deformed ES shapes were computed (Fig. 4). The FE-computed change in LV and RV EF were −5.3% and −0.2%, respectively, along the shape change in ED4 and 3.2% and 3.1%, respectively, along the shape change in ED6. The predicted change in LV and RV EF, based on the correlations between ED shape and SWM and the observed effect on EF (Fig. 1), were −2.8% and −3.0%, respectively along the shape change in ED4 and −3.4% and −2.0%, respectively, along the shape change in ED6.

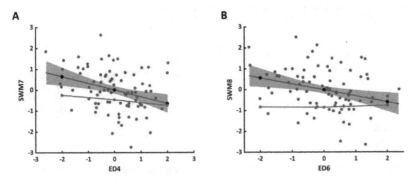

Fig. 5. Comparison of predicted (black circles) and FE-computed (red line and circles) SWM Z-scores for shape perturbations along (A) ED4 $\propto$ SWM7 and (B) ED6 $\propto$ SWM8 using mean material properties. Significant ED shape mode and SWM mode correlations from patient data (blue circles) are represented as linear regression models with the fit (blue line) and 95% confidence intervals (blue shaded area). (Color figure online)

The SWM Z-scores for the FE-computed ED shape perturbations were computed and compared with those predicted from linear regression models fit to patient data. For the correlation between ED4 and SWM7 (Fig. 5A), the FE-computed slope was −0.12

compared to -0.32 in the patient data. For the correlation between ED6 and SWM8 (Fig. 5B), the FE-computed slope was 0.02 compared to -0.29 in the patient data.

For FE-computed models that did not match the predicted SWM, the model contractility was adjusted to produce a better match. For both ED4 and ED6, a more negative Z-score was associated with increased contractility (Fig. 6A and B, respectively).

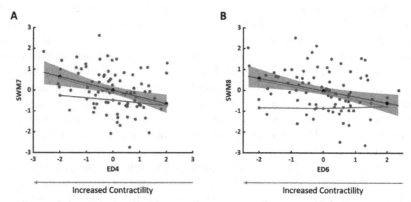

Fig. 6. FE-computed changes in SWM Z-scores for shape perturbations along (A) ED4 $\propto$ SWM7 and (B) ED6 $\propto$ SWM8 with altered contractility (yellow line and circles) overlaid on Fig. 5. The direction of ED shape variation associated with increased contractility is also shown. (Color figure online)

4 Discussion

Patients with rTOF are at risk of developing RV and LV dysfunction as a consequence of adverse ventricular remodeling. While statistical shape modeling has been used to identify features of ventricular remodeling that are associated with function in the rTOF population, it was not clear whether shape features are direct determinants of systolic function. In this study, cardiac biomechanics models were integrated with statistical shape models to test whether clinically significant variations in ventricular shape explain corresponding differences in systolic function or if they are indirect markers of altered myocardial mechanical properties, or both.

4.1 Shape Determinants of Biventricular Function

By correlation analysis, previously identified biventricular ED shape modes that had significant associations with differences in prognosis of patients with rTOF were also significantly correlated with the components of SWM that contributed most to differences in LV and RV EF. Specifically, greater RV apical dilation and less LV dilation were associated with greater systolic displacements near the RV base and apex and the entire LV free wall, and greater RV basal bulging and higher LV conicity were associated with greater systolic displacements around the posterior RV base and the LV base and apex.

The mechanisms of these shape-function relationships were tested using FE models in which observed shape mode differences were simulated without changes in material properties or boundary conditions.

In the case of RV apical dilation and LV dilation, differences in the shape alone partially explained observed differences in systolic function. In terms of global functional metrics, less RV apical dilation and greater LV dilation accounted for more than the expected decrease in LV EF (-5.3% vs. -2.8%) but none of the expected decrease in RV EF (-0.2% vs. -3.0%). In terms of regional measures of SWM, differences in the shape alone explained 38% of the variation in wall motion. Assuming that the remaining correlation could be explained by differences in myocardial material properties, we observed that greater RV apical dilation and less LV dilation were associated with increased contractility. Therefore, greater RV apical dilation may be a surrogate for increased RV basal and apical contractility that may also serve as a compensatory mechanism to prevent LV dilation and preserve LV function. This mechanism may explain why patients with rTOF that had greater RV apical dilation and less LV dilation tended to have better prognoses [15].

In the case of RV basal bulging and LV conicity, differences in shape alone explained the opposite change in systolic function. In terms of global functional metrics, less RV basal bulging and lower LV conicity led to an increase in LV and RV EF (3.2% and 3.1%, respectively) rather than an expected decrease in LV and RV EF (-3.4% and -2.0%, respectively). In terms of regional measures of SWM, changing shape alone produced a slope that was the opposite sign as that of the regression model. When altering myocardial contractility to better match the ED shape and SWM relationship, greater RV basal bulging and higher LV conicity were associated with increased contractility. In order to address this discrepancy, RV basal bulging and LV conicity shape mode scores were correlated with patient-specific LV and RV EFs, providing insight into how these shape changes effect overall differences in global function rather than changes in global function along a specific component of SWM. Through this analysis, less RV basal bulging and lower LV conicity were correlated with higher LV and RV EF. Overall, these results suggest that while RV basal bulging and LV conicity may be surrogate measures for increased contractility in the RV posterior base and LV apex and base, the gain of function in these regions is overshadowed by the loss of function in other regions associated with shape itself. This mechanism may explain why patients with rTOF that had less RV basal bulging and lower LV conicity had better outcomes [15].

4.2 Limitations

There were several simplifications that were employed in our FE analysis of biventricular mechanics. Material properties were assumed to be homogeneous throughout different regions of the heart including the entire RV and the entire LV, rather than including more specific regional heterogeneities that could be the result of scarring or fibrosis in areas with surgical incisions. The myocardial fiber architecture employed was also simplified using a rule-based approximation with a linear transmural gradient rather than using more realistic properties. This was primarily due to the lack of information on ventricular fiber architecture in a representative rTOF patient. Differences in hemodynamics between patients that might have also contributed to variation in shape and SWM were not considered due to the limited clinical availability of this data. Lastly, the FE models employed did not include boundary conditions representative of interactions with the pericardium, atria, and great vessels or include fluid-structure interactions.

5 Conclusions

In this study, statistical shape models were integrated with cardiac biomechanics models to further investigate the degree to which clinically significant shape markers are determinants of systolic function in a cohort of patients with rTOF. In FE simulations of systolic mechanics, variations in ED shape explained differences in SWM to varying degrees. Variation in RV apical dilation and LV dilation partially explained differences in SWM, while variation in RV basal bulging and LV conicity explained the opposite change in SWM. Specifically, greater RV apical dilation and less LV dilation were associated with increased RV basal and apical contractility and may be important determinants of improved function in rTOF. Overall, patients with rTOF may benefit from an atlas-based analysis of biventricular mechanics that can provide insight into mechanisms of pathophysiology and aid long-term clinical management.

6 Competing Interests

ADM and JHO are co-founders of, scientific advisors to, and equity holders in Insilicomed, Inc. ADM is also a co-founder of and scientific advisor to Vektor Medical, Inc. Some of their research grants have been identified for conflict-of-interest management. The authors are required to disclose these relationships in publications acknowledging the grant support; however, the findings reported in this study did not involve the companies in any way and have no relationship with the business activities or scientific interests of either company. The terms of this arrangement have been reviewed and approved by the University of California San Diego in accordance with its conflict-of-interest policies. The rest of the authors do not have any conflict-of-interest.

Funding. Funding was provided by National Institutes of Health R01HL121754 and American Heart Association 19AIML35120034. SG acknowledges National Institutes of Health NHLBI T32HL105373.

References

1. Diller, G.P., et al.: Left ventricular longitudinal function predicts life-threatening ventricular arrhythmia and death in adults with repaired tetralogy of fallot. Circulation **125**(20), 2440–2446 (2012)
2. Orwat, S., et al.: Myocardial deformation parameters predict outcome in patients with repaired tetralogy of Fallot. Heart **102**(3), 209–215 (2016)
3. Kallhovd, S., Sundnes, J., Wall, S.T.: Sensitivity of stress and strain calculations to passive material parameters in cardiac mechanical models using unloaded geometries. Comput. Methods Biomech. Biomed. Eng. 1–12 (2019)

4. Aguado-Sierra, J., et al.: Patient-specific modeling of dyssynchronous heart failure: a case study. Prog. Biophys. Mol. Biol. **107**(1), 147–155 (2011)
5. Krishnamurthy, A., et al.: Patient-Specific Models of Cardiac Biomechanics. J. Comput. Phys. **244**, 4–21 (2013)
6. Medrano-Gracia, P., et al.: Left ventricular shape variation in asymptomatic populations: the Multi-Ethnic Study of Atherosclerosis. J. Cardiovasc. Magn. Reson. **16**, 56 (2014)
7. Farrar, G., et al.: Atlas-Based Ventricular Shape Analysis for Understanding Congenital Heart Disease. Prog. Pediatr. Cardiol. **43**, 61–69 (2016)
8. Gilbert, K., et al.: Independent left ventricular morphometric atlases show consistent relationships with cardiovascular risk factors: A UK Biobank Study. Sci. Rep. **9**(1), 1130 (2019)
9. Mauger, C., et al.: Right ventricular shape and function: cardiovascular magnetic resonance reference morphology and biventricular risk factor morphometrics in UK Biobank. J. Cardiovasc. Magn. Reson. **21**(1), 41 (2019)
10. Mauger, C.A., et al.: Right-left ventricular shape variations in tetralogy of Fallot: associations with pulmonary regurgitation. J. Cardiovasc. Magn. Reson. **23**(1), 105 (2021)
11. Geva, T.: Repaired tetralogy of Fallot: the roles of cardiovascular magnetic resonance in evaluating pathophysiology and for pulmonary valve replacement decision support. J. Cardiovasc. Magn. Reson. **13**, 9 (2011)
12. Broberg, C.S., et al.: Prevalence of left ventricular systolic dysfunction in adults with repaired tetralogy of fallot. Am. J. Cardiol. **107**(8), 1215–1220 (2011)
13. Valente, A.M., et al.: Contemporary predictors of death and sustained ventricular tachycardia in patients with repaired tetralogy of Fallot enrolled in the INDICATOR cohort. Heart **100**(3), 247–253 (2014)
14. Probst, J., et al.: Prevention of sudden cardiac death in patients with Tetralogy of Fallot: Risk assessment and long term outcome. Int. J. Cardiol. **269**, 91–96 (2018)
15. Govil, S., et al., Biventricular Shape Modes Discriminate Pulmonary Valve Replacement in Tetralogy of Fallot Better Than Imaging Indices. Sci. Rep. (2022). (In Review)
16. Fonseca, C.G., et al.: The Cardiac Atlas Project—an imaging database for computational modeling and statistical atlases of the heart. Bioinformatics **27**(16), 2288–2295 (2011)
17. Holzapfel, G.A., Ogden, R.W.: Constitutive modelling of passive myocardium: a structurally based framework for material characterization. Philos. Trans. A Math. Phys. Eng. Sci. **2009**(367), 3445–3475 (1902)
18. Krishnamurthy, A., et al.: Left Ventricular Diastolic and Systolic Material Property Estimation from Image Data: LV Mechanics Challenge. Stat. Atlases. Comput. Models Heart **8896**, 63–73 (2015)
19. Doste, R., et al.: A rule-based method to model myocardial fiber orientation in cardiac biventricular geometries with outflow tracts. Int. J. Numer. Method. Biomed. Eng. **35**(4), e3185 (2019)

Review of Data Types and Model Dimensionality for Cardiac DTI SMS-Related Artefact Removal

Michael Tänzer[1]([envelope]) [iD], Sea Hee Yook[1,2], Pedro Ferreira[1,2] [iD],
Guang Yang[1,2]([envelope]) [iD], Daniel Rueckert[1,3] [iD], and Sonia Nielles-Vallespin[1,2] [iD]

[1] Imperial College London, London, England
{m.tanzer,g.yang}@imperial.ac.uk
[2] Royal Brompton and Harefield Hospital, London, England
[3] Klinikum rechts der Isar, Technische Universität München (TUM),
Munich, Germany

Abstract. As diffusion tensor imaging (DTI) gains popularity in cardiac imaging due to its unique ability to non-invasively assess the cardiac microstructure, deep learning-based Artificial Intelligence is becoming a crucial tool in mitigating some of its drawbacks, such as the long scan times. As it often happens in fast-paced research environments, a lot of emphasis has been put on showing the capability of deep learning while often not enough time has been spent investigating what input and architectural properties would benefit cardiac DTI acceleration the most. In this work, we compare the effect of several input types (magnitude images vs complex images), multiple dimensionalities (2D vs 3D operations), and multiple input types (single slice vs multi-slice) on the performance of a model trained to remove artefacts caused by a simultaneous multi-slice (SMS) acquisition. Despite our initial intuition, our experiments show that, for a fixed number of parameters, simpler 2D real-valued models outperform their more advanced 3D or complex counterparts. The best performance is although obtained by a real-valued model trained using both the magnitude and phase components of the acquired data. We believe this behaviour to be due to real-valued models making better use of the lower number of parameters, and to 3D models not being able to exploit the spatial information because of the low SMS acceleration factor used in our experiments.

Keywords: MRI · Deep learning · Diffusion tensor imaging · Cardiac MRI

1 Introduction

Cardiac diffusion tensor imaging is growing as a novel imaging modality as it is capable of interrogating the microstructure of the beating heart with no invasive

This work was supported in part by the UKRI CDT in AI for Healthcare http://ai4health.io (Grant No. EP/S023283/1), the British Heart Foundation (RG/19/1/34160), and the UKRI Future Leaders Fellowship (MR/V023799/1).
G. Yang, D. Rueckert and S. Nielles-Vallespin—Co-last senior authors.

O. Camara et al. (Eds.): STACOM 2022, LNCS 13593, pp. 123–132, 2022.
https://doi.org/10.1007/978-3-031-23443-9_12

surgery and without the use of any contrast agent [11], making it accessible for patients with reduced kidney functionality [17] or for frequent scans. In clinical research studies, cardiac DTI has been shown to be useful in phenotyping several cardiomyopathies such as hypertrophic cardiomyopathy (HCM) and dilated cardiomyopathy (DCM) by quantitatively analysing the microstructural organisation and orientation of cardiomyocytes within the myocardium.

As cardiac DTI is becoming more and more studied, the use of deep learning-based (DL) approaches applied to it is similarly increasing [4–6,8,14,18,21,22]. Most of the recent work in the field, although, suffer from a common crucial shortcoming: in order to quickly show the potential of deep learning on improving cardiac DTI, many publications don't go beyond applying some well-known general-purpose architecture to the data they have available, often ignoring inherent properties of the acquisition method. As an example, in much of the recently published work, we find a widespread use of out-of-the-box models such as the popular U-Net [16], which is a 2D model designed for real data. MRI data, on the other hand, is complex by definition and the subject of the acquisition is often 3D rather than 2D, making the standard U-Net ill-fitting for the task.

In this work, we compare the effect of making relatively small architectural changes to the popular U-Net model when applied to a general image-to-image task in DTI. Specifically, we choose the task of removing artefacts from cardiac DTI images acquired with an SMS protocol, and we compare the classic 2D magnitude-only U-Net with 3D and complex versions of the same model, with the goal of providing future researchers with a better starting point for their experimental work. To this end, we also make the code for all of our tested models available on GitHub[1].

2   Background

2.1   Cardiac Diffusion Tensor Imaging

Diffusion tensor imaging measures the diffusion of water molecules for every voxel in the imaged tissue and approximates as a 3D tensor. As the free diffusion of water in the tissue is constrained by the shape of cardiac muscle microstructure, studying such tensors has been shown to give us information related to the shape and orientation of the cardiomyocytes in the imaged tissues.

The cardiac diffusion tensor information is commonly visualised and quantified through four per-voxel metric maps: Mean Diffusivity (MD) that quantifies the total diffusion in the voxel (higher corresponds to more diffusion), Fractional Anisotropy (FA) that quantifies the level of organisation of the tissue (higher corresponds to a higher organisation), Helix Angle (HA) and Second Eigenvector (E2) Angle (E2A) that quantify the 3D orientation and shape of the tissue in the voxel [2,10]. These maps have been shown to be a promising tool for phenotyping many cardiac pathology in a clinical setting [3,12,13].

---

[1] https://github.com/Michael-Tanzer/architectures-tanzer-stacom22.

2.2   Simultaneous Multi-slice Acquisition (SMS)

Simultaneous multi-slice techniques have been used with great success to reduce the acquisition time in brain diffusion tensor imaging (DTI) [19]. SMS uses a multi-band excitation pulse to simultaneously excite several 2D slices within the imaged tissue. In SMS, each receiver coil collects a single frequency-domain image for all the excited slices, the signal received is a weighted sum of the signals that would be emitted by exciting each slice individually.

By making use of the redundant information from the receiver coils that surround the tissue we want to image, we can then separate the signal from the excited slices using modified versions of the GRAPPA [7] and SENSE [15] algorithms used for in-plane acceleration. Unfortunately, as the information is often insufficient and as the problem is not fully characterised, the slice separation algorithms introduce artefacts in the separated slices [1]. The artefacts are often referred to as interslice-leakage artefact as they arise when information from one slice erroneously ends up on a different slice. Because of the disposition in space of the MRI coils, the leakage between two slices is more evident when the slices are closer in the imaged tissue.

2.3   Deep Learning in DTI

As cardiac DTI grows in popularity, a lot of work has been published trying to improve the acquisition quality or trying to shorten its long scan-times. Among others, Ferreira et al. [5], Tänzer et al. [21], and Phipps et al. [14] reduced the number of repetitions needed to increase the low SNR by using a de-noising framework to restore high image quality from less acquired data. Schlemper et al. [18] applied a cascade of convolutional neural networks to fill-in k-space entries acquired with a compressed sensing protocol, further reducing the scan times.

Ferreira et al. [6] propose a U-Net for the segmentation of the left ventricle, automating part of the DTI maps computation, Cao et al. [4] show how a GAN model can be used as a de-aliasing model for DTI, and Tian et al. [20] work on self-supervised DTI de-noising is also based on a modified U-Net model.

Most of the examples reported above, despite all being a great contribution to the field of deep-learning-accelerated cardiac DTI, have a 2D real-valued U-Net model at their core. This shows that the choice of data type and dimensionality is not the top priority of many influential publications.

3   Methods

3.1   Complex Neural Networks

As MRI data is inherently complex, we explore the possibility of training using complex data. Traditionally, there are two main ways to achieve this: separating the complex data into non-complex components, or using a model that performs complex operations.

The former is a more straight-forward approach: the complex data is split into its real and imaginary components or into magnitude and phase and then a real-values model is trained using the split data. This can be done in multiple ways: either by treating the components as different channels, or by training two separate models, one for each component. In our comparison we split the data in magnitude and phase as they are more meaningful in the physics of MRI and we use them as separate input channels to a unified model.

The latter option requires more thinking: while some operations naturally extend to the complex domain, some others do not and need to be re-designed. We report the main changes to the used operators in Table 1. In the table $z = a + ib$ where $i = \sqrt{-1}$, c is the convolution layer, and tc is the transpose convolution layer.

Table 1. Neural networks operators and their complex counterparts. Marked with a checkmark those operations that don't need substantial modification to extend their usage to complex space.

Operation		Naive imp	Complex equivalent
Multiplication	Inner product	$\sum_i w_i z_i$	✓
	Convolution	$c(a + ib)$	$c_r(a) - c_i(b) + i(c_i(a) + c_r(b))$
	Trans. conv	$tc(a + ib)$	$tc_r(a) + tc_i(b) + i(tc_i(a) - tc_r(b))$
Activation	Sigmoid	$\frac{1}{1+e^{-(a+ib)}}$	$\mathrm{sigmoid}(a) + i\,\mathrm{sigmoid}(b)$
	ReLU	$\max(0, a + ib)$	$\begin{cases} (|z| + b)\frac{z}{|z|} & \text{if } |z| + b \geq 0 \\ 0 & \text{if } |z| + b < 0 \end{cases}$
	Dropout	$DO(a) + i\,DO(b)$	$DO(a + ib)$
Pooling	Max	$\max_i z_i$	z_k where $\mathrm{argmax}_k\{|z_k|\}$
	Average	$\frac{1}{N}\sum_i z_i$	✓
Normalisation	Batch norm	$\hat{z} = \frac{z - \mathbb{E}[z]}{\sqrt{\mathbb{V}[z] + \epsilon}}$	✓
Loss	Euclidean	$\|Y - Z\|_2^2$	$\|\mathrm{abs}(Y) - \mathrm{abs}(Z)\|_2^2$
	L1	$\|Y - Z\|_1$	$\|\mathrm{abs}(Y) - \mathrm{abs}(Z)\|_1$

Once we have defined these basic operations we can then build a model that uses them and is therefore fit to process complex numbers. The advantage of this approach is that the model makes use of properties of complex numbers during the training instead of letting the network learn a relationship between the input channels like in the previous method.

3.2 Experimental Setting

In order to provide a better understanding on the effect of using a 3D or complex model, we compare all combinations of architectures obtained by modifying the following properties:

- 2D vs 3D: whether the model uses 2D operations or 3D operations.
- Magnitude vs complex data: whether we train the model with the absolute component alone or whether we use the full complex representation of the data. Notice that "complex data" is further split into 1. fully complex models that use complex operations and 2. models that keep standard operations, but use the phase data as a separate input channel. We refer to the former magnitude-only models as "Mag", fully complex models as "Comp", and magnitude-and-phase models as "MagPhs" for brevity.
- All slices vs individual slices: whether the model is trained to correct all SMS-acquired slices simultaneously or whether to correct a single slice at a time. When all the slices are used and the model is 3D, the slices are arranged in the third spatial dimension, while for 2D models the slices are treated as image channels.

This comparison results in 12 combinations of dimensionality (2 types), data (3 types), and input data (2 types). When a single model is referred to, we often use a shorthand: for instance, if we refer to a 2D fully-complex model trained on all the SMS slices, we will shorten it as "2D-All-Comp".

All the models were trained for 200 epochs with the Adam optimiser [9], a learning rate of 0.0003 that was then lowered by a factor of 10 after 100 epochs, a batch size of 16, mean absolute error loss, and residual learning. The data was padded and normalised in the range 0 to 1 and then randomly augmented with random rotation and random vertical and horizontal flipping. All the results were computed on the test set for the epoch in which the validation MAE was lowest.

In order to ensure consistency, we kept the model size and architecture as fixed as possible by choosing a number of parameters, 3 millions, and a general architecture (U-Net with 5 layers with a doubling number of channels in each encoding layer) and subsequently adjusting the initial number of channels. The models have the following starting number of channels:

- 2D Mag and MagPhs: 28
- 3D Mag and MagPhs: 16
- 2D Complex: 20
- 3D Complex: 11

The data used for the training came from 31 ex-vivo swine hearts and was acquired with SMS factor equal to 2 and distance factor equal to 400%. Each heart was scanned in multiple locations to cover as much of the volume as possible. The ground truth images were obtained by scanning each heart again with the same protocol but no SMS acceleration. This results in around 43,000 2D complex slices for the training, 1200 slices for validation and 1200 slices for testing.

3.3   Results Evaluation

When working with DTI data, there are two main components we are interested in evaluating: the acquired images and the DTI maps derived from the images.

The latter are particularly important as they are the main tools a clinician would use in a clinical setting. When evaluating the artefact-removal results of our models we therefore need to take both into account.

To evaluate the image quality we use mostly standard wide-spread metrics:

- Mean Absolute Error (MAE) $\downarrow$: $\frac{1}{nm} \sum_{i=1}^{n} \sum_{j=1}^{m} \left| X^{(i,j)} - Y^{(i,j)} \right|$ where x and y are the predicted and target images respectively. For complex images the MAE is computed with respect to their magnitude and for MagPhs the phase information is not taken into account.
- Peak Signal to Noise Ratio (PSNR) $\uparrow$: $10 \log_{10} \left(\frac{\text{MAX}_X^2}{\text{MSE}} \right)$ where MAX_X is maximum pixel intensity value across the image and MSE is the mean squared error between the predicted image and the target image. In the case of complex images we compute the PSNR on their magnitude and for the MagPhs case we discard the phase data.
- Structural Similarity Index (SSIM) $\uparrow$: SSIM measures the perceived image degradation based on the loss of structure between the output and target images. It is computed as follows:

$$\text{SSIM}(X, Y) = \frac{(2\mu_X \mu_X + c_1)(2\sigma_{XY} + c_2)}{(\mu_X^2 + \mu_Y^2 + c_1)(\sigma_X^2 + \sigma_X^2 + c_2)} \tag{1}$$

Where μ_I represents the mean over I, σ_I the standard deviation over I, σ_{IJ} the covariance, and c_1 and c_2 are fixed scalars used for numerical reasons.

When analysing the DTI maps, we need to distinguish between scalar maps (MD and FA) and angular maps (HA and E2A). While for scalar maps we can use MAE ($\downarrow$) as an error metric, angular maps are defined in the range $[-90°, 90°)$ and they wrap around at the two extrema of this range. For these maps we use the Mean Angle Absolute Error (MAAE, $\downarrow$) defined below instead

$$\text{MAAE}(X, Y) = \frac{1}{NM} \sum_{i=0}^{N} \sum_{j=0}^{M} \begin{cases} \left| X^{(i,j)} - Y^{(i,j)} \right|, & if \ \left| X^{(i,j)} - Y^{(i,j)} \right| < 90° \\ 180° - \left| X^{(i,j)} - Y^{(i,j)} \right|, & \text{otherwise} \end{cases}$$

Moreover, as the DTI maps are not well-defined for the background voxels, we only consider the metrics computed on voxels belonging to the cardiac tissue.

All the results are reported as the median over the means on a per-slice basis and its inter-quartile range as *median [iqr]*. When we perform a statistical significance test we use the Wilcoxon rank test with $P = 0.05$.

4   Results

In Tables 2 and 3 we report the performance metrics for the output images and for the DTI maps on the test set for all our models. The values of MD have been scaled by 10^5 and the values of FA by 10^2 to improve readability. In the table we mark in bold the best result across all models and we underline the second-best result.

We also visually report an example chosen from the test set in Fig. 1.

Table 2. Numerical results related to the DTI maps computed from the AI-processed data. We report MAE for MD and FA and MAAE for HA and E2A. In bold and underlined, respectively, the best and second-best results for each metric.

Run name	HA		E2A		MD ($\times 10^5$)		FA ($\times 10^2$)	
	Median	IQR	Median	IQR	Median	IQR	Median	IQR
2D-ALL-ABS	16.9	5.0	26.1	4.8	5.44	1.51	5.95	1.19
2D-ALL-Comp	18.0	4.6	26.4	4.7	5.50	0.97	**5.41**	1.33
2D-ALL-MagPhs	16.7	2.6	**24.5**	4.4	5.55	1.03	5.86	0.40
3D-ALL-ABS	16.9	4.3	26.4	3.5	5.98	1.26	6.24	0.59
3D-ALL-Comp	19.4	4.8	27.8	5.1	6.19	1.47	6.03	3.26
3D-ALL-MagPhs	**15.0**	3.8	<u>25.5</u>	4.8	<u>4.89</u>	0.67	<u>5.46</u>	1.35
2D-Single-ABS	<u>16.4</u>	4.2	<u>25.5</u>	5.5	5.26	1.28	5.83	0.51
2D-Single-Comp	18.6	4.7	27.3	5.0	5.22	0.95	5.63	2.25
2D-Single-MagPhs	18.9	3.8	27.2	4.7	5.71	1.17	5.93	0.37
3D-Single-ABS	17.8	3.1	27.0	5.4	6.41	1.92	5.76	0.34
3D-Single-Comp	18.9	5.2	27.0	5.3	**4.82**	0.96	5.48	1.12
3D-Single-MagPhs	18.4	3.2	27.9	6.0	6.28	1.88	5.88	0.72

Table 3. Numerical results related to the artefact-removal output of our proposed AI models. The results here refer to the images produced by our model. In bold and underlined, respectively, the best and second-best results for each metric.

Run name	MAE ($\times 10^3$)		PSNR		SSIM	
	Median	IQR	Median	IQR	Median	IQR
2D-ALL-ABS	1.75	0.73	<u>37.2</u>	3.56	0.917	0.021
2D-ALL-Comp	1.82	0.78	36.6	3.26	0.911	0.026
2D-ALL-MagPhs	1.84	0.78	**37.4**	2.85	0.921	0.020
3D-ALL-ABS	1.74	0.62	37.0	2.80	0.920	0.018
3D-ALL-Comp	7.15	3.92	31.1	1.41	0.618	0.012
3D-ALL-MagPhs	1.71	0.71	34.9	3.94	0.922	0.020
2D-Single-ABS	1.68	0.76	37.2	4.52	<u>0.923</u>	0.022
2D-Single-Comp	2.04	0.75	36.2	3.26	0.892	0.024
2D-Single-MagPhs	<u>1.67</u>	0.75	36.6	5.31	**0.925**	0.022
3D-Single-ABS	**1.66**	0.72	27.2	5.46	0.909	0.028
3D-Single-Comp	5.24	4.19	36.2	3.37	0.889	0.020
3D-Single-MagPhs	1.70	0.77	37.1	3.83	<u>0.923</u>	0.026

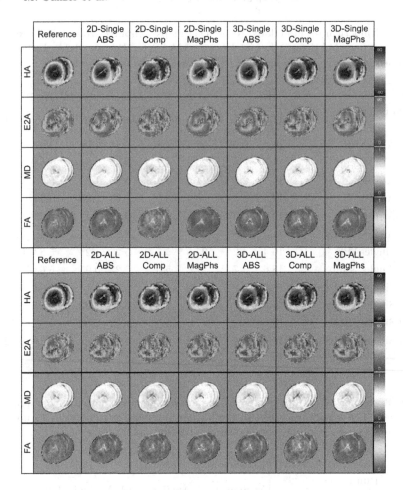

Fig. 1. Example DTI maps for an example from the test set.

5   Discussion

Analysing Tables 2 and 3, we can see how simpler 2D models that only use magnitude data seem to be extremely stable and easy to train, resulting in very good performance overall. Moreover, these models were also faster to train compared to 3D or complex models. 2D MagPhs models also performed remarkably well given how little the architectures differs from the 2D real-values model.

Fully complex and 3D models are significantly slower to train on average due to the higher number of operations needed. Moreover, they are also associated with worse performance especially for the DTI maps results.

There is an important point to be made about the worse performances of the more advanced models: as we aimed to keep the number of parameters fixed, we reduced the number of channels for each layer in the more advanced models, reducing the effective capacity of these models and therefore negatively affecting their performances. It can be hypothesised that these more advanced models would perform better given a higher number of fixed parameters and a longer training time.

Moreover, as our dataset was acquired with SMS factor equal to 2, the 3D architectures have little additional spatial information to exploit when removing the artefacts from the slices. If the acceleration factor were higher, it is likely that better exploiting the 3D spatial information would produce better performances.

Visually, from Fig. 1, we can notice how the AI-derived maps all show extremely similar flaws regardless of the model used (e.g. top-left quadrant of the HA maps), suggesting that 1. the information learned by the models is similar, and 2. almost none of the models is able to overcome the incorrect information present in the SMS version of the images used as input.

6   Conclusion

As cardiac DTI becomes closer and closer to a clinical reality, deep learning is also becoming a vital tool in alleviating some of its downsides such as long scan times and low SNR. In the hurry associated with a new emerging field, many authors prioritise proof-of-concept work to showcase innovative ideas over exploring and comparing known and common options. In this work we lay out the basis of a comparison between input types and model dimensionality and we show how, despite our initial assumption, given a fixed number of parameters and a reasonable train time, 2D models vastly outperform their 3D counterparts and complex-valued networks are not preferable. On the other hand, the seemingly naive use of separate input channels for magnitude and phase data has beneficial performance over discarding the phase information as it is commonly done.

As an advice for future studies, we suggest the use of 2D models and, if available, the of phase information together with the absolute information for their model-development starting point.

References

1. Barth, M., Breuer, F., Koopmans, P.J., Norris, D.G., Poser, B.A.: Simultaneous multislice (SMS) imaging techniques. Magn. Reson. Med. **75**(1), 63–81 (2016). https://onlinelibrary.wiley.com/doi/abs/10.1002/mrm.25897
2. Basser, P.J.: Inferring microstructural features and the physiological state of tissues from diffusion-weighted images. NMR Biomed. **8**(7), 333–344 (1995)
3. Bihan, D.L., et al.: Diffusion tensor imaging: Concepts and applications. J. Magn. Reson. Imaging **13**(4), 534–546 (2001). https://onlinelibrary.wiley.com/doi/abs/10.1002/jmri.1076
4. Cao, Y., et al.: CS-GAN for High-Quality Diffusion Tensor Imaging. Tech. rep., Guizhou University (Sep 2020). https://doi.org/10.21203/rs.3.rs-65572/v1,https://www.researchsquare.com/article/rs-65572/v1
5. Ferreira, P.F., et al.: Accelerating cardiac diffusion tensor imaging with a u-net based model: toward single breath-hold. J. Magn. Reson. Imaging (2022). https://onlinelibrary.wiley.com/doi/abs/10.1002/jmri.28199

6. Ferreira, P.F., et al.: Automating in vivo cardiac diffusion tensor postprocessing with deep learning-based segmentation. Magn. Reson. Med. **84**(5), 2801–2814 (2020). https://onlinelibrary.wiley.com/doi/abs/10.1002/mrm.28294

7. Griswold, M.A., et al.: Generalized autocalibrating partially parallel acquisitions (GRAPPA). Magn. Reson. Med. **47**(6), 1202–1210 (2002). http://doi.wiley.com/10.1002/mrm.10171

8. Karimi, D., Gholipour, A.: Diffusion tensor estimation with transformer neural networks. Artifi. Intell. Med. **130**, 102330 (2022). https://www.sciencedirect.com/science/article/pii/S0933365722000951

9. Kingma, D.P., Ba, J.: Adam: a method for stochastic optimization. arXiv:1412.6980 [cs] (Jan 2017)

10. Kung, G.L., et al.: The presence of two local myocardial sheet populations confirmed by diffusion tensor mri and histological validation. J. Magn. Reson. Imaging **34**(5), 1080–1091 (2011)

11. Mori, S., Zhang, J.: Principles of diffusion tensor imaging and its applications to basic neuroscience research. Neuron **51**(5), 527–539 (2006). https://www.cell.com/neuron/abstract/S0896-6273(06)00634-9

12. Nielles-Vallespin, S., et al.: Assessment of myocardial microstructural dynamics by in vivo diffusion tensor cardiac magnetic resonance. J. Am. College Cardiol. **69**(6), 661–676 (2017). https://linkinghub.elsevier.com/retrieve/pii/S0735109716373272

13. Nielles-Vallespin, S., Scott, A., Ferreira, P., Khalique, Z., Pennell, D., Firmin, D.: Cardiac diffusion: technique and practical applications. J. Magn. Reson. Imaging **52**(2), 348–368 (2020). https://onlinelibrary.wiley.com/doi/abs/10.1002/jmri.26912

14. Phipps, K., et al.: Accelerated in vivo cardiac diffusion-tensor mri using residual deep learning-based denoising in participants with obesity. Radiol. Cardiothoracic Imag. **3**(3), e200580 (2021). https://pubs.rsna.org/doi/abs/10.1148/ryct.2021200580

15. Pruessmann, K.P., Weiger, M., Scheidegger, M.B., Boesiger, P.: SENSE: sensitivity encoding for fast MRI. Magn. Reson. Med. **42**(5), 952–962 (1999)

16. Ronneberger, O., Fischer, P., Brox, T.: U-Net: convolutional networks for biomedical image segmentation. arXiv:1505.04597 [cs] (May 2015)

17. Schlaudecker, J.D., Bernheisel, C.R.: Gadolinium-Associated Nephrogenic Systemic Fibrosis. Am. Family Phys. **80**(7), 711–714 (2009). https://www.aafp.org/afp/2009/1001/p711.html

18. Schlemper, J., et al.: Stochastic deep compressive sensing for the reconstruction of diffusion tensor cardiac MRI. In: Frangi, A.F., Schnabel, J.A., Davatzikos, C., Alberola-López, C., Fichtinger, G. (eds.) MICCAI 2018. LNCS, vol. 11070, pp. 295–303. Springer, Cham (2018). https://doi.org/10.1007/978-3-030-00928-1_34

19. Setsompop, K., et al.: Improving diffusion MRI using simultaneous multi-slice echo planar imaging. NeuroImage **63**(1), 569–580 (2012). https://linkinghub.elsevier.com/retrieve/pii/S1053811912006477

20. Tian, C., Fei, L., Zheng, W., Xu, Y., Zuo, W., Lin, C.W.: Deep Learning on Image Denoising: An overview. arXiv:1912.13171 [cs, eess] (Aug 2020)

21. Tänzer, M., et al.: Faster diffusion cardiac MRI with deep learning-based breath hold reduction. In: Medical Image Understanding and Analysis, pp. 101–115. LNCS. Springer International Publishing, Cham (2022). https://doi.org/10.1007/978-3-031-12053-4_8

22. Weine, J., van Gorkum, R.J.H., Stoeck, C.T., Vishnevskiy, V., Kozerke, S.: Synthetically trained convolutional neural networks for improved tensor estimation from free-breathing cardiac DTI. Comput. Med. Imaging Graph. **99**, 102075 (2022). https://www.sciencedirect.com/science/article/pii/S0895611122000489

Improving Echocardiography Segmentation by Polar Transformation

Zishun Feng[1]([✉]), Joseph A. Sivak[2], and Ashok K. Krishnamurthy[1,3]

[1] Department of Computer Science, University of North Carolina at Chapel Hill,
Chapel Hill, USA
`fzs@cs.unc.edu`
[2] Division of Cardiology, University of North Carolina at Chapel Hill,
Chapel Hill, USA
[3] Renaissance Computing Institute, University of North Carolina at Chapel Hill,
Chapel Hill, USA

Abstract. Segmentation of echocardiograms plays an essential role in the quantitative analysis of the heart and helps diagnose cardiac diseases. In the recent decade, deep learning-based approaches have significantly improved the performance of echocardiogram segmentation. Most deep learning-based methods assume that the image to be processed is rectangular in shape. However, typically echocardiogram images are formed within a sector of a circle, with a significant region in the overall rectangular image where there is no data, a result of the ultrasound imaging methodology. This large non-imaging region can influence the training of deep neural networks. In this paper, we propose to use polar transformation to help train deep learning algorithms. Using the r-θ transformation, a significant portion of the non-imaging background is removed, allowing the neural network to focus on the heart image. The segmentation model is trained on both x-y and r-θ images. During inference, the predictions from the x-y and r-θ images are combined using max-voting. We verify the efficacy of our method on the CAMUS dataset with a variety of segmentation networks, encoder networks, and loss functions. The experimental results demonstrate the effectiveness and versatility of our proposed method for improving the segmentation results.

Keywords: Echocardiography · Segmentation · Polar transformation · Deep learning

1 Introduction

Echocardiography (echo) is a radiation-free and cost-effective imaging modality. Thus, echo is the first-line imaging technique for diagnosing most cardiac diseases. Accurate segmentation of echo images can significantly help the quantitative measurement of the heart and diagnosis of cardiac diseases. For example, segmentation of the left ventricle at end-systolic and end-diastolic frames can be used to calculate ejection fraction (EF), which is an essential cardiac function metric; segmentation of the myocardium is used to calculate wall thickness, which

O. Camara et al. (Eds.): STACOM 2022, LNCS 13593, pp. 133–142, 2022.
https://doi.org/10.1007/978-3-031-23443-9_13

is widely used in cardiac disease diagnosis, such as left ventricular hypertrophy [1–3]. Due to the ultrasound imaging method, a typical echocardiography image is formed within a circular sector, and there is a significant area of the entire rectangular image for which no data are available.

In recent years, deep learning-based segmentation methods have achieved great success in computer vision and medical imaging. Convolutional neural network (CNN) based models, like ResNet, VGG and UNet, have been widely used for medical images analysis in different modalities, such as CT, MRI, and ultrasound.There have been many different kinds of deep learning based echocardiogram analysis applications: 1) view identification [4,5], 2) chamber segmentation [6–9], and 3) disease and abnormality identification[10–12]. For chamber segmentation of echocardiograms specifically Ouyang et al. [7] chose DeepLabV3 [14] to segment the left ventricle for ejection fraction calculation in apical-2-chamber view echocardiograms; Leclerc et al. [6] used UNet [15] to simultaneously segment the left ventricle, myocardium, and left atrium; Liu et al. [8] designed a pyramid local attention architecture for echocardiograms segmentation; Wu et al. [9] proposed a semi-supervised methods for left ventricle segmentation in echocardiography vidoes. All the above-mentioned echocardiogram applications, including classification and segmentation, did not consider the large non-imaging region in the whole echo image, which can influence the training of the networks. Tan et al. [13] applied polar transformations to segment the left ventricle in MRI images. However, the polar transformation was only applied to a small pre-defined region of interest and not the entire imaging area.

In this paper, to address this problem, we propose to use polar transformation to help the segmentation of echo. First, we remove a significant portion of the non-imaging background using r-θ transformation, which helps the network focus on the heart image area. Second, we train the segmentation model on x-y and r-θ spaces simultaneously to let the model capture information from both spaces. Third, during the testing phase, we use max-voting to combine the predictions from the original (x-y) image and the polar (r-θ) image to make the final prediction.

2   Methodology

Our proposed method is shown in Fig. 1 and contains 3 steps: 1) polar transformation that transforms images from the x-y space to r-θ space in order to reduce the non-imaging area, 2) joint model training on both original and polar images and 3) combining original and polar results when testing.

2.1   Polar Transformation of Imaging Region

The ultrasound imaging method results in echocardiography images being formed in a sector of a circle, with large non-imaging regions in the whole rectangular image. The model training can be influenced by this large non-imaging area. To address this problem, we use polar transformation to enlarge the imaging area.

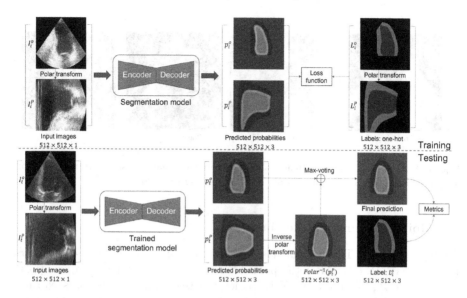

Fig. 1. Training and testing pipeline of the proposed approach. In the training phase, the loss function is calculated in both x-y and r-θ spaces. In the testing phase, performance metrics are only calculated in x-y space.

As shown in Fig. 2, we choose a sector of a circle (orange) in the original echo image, which contains all the imaging areas. We denote original echo image as I^o with size $X \times Y$; and polar image as I^p with size $R \times \Theta$. We use a simple binary segmentation method to identify the center for the sector and the angular extent Θ of the image in polar coordinates. Specifically, we compute the angle between OA and OB, where A and B are the left most point and right most point of the imaging area, and O is the center of the sector. To cover all imaging areas, the radius R is set to be the same as the side of the original image, where $R = Y$. So the value of the polar image I^p at coordinate r-θ can be obtained by:

$$x^* = r \cos(\theta) \tag{1}$$

$$y^* = r \sin(\theta) \tag{2}$$

$$I^p(r, \theta) = I^o(x^*, y^*) \tag{3}$$

where $r \in [0, R], \theta \in [0, \Theta]$. When x^* and y^* are not integer, $I^o(x^*, y^*)$ is calculated by bicubic interpolation.

The polar image label L^p can also be obtained from the original image label L^o by following Eqs. (1)–(2) . However, the label is one-hot coded and the values only contain integers, so we choose nearest interpolation when calculating the polar image labels:

$$L^p(r, \theta) = L^o(round(x^*), round(y^*)) \tag{4}$$

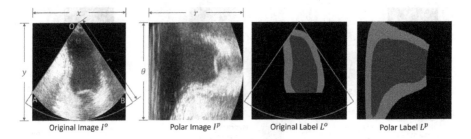

| Original Image I^o | Polar Image I^p | Original Label L^o | Polar Label L^p |

Fig. 2. An example of polar transformation. Before the polar transformation, the region of interest is shown inside the orange lines. (Color figure onlline)

2.2 Joint Training and Testing

Joint Training. The original training set S^o contains N image-label pairs $\{(I_i^o, L_i^o) \mid i \in [1, N]\}$. After the polar transformation, we generate a new polar set $S^p = \{(I_i^p, L_i^p) \mid i \in [1, N]\}$. To let the segmentation networks capture information from both x-y space and r-θ space, we combine these two sets: $S = S^o \cup S^p = \{(I_i^o, I_i^p, L_i^o, L_i^p) \mid i \in [1, N]\}$. So a training batch contains both original and polar images of a sample i in this batch. The pseudo-code is shown below.

Algorithm 1 Joint training and testing

Training:

Input: Original set S^o, network W
Output: Trained network W'
  $S^p \leftarrow Polar(S^o)$
  while training **do**
    $p_i^o \leftarrow W(I_i^o)$
    $p_i^p \leftarrow W(I_i^p)$
    $loss \leftarrow LossFunc(\{L_i^o, L_i^p\}, \{p_i^o, p_i^p\})$
    Update W
  end while
  return Trained network W'

Testing:

Input:
  Test sample I_t^o, L_t^o,
  Trained network W'
Output: Output metrics m
  $I_t^p \leftarrow Polar(I_t^o)$
  $p_t^o \leftarrow W'(I_t^o)$
  $p_t^p \leftarrow W'(I_t^p)$
  $p_t \leftarrow MaxVoting(p_t^o, Polar^{-1}(p_t^p))$
  $m \leftarrow Metric(p_t, L_t^o)$
  return m

Model. In this work, we hope to propose a method that is not bound to a specifically designated network. We choose convolutional neural network (CNN) based segmentation models, which are widely used in medical image analysis, specifically UNet and DeeplabV3+ [16]. Within these segmentation models, we also pick different convolutional neural networks, like ResNet [17] and VGG [18], to be the encoder of the segmentation models.

Loss Function. Similarly, we train the segmentation networks with two different loss functions to evaluate effectiveness: cross-entropy loss and dice loss, are shown below:

$$CrossEntropy = -\sum_{i=1}^{N}\sum_{j=0}^{2} L_{i,j} \log p_{i,j} \tag{5}$$

$$DiceLoss = \sum_{i=1}^{N}(1 - \frac{1}{3}\sum_{j=0}^{2}\frac{2|L_{i,j} \cap p_{i,j}|}{|L_{i,j}| + |p_{i,j}|}) \tag{6}$$

where L and p are the label and the prediction of segmentation networks, i denotes the index of samples, j denotes the index of classes (0 for background, 1 for LV, 2 for wall). Note that the original images and polar images use the same loss function during training.

Testing. In testing, we first transform an original image I_t^o into r-θ space and get the polar image I_t^p. Then, both original and polar images are fed into the trained segmentation model to get predictions p_t^o and p_t^p. To aggregate the predicted probabilities, we use inverse polar transformation to map p_t^p back to x-y space, namely $Polar^{-1}(p_t^p)$. The finally prediction p_t is obtained by max-voting between p_t^o and $Polar^{-1}(p_t^p)$. The final predicted value at (x, y) is calculated below:

$$p_{t,(x,y)} = \begin{cases} p_{t,(x,y)}^o, & max(p_{t,(x,y)}^o) \geq max(Polar^{-1}(p_t^p)_{(x,y)}), \\ Polar^{-1}(p_t^p)_{(x,y)}, & max(p_{t,(x,y)}^o) < max(Polar^{-1}(p_t^p)_{(x,y)}). \end{cases} \tag{7}$$

3 Experiments

3.1 Data

We use the CAMUS dataset to evaluate the effectiveness of our polar transformation method. The dataset contains 450 different patients. For each patient, there are 4 labeled echocardiogram frames: apical 2-chamber (A2C), end-systolic (ES) and end-diastolic (ED), apical 4-chamber (A4C), end-systolic (ES) and end-diastolic (ED). In total, the dataset contains 1800 labeled frames. Our segmentation models predict background, $LV_{chamber}$ and LV_{wall} regions. Additionally, we removed the labels on non-imaging regions, which are not included in the polar transformation. We used 350 patients (1400 images) for training, 50 patients (200 images) for validation, and 50 patients (200 images) for testing. All images and labels were resized to 512×512 for training and testing.

Table 1. The average percentage of each region.

Space	Non-imaging region	Imaging region				Total
		$LV_{chamber}$	LV_{wall}	LV_{all}	Background	
x-y	45.88%	9.27%	9.21%	18.48%	35.64%	54.12%
r-θ	2.10%	19.41%	22.30%	41.71%	56.19%	97.90%

3.2   Experimental Details

Comparison of x-y and r-θ Spaces. To show the effectiveness, we compared our method to 3 other methods: (1) training on x-y space only, (2) training on r-θ space only, (3) separately training on x-y and r-θ space then voting. For fairness, we controlled all methods trained with the same number of images. Specifically, the proposed method was trained for 30 epochs, and the methods (1) and (2) were trained for 60 epochs since the proposed method was trained on 2 spaces, but (1) and (2) were trained only on single space images. The loss function for the joint training also shows that the method has converged around 30 epochs.

Comparison of Segmentation Model Settings. Our base experimental model is a UNet model with ResNet34 encoder and dice loss. To test the effectiveness with different settings, we trained 3 other models: 1) change architecture to DeepLabV3+, 2) change encoder to VGG19, and 3) change loss function to cross-entropy loss. All segmentation models were trained with a batch size of 32 and optimized by the Adam algorithm with a learning rate of $1e^-4$.

We adopted dice similarity coefficient (DSC) as the metric to evaluate the segmentation performance for each region, which can be formulated as Eq. (7). We calculated DSC on 3 regions in x-y space: 1) $LV_{chamber}$, left ventricle outline inside the chamber wall, 2) LV_{wall}, the chamber wall, 3) LV_{all}, outline of the outside chamber that includes the wall.

$$DSC(P,L) = \frac{2\,|\,P \cap L\,|}{|\,P\,|+|\,L\,|} = \frac{2TP}{TP+FP+TP+FN} \tag{8}$$

In this equation, P and L denote prediction and true label, and {TP, FP, TN} denote the number of True Positive, False Positive, False Negative pixels, respectively.

3.3   Experimental Results

We first transformed all original images and labels into r-θ space and calculate the average percentage of each region in the whole image across all images. The resulting statistics are shown in Table 1. From the table, the polar transformation can significantly reduce the non-imaging region in the whole image, and enlarge the size of foreground area and the region of interest.

Table 2. Dice similarity scores of different method with different loss functions, segmentation models and encoder networks. The scores are calculated in x-y space.

Setting	Method	$LV_{chamber}$	LV_{wall}	LV_{all}
UNet-R34-Dice	x-y only	0.9395	0.8877	0.9647
	r-θ only	0.9363	0.8779	0.9613
	x-y+r-θ+voting	0.9405	0.8886	0.9648
	proposed	**0.9418**	**0.8900**	**0.9652**
UNet-R34-CE	x-y only	0.9383	0.8849	0.9635
	r-θ only	0.9358	0.8779	0.9611
	x-y+r-θ+voting	**0.9409**	0.8882	0.9649
	proposed	**0.9409**	**0.8900**	**0.9661**
UNet-V19-Dice	x-y only	0.9386	0.8843	0.9629
	r-θ only	0.9358	0.8716	0.9579
	x-y+r-θ+voting	0.9397	0.8846	0.9630
	proposed	**0.9400**	**0.8857**	**0.9634**
DL-R34-Dice	x-y only	0.9389	0.8869	0.9644
	r-θ only	0.9370	0.8782	0.9621
	x-y+r-θ+voting	0.9415	0.8892	0.9654
	proposed	**0.9426**	**0.8929**	**0.9663**

CE: cross-entropy loss; Dice: dice loss;
DL: DeepLabV3+; R34: Resnet34; V19: VGG19.

We show our segmentation results in Table 2, with each row showing a different model choice. In each row, the proposed joint training method achieves the best results, since the proposed method can learn information from both spaces. Training on polar images only does not achieve better results compared with training on original images only; we speculate that this is because the supervision is in r-θ space, but the evaluation metric is in x-y space. Therefore, errors are accumulated when transforming from r-θ space to x-y space. Comparing "x-y only" and "r-θ only" with "x-y+r-θ+voting" results shows that the models trained on different spaces can capture more information. The voting of two single-space-trained models aggregated information from different spaces and improved the segmentation results. But the two models were trained separately, so they could not extract information as well as the jointly trained model.

We also list three groups of results with different settings: changing loss function to cross-entropy, changing encoder to VGG19, and changing segmentation model architecture to DeeplabV3+. Within each group, the proposed method outperformed all other methods, which demonstrated that the proposed method was effective with different loss functions, encoder networks, and segmentation model architectures.

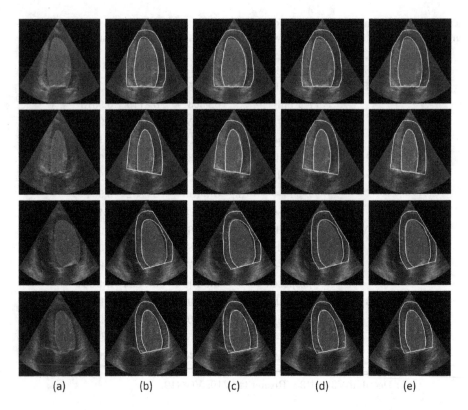

Fig. 3. Segmentation results of one subject with DL-R34-Dice setting. From top to bottom: A2C-ED, A2C-ES, A4C-ED, A4C-ES. (a) Ground-truth. (b) x-y only. (c) r-θ only. (d) x-y+r-θ+voting. (e) Propsed. The yellow contour denotes ground-truth in (b)-(e). $LV_{chamber}$: green. LV_{wall}: blue. LV_{all}: blue+green. (Color figure online)

A visualization of the segmentation results is shown in Fig. 3. The qualitative results show the effectiveness of our method, which is able to give better predictions for pixels near the region boundary.

4   Conclusion

In this paper, we have proposed a new polar transformation based method to improve echocardiography segmentation performance. The polar transformation helps the segmentation models focus more on the imaging region of the image, and the joint training on the original and polar image lets the models capture information from x-y and r-θ spaces. The max-voting aggregates prediction from 2 spaces and achieves better performance. The experimental results show our method can effectively improve the performance with different segmentation models, encoders and loss functions.

Acknowledgments. This work was funded by NSF Grant 1633295 BIGDATA: F: Collaborative Research: From Visual Data to Visual Understanding.

References

1. Troy, B.L., et al.: Measurement of left ventricular wall thickness and mass by echocardiography. Circulation **45**(3), 602–611 (1972)
2. Devereux, R.B., et al.: Echocardiographic assessment of left ventricular hypertrophy: comparison to necropsy findings. Am. J. Cardiol. **57**(6), 450–458 (1986)
3. Lang, R.M., et al.: Recommendations for cardiac chamber quantification by echocardiography in adults: an update from the american society of echocardiography and the european association of cardiovascular imaging. Euro. Heart J. Cardiovas. Imaging **16**(3), 233–271 (2015)
4. Madani, A., et al.: Fast and accurate view classification of echocardiograms using deep learning. NPJ Digit. Med. **1**(1), 1–8 (2018)
5. Østvik, A., et al.: Real-time standard view classification in transthoracic echocardiography using convolutional neural networks. Ultrasound Med. Biol. **45**(2), 374–384 (2019)
6. Leclerc, S., et al.: Deep learning for segmentation using an open large-scale dataset in 2D echocardiography. IEEE TMI **38**(9), 2198–2210 (2019)
7. Ouyang, D., et al.: Video-based AI for beat-to-beat assessment of cardiac function. Nature **580**(7802), 252–256 (2020)
8. Liu, F., et al.: Deep pyramid local attention neural network for cardiac structure segmentation in two-dimensional echocardiography. Med. Image Anal. **67**, 101873 (2021)
9. Wu, H., et al.: Semi-supervised segmentation of echocardiography videos via noise-resilient spatiotemporal semantic calibration and fusion. Med. Image Anal. **78**, 102397 (2022)
10. Madani, A., et al.: Deep echocardiography: data-efficient supervised and semi-supervised deep learning towards automated diagnosis of cardiac disease. NPJ Digit. Med. **1**(1), 1–11 (2018)
11. Zhang, J., et al.: Fully automated echocardiogram interpretation in clinical practice: feasibility and diagnostic accuracy. Circulation **138**(16), 1623–1635 (2018)
12. Feng, Zi., et al.: Two-stream attention spatio-temporal network for classification of echocardiography videos. In: 2021 IEEE 18th International Symposium on Biomedical Imaging (ISBI), pp. 1461–1465. IEEE (2021)
13. Tan, L.K., et al.: Fully automated segmentation of the left ventricle in cine cardiac MRI using neural network regression. J. Magn. Reson. Imaging **48**(1), 140–152 (2018)
14. Chen, L.-C., et al. Rethinking atrous convolution for semantic image segmentation. arXiv preprint arXiv:1706.05587 (2017)
15. Ronneberger, O., Fischer, P., Brox, T.: U-Net: convolutional networks for biomedical image segmentation. In: Navab, N., Hornegger, J., Wells, W.M., Frangi, A.F. (eds.) MICCAI 2015. LNCS, vol. 9351, pp. 234–241. Springer, Cham (2015). https://doi.org/10.1007/978-3-319-24574-4_28
16. Chen, L.-C., Zhu, Y., Papandreou, G., Schroff, F., Adam, H.: Encoder-decoder with atrous separable convolution for semantic image segmentation. In: Ferrari, V., Hebert, M., Sminchisescu, C., Weiss, Y. (eds.) ECCV 2018. LNCS, vol. 11211, pp. 833–851. Springer, Cham (2018). https://doi.org/10.1007/978-3-030-01234-2_49

17. He, K., et al.: Deep residual learning for image recognition. In: Proceedings of the IEEE Conference on Computer Vision and Pattern Recognition (2016)
18. Simonyan, K., Zisserman, A.: Very deep convolutional networks for large-scale image recognition. arXiv preprint arXiv:1409.1556 (2014)

Spatiotemporal Cardiac Statistical Shape Modeling: A Data-Driven Approach

Jadie Adams[1,2(✉)], Nawazish Khan[1,2], Alan Morris[1,2], and Shireen Elhabian[1,2]

[1] Scientific Computing and Imaging Institute, University of Utah,
Salt Lake City, UT, USA
{jadie,nawazish.khan,amorris,shireen}@sci.utah.edu
[2] School of Computing, University of Utah, Salt Lake City, UT, USA

Abstract. Clinical investigations of anatomy's structural changes over time could greatly benefit from population-level quantification of shape, or spatiotemporal statistic shape modeling (SSM). Such a tool enables characterizing patient organ cycles or disease progression in relation to a cohort of interest. Constructing shape models requires establishing a quantitative shape representation (e.g., corresponding landmarks). Particle-based shape modeling (PSM) is a data-driven SSM approach that captures population-level shape variations by optimizing landmark placement. However, it assumes cross-sectional study designs and hence has limited statistical power in representing shape changes over time. Existing methods for modeling spatiotemporal or longitudinal shape changes require predefined shape atlases and pre-built shape models that are typically constructed cross-sectionally. This paper proposes a data-driven approach inspired by the PSM method to learn population-level spatiotemporal shape changes directly from shape data. We introduce a novel SSM optimization scheme that produces landmarks that are in correspondence both across the population (inter-subject) and across time-series (intra-subject). We apply the proposed method to 4D cardiac data from atrial-fibrillation patients and demonstrate its efficacy in representing the dynamic change of the left atrium. Furthermore, we show that our method outperforms an image-based approach for spatiotemporal SSM with respect to a generative time-series model, the Linear Dynamical System (LDS). LDS fit using a spatiotemporal shape model optimized via our approach provides better generalization and specificity, indicating it accurately captures the underlying time-dependency.

Keywords: Statistical shape modeling · Cardiac dynamics · Statistical morphology analysis

1 Introduction

Clinical studies that track dynamic or evolving organ shape within and across subjects in a cohort are critical to cardiac research. Statistical shape modeling

---

J. Adams and N. Khan—Contributed equally.

O. Camara et al. (Eds.): STACOM 2022, LNCS 13593, pp. 143–156, 2022.
https://doi.org/10.1007/978-3-031-23443-9_14

(SSM) from medical image data is a valuable tool in such studies, as it allows for population-level quantification and analysis of anatomical shape variation. However, well-established approaches for SSM [1,26,28] assume static shape and are incapable of directly modeling a population of sequences of shape over time (i.e., spatiotemporal data), preventing their use in dynamic or longitudinal analysis. In SSM, shape is either represented explicitly as dense sets of correspondence points [27,32,34], or implicitly via deformation fields (coordinate transformations in relation to a predefined atlas) [6,24]. Progress toward spatiotemporal SSM has been made using both representations. Concerning the deformation-based approach, early work applied pure regression, relying on either a cross-sectional atlas for correlation analysis [23] or techniques for estimating time-varying deformations [7,13,18,33]. Such approaches average shape evolution's without considering the inter-subject variability. Works such as [12,14] require predefined spatiotemporal atlas of normative anatomical changes to quantify patient-specific disease-relevant changes.

The explicit landmark or correspondence-based approach represents shapes in the form of point distribution models (PDM). In this paper, we focus on PDM because it is more intuitive than implicit representations, more easily interpreted by clinicians, and can readily be visualized. Automatic PDM construction has been formulated as an optimization problem via metrics such as entropy [4] and minimum description length [10], as well as using a parametric representation of the surface using spherical harmonics, assuming a sphere template [31]. The particle-based shape modeling (PSM) approach [4] is a data-driven SSM approach for generating PDMs that does not require an initial atlas and learns directly from shape data by optimizing the landmarks placement. PSM has proven effective in quantifying group differences and in downstream tasks such as pathology detection and disease diagnosis [2,3,16,19].

However, PSM, like other well-established SSM approaches, assumes static shape and is incapable of directly modeling a population of sequences of shape over time (i.e., spatiotemporal data). Many methods have been proposed for the statistical analysis of cross-sectional time-series data [11,20,21], including a PSM approach, where a time-agnostic PDM is fit then regressed over time [8]. These approaches either do not contain repeated subject measurements or incorrectly assume shape samples are drawn independently and do not consider the inherent correlation of shapes from the same sequence, confounding the population analysis [15,17]. In other existing methods, an auxiliary time-dependent model such as mixed-effects model [9] is applied to a PDM. However, mixed-effects models do not explicitly model dynamically changing anatomies with spatiotemporal movement. An image-based approach has been proposed for estimating organ segmentation and functional measurements over time [25]. Such an approach could be used for spatiotemporal SSM by first generating a PDM for a single time point across subjects, then independently propagating the correspondence points across individual time sequences using image-based deformable registration. While this technique incorporates time-dependency into the PDM, it relies on image-based features alone to do so and thus is prone to miscorrespondence. We use this approach as a baseline for comparison.

We propose a novel entropy-based PSM optimization objective that disentangles subject and time-dependencies to provide more accurate spatiotemporal SSM. This technique encourages maximal inter-subject shape correspondence across the population and temporal intra-subject correspondence across time points. We demonstrate that the PDM resulting from proposed method effectively captures dynamic shape change over time in a population of left atrium. This use case exemplifies dynamic motion over a short discretized time interval, however our method could also be applied to longitudinal studies where sparse observations are consistently made over long periods of time. The existing literature lacks analysis metrics for quantifying how well time-dependency is captured. To address this, we employ a generative time-series model for verification - the linear dynamical system (LDS). LDS is a state space model that explicitly represents the underlying evolving process of dynamically changing anatomies over time. LDS enables PCA-like analysis over time using linear transitions and Gaussian latent variables [22, 29, 30]. When fit to time-series data via the Expectation-Maximization (EM) algorithm, LDS captures the underlying time evolution and measurement processes and can be used to infer future and missing time points. We demonstrate that LDS fit to a PDM generated via our approach is more accurate in terms of generalization, specificity, and inference; suggesting our method accurately captures the dynamics of shape over time.

2 Methods

2.1 Notation

Given a cohort of N subjects, each subject has a time-sequence of $T-$shapes, where the temporal frames are consistent across subjects. Each shape is represented by a set of $d-$dimensional points (or particles). In this work, shape is segmented from volumetric images, so $d = 3$. We consider two forms of random variables: configuration and shape space variables, where the former captures sample-specific geometry and the latter describes population-level shape statistics. The configuration space variable $\mathbf{X}_{n,t}$ represents the particle position on the $n-$th subject at the $t-$th time point, where $n \in [1, 2, \ldots N]$ and $t \in [1, 2, \ldots T]$. $M-$realizations of this random variable defines the point set (or PDM) of the $n, t-$shape, $\mathbf{x}_{n,t} = [\mathbf{x}_{n,t}^1, \mathbf{x}_{n,t}^2, \ldots, \mathbf{x}_{n,t}^M] \in \mathbb{R}^{dM}$ ($\mathbf{x}_{n,t}^m \in \mathbb{R}^d$), where $\mathbf{x}_{n,t}^m$ is the vector of the $\{x, y, z\}$ coordinates of the $m-$th particle. Generalized Procrustes alignment is used to estimate a rigid transformation matrix $\mathbf{T}_{n,t}$ that can transform the particles in the local coordinate $\mathbf{x}_{n,t}^m$ in the configuration space to the world common coordinate $\mathbf{z}_{n,t}^m$ in the shape space such that $\mathbf{z}_{n,t}^m = \mathbf{T}_{n,t}\mathbf{x}_{n,t}^m$. In the cross-sectional PSM formulation, shape distribution in the shape space is modeled by a single random variable $\mathbf{Z} \in \mathbb{R}^{dM}$ that is assumed to be Gaussian distributed and only captures inter-subject variations. We define two different shape space random variables, one represents shapes across subjects at a specific time point t (i.e., inter-subject variable) and is denoted $\mathbf{Z}_t$ and the other represents shape across time for a specific subject n (i.e., intra-subject variable) and is denoted $\mathbf{Z}_n$.

2.2   Proposed Spatiotemporal Optimization Scheme

We build on the PSM approach for optimizing population-specific PDMs. Vanilla PSM assumes cross-sectional data (denoted hereafter "cross-sectional PSM") and optimizes particle positions for observations at a single point in time using an entropy-based scheme [4,5]. Here intuitively $\mathbf{X}_n = \mathbf{X}_{n,t=1}$ and $\mathbf{Z} = \mathbf{Z}_{t=1}$. The optimization objective to be minimized is then defined as:

$$\mathcal{Q}_{cross-sectional} = \alpha H(\mathbf{Z}) - \sum_{n=1}^{N} H(\mathbf{X}_n) \tag{1}$$

where H is the differential entropy of the respective random variable and α is a relative weighting parameter that defines the contribution of the correspondence objective $H(\mathbf{Z})$ to the particle optimization process. In this work α was experimentally tuned to be initialized as $\alpha = 100$ then gradually decreased during optimization until $\alpha = 0.1$ for both the proposed and comparison method. Minimizing this objective balances two terms. The first encourages a compact distribution of samples in the shape space, ensuring maximal correspondence between particles across shapes (i.e., lower model complexity). The second encourages maximal uniformly-distributed spread of points across individual shapes so that shape is faithfully represented (i.e., better geometric accuracy).

We propose a novel spatiotemporal optimization equation that disentangles the shape space entropy for $\mathbf{Z}_t$ and $\mathbf{Z}_n$. It is defined as follows:

$$\mathcal{Q} = \alpha \left(\sum_{t=1}^{T} H(\mathbf{Z}_t) + \sum_{n=1}^{N} H(\mathbf{Z}_n) \right) - \sum_{n=1}^{N} \sum_{t=1}^{T} H(\mathbf{X}_{n,t}) \tag{2}$$

The first term encourages intra-subject correspondence across time points, the second inter-subject correspondence across sequences, and the third retains geometric accuracy across subjects and time points. The cost function is optimized via gradient descent. To find the correspondence point updates, we must take the derivative of $H(\mathbf{Z}_n)$ and $H(\mathbf{Z}_t)$ with respect to particle positions. Similarly to the cross-sectional PSM formulation [4], we model $p(\mathbf{Z}_t)$ and $p(\mathbf{Z}_n)$ parametrically as Gaussian distributions with covariance matrices $\boldsymbol{\Sigma}_t$ and $\boldsymbol{\Sigma}_n$, respectively. These covariance matrices are directly estimated from the data. The entropy terms are then given by:

$$H(\mathbf{Z}_*) \approx \frac{1}{2} \log \boldsymbol{\Sigma}_* = \frac{1}{2} \sum_{i=1}^{dM} \log \lambda_{*,i} \tag{3}$$

where $*$ represents either t or n and $\lambda_{*,i}$ are the eigenvalues of $\boldsymbol{\Sigma}_*$. We estimate the covariances from the data (closely following [4]) and find:

$$\frac{-\partial H(\mathbf{Z}_*)}{\partial \mathbf{X}} \approx \mathbf{Y}_*(\mathbf{Y}_*^\top \mathbf{Y}_* + \alpha \mathbf{I})^{-1} \tag{4}$$

where the respective $\mathbf{Y}_*$ matrices denote the matrix of points minus the sample mean for the ensemble and the regularization α accounts for the possibility of diminishing gradients (see [4] for more detail). Combining Eq. 4 with the shape-based updates explained in [4] we get an update for each point.

2.3   Image-Based Comparison Method

As a baseline for comparison, we consider the approach presented in "An Image-based Approach for 3D Left Atrium Functional Measurements" [25]. In this work, anatomic segmentations from high-resolution Magnetic Resonance Angiographies (MRA) were registered and propagated through pairwise deformable registrations to cover the cardiac cycle. We applied this to spatiotemporal SSM by fitting a PDM to a single corresponding time-point across the cohort, cross-sectionally, then individually propagating particles across time points for each subject using the image-to-image deformable registration transforms. While this approach provides intra- and inter-subject correspondences, the correspondence distribution across time points involves only one subject at a time and is unable to be optimized for shape statistics across the cohort.

2.4   LDS Evaluation Metrics

A time-series generative model is required to evaluate how well a spatiotemporal PDM captures the underlying time dependencies. Here, we use a Linear Dynamic System (LDS). LDS is equivalent to a Hidden Markov Model (HMM) except it employs linear Gaussian dynamics and measurements and uses a continuous latent/state variable instead of discrete (Fig. 1). The LDS state representation, $\mathbf{S}$, can be fit to the PDM, along with model parameters, via the expectation maximization (EM) algorithm. We provide a more detailed explanation of the LDS model and EM algorithm in the Appendix A. Generalization and specificity of the fit LDS model are used to determine if the temporal dynamics are accurately portrayed by the PDM.

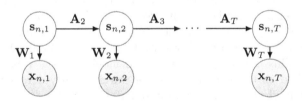

Fig. 1. Time-varying LDS model

Partial and Full-Sequence Reconstruction. If the time-dependency is captured, an LDS model fit on a subset of the PDM's should generalize to held out examples (via multi-fold validation). The reconstruction error is calculated as the root mean square error (RMSE) between unseen observation sequences $\mathbf{x}_{n,t}$ and those reconstructed from the state/latent space by the LDS model $\hat{\mathbf{x}}_{n,t}$:

$$\text{RMSE} = \sqrt{\frac{1}{NTdM} \sum_{n=1}^{N} \sum_{t=1}^{T} \sum_{m=1}^{M} \sum_{i=1}^{d} \left(\mathbf{x}_{n,t}^{m_i} - \hat{\mathbf{x}}_{n,t}^{m_i} \right)^2} \tag{5}$$

Here $\hat{x}_{n,t}$ is found by estimating the state values (using the E-step (A.2)) then applying the LDS observation equation (Eq. A.1.2) reconstruct observations. When provided a partial sequence, LDS should also be able to reconstruct the missing points. The LDS state equation (Eq. A.1.1) estimates the state values for randomly masked missing time points. The partial observations are then reconstructed in the same manner and compared to the full unmasked sequence.

Specificity. The specificity of the fit LDS model is another indication of how well the shape model captures the time-dependency. The fit LDS model should be specific to the training set, meaning it only represents valid instances of shape sequences. To test this, we sample new observations from the LDS model, then compute the RMSE between these samples and the closest training sequences.

3   Results

3.1   4D Left Atrium Data

We performed experimental analysis on a cohort comprised of 3D LGE and stacked CINE CMR scans through the left atrium for 28 patients presenting with atrial fibrillation between 2019 and 2020. Average patient age was 64.9 years with 15 male and 13 female. Scans were captured for each patient before and after a RF cardiac ablation procedure for a total of 56 scans. Each CINE scan contained 25 consistent time points covering the cardiac cycle. The temporal dimension was normalized at time of acquisition to cover one heart beat for each patient. The number of milliseconds covered by the temporal dimension thus varies by patient. The 3D LGE images were manually segmented by a cardiac imaging expert and this segmentation was matched to the closest CINE time-point based on CMR trigger time. The segmentation was then transformed to each time point through time point to time point deformable registrations to create a full 3D segmentation sequence [25]. This dataset is selected to demonstrate the robustness of the proposed method as the left atrium shape varies greatly across patients and atrial fibrillation effects the dynamics in differing ways (Fig. 2).

3.2   Evaluation

We leverage ShapeWorks [28], the open source implementation of the PSM method, as a starting point for out implementation. ShapeWorks utilizes particle splitting, constraining the number of particles to be a power of two. We chose to use 256 particles for both the proposed and comparison PDM, as this is the smallest number that is dense enough to accurately represent the geometry of the left atrium. Figure 2 displays the part of the PDM generated via the proposed method on three subjects at selected time points, illustrating both intra- and inter-subject correspondence. The subset of time points selected for display were chosen to be meaningful points in the cardiac cycle using left atrium volume.

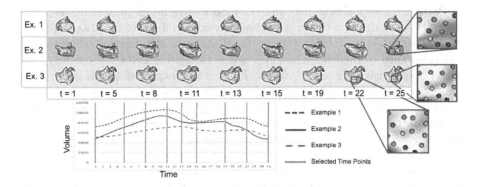

Fig. 2. PDM Examples Correspondence points for three subjects on the ground truth meshes (anterior view). Color denotes correspondence (see zoomed in boxes). The plot of left atrium volume over time shows the time points selected for display.

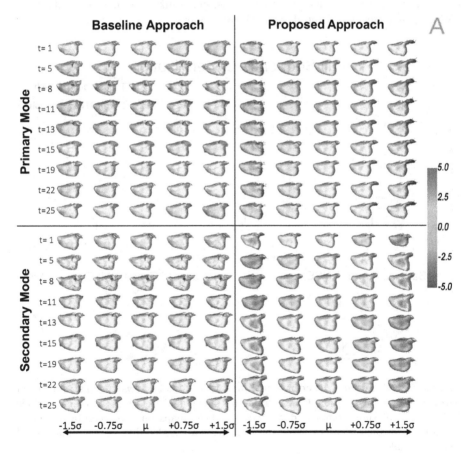

Fig. 3. Modes of Variation Anterior view of Left Atrium across the primary and secondary mode of variation. Heat maps show the distance from mean shape.

Particle correspondence captures the modes of variation present in the population. Figure 3 displays the primary and secondary modes of variation captured by the two shape models. In the proposed shape model, the primary mode of variation captures the left atrium appendage elongation and the secondary captures the sphericity of the left atrium. This is to be expected as this is the main source of variation across patients and time. The shape variation captured in the modes of the baseline approach PDM has smaller magnitude and is more difficult to interpret, suggesting weaker correspondence. Additionally, we observe the temporal dynamics are more smoothly captured by the proposed approach shape model. This example illustrates the utility of our approach as a tool for statistically quantifying and visualizing population variation.

We employ the LDS model to analyze how well the PDM's capture the underlying dynamics. We fit LDS models with the same initial parameters latent dimension $L = 64$ to both PDM's for 50 EM iterations. To ensure a fair comparison, training was done on a version of the correspondence particles ($d = 3$) scaled uniformly to a range of 0 to 1. Thus, generalization and specificity evaluation metrics are interpreted on a relative scale.

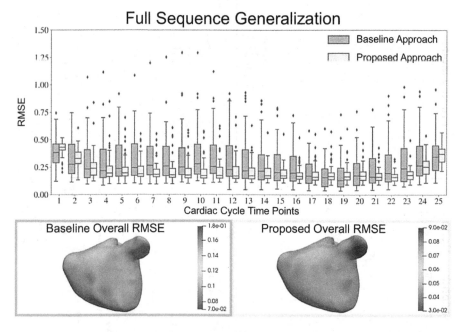

Fig. 4. Full Sequence Reconstruction The box plot shows the distribution of particle-wise reconstruction RMSE at each time point. Below the average particle RMSE over all time points is displayed as a heat maps on a representative mesh, illustrating regional accuracy. Note the heat maps are scaled differently to make local changes visible.

Figure 4 displays the distribution of full sequence reconstruction error across time points. We can see that LDS fit to the proposed PDM more accurately reconstructs held out sequences. The partial sequence generalization results in Fig. 5 similarly demonstrates that LDS fit to our PDM more accurately infers missing time points within a sequence and notably performs better with a larger percent missing. Finally, the specificity results, shown in Fig. 6 demonstrate that our approach leads to a more specific LDS model.

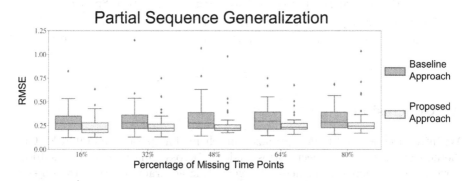

Fig. 5. Partial Sequence Reconstruction The box plot shows the distribution of error with various percentages of missing time points.

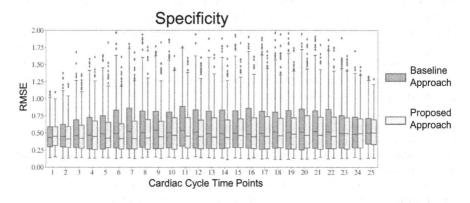

Fig. 6. Specificity The box plot shows the distribution of particle-wise RMSE between sampled and closest true particle sets at each time point.

Table 1 provides the overall RMSE with standard deviation for these metrics generated by 5-fold cross-validation experiments on the LA data. Lower reconstruction error on held out sequences (full or partial) indicates the LDS model has better generalization, suggesting the underlying shape dynamics are better represented. Regarding specificity, a lower RMSE indicates that the generated sequences sampled from LDS are closer to the training sequences, suggesting the time-dependency is accurately described by the PDM. Thus these metrics support the conclusion that our PSM technique is capable of accurately capturing the temporal projection of shapes.

Table 1. Overall RMSE for LDS metrics.

Approach	Full-seq. generalization	Partial-seq. generalization	Specificity
Baseline	0.345 ± 0.065	0.329 ± 0.233	0.576 ± 0.357
Proposed	0.243 ± 0.035	0.270 ± 0.204	0.519 ± 0.296

4   Conclusion

We introduced a novel PDM optimization scheme for spatiotemporal SSM. Our method ensures intra-subject correspondence across time points, inter-subject correspondence across sequences, and geometric accuracy across both subjects and time points. We demonstrated the efficacy of our method in capturing population variability in the left atrium throughout the cardiac cycle. Our method outperformed a comparison image-based approach for generating a spatiotemporal PDM in terms of capturing shape variation and the underlying time-dependency, as demonstrated via LDS generalization and specificity. In future analysis, spatiotemporal SSM could be used to analyze group differences in the cardiac cycle of patients before and after the ablation procedure. Another consideration is that while our optimization objective disentangles subject and time correspondence, it does not fully capture their interaction as a time-sequence generative modeling approach would. In future work, such a generative model could be integrated into the optimization scheme so that the modeled distribution informs particle position updates.

Acknowledgements. This work was supported by the National Institutes of Health under grant numbers NIBIB-U24EB029011, NIAMS-R01AR076120, NHLBI-R01HL135568, and NIBIB-R01EB016701. The content is solely the responsibility of the authors and does not necessarily represent the official views of the National Institutes of Health. The authors would like to thank the University of Utah Division of Cardiovascular Medicine for providing left atrium MRI scans and segmentations from the Atrial Fibrillation projects.

Appendix A   Linear Dynamical System

A.1   Time-Variant Model

A linear dynamical system (LDS) is a generative model capable of modeling time-series observations (Fig. 1). LDS is state space model that differs from a Hidden

Markov Model (HMM) in two ways: (1) it employs linear Gaussian dynamics and measurements and (2) it uses a continuous latent/state variable instead of discrete. LDS makes use of a latent representation ($\mathbf{S}$) and is defined by Eqs. 6–8.

$$\text{State Equation:} \quad \mathbf{s}_{n,t} = \mathbf{A}_t \mathbf{s}_{n,t-1} + \boldsymbol{\epsilon}^s_{n,t} \quad \text{where } \boldsymbol{\epsilon}^s_{n,t} \in \mathcal{N}(0, \boldsymbol{\Sigma}^s) \quad (6)$$

$$\text{Observation Equation:} \quad \mathbf{x}_{n,t} = \mathbf{W}_t \mathbf{s}_{n,t} + \boldsymbol{\epsilon}^x_{n,t} \quad \text{where } \boldsymbol{\epsilon}^x_{n,t} \in \mathcal{N}(0, \boldsymbol{\Sigma}^x) \quad (7)$$

$$\text{Prior Equation:} \quad \mathbf{s}_{n,1} = \boldsymbol{\mu}_0 + \boldsymbol{\varphi}_{n,0} \quad \text{where } \boldsymbol{\varphi}_{n,0} \in \mathcal{N}(0, \mathbf{V}_0) \quad (8)$$

Here $\mathbf{s}_{n,t} \in \mathbb{R}^L$ is the state (latent) vector, $\mathbf{A}_t \in \mathbb{R}^{L \times L}$ is the transition matrix, $\boldsymbol{\Sigma}^s \in \mathbb{R}^{L \times L}$ is the state covariance matrix, $\mathbf{W}_t \in \mathbb{R}^{dM \times L}$ is the loading/observation-system matrix, $\boldsymbol{\Sigma}^x \in \mathbb{R}^{dM \times dM}$ is the observation covariance matrix, $\boldsymbol{\mu}_0 \in \mathbb{R}^L$ is the latent prior mean, and $\mathbf{V}_0 \in \mathbb{R}^{L \times L}$ is the latent prior covariance matrix. Thus the time-variant LDS parameters are: $\boldsymbol{\theta} = \{\{\mathbf{A}_t, \mathbf{W}_t\}_{t=1}^T, \boldsymbol{\Sigma}^s, \boldsymbol{\Sigma}^x, \boldsymbol{\mu}_0, \mathbf{V}_0\}$.

In our case, $\mathbf{s}_{n,t}$ is the L dimensional latent representation of the particle set for the nth subject at time t ($\mathbf{s}_{n,t} \in \mathbb{R}^L$).

A.2   EM Algorithm

The latent states and parameters of LDS can be fit using the EM algorithm. The formulation for time-variant LDS with multiple observations is outlined below.

E-Step. In the E-step we run the inference algorithm to determine the posterior distribution of the latent variables $p(\mathbf{S}|\mathbf{X}, \boldsymbol{\theta}^{old})$. This step is composed of two parts: Kalman filtering and RTS smoothing.

1. Kalman Filtering
   Prediction Step: We get the predicted distribution (Eq. 9) by solving Eqs. 10 and 11.

$$P(\mathbf{s}_{n,t}|\mathbf{x}_{n,1:t-1}) \mathcal{N}(\mathbf{s}_{n,t}|\boldsymbol{\mu}^{1:t-1}_{n,t}, \mathbf{V}^{1:t-1}_t) \quad (9)$$

$$\boldsymbol{\mu}^{1:t-1}_{n,t} = \mathbf{A}_t \boldsymbol{\mu}^{1:t-1}_{n,t-1} \quad (10)$$

$$\mathbf{V}^{1:t-1}_t = \mathbf{A}_t \mathbf{V}^{1:t-1}_{t-1}(\mathbf{A}_t)^\top + \boldsymbol{\Sigma}^s \quad (11)$$

Measurement Step: We get the filtered distribution (Eq. 12) by solving 13 and 14.

$$p(\mathbf{s}_{n,t}|\mathbf{x}_{n,1:t}) = \mathcal{N}(\mathbf{s}_{n,t}|\boldsymbol{\mu}^{1:t}_{n,t}, \mathbf{V}^{1:t}_t) \quad (12)$$

$$\boldsymbol{\mu}^{1:t}_{n,t} = \boldsymbol{\mu}^{1:t-1}_{n,t} + \mathbf{K}_t(\mathbf{r}_{n,t}) \quad (13)$$

$$\mathbf{V}^{1:t}_t = \mathbf{V}^{1:t-1}_t - \mathbf{K}_t \mathbf{W}_t \mathbf{V}^{1:t-1}_t \quad (14)$$

Where $\mathbf{r}_{n,t} = \mathbf{x}_{n,t} - \hat{\mathbf{x}}_{n,t}$ and $\hat{\mathbf{x}}_{n,t} = \mathbf{W}_t \boldsymbol{\mu}^{1:t-1}_{n,t}$ and the Kalman Gain matrix is defined as $\mathbf{K}_t = \mathbf{V}^{1:t-1}_t(\mathbf{W}_t)^\top(\mathbf{P}_t)^{-1}$ where $\mathbf{P}_t = \mathbf{W}_t \mathbf{V}^{1:t-1}_t(\mathbf{W}_t)^\top + \boldsymbol{\Sigma}^x$. To remove the risk of computational round-off error causing negative diagonal values, we use an equivalent form of Eq. 14, called the Joseph form:

$$\mathbf{V}^{1:t}_t = [\mathbf{I} - \mathbf{K}_t \mathbf{W}_t]\mathbf{V}^{1:t-1}_t[\mathbf{I} - \mathbf{K}_t \mathbf{W}_t]^\top + \mathbf{K}_t \boldsymbol{\Sigma}^x \mathbf{K}^\top_t \quad (15)$$

2. RTS Smoothing

We get the smoothed posterior distribution (Eq. 16) by solving Eqs. 17 and 18.

$$p(\mathbf{s}_{n,t}|\mathbf{x}_{n,1:T}) = \mathcal{N}(\mathbf{s}_{n,t}|\boldsymbol{\mu}_{n,t}^{1:T}, \mathbf{V}_t^{1:T}) \tag{16}$$

$$\boldsymbol{\mu}_{n,t}^{1:T} = \boldsymbol{\mu}_{n,t}^{1:t} + \mathbf{J}_t(\boldsymbol{\mu}_{n,t+1}^{1:T} - \boldsymbol{\mu}_{n,t+1}^{1:t}) \tag{17}$$

$$\mathbf{V}_t^{1:T} = \mathbf{V}_t^{1:t} + \mathbf{J}_t(\mathbf{V}_{t+1}^{1:T} - \mathbf{V}_{t+1}^{1:t})(\mathbf{J}_t)^\top \tag{18}$$

Where the backward Kalman gain matrix is defined as:

$$\mathbf{J}_t = \mathbf{V}_t^{1:t}(\mathbf{A}_{t+1})^\top(\mathbf{V}_{t+1}^{1:t})^{-1}. \tag{19}$$

M-Step. Now that we can estimate $p(\mathbf{S}|\mathbf{X}, \boldsymbol{\theta}^{old})$, we optimize $\boldsymbol{\theta}^{new}$ by maximizing $\mathcal{Q}$:

$$Q(\boldsymbol{\theta}^{new}, \boldsymbol{\theta}) = \mathbb{E}_{\mathcal{Z}|\mathcal{X},\theta}[\ln p(\mathcal{X}, \mathcal{Z}|\boldsymbol{\theta})] = \frac{1}{N}\sum_{n=1}^{N}\mathbb{E}_{\mathbf{S}_n|\mathbf{X}_n,\theta}[\ln p(\mathbf{X}_n, \mathbf{S}_n|\boldsymbol{\theta})] \tag{20}$$

Where the log likelihood function for a single sample n is:

$$\ln p(\mathbf{X}_n, \mathbf{S}_n|\theta) = \ln p(\mathbf{s}_{n,1}|\boldsymbol{\mu}_0, \mathbf{V}_0) + \sum_{t=2}^{T}\ln p(\mathbf{s}_{n,t}|\mathbf{s}_{n,t-1}, \mathbf{A}_t, \boldsymbol{\Sigma}^s) \tag{21}$$

$$+ \sum_{t=1}^{T}\ln p(\mathbf{x}_{n,t}|\mathbf{s}_{n,t}, \mathbf{W}_t, \boldsymbol{\Sigma}^x)$$

References

1. Aramis Lab Brain and Spine Institute: Deformetrica. www.deformetrica.org
2. Atkins, P.R., et al.: Quantitative comparison of cortical bone thickness using correspondence-based shape modeling in patients with cam femoroacetabular impingement. J. Orthopaedic Res. **35**(8), 1743–1753 (2017)
3. Bhalodia, R., Dvoracek, L.A., Ayyash, A.M., Kavan, L., Whitaker, R., Goldstein, J.A.: Quantifying the severity of metopic craniosynostosis: a pilot study application of machine learning in craniofacial surgery. J. Craniofacial Surg. **31**, 697 (2020)
4. Cates, J., Fletcher, P.T., Styner, M., Shenton, M., Whitaker, R.: Shape modeling and analysis with entropy-based particle systems. In: Karssemeijer, N., Lelieveldt, B. (eds.) IPMI 2007. LNCS, vol. 4584, pp. 333–345. Springer, Heidelberg (2007). https://doi.org/10.1007/978-3-540-73273-0_28
5. Cates, J., Elhabian, S., Whitaker, R.: Shapeworks: particle-based shape correspondence and visualization software. In: Statistical Shape and Deformation Analysis, pp. 257–298. Elsevier (2017)
6. Cootes, T.F., Twining, C.J., Taylor, C.J.: Diffeomorphic statistical shape models. In: BMVC, pp. 1–10. Citeseer (2004)
7. De Craene, M., Camara, O., Bijnens, B.H., Frangi, A.F.: Large diffeomorphic FFD registration for motion and strain quantification from 3D-US sequences. In: Ayache, N., Delingette, H., Sermesant, M. (eds.) FIMH 2009. LNCS, vol. 5528, pp. 437–446. Springer, Heidelberg (2009). https://doi.org/10.1007/978-3-642-01932-6_47

8. Datar, M., Cates, J., Fletcher, P.T., Gouttard, S., Gerig, G., Whitaker, R.: Particle based shape regression of open surfaces with applications to developmental neuroimaging. In: Yang, G.-Z., Hawkes, D., Rueckert, D., Noble, A., Taylor, C. (eds.) MICCAI 2009. LNCS, vol. 5762, pp. 167–174. Springer, Heidelberg (2009). https://doi.org/10.1007/978-3-642-04271-3_21

9. Datar, M., et al.: Mixed-effects shape models for estimating longitudinal changes in anatomy. In: Durrleman, S., Fletcher, T., Gerig, G., Niethammer, M. (eds.) STIA 2012. LNCS, vol. 7570, pp. 76–87. Springer, Heidelberg (2012). https://doi.org/10.1007/978-3-642-33555-6_7

10. Davies, R.H., Twining, C.J., Cootes, T.F., Waterton, J.C., Taylor, C.J.: A minimum description length approach to statistical shape modeling. IEEE Trans. Med. Imaging **21**(5), 525–537 (2002)

11. Davis, B.C., Fletcher, P.T., Bullitt, E., Joshi, S.: Population shape regression from random design data. Int. J. Comput. Vision **90**(2), 255–266 (2010)

12. Durrleman, S., Pennec, X., Trouvé, A., Braga, J., Gerig, G., Ayache, N.: Toward a comprehensive framework for the spatiotemporal statistical analysis of longitudinal shape data. Int. J. Comput. Vision **103**, 22–59 (2012)

13. Fishbaugh, J., Durrleman, S., Gerig, G.: Estimation of smooth growth trajectories with controlled acceleration from time series shape data. In: Fichtinger, G., Martel, A., Peters, T. (eds.) MICCAI 2011. LNCS, vol. 6892, pp. 401–408. Springer, Heidelberg (2011). https://doi.org/10.1007/978-3-642-23629-7_49

14. Fishbaugh, J., Prastawa, M., Durrleman, S., Piven, J., Gerig, G.: Analysis of longitudinal shape variability via subject specific growth modeling. In: Ayache, N., Delingette, H., Golland, P., Mori, K. (eds.) MICCAI 2012. LNCS, vol. 7510, pp. 731–738. Springer, Heidelberg (2012). https://doi.org/10.1007/978-3-642-33415-3_90

15. Fitzmaurice, G.M., Ravichandran, C.: A primer in longitudinal data analysis. Circulation **118**(19), 2005–2010 (2008)

16. Gaffney, B.M., Hillen, T.J., Nepple, J.J., Clohisy, J.C., Harris, M.D.: Statistical shape modeling of femur shape variability in female patients with hip dysplasia. J. Orthopaedic Res.h® **37**(3), 665–673 (2019)

17. Gerig, G., Fishbaugh, J., Sadeghi, N.: Longitudinal modeling of appearance and shape and its potential for clinical use (2016)

18. Grenander, U., Srivastava, A., Saini, S.: A pattern-theoretic characterization of biological growth. IEEE Trans. Med. Imaging **26**(5), 648–659 (2007)

19. Harris, M.D., Datar, M., Whitaker, R.T., Jurrus, E.R., Peters, C.L., Anderson, A.E.: Statistical shape modeling of cam femoroacetabular impingement. J. Orthopaedic Res. **31**(10), 1620–1626 (2013). https://doi.org/10.1002/jor.22389

20. Hart, G., Shi, Y., Zhu, H., Sanchez, M., Styner, M., Niethammer, M.: Dti longitudinal atlas construction as an average of growth models. Miccai Stia (2010)

21. Khan, A.R., Beg, M.F.: Representation of time-varying shapes in the large deformation diffeomorphic framework. In: 2008 5th IEEE International Symposium on Biomedical Imaging: From Nano to Macro, pp. 1521–1524. IEEE (2008)

22. Kim, C.J.: Dynamic linear models with markov-switching. J. Econometrics **60**(1), 1 – 22 (1994). https://doi.org/10.1016/0304-4076(94)90036-1, http://www.sciencedirect.com/science/article/pii/0304407694900361

23. Mansi, T., et al.: A statistical model of right ventricle in tetralogy of fallot for prediction of remodelling and therapy planning. In: Yang, G.-Z., Hawkes, D., Rueckert, D., Noble, A., Taylor, C. (eds.) MICCAI 2009. LNCS, vol. 5761, pp. 214–221. Springer, Heidelberg (2009). https://doi.org/10.1007/978-3-642-04268-3_27

24. Miller, M.I., Younes, L., Trouvé, A.: Diffeomorphometry and geodesic positioning systems for human anatomy. Technology **2**(01), 36–43 (2014)
25. Morris, A., Kholmovski, E., Marrouche, N., Cates, J., Elhabian, S.: An image-based approach for 3D left atrium functional measurements. In: 2020 Computing in Cardiology, pp. 1–4. IEEE (2020)
26. Neuro Image Research and Analysis Laboratories at University of North Carolina at Chapel Hill: Spharm-pdm. https://www.nitrc.org/projects/spharm-pdm/
27. Sarkalkan, N., Weinans, H., Zadpoor, A.A.: Statistical shape and appearance models of bones. Bone **60**, 129–140 (2014)
28. Scientific Computing and Imaging Institute at the University of Utah: Shapeworks. https://www.sci.utah.edu/software/shapeworks.html
29. Shumway, R.H., Stoffer, D.S.: An approach to time series smoothing and forecasting using the em algorithm. J. Time Series Anal. **3**(4), 253–264 (1982). https://doi.org/10.1111/j.1467-9892.1982.tb00349.x, https://onlinelibrary.wiley.com/doi/abs/10.1111/j.1467-9892.1982.tb00349.x
30. Shumway, R.H., Stoffer, D.S.: Dynamic linear models with switching. J. Am. Stat. Assoc. **86**(415), 763–769 (1991). http://www.jstor.org/stable/2290410
31. Styner, M., et al.: Framework for the statistical shape analysis of brain structures using spharm-pdm (2006)
32. Thompson, D.: On Growth and Form. Cambridge University Press, Cambridge (1917)
33. Trouvé, A., Vialard, F.X.: Shape splines and stochastic shape evolutions: a second order point of view. Quart. Appl. Math., 219–251 (2012)
34. Zachow, S.: Computational planning in facial surgery. Facial Plast. Surg. **31**(05), 446–462 (2015)

Interpretable Prediction of Post-Infarct Ventricular Arrhythmia Using Graph Convolutional Network

Buntheng Ly[1]([✉]), Sonny Finsterbach[2], Marta Nuñez-Garcia[3], Pierre Jais[3], Damien Garreau[4], Hubert Cochet[3], and Maxime Sermesant[1]

[1] Inria, Université Côte d'Azur, Epione Team, Sophia Antipolis, France
{buntheng.ly,maxime.sermesant}@inria.fr
[2] CHU Bordeaux, Université de Bordeaux, Bordeaux, France
[3] IHU Liryc, Université de Bordeaux, Bordeaux, France
[4] Université Côte d'Azur, Nice, France

Abstract. Heterogeneity of left ventricular (LV) myocardium infarction scar plays an important role as anatomical substrate in ventricular arrhythmia (VA) mechanism. LV myocardium thinning, as observed on cardiac computed tomography (CT), has been shown to correlate with LV myocardial scar and with abnormal electrical activity. In this project, we propose an automatic pipeline for VA prediction, based on CT images, using a Graph Convolutional Network (GCN). The pipeline includes the segmentation of LV masks from the input CT image, the short-axis orientation reformatting, LV myocardium thickness computation and mid-wall surface mesh generation. An average LV mesh was computed and fitted to every patient in order to use the same number of vertices with point-to-point correspondence. The GCN model was trained using the thickness value as the node feature and the atlas edges as the adjacency matrix. This allows the model to process the data on the 3D patient anatomy and bypass the "grid" structure limitation of the traditional convolutional neural network. The model was trained and evaluated on a dataset of 600 patients (27% VA), using 451 (3/4) and 149 (1/4) patients as training and testing data, respectively. The evaluation results showed that the graph model (81% accuracy) outperformed the clinical baseline (67%), the left ventricular ejection fraction, and the scar size (73%). We further studied the interpretability of the trained model using LIME and integrated gradients and found promising results on the personalised discovering of the specific regions within the infarct area related to the arrhythmogenesis.

Keywords: Graph neural network · Ventricular arrhythmia · Interpretable AI · Cardiac CT

1 Introduction

VA is the an abnormal heart rhythm most observed leading to the sudden cardiac death (SCD), ranking among the highest causes of mortality in the

O. Camara et al. (Eds.): STACOM 2022, LNCS 13593, pp. 157–167, 2022.
https://doi.org/10.1007/978-3-031-23443-9_15

developed countries [6]. The current gold standard predictor of SCD is still the left ventricular ejection fraction (LVEF), despite a majority of SCDs occurred in patients with preserved LVEF (>35%). The post-infarction scar on the LV myocardium (LVMYO) is recognised as the anatomical substrate of the VA mechanism. Nonetheless, identifying the precise arrhythmogenesis characteristics of the scar regions is still challenging, thus limiting its application in the clinical practice. Recent study has shown that the LVMYO scar can be located through the analysis of the LV wall thinning in CT imaging [5], as the alternative to the late gadolinium enhancement (LGE) MRI. CT imaging is generally more accessible in terms of machine availability and patient compatibility (i.e. patients with metallic device) compared to MRI. For image processing, CT imaging also have higher image resolution and quantitative image intensity consistency between manufacturers, thus reducing the potential error in the automatic processing pipeline.

In recent years, Deep Learning (DL), notably the convolution neural network (CNN), has made a remarkable impact in medical image processing, ranging from semantic segmentation to diagnosis and outcome prediction. Although, the grid-like property of the CNN can limit its efficacy on the organs with specific 3D anatomy, such as the LVMYO. Our previous work proposed transforming the CT input image into the 2D bullseye representation of the LV thickness to remove the blank voxels of the endocardium [4]. However, the flattening of the 3D led to the distortion of the 3D heterogeneity of the thickness map, impacting the subsequent fitting of the DL model. On the other hand, graph neural networks have gained more applications in medical image processing and bioinformatics [16], thanks to its adaptability to the specific data geometry.

In this study, we investigated the graph-level classification task using a model built with graph convolutional network (GCN) layers [3], and the LV thickness map as input. Moreover, we studied the interpretability of the graph model using LIME [8] and integrated gradients [11] to locate the specific regions contributed to the VA prediction and to gain better insight of the explicit arrhythmogenic regions from the model perspective.

2   Method

Our image processing pipeline formulates the 3D CT scan into a graph input. The processing steps were image segmentation, short-axis (SAX) reorientation, LVMYO thickness calculation, LV mid-wall meshing and average mesh fitting. From there, a graph classification network was built using the atlas mesh and the thickness value as input. The image processing pipeline and the model architecture are described in Sect. 2.1 and Sect. 2.2.

2.1   Image Processing

The 3D ventricular masks were segmented from the input CT image using a Dual-UNet segmentation model, composing of a "coarse" UNet [9] for ROI segmentation and a "refine" UNet for the cropped input segmentation. The required masks included the epicardium and endocardium of the LV and the epicardium of the RV. Using the LV and RV mask as landmarks, automatic SAX view reorientation was done using the method described in [7]. The SAX ventricular masks were then resampled isotropically at 0.5 mm. The LV wall thickness and the mid-wall points were calculated using the Eulerian partial differential method as proposed by [15]. The thickness value was normalised by clipping at 10 mm then divided by 10. Then, the surface mid-wall mesh was generated from the mid-wall points using marching cube algorithm and uniformly remeshed using appoximated centroid voronoi diagrams (ACVD) method[13]. Finally, an average mid-wall mesh was generated and fitted to every patient based on the large deformation diffeomorphic metric mapping (LDDMM) framework. The atlas fitting was done using the deterministic atlas function provided by the Deformetrica software [1]. Rigid registration of the meshes was done prior to atlas fitting. As the mesh rotation and size were already registered in the SAX reorientation step, we translated the centre of mass of all the meshes toward a reference from a randomly selected mesh. At inference, the atlas mesh could be directly registered to the new mesh using affine registration. The image process steps are shown in Fig. 1.I.

2.2   Graph Convolutional Network Model

We built the model using the GCN layer [3], which used the following convolutional operation:

$$X' = \hat{D}^{-\frac{1}{2}} \hat{A} \hat{D}^{\frac{1}{2}} X W + b, \tag{1}$$

where $X' \in \mathbb{R}^{N \times F}$ is the output of N nodes and F filters, $X \in \mathbb{R}^{N \times C}$ is the input of C features, $W \in \mathbb{R}^{C \times F}$ is the weights, b is the bias, $\hat{D} \in \mathbb{R}^{N \times N}$ the degree matrix, and $\hat{A} \in \mathbb{R}^{N \times N}$ is the self-loop adjacency matrix ($\hat{A} = I_N + A$).

We considered the LV mid-wall mesh as an indirect graph $\mathcal{G} = (\mathcal{V}, \mathcal{E})$ with N nodes $v_i \in \mathcal{V}$ and edges $(v_i, v_j) \in \mathcal{E}$. We used the thickness value of each node v_i as the input feature $X \in \mathbb{R}^{N \times 1}$. Since the input were uniformly remeshed, no edge feature was use. The adjacency matrix A was extracted from the atlas mesh edges and used for every inputs. The LV mesh could then be represented as a graph $\mathcal{G} = (X, A)$.

Index Pooling. We employed the pooling method as proposed by [10], where a coarse graph was generated to pre-define the corresponding pooling index and the new adjacency matrix during the model construction. In our case, the coarse graph was generated using the ACVD uniform meshing, with reduced number of points, and the corresponding pooling indexes were defined using k-dimensional tree nearest neighbour search (KD-tree NN). In this project, we built the index pooling layer using max pooling method. The graph coarsening and pooling indexes searching are computationally expensive, which is unfit for online processing. This pooling method is more suitable with a point-to-point corespondent graph input, as the pooling indexes of the atlas mesh can be applied to all the input graphs. The index pooling steps are shown in Fig. 1.III.

Model Architecture. The model architecture is shown in Fig. 1.II. Starting with an input graph of N nodes, the model was built using two consecutive GCN layers, followed by an index pooling layer, reducing the aggregated graph to N_{dwn} nodes. The pooling output was then passed through three more GCN layers, before feeding to the fully-connected network (FCN) classification block. We set the filter size to 64 for all the GCNs, except the last GCN, where the filter was set to 1. The classification block was built with 4 fully-connected layers with [128, 64, 32, 2] units, respectively. We applied linear rectifier activation the first 3 layers and the softmax activation for the output layer. The model was trained using the binary cross-entropy loss. The model was built using tools provided by Spektral[1] (GCN layer) and Tensorflow[2].

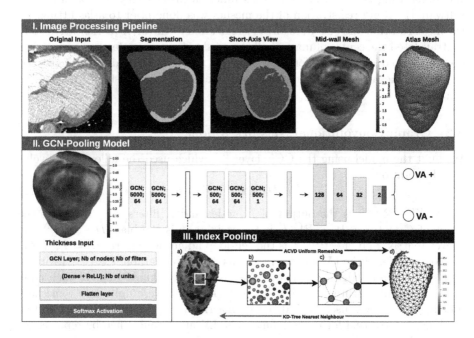

Fig. 1. Image processing pipeline, GCN-Pooling architecture and Index Pooling layer. The thickness input mesh was the result after fitting the average mesh to the mid-wall mesh, ensuring the point-to-point correspondence of every input. *Index Pooling:* a). pooling patches (from KD-tree NN); b). original (smaller spheres) and pooling nodes (bigger spheres); c & d). coarse graph.

---

[1] https://github.com/danielegrattarola/spektral.
[2] https://www.tensorflow.org/.

2.3   Interpretability Study

We studied the interpretability of the graph neural network by adapting to our context two existing methods: Local Interpretable Model-agnostic Explanation (LIME, [8]), and integrated gradients [11].

LIME. To summarize briefly, in the case of image data LIME computes a local surrogate linear model based on the absence or presence of superpixels. Our use-case is similar if we consider the input graph $\mathcal{G} = (X, A)$ as an image with pixel values defined as the thickness. Following [2], our method can be described as:

1. segment the mesh into d patches;
2. randomly *turn on/off* the patches to create new inputs $\mathcal{G}_1 = (X_1, A), \ldots,$ $\mathcal{G}_n = (X_n, A)$;
3. run the model prediction on all the new inputs $y_i = f(\mathcal{G}_i)$;
4. compute $\hat{\beta}_n$ by fitting the y_is to a local weighted surrogate model.

We used two segmentation methods in step 1. The first is *quick shift* [14], as is default for the image version of LIME, which bases the segmentation on both the nodes 3D coordinates and the thickness value[3]. We scaled the magnitude of the node coordinates and thickness value using scikit-learn standard feature scaler[4]. The quickshift algorithm was then applied with the distance threshold of 0.2. The second is *grid segmentation* with KD-tree NN, which is based only on the nodes 3D coordinates. The mesh was first downsampled to 30 mm radius using voxel downsampling, before running the KD-Tree search, which generated a quasi uniform grid segmentation of the input mesh. Turning the patches i *on* meant replacing the original value of patch with the replacement value, and vice versa. To study the case of VA+ classification, the replacement value was set to 1 to represent the normalised healthy LVMYO, under the hypothesis that the completely healthy LV would be classified as VA-.

Integrated Gradients. Integrated gradients is computed as the integral of back-propagated gradients of the straightline path from the baseline input x' to the original input x. To adapt the method to the graph network, the integrated gradients along the dimension i^{th} for a model $f(\mathcal{G})$ with the input graph $\mathcal{G} = (X, A)$ and baseline graph $\mathcal{G}' = (X', A)$ was calculated as:

$$\text{IG}_i(\mathcal{G} = (X, A)) := (X_i - X'_i) \times \int_{\alpha=0}^{1} \frac{\partial f(X' + \alpha(X - X'), A)}{\partial X_i} d\alpha, \quad (2)$$

where $\alpha \in [0, 1]$ denotes the step coefficients between the baseline input ($\alpha = 0$) to the input ($\alpha = 1$). We set the number of step to 50. We set the baseline feature X' the same way as the replacement input in LIME by changing the thickness feature value to 1.

---

[3] https://github.com/Nick-Ol/MedoidShift-and-QuickShift.
[4] https://scikit-learn.org/.

3   Experimental Setup

We studied our method using a retrospective dataset of 600 patients, collected between 2010 and 2020. The dataset included the CT images captured at diastolic of patients with history myocardial infarction (MI) at least 1 month before the scan date and without any history of LV surgery. The dataset included 27.5% of patients labelled as VA+. The VA inclusion criteria were any episode of ventricular tachycardia (VT) or fibrillation (VF), and aborted cardiac arrest.

The total population was randomly divided at 1 : 4 ratio, resulting in 451 and 149 patients for training and testing, respectively. Two-sample t-test was used on all the available clinical characteristics to ensure that there was no significant difference between the training and testing population. The available characteristics in the dataset were age, gender, hypertension, diabetes, dyslipidaemia, smoking, LVEF, scar size and scar age. The scar size was calculated as the area on the LV surface with $< 5cm$ thickness, and the scar age was calculated as the delay between the MI and the scan date.

3.1   Baseline Models

DL Model. We studied the performance of the GCN-Pooling model against the baseline DL models using 10-fold cross-validation on the training dataset.

Using the 2D thickness bullseye prediction pipeline proposed by [4], we tested two models: the conditional variational autoencoder classification (CVAE-Class) model and the direct CNN classification model (using the same layers as the CVAE-Class without the decoder). To test the performance of the traditional CNNs, we also run the cross-validation test with the 3D LV wall mask as input, using the 3D variant of the VAE-Class (without the condition) and direct CNN model. We used the SAX oriented LV wall masks to ensured the rotation and the voxel spacing consistency between inputs.

For the graph input, we generated two sets of LV mid-wall meshes at 500 and 5000 nodes using the pipeline described in Sect. 2.1. With the atlas-fitted inputs, we run cross-validation on the GCN-Pooling, GCN (without index pooling) and FCN model. The baseline GCN model was built with the same layers as the GCN-Pooling, and the FCN model was built the same layers as the classifier block (4-consecutive fully-connected layers). As ablation study, we also run cross-validation on the GCN and FCN model using the rigid registered remesh inputs without the atlas fitting.

Clinical Baseline. Finally, we compared the best DL model against the clinical characteristics. With training dataset, we trained the best DL model and run univariate analysis to find the significant variables associated with VA. Then, the optimal cut-off values were calculated with the receiver operator curve (ROC) analysis [12] using the following equation:

$$\text{Cutoff}_{\text{Optimal}} = \min(|\text{AUC} - \text{Spe}_c| + |\text{AUC} - \text{Sen}_c|). \tag{3}$$

We then compared the generalisability of the DL model and the cutoff values on the testing dataset.

4   Results

The Table 1 shows the cross-validation results of the DL models. The GCN-Pooling model outperformed the baselines models with the cross-validate accuracy of 0.818 ± 0.06. It outperformed the 2D and 3D input models (both CVAE-Class and CNN-Class), thus proving that the LV wall thickness could be formulated as graph to accurately predict the presence/absence of VA. The 3D models also underperformed compared to the 2D model, which further demonstrated the limit of the 3D CNNs in the classification task of the 3D anatomy such as the LV wall.

For the FCN and GCN models, the point-to-point correspondence was crucial for the models optimisation, as proven by the plummet in the accuracy of the models trained with non atlas fitting inputs. Without the index pooling, the 500-node models (FCN and GCN) outperformed the 5000-node models, however their accuracy could only reach 0.78. The prediction accuracy was increased for the 500-node models with additional node coordinates as the input feature. By adding the index pooling layer, the GCN-Pooling was able to properly learn the more complex heterogeneity of the 5000-node inputs and achieved higher prediction accuracy.

The univariate analysis on the training population clinical characteristics resulted in 4 significant variables ($p < 0.01$): LVEF, gender, scar size and scar age. The optimal cutoff values calculated from the dataset were 43%, $65cm^2$ and $132months$ for the LVEF, scar size and scar age, respectively.

The evaluation results on the testing dataset are shown in Table 2. The 5000-node GCN-Pooling model achieved the highest prediction accuracy at 0.812, followed by the scar-based markers at 0.725 (scar size) and 0.689 (scar age). The LVEF had the lowest performance at 0.671 accuracy.

4.1   Model Interpretability

The Fig. 2 shows the output coefficients of LIME and integrated gradients on a true positive prediction from the testing population. We could note the consistency of the high coefficient (blue) regions between the two methods. The high coefficient regions were also faithful among LIME outputs when the different segmentation methods were used.

We observed that the regions with prominent coefficient correlated strongly with the thinning regions of the LVMYO. This validates the hypothesis that the post-infarction LV thinning is associated with the presence of VA. On top of that, a closer inspection of the top coefficient patches showed that the GCN-Pooling model did not base its VA+ prediction on the entire thinning regions. Rather, higher coefficients were focused on smaller distinctive regions within or adjacent to the scar regions, as indicates by the yellow ellipsoids in the Fig. 2.

Table 1. 10-fold cross-validation of the DL baseline models. The results were diplayed as *mean* (±*std*). (*) Non average fitted input, (**) Non average fitted input with position and thickness features. **Bold row**: best accuracy model.

Model	Input Shape	Accuracy	Sensitivity	Specificity
CNN-class	256×256	0.775 (±0.05)	0.616 (±0.10)	0.834 (±0.06)
	$240 \times 240 \times 288$	0.759 (±0.06)	0.508 (±0.19)	0.853 (±0.05)
(C)VAE-Class	256×256	0.788 (±0.06)	0.691 (±0.18)	0.825 (±0.05)
	$240 \times 240 \times 288$	0.775 (±0.06)	0.457 (±0.18)	0.875 (±0.03)
FCN	$N = 500*$	0.604 (±0.04)	0.525 (±0.15)	0.643 (±0.09)
	$N = 500**$	0.700 (±0.06)	0.266 (±0.17)	0.862 (±0.05)
	$N = 500$	0.770 (±0.06)	0.716 (±0.18)	0.790 (±0.07)
	$N = 5000*$	0.706 (±0.07)	0.250 (±0.11)	0.878 (±0.09)
	$N = 5000**$	0.743 (±0.03)	0.120 (±0.09)	0.975 (±0.03)
	$N = 5000$	0.761 (±0.12)	0.775 (±0.17)	0.756 (±0.17)
GCN	$N = 500*$	0.701 (±0.05)	0.626 (±0.09)	0.742 (±0.11)
	$N = 500**$	0.718 (±0.07)	0.766 (±0.17)	0.700 (±0.09)
	$N = 500$	0.784 (±0.08)	0.716 (±0.17)	0.809 (±0.09)
	$N = 5000*$	0.679 (±0.12)	0.633 (±0.17)	0.696 (±0.21)
	$N = 5000**$	0.677 (±0.09)	0.566 (±0.20)	0.718 (±0.16)
	$N = 5000$	0.779 (±0.08)	0.716 (±0.19)	0.803 (±0.10)
GCN-Pooling	$N = 500$	0.784 (±0.05)	0.808 (±0.07)	0.775 (±0.06)
	$N = 5000$	**0.818 (±0.06)**	**0.766 (±0.10)**	**0.837 (±0.07)**

Table 2. Evaluation results on the testing population of the GCN-Pooling model and the clinical baselines.

	Accuracy	Sensitivity	Specificity
GCN-Pooling ($N = 5000$)	**0.812**	**0.780**	**0.824**
Scar Size $> 65.49\,cm^2$	0.725	0.683	0.741
Scar Age $> 132 months$	0.698	0.839	0.640
LVEF $< 43\%$	0.671	0.707	0.657

5   Discussion

Although, the testing dataset were not used during the designing and tuning of the model, the cross-validation and evaluation results were still based on mono-centre dataset. The retrospective nature of dataset could also introduce additional biases relating to the center imaging guideline, which skewed toward a selected population. Therefore, the current study population might not be the realistic representation of the general public.

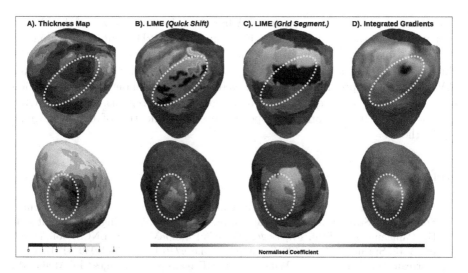

Fig. 2. Interpretability studies of the GCN-Pooling model on a true positive prediction case from testing population. The yellow ellipsoids highlight the regions with consistently high coefficient among different methods. First row: inferior LV view; second row: apex view. (Color figure online)

The interpretablity analysis only provided a perspective into the relation between the model prediction and the input. The high coefficient regions could be read as highly arrhythmogenic, simply because they were the regions allowing the model to classify the input as VA+. While the choice of the segmentation method in LIME is not trivial, we focused the analysis only on two segmentation methods, to limit the scope of this study. The quick shift method was chosen as it incorporates both physical coordinates and the thickness values of the node and was parameterised to generate small segments (by setting smaller threshold). On contrast, the grid segmentation method does not consider the thickness value and was parameterised to generate larger segments (using larger neighbour searching radius). Therefore, the consistency of the two LIME methods, as well as the integrated gradients, further solidify the tolerance of our interpretability approaches toward superpixel segmentation bias. Nevertheless, limiting by the retrospective aspect of the available data, the current analysis of the interpretability outputs were restricted to hypothesising.

A prospective and multi-centre dataset is necessary to confirm the generalisability of the prediction pipeline and the validity of the detected arrhythmogenic site.

6   Conclusion

We proposed a novel automatic pipeline for VA prediction using the cardiac CT images. Formulating the LV thickness map into a graph input allowed the model to perform an accurate prediction of VA based on the 3D anatomy of the LV and achieved a better score then the model trained on the 2D input. The evaluation on the testing dataset showed that the GCN-Pooling model outperformed the clinical markers, especially comparing to the current gold standard predictor (LVEF). The interpretabiliy studies proved that the model derived its prediction from the very specific regions within the greater scar regions, an encouraging evidence for the future work on the personalised identification of arrhythmogenic site of the LV.

References

1. Bône, A., Louis, M., Martin, B., Durrleman, S.: Deformetrica 4: An Open-Source Software for Statistical Shape Analysis. In: Reuter, M., Wachinger, C., Lombaert, H., Paniagua, B., Lüthi, M., Egger, B. (eds.) ShapeMI 2018. LNCS, vol. 11167, pp. 3–13. Springer, Cham (2018). https://doi.org/10.1007/978-3-030-04747-4_1
2. Garreau, D., Mardaoui, D.: What does LIME really see in images? In: Meila, M., Zhang, T. (eds.) Proceedings of the 38th International Conference on Machine Learning. Proceedings of Machine Learning Research, PMLR (2021)
3. Kipf, T.N., Welling, M.: Semi-supervised classification with graph convolutional networks. In: 5th International Conference on Learning Representations, ICLR 2017 - Conference Track Proceedings (2017)
4. Ly, B., Finsterbach, S., Nuñez-Garcia, M., Cochet, H., Sermesant, M.: Scar-related ventricular arrhythmia prediction from imaging using explainable deep learning. In: Ennis, D.B., Perotti, L.E., Wang, V.Y. (eds.) FIMH 2021. LNCS, vol. 12738, pp. 461–470. Springer, Cham (2021). https://doi.org/10.1007/978-3-030-78710-3_44
5. Mahida, S., et al.: Cardiac Imaging in Patients with Ventricular Tachycardia. Circulation (2017)
6. Nielsen, J.C., et al.: European heart rhythm association (ehra)/heart rhythm society (hrs)/asia pacific heart rhythm society (aphrs)/latin american heart rhythm society (lahrs) expert consensus on risk assessment in cardiac arrhythmias: use the right tool for the right outcome, in the right population. Europace (2020)
7. Nuñez-Garcia, M., Cedilnik, N., Jia, S., Sermesant, M., Cochet, H.: automatic multiplanar CT Reformatting from trans-axial into left ventricle short-axis view. In: STACOM 2020–11th International Workshop on Statistical Atlases and Computational Models of the Heart, Lima, Peru (Oct 2020)
8. Ribeiro, M.T., Singh, S., Guestrin, C.: "Why should i trust you?": explaining the predictions of any classifier. In: Proceedings of the 22nd ACM SIGKDD International Conference on Knowledge Discovery and Data Mining, KDD 2016. Association for Computing Machinery, New York (2016)
9. Ronneberger, O., Fischer, P., Brox, T.: U-Net: convolutional networks for biomedical image segmentation. In: Navab, N., Hornegger, J., Wells, W.M., Frangi, A.F. (eds.) MICCAI 2015. LNCS, vol. 9351, pp. 234–241. Springer, Cham (2015). https://doi.org/10.1007/978-3-319-24574-4_28
10. Simonovsky, M., Komodakis, N.: Dynamic edge-conditioned filters in convolutional neural networks on graphs. In: Proceedings - 30th IEEE Conference on Computer Vision and Pattern Recognition, CVPR 2017 (2017)

11. Sundararajan, M., Taly, A., Yan, Q.: Axiomatic attribution for deep networks. In: Precup, D., Teh, Y.W. (eds.) Proceedings of the 34th International Conference on Machine Learning. Proceedings of Machine Learning Research, vol. 70, pp. 3319–3328. PMLR (06–11 Aug 2017)

12. Unal, I.: Defining an optimal cut-point value in ROC analysis: An Alternative Approach. In: Computational and Mathematical Methods in Medicine (2017)

13. Valette, S., Chassery, J.M.: Approximated centroidal voronoi diagrams for uniform polygonal mesh coarsening. In: Computer Graphics Forum (2004)

14. Vedaldi, A., Soatto, S.: Quick shift and kernel methods for mode seeking. In: Forsyth, D., Torr, P., Zisserman, A. (eds.) ECCV 2008. LNCS, vol. 5305, pp. 705–718. Springer, Heidelberg (2008). https://doi.org/10.1007/978-3-540-88693-8_52

15. Yezzi, A.J., Prince, J.L.: An Eulerian PDE approach for computing tissue thickness. IEEE Trans. Med. Imaging (2003)

16. Zhang, X.M., Liang, L., Liu, L., Tang, M.J.: Graph neural networks and their current applications in bioinformatics. Front. Genet. (2021)

Unsupervised Echocardiography Registration Through Patch-Based MLPs and Transformers

Zihao Wang, Yingyu Yang$^{(\boxtimes)}$, Maxime Sermesant, and Hervé Delingette

Université Côte d'Azur, Inria Epione Team, Sophia Antipolis, France
{zihao.wang,yingyu.yang,maxime.sermesant,herve.delingette}@inria.fr

Abstract. Image registration is an essential but challenging task in medical image computing, especially for echocardiography, where the anatomical structures are relatively noisy compared to other imaging modalities. Traditional (non-learning) registration approaches rely on the iterative optimization of a similarity metric which is usually costly in time complexity. In recent years, convolutional neural network (CNN) based image registration methods have shown good effectiveness. In the meantime, recent studies show that the attention-based model (e.g., Transformer) can bring superior performance in pattern recognition tasks. In contrast, whether the superior performance of the Transformer comes from the long-winded architecture or is attributed to the use of patches for dividing the inputs is unclear yet. This work introduces three patch-based frameworks for image registration using MLPs and transformers. We provide experiments on 2D-echocardiography registration to answer the former question partially and provide a benchmark solution. Our results on a large public 2D-echocardiography dataset show that the patch-based MLP/Transformer model can be effectively used for unsupervised echocardiography registration. They demonstrate comparable and even better registration performance than a popular CNN registration model. In particular, patch-based models better preserve volume changes in terms of Jacobian determinants, thus generating robust registration fields with less unrealistic deformation. Our results demonstrate that patch-based learning methods, whether with attention or not, can perform high-performance unsupervised registration tasks with adequate time and space complexity.

Keywords: Unsupervised registration · MLP · Transformer · Echocardiography

1 Introduction

Image registration is essential for clinical usage; for example, the registration of cardiac images between end-diastole and end-systole is meaningful in myocardium deformation analysis. Non-rigid echocardiography image registration is one of the most challenging image registration tasks, as finding the defor-

---

Z. Wang and Y. Yang—These authors contributed equally to this work.

O. Camara et al. (Eds.): STACOM 2022, LNCS 13593, pp. 168–178, 2022.
https://doi.org/10.1007/978-3-031-23443-9_16

mation field between noisy images is a highly nonlinear problem in the absence of ground truth deformation. Specifically, various image registration problems require the mapping between moving and fixed images to be folding free [2,7,32]. Traditional non-learning approaches rely on the optimizing similarity metrics to measure the matching quality between image pairs [10,26,31]. With the rapid promotion of deep learning, various frameworks of convolutional neural networks (CNN) have been introduced in image registration and have shown impressive performance in many research works.

We consider a 2D non-rigid machine learning-based image registration task in this work. With two given images: $I_{fix}^N, N \in \mathbb{R}$ and $I_{mov}^N, N \in \mathbb{R}$, we want to learn a model $\mathcal{T}_\omega(I_{mov}, I_{fix}) \rightarrow \phi(\theta)$ that generates a constrained transformation $\phi(\theta)$ based on a similarity measurement $\mathcal{M}$ to warp the moving image by minimising the loss function:

$$L = \arg\min_\theta \mathcal{M}(I_{fix}, I_{mov} \circ \phi(\theta)) + \lambda \mathcal{C}(\phi(\theta)) \tag{1}$$

where the transformation ϕ is parameterised by the parameter θ and constrained by a regularisation term $C(\phi(\theta))$ to ensure ϕ to be a spatially smooth transformation. However, iterative optimization of Eq. 1 is very time-consuming, whereas a well-trained CNN does not need any iterative minimization of the loss function at test time. This advantage drives researchers' attention to learning-based registration. Learning-based registration methods can be categorised into supervised and unsupervised registration approaches.

Supervised Registration. The supervised learning registration methods [6,27,33] are primarily trained on a ground-truth training set for which simulated or registered displacement fields are available. The training dataset is usually generated with traditional registration frameworks or by generating artificial deformation fields as ground truth for warping the moving images to get the fixed images. [28,34]. One of the limitations of the supervised registration approaches is the registration quality, which is highly influenced by the nature of the training set of deformation map, although the requirement in terms of training set can be partially alleviated by using weakly-supervised learning [4,5,12,14,15].

Unsupervised Registration. In unsupervised registration [4,9,13,17,24], we rely on a similarity measure and regularisation to optimize the neural network for learning the transformations between the fixed and moving images. Usually, a CNN is used directly for warping the moving images, which is then compared to the fixed-image with the similarity loss. The displacement field can also be obtained from a generative adversarial neural network, which introduces a discriminator neural network for assessing the generated deformation field quality. [11,23,29,36].

Multi-layer Perceptron and Transformers MLP is one of the most classical neural networks and consists of a stack of linear layers along with non-linear activation [30]. For several years, CNN has been widely used due to its performance on vision tasks and its computation efficiency [19]. Recently, several alternatives to the CNN have been proposed such as Vision Transformer (ViT) [1] or MLP-Mixer [16] which demonstrated comparable or even better performance than CNN on classification or detection tasks. There are currently intense discussions in the community of whether patching, attention or simple MLP play the most important role in such good performance.

In this paper, we propose three MLP/Transformer based models for echocardiography image registration and compare them with one representative CNN model in unsupervised echocardiography image registration. There are already works using transformers to register medical images, such as TransMorph [8] and Dual Transformer [35], but they are mostly restricted to high signal-to-noise medical images, such as MRI images and CT images. While ultrasound images (2D) are actually the most popular imaging modality in the real world. Our inspiration not only comes from the trending debate over Transformer and MLP, but also stems from the intuition that patch-based learning methods share similar logic to traditional block matching method for cardiac tracking.

Our contributions are two-folded. First, we show the effectiveness of patch-based MLP/Transformer models in medical image registration compared with a CNN-based registration model. Second, we conduct a thorough ablation study of the influence of different structures (MLP, MLP-Mixer, Transformer) and different scales (single scale or multiple scales). Our results provide empirical support to the observation that the attention mechanism may not be the only key factor in the SOTA performances. [21, 25], at least in the field of unsupervised image registration.

2 Methodology

2.1 Diffeomorphic Registration

We estimate a diffeomorphic transform between images, which preserves topology and is folding-free. Our model generates stationary velocity field $v(\theta)$ [3] instead of generating displacement maps, thanks to an integration layer applied to the velocity field leading to diffeomorphism $\phi(\theta)$. Formally, the diffeomorphic transformation ϕ is solution to a differential equation related to the predicted (stationary) velocity field V [9]: $\frac{\partial \phi_t}{\partial t} = v(\phi_t); \phi_{t=0} = Id$. In a stationary velocity field setting, the transformation ϕ is defined as the exponential of the velocity field $\phi = \exp(v)$ [2]. The integration (exponential) layer applies the scaling and squaring method to approximate the diffeomorphic transform [18]. The obtained transformation ϕ is then used by a spatial transform layer to deform the image.

2.2 Proposed Frameworks

Given two images, I_{fix} and I_{mov}, we would to estimate the transformation $\phi(\theta)$ that transforms the moving image to the fixed image so that $I_{fix} \approx I_{mov} \circ \phi(\theta)$. We approximate the ideal $\phi(\theta)$ by the following proposed frameworks.

The following three propositions are all based on patch-wise manipulations and share a similar general architecture. As shown in Fig. 1, I_{fix} and I_{mov} are both processed by an identical feature extractor (green block) separately. The two feature maps are then passed through the cross feature block (blue block). After two linear layers, we obtain their corresponding velocity field. The velocity field passes through an integration layer and we obtain the final displacement field by upsampling it to the original image size.

Pure MLP Registration Framework. The same MLP block (Block I in Fig. 1) are used for feature extractor and the cross feature block in this model. The outputs from two separate feature extractor (shared weights) are added together before feed into the cross feature block. We note this model **PureMLP** for abbreviation in the following paper.

MLP-Mixer Registration Framework. The MLP-Mixer registration framework is very similar to the former Pure MLP framework. The only difference is that the three MLP blocks used for separate feature extraction and cross feature processing are replaced by MLP-Mixer blocks [16] (Block II in Fig. 1). The MLP-Mixer block has an identical structure as the MLP block, but with feature map transpose to obtain channel-wise feature fusion (the red cell of Block II in Fig. 1). We note this model **MLPMixer** for abbreviation in the following paper.

Swin-Transformer Registration Framework. This model uses the MLP block (Block I in Fig. 1) to first extract patch based features for both I_{fix} and I_{mov}. For cross feature block, we adapt the recent Swin Transformer [22] to do the cross-patch attention locally (Swin Block in Fig. 1). Our Swin block accepts feature input from both images (I_{fix} and I_{mov}), where key K and value V are normalised I_{fix} features while query Q comes from normalised I_{mov} features. Swin block calculates the cross-attention within a pre-defined window region. We perform normal window partition for I_{fix} features and one normal partition, one shifted partition for I_{mov} features. The cross-attention under the two types of window partition configurations are summed together before feeding to the final linear layers. Due to the page limit, we invite interested readers to refer to [22] for detailed description of Swin transformer mechanism. We note this model **SwinTrans** for abbreviation in the following paper.

2.3   Multi-scale Features

In order to enforce different reception fields for patch-based models, we decide to combine multi-scale models together. This is accomplished by adopting models of different patch sizes together. It is quite similar to how CNN achieves this goal, by applying larger kernel or adding pooling layers. In particular, our multi-scale model consists of several parallel independent single-scale child models. The output of each child model is upsampled and then combined together to form the final estimation of velocity field $v(\theta)$

$$v(\theta) = \sum_{C=1}^{N} \omega_C Out_C \tag{2}$$

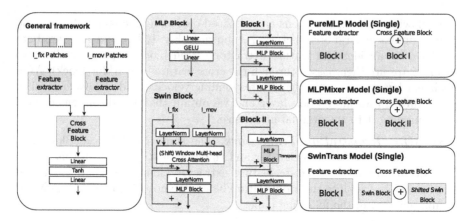

Fig. 1. The detailed composition of proposed three frameworks. Here we only show single-scale models. Please read Sect. 2.2 for more description.

where Out_C is the output of child model C. The final velocity field v is then passed to calculate the final transform ϕ as depicted in former subsections.

3   Experiments and Results

3.1   Dataset

To evaluate the effectiveness of our unsupervised registration models, we use a publicly accessible 2D echocardiography dataset CAMUS[1]. This dataset consists of 500 patients, each has 2D apical 4-chamber (A4C) and 2-chamber (A2C) view sequences. Manual annotation of cardiac structures (left endocardium, left epicardium and left atrium) were acquired by expert cardiologists for each patient in each view, at end-diastole (ED) and end-systole (ES) [20]. The structure annotations of 450 patients are public available while that of the other 50 patients unreleased. In total we have 1000 pairs of ED/ES images and we randomly split (still considering age and image quality distribution) the 900 pairs (with annotations) into training (630), validation(90) and test data (180). The 100 pairs (without annotations) are included into the training set (730).

3.2   Implementation

We compare our proposed three models with a very popular CNN registration model VoxelMorph [4]. To be consistent with our setting, we make use of the diffeomorphic version of VoxelMorph model (we use the abbreviation **Vxm** in the following paper). We train all the models with input images resized to 128×128 pixels and use an Adam optimiser (learning rate = 0.0001). We set training epoch to be 500 and training is early stopped when there is no improvement on

---

[1] https://www.creatis.insa-lyon.fr/Challenge/camus/.

validation set over 30 epochs. Our codes are available (https://gitlab.inria.fr/epione/mlp_transformer_registration).

Loss Function. In order to enforce the diffeomorphic property of our registration model, we apply a symmetric loss function for all the unsupervised models:

$$\arg \min L = L_{mse}(\hat{\phi}(I_{move}), I_{fix}) + L_{mse}(\hat{\phi}^{-1}(I_{fix}), I_{move}) + \lambda * L_{diff}(\hat{\phi}) \quad (3)$$

where $\hat{\phi}^{-1}$ is the inverse of $\hat{\phi}$ and L_{diff} is a diffusion regularizer for smoothness $L_{diff} = \int ||\nabla_x \phi + \nabla_y \phi||^2$ and set $\lambda = 0.01$ according to [4].

Data Augmentation. In order to improve model generalisation and avoid outfitting, we apply the same random data augmentation tricks for each image pair during training phase. The following augmentation techniques: rotation, cropping and resizing, brightness adjustment, contrast change, sharpening, blurring and speckle noise addition are conducted with a probability of 0.5 separately. No augmentation is applied during validation nor test phase.

3.3 Experiments

Multi-scale Models. (abbreviation: model name + _M) we apply three child models for PureMLP, MLPMixer and SwinTrans (with patch size of 4×4, 8×8 and 16×16 respectively). For child model of size 4×4, 8×8 and 16×16 in SwinTrans, we set the number of window size to be 8, 4 and 2, the number of heads to be 32, 16 and 8 respectively. The dimension of patch-embedding is set to be 128 for all patch-based methods. ω_C in Eq. 2 is set to be 0.5, 0.3, 0.2 for child model with patch size of 4×4, 8×8, 16×16 separately.

Single-Scale Models. (abbreviation: model name + _S) we run single-scale models for PureMLP, MLPMixer and SwinTrans three proposed frameworks (using patch size of 4×4 pixels). The same configuration is set as for child model with patch-size 4×4 in multi-scale models.

3.4 Results

Since our SwinTrans model relies mostly on features of I_{fix} (with skip-connection of I_{fix} features), we report only the metrics related to the transformation $\phi(\theta)$ that $I_{fix} \approx I_{move} \circ \phi(\theta)$. For CAMUS test dataset, we report the Dice score, Hausdorff distance (HD) and mean surface distance (MSD) between ground truth ED mask and transformed ES mask and the Jacobian determinant in the area of myocardium region.

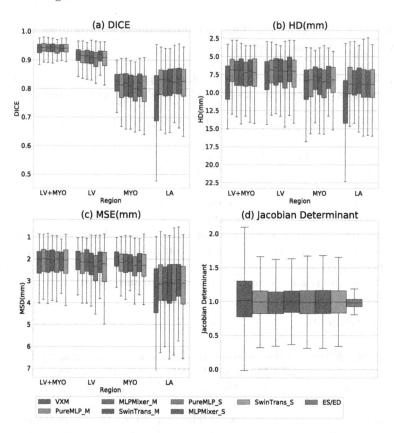

Fig. 2. Comparison of evaluation metrics (Dice score, Hausdorff distance (HD), mean surface distance (MSD) and Jacobin determinant) on test dataset of CAMUS. The Jacobin determinant is only computed in the myocardium region. Except Jacobin determinant figure, the higher the boxplot is in the figure, the better performance it will be.

Evaluation on CAMUS Dataset. From Fig. 2 we can observe that on CAMUS test dataset, almost all the proposed models, no matter it is multi-scale or single-scale, no matter what kind of sub-block it contains (MLP or Transformer or MLP-Mixer), have achieved comparable performance than the CNN model (Vxm), in particular for the whole left ventricle and left atrium registration. In addition, the distribution of Jacobin determinant shows that our patch based methods tend to generate more plausible transform, i.e. closer to real ES/ED myocardium area change. This is consistent with the Hausdorff distance results, which indicates that while preserving comparable registration performance, patch-based methods are more resistant to estimation of false large deformation (see the atrium and myocardium region of example in Fig. 3). What's more, single-scale models and multi-scale models have similar performance. With single-sized patches, we are already capable to let feature infor-

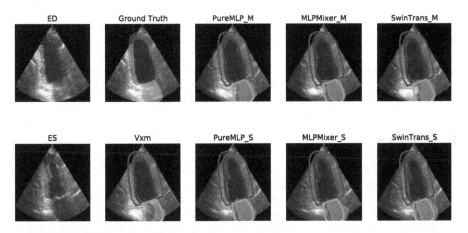

Fig. 3. The same registration example on CAMUS test data with transformed ES masks. Colourful patches are corresponding estimations while bold contours are the ground truth (Yellow: left atrium, Purple: left ventricle, Green: myocardium). (Color figure online)

mation flow through the whole image area and estimate registration transform efficiently (see time and space complexity in Table 1).

Table 1. Time and space complexity between different models (evaluated on a GTX 2080Ti)

Model	GPU memory	Train time (s/pair)	Test time (s/pair)
Vxm	1365 MiB	0.020	0.0047
PureMLP_S	1411 MiB	0.020	0.0038
MLPMixer_S	1447 MiB	0.020	0.0038
SwinTrans_S	1479 MiB	0.029	0.0047

4   Conclusion

In summary, we propose three novel patch-based registration architectures using only MLPs and Transformers. We show that our single and multi-scale models perform similarly and even better to CNN-based registration frameworks on a large echocardiography dataset. The three proposed models demonstrate similar performance among themselves. Our experiments show that patch-based models using MLP/Transformer can perform 2D medical image registration. We shared a similar conclusion with previous works [21,25] that the success of Transformer in vision tasks cannot be simply attributed to the attention mechanism, at least in image registration task. Future works will concentrate on the application of MLP/Transformer in time-series motion tracking.

Acknowledgements. This work has been supported by the French government through the National Research Agency (ANR) Investments in the Future with 3IA Côte d'Azur (ANR-19-P3IA-0002) and by Inria PhD funding.

References

1. Dosovitskiy, A., et al.: An image is worth 16x16 words: Transformers for image recognition at scale. ArXiv abs/ arXiv: 2010.11929 (2021)
2. Arsigny, V., Commowick, O., Pennec, X., Ayache, N.: A log-euclidean framework for statistics on diffeomorphisms. In: Larsen, R., Nielsen, M., Sporring, J. (eds.) MICCAI 2006. LNCS, vol. 4190, pp. 924–931. Springer, Heidelberg (2006). https:// doi.org/10.1007/11866565_113
3. Ashburner, J., Friston, K.J.: Voxel-based morphometry-the methods. Neuroimage **11**(6), 805–821 (2000)
4. Balakrishnan, G., et al.: Voxelmorph: A learning framework for deformable medical image registration. IEEE Trans. Med. Imaging **38**(8), 1788–1800 (2019)
5. Blendowski, M., et al.: Weakly-supervised learning of multi-modal features for regularised iterative descent in 3d image registration. Med. Image Anal. **67**, 101822 (2021)
6. Cao, X., et al.: Deformable image registration based on similarity-steered CNN regression. In: Descoteaux, M., Maier-Hein, L., Franz, A., Jannin, P., Collins, D.L., Duchesne, S. (eds.) MICCAI 2017. LNCS, vol. 10433, pp. 300–308. Springer, Cham (2017). https://doi.org/10.1007/978-3-319-66182-7_35
7. Cao, Y., et al.: Large deformation diffeomorphic metric mapping of vector fields. IEEE Trans. Med. Imaging **24**(9), 1216–1230 (2005)
8. Chen, J., Frey, E.C., He, Y., Segars, W.P., Li, Y., Du, Y.: Transmorph: Transformer for unsupervised medical image registration. arXiv preprint arXiv:2111.10480 (2021)
9. Dalca, A.V., et al.: Unsupervised learning of probabilistic diffeomorphic registration for images and surfaces. Med. Image Anal. **57**, 226–236 (2019)
10. Davatzikos, C.: Spatial transformation and registration of brain images using elastically deformable models. Comput. Vis. Image Underst. **66**(2), 207–222 (1997)
11. Debayle, J., Presles, B.: Rigid image registration by general adaptive neighborhood matching. Pattern Recogn. **55**, 45–57 (2016)
12. Ferrante, E., et al.: Weakly supervised learning of metric aggregations for deformable image registration. IEEE J. Biomed. Health Inform. **23**(4), 1374–1384 (2019)
13. Hering, A., et al.: Cnn-based lung ct registration with multiple anatomical constraints. Med. Image Anal. **72**, 102139 (2021)
14. Hu, Y., et al.: Label-driven weakly-supervised learning for multimodal deformable image registration. In: 2018 IEEE 15th International Symposium on Biomedical Imaging (ISBI 2018), pp. 1070–1074 (2018)
15. Hu, Y., et al.: Weakly-supervised convolutional neural networks for multimodal image registration. Med. Image Anal. **49**, 1–13 (2018)
16. Ilya, et al.: Mlp-mixer: An all-mlp architecture for vision. CoRR abs/ arXiv: 2105.01601 (2021)

17. Krebs, J., et al.: Learning a probabilistic model for diffeomorphic registration. IEEE Trans. Med. Imaging **38**(9), 2165–2176 (2019)
18. Krebs, J., Mansi, T., Mailhé, B., Ayache, N., Delingette, H.: Unsupervised probabilistic deformation modeling for robust diffeomorphic registration. In: Stoyanov, D., et al. (eds.) DLMIA/ML-CDS -2018. LNCS, vol. 11045, pp. 101–109. Springer, Cham (2018). https://doi.org/10.1007/978-3-030-00889-5_12
19. Krizhevsky, A., et al.: Imagenet classification with deep convolutional neural networks. In: Pereira, F., Burges, C.J.C., Bottou, L., Weinberger, K.Q. (eds.) Advances in Neural Information Processing Systems, vol. 25. Curran Associates, Inc. (2012)
20. Leclerc, S., Smistad, E., et al.: Deep learning for segmentation using an open large-scale dataset in 2d echocardiography. IEEE Trans. Med. Imaging **38**(9), 2198–2210 (2019)
21. Liu, H., et al.: Pay attention to mlps (2021)
22. Liu, Z., et al.: Swin transformer: Hierarchical vision transformer using shifted windows. In: International Conference on Computer Vision (ICCV) (2021)
23. Mahapatra, D., Ge, Z.: Training data independent image registration using generative adversarial networks and domain adaptation. Pattern Recogn. **100**, 107109 (2020)
24. Mansilla, L., et al.: Learning deformable registration of medical images with anatomical constraints. Neural Netw. **124**, 269–279 (2020)
25. Melas-Kyriazi, L.: Do you even need attention? a stack of feed-forward layers does surprisingly well on imagenet (2021)
26. Oliveira, F.P., Tavares, J.M.R.: Medical image registration: a review. Comput. Methods Biomech. Biomed. Eng. **17**(2), 73–93 (2014)
27. Rohé, M.-M., Datar, M., Heimann, T., Sermesant, M., Pennec, X.: SVF-Net: learning deformable image registration using shape matching. In: Descoteaux, M., Maier-Hein, L., Franz, A., Jannin, P., Collins, D.L., Duchesne, S. (eds.) MICCAI 2017. LNCS, vol. 10433, pp. 266–274. Springer, Cham (2017). https://doi.org/10.1007/978-3-319-66182-7_31
28. Sokooti, H., de Vos, B., Berendsen, F., Lelieveldt, B.P.F., Išgum, I., Staring, M.: Nonrigid image registration using multi-scale 3d convolutional neural networks. In: Descoteaux, M., Maier-Hein, L., Franz, A., Jannin, P., Collins, D.L., Duchesne, S. (eds.) MICCAI 2017. LNCS, vol. 10433, pp. 232–239. Springer, Cham (2017). https://doi.org/10.1007/978-3-319-66182-7_27
29. Tanner, C., et al.: Generative adversarial networks for mr-ct deformable image registration (2018)
30. Van Der Malsburg, C.: Frank rosenblatt: Principles of neurodynamics: Perceptrons and the theory of brain mechanisms. In: Palm, G., Aertsen, A. (eds.) Brain Theory, pp. 245–248. Springer, Berlin (1986).https://doi.org/10.1007/978-3-642-70911-1_20
31. Vercauteren, T., Pennec, X., Perchant, A., Ayache, N.: Non-parametric diffeomorphic image registration with the demons algorithm. In: Ayache, N., Ourselin, S., Maeder, A. (eds.) MICCAI 2007. LNCS, vol. 4792, pp. 319–326. Springer, Heidelberg (2007). https://doi.org/10.1007/978-3-540-75759-7_39
32. Vercauteren, T., Pennec, X., Perchant, A., Ayache, N.: Symmetric log-domain diffeomorphic registration: a demons-based approach. In: Metaxas, D., Axel, L., Fichtinger, G., Székely, G. (eds.) MICCAI 2008. LNCS, vol. 5241, pp. 754–761. Springer, Heidelberg (2008). https://doi.org/10.1007/978-3-540-85988-8_90

33. Wu, G., et al.: Scalable high-performance image registration framework by unsupervised deep feature representations learning. IEEE Trans. Biomed. Eng. **63**(7), 1505–1516 (2016)
34. Yang, X., Kwitt, R., Styner, M., Niethammer, M.: Quicksilver: Fast predictive image registration - a deep learning approach. Neuroimage **158**, 378–396 (2017)
35. Zhang, Y., Pei, Y., Zha, H.: Learning dual transformer network for diffeomorphic registration. In: de Bruijne, M., et al. (eds.) MICCAI 2021. LNCS, vol. 12904, pp. 129–138. Springer, Cham (2021). https://doi.org/10.1007/978-3-030-87202-1_13
36. Zheng, Y., et al.: Symreg-gan: Symmetric image registration with generative adversarial networks. IEEE Trans. Pattern Anal. Mach. Intell. 1–1 (2021)

Sensitivity Analysis of Left Atrial Wall Modeling Approaches and Inlet/Outlet Boundary Conditions in Fluid Simulations to Predict Thrombus Formation

Carlos Albors[1]([✉]), Jordi Mill[1], Henrik A. Kjeldsberg[2], David Viladés Medel[3], Andy L. Olivares[1], Kristian Valen-Sendstad[2], and Oscar Camara[1]

[1] Sensing in Physiology and Biomedicine (PhySense), Department of Information and Communication Technologies, Universitat Pompeu Fabra, 08018 Barcelona, Spain
carlos.albors@upf.edu
[2] Department of Computational Physiology, Simula Research Laboratory AS, Kristian Augusts Gate 23, 0164 Oslo, Norway
[3] Cardiac Imaging Unit, Cardiology Department, Hospital Santa Creu i Sant Pau, Universitat Autònoma de Barcelona, Sant Antoni Maria Claret 167, 08025 Barcelona, Spain

Abstract. In-silico fluid simulations of the left atria (LA) in atrial fibrillation (AF) patients can help to describe and relate patient-specific morphologies and complex flow haemodynamics with the pathophysiological mechanisms behind thrombus formation. Even in AF patients, LA wall motion plays a non-negligible role in LA function and blood flow patterns. However, obtaining 4D LA wall dynamics from patient-specific data is not an easy task due to current image resolution limitations. Therefore, several approaches have been proposed in the literature to include left atrial wall motion in fluid simulations, being necessary to benchmark them in relation to their estimations on thrombogenic risk. In this work, we present results obtained with computational fluid dynamic simulations of LA geometries from a control and an AF patient with different left atrial wall motion approaches: 1) assuming rigid walls; 2) with a passive movement of the mitral valve annulus plane from a dynamic mesh approach based on springs (DM-SB); and 3) imposing LA wall deformation extracted from dynamic computed tomography (DM-dCT) images. Different strategies for the inlet/outlet boundary conditions were also tested. The DM-dCT approach was the one providing simulation results closer to velocity curves extracted from the reference echocardiographic data, whereas the rigid wall strategy over-estimated the risk of thrombus formation.

Keywords: Computational fluid dynamics · Atrial fibrillation · Left atrial appendage · Dynamic mesh · Thrombus formation · Boundary conditions

O. Camara et al. (Eds.): STACOM 2022, LNCS 13593, pp. 179–189, 2022.
https://doi.org/10.1007/978-3-031-23443-9_17

1   Introduction

Stroke is the main cause of morbidity and mortality from cardiovascular diseases. A primary risk factor is atrial fibrillation (AF), cataloged as the most frequent cardiac arrhythmia in clinical practice and responsible for 15–25% of all ischemic strokes [11]. More than 99% of AF-related strokes originates in an ear-shaped cavity of great morphological variability located in the left atrium (LA), the left atrial appendage (LAA) [3]. Although pharmacological therapies, e.g., oral anti-coagulation (OACS), are first-line therapy for AF patients, left atrial appendage occlusion (LAAO) or excision are also available treatments to prevent and revert the formation process for patients with OAC contra-indications. However, a successful LAAO device implantation requires substantial clinician's experience due to the variability in LAA morphologies and the wide range of device settings to personalise (e.g., design, size, position). Sub-optimal LAAO device choices can lead to abnormal events at follow-up such as device-related thrombosis (DRT), which happens in 2–5% of cases [1,5]. Identifying the LA morphological and haemodynamics characteristics that led to DRT is key to optimise LAAO-based therapy.

Imaging techniques are the gold standard on clinical routine to study LA haemodynamics, specifically Doppler echocardiography images. However, the limited spatial resolution available in these images, representing the flow in one plane and at one time, is not enough to characterise the 4D nature of blood flow patterns in the LA. 4D flow magnetic resonance imaging (4D flow MRI) allows better blood flow characterization, but as a novel modality, it still provides limited LA information [12]. As a potential alternative, in-silico computational fluid simulations (CFD) can help to describe and relate patient-specific LA/LAA morphology and complex hemodynamics to understand the mechanism behind thrombus formation.

Even with the increase in the LA/LAA fluid modeling studies in the literature (see [15] for a recent review), no consensus on optimal boundary conditions have been reached, both for inlet/outlet configurations (i.e., based on pressure and velocities) and LA wall behaviour. The latter is particularly controversial since some researchers [2,8] have assumed rigid LA walls, arguing that LA movement is very limited in AF patients. However, even in those patients, passive movement of the LA induced by the left ventricle (LV) still occurs. Therefore, dynamic mesh approaches using generic mitral valve ring plane excursions [15] or overall LA sinusoidal motion [13] have also been proposed. More advanced strategies have also been used including the extraction of patient-specific LA wall motion from dynamic computed tomography (dCT) images [9,10,17] and fluid-structure interaction (FSI) modelling [4]. Unfortunately, obtaining dCT images of LAAO patients is rarely available in clinical; in parallel, FSI is usually associated to high computational costs and it is uncertain how to define LA wall material properties.

Accordingly, we present in this work a sensitivity analysis on how to incorporate LA wall behaviour in fluid simulations of the left atria and determine the optimal strategy to accurately predict the risk of thrombus formation after LAAO device implantation. The tested strategies were the following: 1) rigid wall assumption; 2) dynamic mesh with a spring-based method (DM-DB); and 3) dynamic mesh guided by patient-specific dCT data (DM-dCT). Experiments were achieved using dCT imaging from a control and an AF patient, from which echocardiographic data was also available. Several inlet/outlet BC configurations were also studied.

2   Material and Methods

2.1   Clinical Data and 3D Model Generation

Two retrospective cardiac-gated computed tomography angiography images, one from a healthy individual and an AF patient, were provided by Hospital de la Santa Creu i Sant Pau (Barcelona, Spain), after approval by the institutional Ethics Committee and informed consent from patients. The images were acquired with a Somatom Force (Siemens Healthineers, Erlangen, Germany) using a biphasic contrast injection protocol defining a slice thickness of 2 mm without overlapping. In the control case, the full cardiac phase reconstruction was every 1% of the cardiac cycle except for the 34%–48% interval, which was not acquired. For the AF patient, images were acquired at every 5% of the cardiac cycle (0 to 99% of R-R interval). Both LA were segmented from dCT images using semi-automatic tools available in Slicer 4.10.11[1]. 3D surface meshes were created from binary marks out of the segmentation with the flying edges algorithm available in Slicer. Then, volumetric LA geometries were generated from the 3D surface ones using the Gmsh 4.5.4 software[2], with a total of 12×10^5 elements.

2.2   Computational Fluid Dynamic Simulations

Modelling Strategies for Left Atrial Wall Motion. The different approaches to incorporate LA wall motion in fluid simulations for the AF patients were the following: 1) rigid walls; 2) a dynamic mesh method with a spring-based approach (DM-SB) to apply a passive mitral valve (MV) annulus plane movement from literature [18]; and 3) a patient-specific dynamic mesh extracted from the patient's dCT scan (DM-dCT). In the healthy case, only rigid walls and DM-dCT strategies were tested since the second scenario, with only a passive MV movement does not have physiological sense in a non-disease heart.

To extract LA wall motion from the dCT data the first step was to perform the registration of the images acquired every 5% in the AF and 1% in the control case of the cardiac cycle. The image registration step was performed using a concatenation of different models using the ANTs software[3]: (1) rigid registration (i.e., only rotation and translation); (2) affine registration (i.e., adding scaling and shearing); and (3) a symmetric normalization registration (i.e., adding a free-form deformation-based deformable transformation with mutual information as

---

[1] https://www.slicer.org/.
[2] https://gmsh.info/.
[3] http://stnava.github.io/ANTs/.

optimization metric). The reference mesh always was the cardiac frame starting from ventricular systole (frame 0, i.e., 0% of cardiac cycle). Therefore, 19 and 87 image registrations were performed for the AF and control case, respectively; i.e., in AF from frame 0 to 5, from frame 0 to frame 10 and so on, until reaching the last time-frame of the dCT. For each registration, the output was a matrix of displacements for each image voxel. In addition, the image corresponding to the first time-frame was manually segmented, generating a corresponding 3D surface mesh. Subsequently, displacement matrices extracted from the image registration step were applied to the 3D surface mesh using an piece-wise cubic Hermite interpolating polynomial (PCHIP), displacing the nodes to the following time-frame. MATLAB R2018a (Mathworks, Natick, MS, USA) was used for the image processing and mesh warping steps described above.

The resulting 3D surface mesh and set of displacements at each time-step was then mapped to simulation time in an Ansys file format to be applied as a boundary condition for the LA wall movement. In each simulation time step, a local remeshing process of faces and cells was performed with the proprietary Ansys algorithm lasting 4 min in the DM-dCT approach and 10 min in DM-SB, checking for negative cell volumes and the improvement of the skewness value for mesh quality. Since the nodes were given in DM-dCT, no resolution change has been experienced in the remeshing process. However, changes in mesh resolution of 3–6% were observed in the second modelling strategy.

Boundary Conditions and Simulation Setup. Two boundary condition settings of the inlets/outlets (pulmonary veins/mitral valve) were tested in each of the analysed LA geometrical models. First, a patient-specific velocity profile at the pulmonary veins (PV) was estimated from the derivative of left atrial and left ventricular volume changes measured from the dCT segmented images. In the first BC scenario (Configuration 1), the mitral valve was modelled as a wall during ventricular systole, representing its closing, and a constant pressure value (7 and 8 mmHg for healthy and AF cases, respectively, following [16]) at ventricular diastole, simulating its opening.

In the second BC configuration (Configuration 2), a generic pressure curve was defined at the PV (in sinus rhythm and with AF for the healthy and diseased cases, respectively). A patient-specific velocity profile was defined at the mitral valve, also derived from dCT-derived volume changes of the LA and LV.

To define the passive motion of the mitral annulus in the DM-SB scheme, a displacement function from literature [18] was imposed in the MV annulus plane, describing the longitudinal excursion of the MV. Then, a spring-based dynamic solution of the CFD solver was employed to ensure motion diffusion through the LA wall geometry.

Blood characteristics were set to 1060 kg/m^3 density and the viscosity to 0.0035 Pa/s. Three cardiac cycles were simulated, with a time-step of 0.01 s and 100 steps per beat, according to the heart rate (HR) of each patient. Flow simulations were performed with the Computational Fluid Dynamics solver available in Ansys Fluent 19 R3 (ANSYS Inc, USA), with the following computational resources: Intel i9-9900K CPU, RTX 2080 Ti GPU with 64 Gb RAM. Post-processing and visualization of simulation results were conducted with ParaView

5.8^4. Computational times ranged between 40–48 h depending on the wall scheme and BC configuration selected.

2.3   In-silico Haemodynamic Indices

The simulated blood flow patterns were qualitatively assessed together with average flow velocities in key LA regions (PV, MV, ostium). Pressure distribution in the LA was also analysed. Moreover, the endothelial cell activation potential (ECAP) was estimated since it has demonstrated to identify regions with high risk of thrombus formation [14]. The mentioned indices were calculated on the third simulated cardiac beat to avoid convergence problems. Additionally, simulated velocity profiles were compared with the estimated from dCT processing in those LA regions where velocity has not been imposed as BC, using the root mean square error (RMSE).

3   Results

The LA displacement fields extracted from dCT images were consistent with patient's conditions, with a MV annular longitudinal excursion of 3 and 9 mm for the AF and healthy case, respectively. Resulting volume change curves from dCT images displayed a volume increase at the maximum diastolic frame of 39 (t = 0.35 s) and 11 (t = 0.45 s) mL for the healthy and AF case, respectively; with slight differences of 4 mL and 3 mL in the same diastolic time frame compared to the volume curves extracted from segmentation in both cases (43 mL and 14 mL for the healthy and AF cases, respectively).

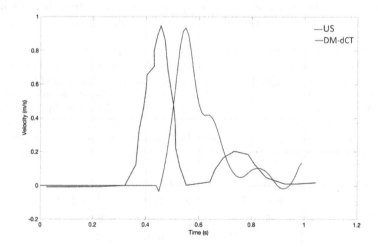

Fig. 1. Velocity curves at the mitral valve (MV) in the AF patient from ultrasound (US) images (blue), and estimated from the left atrial movement derived from dynamic computed tomography (DM-dCT) data (red). (Color figure online)

---

4 https://www.paraview.org/.

Figure 1 shows the similarity between the velocity curve in the MV extracted from the dCT and the US data in the AF patient. The velocity magnitudes were in the same range (maximum of 0.94 m/s and 0.93 m/s for US and dCT, respectively). A very small "A-wave" was detected on the US image, even smaller on dCT, with some flow regurgitation (red line in Fig. 1), being a consistent AF behavior. Averaging the results in the two analysed cases, the DM-dCT approach to incorporate LA wall motion into fluid simulations was the one providing velocity curves closer to the ones derived from dynamic imaging (see Fig. 3): RMSE = 0.09 for DM-dCT; RMSE > 0.12 for the remaining strategies (e.g., RMSE of 0.05 and 0.14 for DM-dCT and rigid walls in the control case, respectively). Qualitatively, only the DM-dCT approach faithfully represented the A wave in the control case, as well as correct S waves in the velocity curves at the PV (i.e., a large S1 and a tiny S waves at the healthy and AF cases, respectively, in Fig. 3).

Figure 2 (first two rows) shows how fluid simulations in the control case presented higher velocities and better washout in the LAA than the AF case, as expected. Larger variations in LAA velocities provided by the different BC and LA wall strategies were obtained at the deeper part of the LAA, while they were quite similar at the ostium. The DM-dCT was the approach providing higher blood flow peak velocities (>0.15 m/s even in distal LAA lobes in the healthy case), while the DM-SB and rigid wall strategies gave similar results since the passive MV excursion in the DM-SB did not result in strong motion of the LAA.

As for inlet/outlet BC configurations, defining velocity at the PV (Configuration 1) provided a better blood washout and higher velocities in the LAA and ostium than Configuration 2 BC set-up in the AF patient, while it was the opposite for the control case (see Table 1). Velocity values simulated with the different LA wall motion strategies were quite similar, without a clear distinctive pattern, being less relevant that the inlet/outlet configuration.

Table 1. Average ostium velocities (m/s) during the whole cardiac cycle in a control and an AF patient with the different evaluated boundary conditions and left atrial wall motion approaches. C1 (configuration 1): velocity profile at the inlet pulmonary veins (PV) and pressure constant values at the outlet mitral valve (MV). C2 (configuration 2): pressure at the inlet PV and velocities at the MV outlet. DM-dCT and DM-SB: dynamic mesh approaches guided by dynamic computed tomography images and spring-based method, respectively.

	Control		Atrial fibrillation		
	Rigid	DM-dCT	Rigid	DM-SB	DM-dCT
C1	0.128	0.1495	0.078	0.082	0.08
C2	0.159	0.164	0.084	0.074	0.078

Blood flow patterns were also qualitatively analysed to identify if fluid simulations could reproduce the two main vortices in the LA at end-systolic and

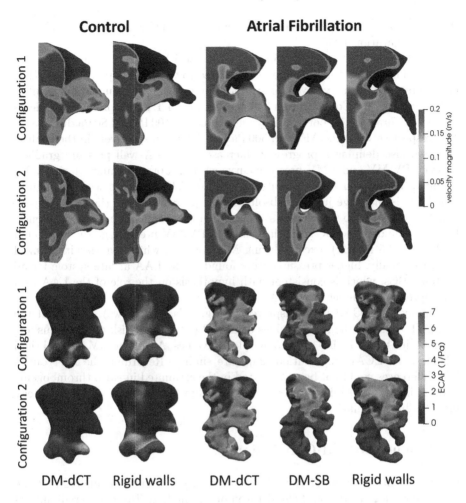

Fig. 2. First two rows: Blood velocity patterns in the left atria (LA) during early diastole (t = 0.6 s) in a control and an AF patient with the different evaluated boundary conditions and left atrial wall motion approaches. Last two rows: Endothelial cell activation (ECAP) maps in a control and an AF patient with the different evaluated boundary conditions and left atrial wall motion approaches. BC configuration 1: velocity profile at the inlet pulmonary veins (PV) and pressure constant values at the outlet mitral valve (MV). BC configuration 2: pressure at the inlet PV and velocities at the MV outlet. DM-dCT and DM-SB: dynamic mesh approaches guided by dynamic computed tomography images and spring-based method, respectively.

mid-diastolic (ventricular) phases reported in 4D flow MRI studies [7]. The DM-dCT approach succeeded to capture both vortices in the control case, with the remaining strategies only reproduced the end-diastolic one (only this one could be seen for all approaches in the AF case). Furthermore, abnormal PV regurgitation was mainly observed in fluid simulations with the DM-SB and rigid wall strategies to fulfill mass conservation law during ventricular systole, especially with Configuration 1 BC.

The DM-dCT strategy provided the largest PV-MV gradients due to its associated larger LA volume increase compared to the other approaches, which is better observed during ventricular systole. Comparing the evaluated inlet/outlet BC configurations, Configuration 2 generated higher PV-MV pressure gradients than Configuration 1 in both healthy and AF cases (100 Pa and 80 Pa during systole, respectively, for the AF case; 500 Pa and 200 Pa, respectively in the healthy subject), also defining a progressive increase of the LA wall pressure gradient between PV-MV and a MV pressure curve without abrupt changes during valve opening. The reason for instabilities and unrealistic patterns with Configuration 1 BC is the large jump in absolute value pressures when the MV opens (8 mmHg) compared to ventricular systole, when there is not a pressure reference. However, all analysed BC scenarios achieved PV-MV pressure gradients around 70 Pa (i.e., 0.5 mmHg), except during MV opening, which agrees with clinical data. Regionally, higher pressures were found in the LAA at late systole for all analysed BCs, which is consistent with hypothesis on the role of the LAA as a decompression chamber of the LA.

Finally, the ECAP maps displayed in Fig. 2 (last two rows) shows that the DM-dCT achieved the lowest values of ECAP (i.e., lowest risk of thrombus formation) due to the higher blood flow velocity values. As for the DM-SB and rigid schemes, the ECAP distribution maps are similar. Regarding the inlet/outlet BC, differences could not be appreciated in the estimated in-silico thrombogenic indices, being more relevant how to incorporate the LA wall motion strategy into the fluid simulations.

4   Discussion and Conclusions

Computational fluid simulations can provide useful insight to understand the patho-physiological factors behind thrombus formation. However, credibility is a cornerstone of in-silico models requiring personalization, verification and validation with data from different sources (e.g. in-vivo, ex-vivo, etc.), following standards such as the V&V40 guidelines [19]. Comparing different modelling options are then key to establish best practices.

In fluid simulations of the LA, the most important consideration is having a good synchronisation between the LA wall movement, anatomy, and volume changes, with the inlet/outlet parameters (i.e., velocities and pressures) selected at the pulmonary veins and mitral valve, respectively. Left atrial wall motion plays a non-negligible role in simulation scenarios, even in AF patients, although the assumption of rigid walls has been used for years to study them.

Our experiments have demonstrated that incorporating LA wall motion from patient-specific dCT images into fluid simulations is a better strategy than DM-SB and rigid ones, in agreement with recent literature [9]. The DM-dCT was the only LA wall motion strategy able to represent a small S-wave in the PV velocity curves in both patients, which is normally neglected in CFD simulations but widely reported in clinical studies due to its relationship with several pathologies. In addition, the (largest) LA volume increase defined by the DM-dCT motion at the ventricular systolic phase (MV closure) facilitated the fulfilment of mass conservation law with Configuration 1 BC (i.e., velocities imposed at inlet PV). On the other hand, DM-SB and rigid wall strategies generated non-physiological PV regurgitation to compensate mass imbalance, especially the latter.

One of the main advantages of dCT data is that velocity curves can also be extracted (through mass conservation law) and used them as patient-specific BCs. In our experiments, the resulting dCT velocity curve at the MV was similar to the one measured with ultrasound data (Fig. 1), with equivalent maximum values and with a reduced A wave due to AF. The observed differences in Fig. 1 could be explained by varying patient heartbeat, time between acquisitions, or different positions of velocity measurement in the two modalities. As for the simulated PV velocity curve, it was similar to the one directly derived from dCT data. However, none of them succeeded on properly replicating the complex ventricular systole dynamics at the PV, as already noticed in previous works [10].

Although there were not many global differences in simulated blood flow patterns with the DM-dCT approach (e.g., similar blood flow velocity magnitudes at the ostium), we saw that it led to a better LAA washing and more blood recirculations. Nevertheless, none of the AF tested BC configurations was able to replicate the systolic vortex reported in the literature [6], which could be related to velocity curve inaccuracies.

Combined with the dynamic mesh approach for LA wall motion modelling, Configuration 2 of BC (i.e., imposing velocity at the MV), provided higher ostium velocities in the control case and slightly slower in AF, but more coherent and smooth pressure map distribution in both patients over the cardiac cycle. The DM-SB and rigid wall approaches generated some unrealistic pressure peaks, the latter not creating any pressure differential during ventricular systole. Finally, the thrombogenic ECAP index confirmed the over-estimation of thrombus risk when patient-specific wall motion is not considered due to low time wall shear stress values induced by low/non-existent flow velocities.

Despite the advantages of including LA wall motion from patient-specific dCT data, it is not obvious to acquire such images on a regular basis, being complicated to create large cohorts. Moreover, dCT data still does not fully capture ventricular systole dynamics. Alternatives need to be investigated in the future to transport knowledge on LA wall motion from dCT images available on a small dataset of cases onto each LA to be modelled or to recover as much LA wall motion as possible from 2D-3D echocardiographic data.

Acknowledgement. This project has received funding from the European Union's Horizon 2020 research and innovation programme under grant agreement No 101016496 (SimCardioTest).

A   Appendix

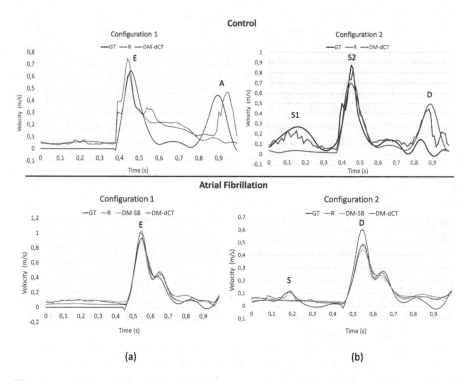

Fig. 3. Validation of the different boundary condition (BC) scenarios in a control and an AF patient. (a) and (b) represents the velocity curves at the mitral valve (MV) and the left superior pulmonary vein (LSPV), respectively, provided by the different approaches for incorporating left atrial wall motion into fluid simulations. Configuration 1: velocity profile at the inlet pulmonary veins (PV) and pressure constant values at the MV outlet. Configuration 2: pressure at the inlet PV and velocities at the MV outlet. GT: Velocity curve extracted from dynamic computed tomography taken as a ground-truth for comparison purposes. R: rigid walls. DM-dCT and DM-SB: dynamic mesh approaches guided by dynamic computed tomography images and spring-based method, respectively.

References

1. Aminian, A., et al.: Incidence, characterization, and clinical impact of device-related thrombus following left atrial appendage occlusion in the prospective global AMPLATZER amulet observational study. JACC: Cardiovasc. Interv. **12**(11), 1003–1014 (2019)
2. Bosi, G.M., et al.: Computational fluid dynamic analysis of the left atrial appendage to predict thrombosis risk. Front. Cardiovasc. Med. **5**, 34 (2018)

3. Cresti, A., et al.: Prevalence of extra-appendage thrombosis in non-valvular atrial fibrillation and atrial flutter in patients undergoing cardioversion: a large transoesophageal echo study. EuroIntervention **15**(3), e225–e230 (2019)
4. Fang, R., Li, Y., Zhang, Y., Chen, Q., Liu, Q., Li, Z.: Impact of left atrial appendage location on risk of thrombus formation in patients with atrial fibrillation. Biomech. Model. Mechanobiology **20**(4), 1431–1443 (2021). https://doi.org/10.1007/s10237-021-01454-4
5. Fauchier, L., et al.: Device-related thrombosis after percutaneous left atrial appendage occlusion for atrial fibrillation. J. Am. Coll. Cardiol. **71**(14), 1528–1536 (2018)
6. Fyrenius, A., Wigström, L., Ebbers, T., Karlsson, M., Engvall, J., Bolger, A.F.: Three dimensional flow in the human left atrium. Heart **86**(4), 448–455 (2001)
7. Garcia, J., et al.: Left atrial vortex size and velocity distributions by 4d flow MRI in patients with paroxysmal atrial fibrillation: Associations with age and CHA2DS2-VASc risk score. J. Magn. Reson. Imaging **51**(3), 871–884 (2020)
8. García-Isla, G., et al.: Sensitivity analysis of geometrical parameters to study haemodynamics and thrombus formation in the left atrial appendage. Int. J. Numer. Methods Biomed. Eng. **34**(8), e3100 (2018)
9. García-Villalba, M., et al.: Demonstration of patient-specific simulations to assess left atrial appendage thrombogenesis risk. Front. Physiol. **12**, 596596 (2021)
10. Gonzalo, A., et al.: Non-newtonian blood rheology impacts left atrial stasis in patient-specific simulations. Int. J. Numer. Methods Biomed. Eng., e3597 (2022)
11. Haeusler, K.G., et al.: Expert opinion paper on atrial fibrillation detection after ischemic stroke. Clin. Res. Cardiol. **107**(10), 871–880 (2018). https://doi.org/10.1007/s00392-018-1256-9
12. Markl, M., et al.: Assessment of left atrial and left atrial appendage flow and stasis in atrial fibrillation. J. Cardiovasc. Magn. Reson. **17**(1), 1–2 (2015)
13. Masci, A., et al.: A proof of concept for computational fluid dynamic analysis of the left atrium in atrial fibrillation on a patient-specific basis. J. Biomech. Eng. **142**(1) (2020)
14. Mill, J., et al.: Patient-specific flow simulation analysis to predict device-related thrombosis in left atrial appendage occluders. REC: Interv. Cardiol. **3**(4), 278–85 (2021)
15. Mill, J., et al.: Sensitivity analysis of in silico fluid simulations to predict thrombus formation after left atrial appendage occlusion. Mathematics **9**(18), 2304 (2021)
16. Nagueh, S.F., et al.: Recommendations for the evaluation of left ventricular diastolic function by echocardiography. Eur. J. Echocardiogr. **10**(2), 165–193 (2009)
17. Otani, T., Al-Issa, A., Pourmorteza, A., McVeigh, E.R., Wada, S., Ashikaga, H.: A computational framework for personalized blood flow analysis in the human left atrium. Ann. Biomed. Eng. **44**(11), 3284–3294 (2016)
18. Veronesi, F., et al.: Quantification of mitral apparatus dynamics in functional and ischemic mitral regurgitation using real-time 3-dimensional echocardiography. J. Am. Soc. Echocardiogr. **21**(4), 347–354 (2008)
19. Viceconti, M., Pappalardo, F., Rodriguez, B., Horner, M., Bischoff, J., Tshinanu, F.M.: In silico trials: verification, validation and uncertainty quantification of predictive models used in the regulatory evaluation of biomedical products. Methods **185**, 120–127 (2021)

APHYN-EP: Physics-Based Deep Learning Framework to Learn and Forecast Cardiac Electrophysiology Dynamics

Victoriya Kashtanova[1,2]([✉]), Mihaela Pop[1,3], Ibrahim Ayed[4,5],
Patrick Gallinari[4,6], and Maxime Sermesant[1,2]

[1] Inria, Université Côte d'Azur, Nice, France
{victoriya.kashtanova,maxime.sermesant}@inria.fr
[2] 3IA Côte d'Azur, Sophia Antipolis, France
[3] Sunnybrook Research Institute, Toronto, Canada
[4] Sorbonne University, Paris, France
[5] Theresis Lab, Thales, France
[6] Criteo AI Lab, Paris, France

Abstract. Biophysically detailed mathematical modeling of cardiac electrophysiology is often computationally demanding, for example, when solving problems for various patient pathological conditions. Furthermore, it is still difficult to reduce the discrepancy between the output of idealized mathematical models and clinical measurements, which are usually noisy. In this paper, we propose a fast physics-based deep learning framework to learn cardiac electrophysiology dynamics from data. This novel framework has two components, decomposing the dynamics into a physical term and a data-driven term, respectively. This construction allows the framework to learn from data of different complexity. Using 0D in silico data, we demonstrate that this framework can reproduce the complex dynamics of transmembrane potential even in presence of noise in the data. Additionally, using ex vivo 0D optical mapping data of action potential, we show the ability of our framework to identify the relevant physical parameters for different heart regions.

Keywords: Physics-based learning · Deep learning · Electrophysiology · Simulations

1 Introduction

Multi-physics phenomena involved in cardiac function can be studied using mathematical models of different complexity. Among them, several electrophysiology (EP) models can accurately reproduce the electrical behaviour of the heart at cellular, tissue or organ level. For instance, in order to describe the dynamics of transmembrane voltage, current and different ionic concentrations in the cardiac cell, detailed biophysical models such as the Ten Tusscher-Panfilov model [21,22] have been proposed. However, these models are intricate and computationally

© The Author(s), under exclusive license to Springer Nature Switzerland AG 2022
O. Camara et al. (Eds.): STACOM 2022, LNCS 13593, pp. 190–199, 2022.
https://doi.org/10.1007/978-3-031-23443-9_18

expensive, and have numerous hidden variables which are nearly impossible to measure all, making the model parameters difficult to personalise.

Alternatively, one can use phenomenological models that involve simplified descriptions derived from biophysical models. Examples of such models are the FitzHugh-Nagumo, Aliev-Panfilov, and Mitchell-Schaeffer models [1,6,12–14], which all employ fewer variables and parameters, and are therefore especially useful for rapid computational modelling of wave propagation at tissue and organ level. However, these are less realistic and, consequently, need a complementary mechanism to enable their fitting to measured data. Machine learning and in particular deep learning (DL) approaches could help provide such correction mechanisms. Thus, the combination of rapid phenomenological models and machine learning components may allow the development of rapid and accurate models of transmembrane dynamics.

For this reason, researchers have started to use coupled physico-statistical approaches for cardiac electrophysiology simulations with a high precision and at low cost. For example, one group designed a neural network that approximates the FitzHugh-Nagumo model [5], others used a physics-informed neural network for cardiac activation mapping accounting for underlying wave propagation dynamics [20], while another group proposed an approach to create a nonlinear reduced order model with the help of deep learning algorithms (DL-ROM) designed for cardiac EP simulations [7]. Lastly, other researchers presented a physics-informed neural network for accurate simulation of action potential and correct estimation of model parameters [9]. However, the majority of these coupled approaches is based on high-fidelity physical models. Fitting those to the data could be not only computationally expensive, but also difficult to properly manage large discrepancies between simulated and real data.

To alleviate this limitation, here we propose to use the APHYN-EP framework, a DL framework based on a fast low-fidelity (or incomplete) physical model, inspired by [23]. This framework has two components, decomposing the dynamics into a physical term and a data-driven term, respectively. Notably, the data-driven deep learning component is specifically designed to capture only the information that cannot be modeled by the incomplete physical model. In our previous paper [10] we showed that this framework is able to reproduce with good precision the 2D dynamics simulated by the Ten Tusscher - Noble - Noble - Panfilov ionic model. But we applied the framework only on simulated data and on the depolarisation part of a cardiac action potential cycle, neglecting the repolarisation part.

In this paper, we specifically explore the capability of APHYN-EP framework to learn the full cardiac action potential cycle (i.e., depolarisation and repolarisation phases). Using 0D *in silico* data, we demonstrate that our framework can reproduce the complex dynamics of transmembrane potential even in presence of noise in the data. Additionally, using *ex vivo* 0D epicardial optical fluorescence mapping data, we show the capability of this framework to identify relevant physical parameters for different anatomical zones of the heart.

2   Learning Framework

In order to learn the cardiac electrophysiology dynamics (X_t), here we solve an optimization problem via our physics-based data-driven APHYN-EP framework. This particular framework combines a physical model (F_p) representing an incomplete description of the underlying phenomenon with a neural network (F_{dl}), where the latter complements the physical model by capturing the information that cannot be modeled by the physics-described component:

$$\min_{F_p \in \mathcal{F}_p, F_{dl} \in \mathcal{F}_{dl}} \|F_{dl}\| \text{ subject to } \forall X \in \mathcal{D}, \forall t, \frac{dX_t}{dt} = F(X_t) = (F_p + F_{dl})(X_t). \quad (1)$$

Our incomplete physical model is the two-variable (v, h) model by [12] for cardiac EP simulations (2). The variable v represents a normalised ($v \in [0,1]$) dimensionless transmembrane potential while the "gating" variable h controls the repolarisation phase (i.e., the return to initial resting state):

$$\begin{aligned} \frac{dv}{dt} &= \frac{hv^2(1-v)}{\tau_{\text{in}}} - \frac{v}{\tau_{\text{out}}} + J_{\text{stim}}(t) \\ \frac{dh}{dt} &= \begin{cases} \frac{1-h}{\tau_{\text{open}}} & \text{if } v < v_{\text{gate}} \\ \frac{-h}{\tau_{\text{close}}} & \text{if } v > v_{\text{gate}} \end{cases} \end{aligned} \quad (2)$$

where J_{stim} is a transmembrane potential activation function, which is equal to 1 during stimulation time (t_{stim}). This physical model has been successfully used in patient-specific modelling [19], covering general EP dynamics. Furthermore, in contrast to very detailed ionic/cellular models, this model is flexible in terms of spatial and temporal steps. Thus, assuming the initial conditions for this system (2) $v(t=0) = 0$ and $h(t=0) = 1$ we can compute an approximation of h for any time point t by employing a simple integration scheme.

The framework's deep learning component (F_{dl}) is a Residual Network (ResNet) [8], because it can accurately reproduce a complex dynamics due to residual connections in its architecture. The choice of data-driven component is not limited by ResNet architecture. We performed preliminary experiments with other types of networks (out of scope for this paper), but overall the ResNet model was the most stable along the different simulations.

In APHYN-EP framework the physical and the data-driven components are trained simultaneously using automatic differentiation tools provided by the Pytorch library [17]. Here, the Loss function $(\mathcal{L})$ for training consists of 2 parts: trajectory-based loss and loss on norm of F_{dl}, being represented as following:

$$\mathcal{L} = \lambda * \mathcal{L}_{\text{traj}} + \|F_{dl}\| = \lambda * \sum_{i=1}^{N} \sum_{h=1}^{T/\Delta t} \|X_{h\Delta t}^{(i)} - \tilde{X}_{h\Delta t}^{(i)}\| + \|F_{dl}\|$$

where each state $\tilde{X}_{h\Delta t}^{(i)} = \int_{X_0^{(i)}}^{X_0^{(i)} + h\Delta t} (F_p^{\theta_p} + F_a^{\theta_a})(X_s) \, dX_s$ is calculated from the initial state $X_0^{(i)}$ via a differentiable ODE solver [3,4]. The role of the λ coefficient is to balance the two parts of the loss. During training, we use λ in the dynamic mode as $\lambda_{j+1} = \lambda_j + \gamma \mathcal{L}_{\text{traj}}(\theta_{j+1})$ (j is an epoch number) to artificially increase the importance of $\mathcal{L}_{\text{traj}}$ at the beginning of training and gradually decrease it, changing the focus of optimization on norm of F_{dl}. This training algorithm is described in detail in [10,23].

3   Experimental Settings

To test the performance of APHYN-EP framework, we chose two types of experiments. First, using *in silico* data, we tested the ability of the framework to reproduce the complex dynamics of transmembrane potential even in the presence of noise in the data. Second, using *ex vivo* optical mapping data, we showed that our framework can identify the relevant physical parameters for different anatomical zones of the unhealthy heart. The details of data collection and training settings used for the experiments are included below.

3.1   *In Silico* Data

Data Collection. To evaluate our method, we used a dataset of transmembrane potential activation simulated with a monodomain reaction-diffusion equation and the Ten Tusscher – Panfilov ionic model [22], which represents 12 different transmembrane ionic currents. The simulations were performed with finetwave software[1], described in detail in the second chapter of [15]. The computational domain represents a 2D slab of cardiac tissue (isotropic), with 24×24 elements in size. One stimulation was applied for 1 ms in the selected area (left top corner for training and near the center for validation datasets, see Fig. 1) for transmembrane potential activation. Simulations were conducted for the first 350 ms of a heart beat, to achieve full depolarization-repolarization cycle (note that videos showing simulated dynamics across time is available online[2]). Next, we saved a time sequence for each pixel of the cardiac slab in separate files, creating two databases: for training and validation, respectively. To simplify the workflow for our framework, we removed the section of time sequences where the potential is equal to zero (Fig. 1(c)). To test the ability of the framework to operate with noisy data we add to each time sequence a 5% random noise with normal Gaussian distribution.

The data simulated via the Ten Tusscher – Panfilov model with added noise were considered here as the ground truth. The objective was then to learn the complex dynamics generated via this model with the APHYN-EP framework, by combining a simplified physics description with a deep learning component. This should result in a low computational cost surrogate model of the computationally intensive biophysically detailed Ten Tusscher – Panfilov model.

---

[1] https://github.com/TiNezlobinsky/Finitewave.
[2] https://doi.org/10.6084/m9.figshare.21648752.v3.

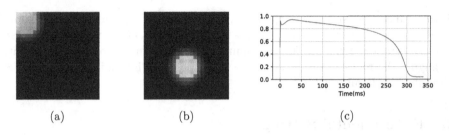

Fig. 1. (a,b) 2D cardiac slab with transmembrane potential activation (yellow) and resting phase (dark blue), for train and validation dataset, respectively. (c) Typical temporal sequence for the experiment (without noise), normalised amplitude of the transmembrane potential on the Y-axis and time (ms) on the X-axis. (Color figure online)

Training Settings. The physical model (F_p) of Eq. (2) was implemented with a standard finite-difference scheme. We estimated only τ_{in}, τ_{out} and τ_{close} as unknown parameters in (2), since they control the major part of model dynamics and thus the main difference between the Mitchell-Schaeffer and the Ten Tusscher – Panfilov models in our simulations. The initial Mitchell-Schaeffer model parameters were taken as in the original paper [12]: $\tau_{in} = 0.3$, $\tau_{out} = 6$, $\tau_{open} = 120$, $\tau_{close} = 150$, $v_{gate} = 0.13$ and $t_{stim} = 1$. For the deep learning component (F_{dl}) of the framework, we used a ResNet with 8 filters at the initial stage and 3 intermediary blocks, and started with a re-weighted orthogonal initialisation for its parameters.

We used a time resolution of 0.1 ms to compute the forecast given by APHYN-EP framework. The training was performed using a horizon of 350 ms. We trained the framework until full model convergence (about 100 epochs) using an ADAM optimiser [11] with initial learning rate of 10^{-3}. The hyper-parameters λ_0 and γ of the algorithm were set to 1 and 10, respectively.

It is important to notice that, while the framework takes a few hours to train (about 2 h on Nvidia Quadro M2200 GPU), once this is done the inference is quick (less than 10 s to compute 350 ms of forecasting) and does not require any re-calibration.

3.2 *Ex Vivo* Data

Data Collection. We tested APHYN-EP framework performance on *ex vivo* datasets from optical fluorescence imaging of action potential. Briefly, the optical signals were recorded *ex vivo* on an heart explanted from a juvenile swine (25 kg in weight). The heart was attached to a Langendorff perfusion system and a voltage sensitive dye (i.e., di-4-ANEPPS) was injected into the perfusate. In order to avoid cardiac motion artifacts during the electrophysiological recordings, the heart contraction was suppressed by a bolus of saline and Cytochalasin D, an electro-mechanical uncoupler. All optical images were acquired epicardially using a high-speed CCD camera (MICAM02, BrainVision Inc. Japan), with high-temporal resolution (3.7 ms) as well as a high-spatial resolution (i.e., pixel size 0.7 mm x 0.7 mm). The action potential was then derived at each pixel from the relative change in the intensity of fluorescence signal. The experiments are described in more detail in [18]. The recorded signals were exported from the BV-Ana acquisition software and further signal analysis was performed using in-house scripts written in Matlab and Python.

For this experiment, we used a heart with an ischaemic region. We manually selected two rectangular regions of interest (ROIs) with different action potential dynamics across time (see Fig. 2). Next, we normalised the optical signal, to obtain a [0, 1] min/max interval for transmembrane potential, while keeping the noise in the data. We took a first full cardiac cycle and removed the parts with zero potential, keeping only time-sequences of 300 ms per experiment. Then, we saved a time sequence for each pixel from each ROI in separate files, creating two databases (ROI A and ROI B) containing each about 10 and 5 time-sequences for training and validation respectively.

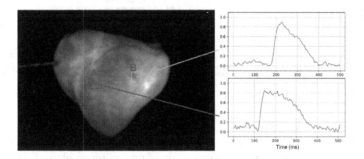

Fig. 2. Example of optical mapping data (tracings of denoised action potential waves) recorded ex vivo in a porcine heart. ROI B represents an ischaemic region characterized by a shorten action potential duration (APD) compared to the normal APD recorded in ROI A.

The optical data were considered here as the ground truth. Our specific objective was to learn the complex dynamics of measured action potential, and then to identify the relevant physical parameters for different parts of the heart.

Training Settings. The training settings for this experiment were similar to the ones described in Sect. 3.1, except for training horizon (300 ms) and parameter γ, which was set to 1 for better equilibrium.

4   Results

4.1   *In Silico* Data

Results demonstrated that APHYN-EP framework was able to accurately reproduce several key features, the wave morphology and the electrical conduction properties of the transmembrane potential generated by the Ten Tusscher – Panfilov model (see Fig. 3), even in presence of noise in data. Figure 3(a) shows that in absence of parameter τ_{close} (controlling the repolarisation), the data-driven component (ResNet model) completed the dynamics generated by physical component. In the presence of τ_{close} the error of dynamics learned by data-driven component is minimal (Fig. 3(b)). The predicted dynamic was generated via an Euler integration scheme, by assimilating only one first measurement of the transmembrane potential dynamics. The framework showed robust resistance to noise in the data, and, as a result, it neglected rapid changes in transmembrane potential activation (see Fig. 3, first 40 ms).

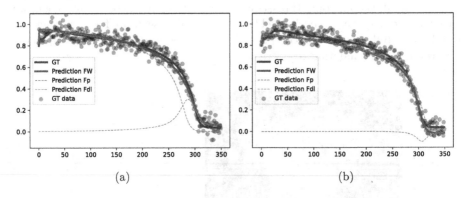

Fig. 3. Validation results of trained framework with learning of (a) 2 (τ_{in} and τ_{out}) and (b) 3 (τ_{in}, τ_{out} and τ_{close}) physical parameters. Ground truth (GT), prediction of the framework (Prediction FW), decomposition of prediction on physical (F_p) and DL (F_{dl}) parts.

4.2   *Ex Vivo* Data

Using optical imaging mapping data, our APHYN-EP framework was able to reproduce the observed action potential dynamics for different ROIs within the heart, identifying the 3 major physical dynamics parameters (τ_{in}, τ_{out} and τ_{close}). Figure 4 demonstrates that the framework correctly estimated the difference in value of parameter τ_{close} which increased APD or shortened it, respectively.

Table 1 shows the mean squared error (MSE) results for our framework forecasting on train and validation data samples. To calculate this error, for each data sample, we fed the framework with only one initial measurement, then let it predict 300 ms forward without any additional information. It can be seen that the obtained MSE is small for both ROI and, despite the use of a limited dataset for training, the APHYN-EP framework achieves forecasting the dynamics with good accuracy for new data samples from the validation dataset. We also added for comparison a baseline model corresponding to the "incomplete"

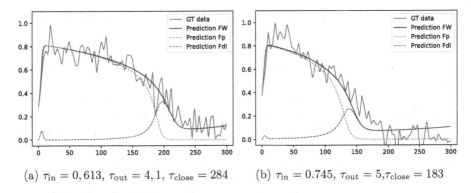

(a) $\tau_{\text{in}} = 0,613$, $\tau_{\text{out}} = 4, 1$, $\tau_{\text{close}} = 284$ (b) $\tau_{\text{in}} = 0.745$, $\tau_{\text{out}} = 5$, $\tau_{\text{close}} = 183$

Fig. 4. Validation results of the framework trained on: (a) ROI A data and (b) ROI B data. Ground truth (GT) data, prediction of the framework (Prediction FW), decomposition of prediction on physical (F_p) and DL (F_{dl}) parts.

physical model trained alone on the same dataset as APHYN-EP framework, described in 3.2. As we can observe, our framework outperform the physical model for every dataset, while the contribution of F_{dl} component is still minimal.

Table 1. Mean-squared error (MSE) of normalised transmembrane potential forecasting (forecasting horizon of 300 ms).

	ROI A		ROI B	
	Train	Validation	Train	Validation
APHYN-EP	$9.12 * 10^{-3}$	$5.72 * 10^{-3}$	$1 * 10^{-2}$	$8 * 10^{-3}$
framework ($\lVert F_{dl} \rVert^2$)	$(1.6 * 10^{-4})$	$(8 * 10^{-5})$	$(8 * 10^{-5})$	$(7 * 10^{-5})$
Physical model only	$1.4 * 10^{-2}$	$1 * 10^{-2}$	$1.45 * 10^{-2}$	$9.3 * 10^{-3}$

It is important to notice that we cannot use the Ten Tusscher – Panfilov ionic model [22] applied on *ex vivo* data as a baseline model due its parameter personalisation difficulties [2]. Despite the expectations and suggestions indicated in [2,16], to date, these difficulties have not been overcome. That is because complex biophysical models have too many parameters and instabilities to enable robust parameter estimation from such data.

5   Discussion and Conclusion

In this work, we presented a learning framework that is able to learn the considered dynamics. Overall, our results suggest that automated learning of cardiac electrophysiology dynamics is feasible and has great potential. We showed that

APHYN-EP framework can learn and forecast the EP data of different origin (i.e., data simulated via the Ten Tusscher – Panfilov ionic model, as well as real optical mapping data), even in presence of noise in the data. This framework may be useful for applications concerning fast parameterization of computational heart models.

The main advantage of our proposed framework is its coupled architecture, which allows us to use a simplified low-fidelity electrophysiological model as a physical component of the framework. Such framework opens up possibilities in order to introduce prior knowledge in deep learning approaches through explicit equations, as well as to correct physical model errors from data.

However, we acknowledge that some limitations exist in our model. For instance, due to the data-driven architecture of the framework, its training is not regular. This could lead to a local minimum for the parameters of the physical component and additional involvement of the deep learning component. Several solutions for this problem may be given by: a more advanced training protocol; rigid boundaries on physical model parameters; or, performing ablation studies including deep learning component architectural changes, which will be addressed in our future work.

Acknowledgements. This work has been supported by the French government, through the 3IA Côte d'Azur Investments in the Future project managed by the National Research Agency (ANR) with the reference number ANR-19-P3IA-0002 and through the "Research and Teaching chairs in artificial intelligence (AI Chairs)" funding for DL4Clim project. The authors are grateful to the OPAL infrastructure from Université Côte d'Azur for providing resources and support, and to Dr. A. Panfilov and his PhD student T. Nezlobinsky for showing us their software FiniteWave to easily simulate the EP data.

References

1. Aliev, R.R., Panfilov, A.V.: A simple two-variable model of cardiac excitation. Chaos, Solitons Fractals **7**(3), 293–301 (1996)
2. Camara, O., et al.: Inter-model consistency and complementarity: learning from ex-vivo imaging and electrophysiological data towards an integrated understanding of cardiac physiology. Prog. Biophys. Mol. Biol. **107**(1), 122–133 (2011). Experimental and Computational Model Interactions in Bio-Research: State of the Art
3. Chen, R.T.Q., Amos, B., Nickel, M.: Learning neural event functions for ordinary differential equations. ICRL (2021)
4. Chen, R.T.Q., Rubanova, Y., Bettencourt, J., Duvenaud, D.: Neural ordinary differential equations. In: Advances in Neural Information Processing Systems (2018)
5. Court, S., Kunisch, K.: Design of the monodomain model by artificial neural networks. arXiv preprint:2107.03136 (2021)
6. FitzHugh, R.: Impulses and physiological states in theoretical models of nerve membrane. Biophys. J . **1**(6), 445–466 (1961)
7. Fresca, S., Manzoni, A., Dedè, L., Quarteroni, A.: POD-enhanced deep learning-based reduced order models for the real-time simulation of cardiac electrophysiology in the left atrium. Front. Physiol. **12** (2021)

8. He, K., Zhang, X., Ren, S., Sun, J.: Deep residual learning for image recognition. In: IEEE Conference CVPR, pp. 770–778 (2016)
9. Herrero Martin, C., et al.: EP-PINNs: Cardiac electrophysiology characterisation using physics-informed neural networks. Front. Cardiovasc. Med. **8**, 2179 (2022)
10. Kashtanova, V., Ayed, I., Arrieula, A., Potse, M., Gallinari, P., Sermesant, M.: Deep learning for model correction in cardiac electrophysiological imaging. In: MIDL 2022-Medical Imaging with Deep Learning (2022)
11. Kingma, D.P., Ba, J.: Adam: A method for stochastic optimization. arXiv preprint:1412.6980 (2014)
12. Mitchell, C.C., Schaeffer, D.G.: A two-current model for the dynamics of cardiac membrane. Bull. Math. Biol. **65**(5), 767–793 (2003)
13. Nagumo, J., Arimoto, S., Yoshizawa, S.: An active pulse transmission line simulating nerve axon. Proc. IRE **50**(10), 2061–2070 (1962)
14. Nash, M.P., Panfilov, A.V.: Electromechanical model of excitable tissue to study reentrant cardiac arrhythmias. Prog. Biophys. Mol. Biol. **85**(2), 501–522 (2004)
15. Nezlobinsky, T.: Software for cardiac modelling and patch clamp experiments, and its application to study conduction in the presence of fibrosis. Ph.D. thesis, Ghent University (2021)
16. Niederer, S., Mitchell, L., Smith, N., Plank, G.: Simulating human cardiac electrophysiology on clinical time-scales. Front. Physiol. **2**, 14 (2011)
17. Paszke, A., et al.: Pytorch: an imperative style, high-performance deep learning library. In: Advances in Neural Information Processing Systems, vol. 32 (2019)
18. Pop, M., et al.: Fusion of optical imaging and MRI for the evaluation and adjustment of macroscopic models of cardiac electrophysiology: a feasibility study. Med. Image Anal. **13**(2), 370–380 (2009). Includes Special Section on Functional Imaging and Modelling of the Heart
19. Relan, J., et al.: Coupled personalization of cardiac electrophysiology models for prediction of ischaemic ventricular tachycardia. Interface Focus **1**(3), 396–407 (2011)
20. Sahli Costabal, F., Yang, Y., Perdikaris, P., Hurtado, D.E., Kuhl, E.: Physics-informed neural networks for cardiac activation mapping. Front. Phys. **8**, 42 (2020)
21. Ten Tusscher, K.H.W.J., Noble, D., Noble, P.J., Panfilov, A.V.: A model for human ventricular tissue. Am. J. Physiol. - Heart Circ. Physiol. **286**, H1573–H1589 (2004)
22. Ten Tusscher, K.H.W.J., Panfilov, A.V.: Alternans and spiral breakup in a human ventricular tissue model. Am. J. Physiol. - Heart Circ. Physiol. **291**(3), H1088–H1100 (2006)
23. Yin, Y., et al.: Augmenting physical models with deep networks for complex dynamics forecasting. In: International Conference on ICRL (2021)

Unsupervised Machine Learning Exploration of Morphological and Haemodynamic Indices to Predict Thrombus Formation in the Left Atrial Appendage

Marta Saiz-Vivó[1]([✉])[iD], Jord Mill[1][iD], Josquin Harrison[2],
Guillermo Jimenez-Pérez[1][iD], Benoit Legghe[3], Xavier Iriart[3][iD],
Hubert Cochet[3][iD], Gemma Piella[1][iD], Maxime Sermesant[2][iD],
and Oscar Camara[1][iD]

[1] Physense, BCN Medtech, Department of Information and Communication
Technologies, Universitat Pompeu Fabra, Barcelona, Spain
{marta.saiz,jordi.mill,gemma.piella,oscar.camara}@upf.edu,
guillermo@jimenezperez.com
[2] Inria, Université Côte d'Azur, Epione Team, Sophia Antipolis, Valbonne, France
{josquin.harrison,maxime.sermesant}@inria.fr
[3] Hôpital de Haut-Lévêque, Bordeaux, France
{xavier.iriart,hubert.cochet}@chu-bordeaux.fr

Abstract. Atrial Fibrillation (AF) is the most common cardiac arrhythmia, and it is associated with an increased risk of embolic stroke. It is known that AF-related thrombus formation occurs predominantly in the left atrial appendage (LAA). However, it is still unknown the structural and functional characteristics of the left atria (LA) that promote low velocities and stagnated blood flow, thus a high risk of thrombogenesis. In this work, we investigated morphological and in-silico haemodynamic indices of the LA and LAA with unsupervised machine learning (ML) techniques, to identify the most relevant features that could subsequently be used to generate thrombus prediction models. A fully automatic pipeline was implemented to extract multiple morphological parameters from a 3D mesh of a LA. Morphological parameters were then combined with particle flow parameters from in-silico fluid simulations. Unsupervised multiple kernel learning (MKL) was used for dimensionality reduction, resulting in a latent space positioning patients based on feature similarity. Clustering applied to the MKL output space estimated clusters with different proportion of thrombus cases. The cluster with the highest risk of thrombus formation was characterised by high values of LAA height, tortuosity and ostium perimeter, as well as total number of flow particles in the LAA and low angle between the LAA and the left superior pulmonary vein, proving the usefulness of unsupervised ML techniques to extract knowledge from the data, and early identify AF patients at higher risk of thrombus formation.

O. Camara et al. (Eds.): STACOM 2022, LNCS 13593, pp. 200–210, 2022.
https://doi.org/10.1007/978-3-031-23443-9_19

Keywords: Atrial fibrillation · Left atrial appendage · Unsupervised machine-learning · Thrombus

1 Introduction

Atrial Fibrillation (AF) is the most common type of cardiac arrhythmia, being an important risk factor for stroke. In patients with non-valvular AF, it is estimated that 99% of thrombus are formed in the left atrial appendage (LAA) [6]. Previous studies have reported the LAA morphology as a risk factor for thrombus formation [7]. Thus, Di Base et al. [7] classified the LAA morphology in qualitative groups, reporting that patients with non-chicken wing morphologies had an increased risk of thrombus. However, qualitative classifications are prone to inter-observer discrepancies and may lack morphological details. As a result, recent studies have sought to quantify the LAA shape through the extraction of objective morphological parameters such as LAA ostium (i.e., interface between the LA and the LAA) dimensions or the LAA bending angle [19], among others. More recently, the LA shape and pulmonary veins (PV) configuration have also been studied as markers for thrombus prediction [13]. One of the main problems of these studies is that not all morphological features can be automatically extracted from the patient images, hampering the analysis of large datasets.

Furthermore, LA haemodynamics play a significant role in thrombus formation [4,14]. The Virchow's triad describes the main thromboembolic events as: hypercoagulability, hemodynamic changes, and endothelial injury dysfunction [18]. In AF patients, there is a decrease in LAA contractility function, which may lead to low LAA blood flow velocity, blood stagnation and LAA dilation [5]. Recently, Pons et al. [15] performed a statistical analysis combining morphological indices with LA haemodynamic indices derived from Computational Fluid Dynamics (CFD). More recently, Fang et al. [8] applied particle-based modeling to study blood stagnation in LAA. However, few studies have applied advanced machine learning (ML) techniques to identify the most relevant morphological and haemodynamic features related to thrombus formation in AF patients. Harrison et al. [11] built a latent space merging multiple heterogeneous datasets, finding interesting links between the LAA and PV positions and orientations with the risk of thrombus formation, but without considering LA haemodynamics.

Unsupervised machine learning techniques such as similarity-based dimensionality reduction has previously been used to learn patterns from unlabeled data, through the integration of multiple features and have been previously employed in cardiac feature extraction [16]. In this work, we investigated the relationship between LA morphological/haemodynamic parameters and the risk of thrombus by identifying patient subgroups in a low-dimensional and interpretable output space, which could ultimately be used for the generation of ML-based thrombus prediction models. Specifically, a set of LA, LAA and left superior PV morphological features were combined with particle flow in-silico indices through an unsupervised ML dimensionality reduction technique known as Multiple Kernel Learning (MKL). Furthermore, the LAA-based morphological parameter extraction was fully automatized aiming for a future implementation in the clinical practice.

2   Materials and Methods

2.1   Data

The dataset used in this work was provided by Hospital Haut-Lévêque (Bordeaux, France) and included high-quality pre-procedural computed tomography (CT) scans from AF patients that underwent a left atrial appendage occlusion (LAAO) procedure. After the automatic morphology extraction pipeline, a total of 121 patients were included in this study. The cardiac scan acquisition was performed on a 64-slice dual source CT system (Siemens Definition, Siemens Medical Systems, Forcheim, Germany) with a biphasic injection protocol: 1 mL/kg of Iomeprol 350 mg/mL (Bracco, Milan, Italy) at the rate of 5 mL/s, followed by a 1 mL/kg flush of saline at the same rate. The CT images provided had 512×512 pixel matrix dimensions and 169 to 979 number of slices with $\sim 0.5 \times 0.5 \times 0.5$ mm pixel spacing. Informed consent was provided by all patients and the study was approved by the Institutional Ethics Committee.

With respect to thrombus formation, 51 cases in the database had history of LAA-related thrombus, and 70 cases were controls. Furthermore, a classification combining information of LAA shape & LAA-LSPV alignment was provided since it was shown to be relevant for thrombus prediction [13]. The first three groups categorized LA geometries by degrees of alignment: the J group (n = 54), with subjects having a LAA-LSPV alignment; the A group (n = 15), with LAA-LSPV misalignment; the NA group (n = 28), with moderate LAA-LSPV misalignments. Finally, the CW group (n = 24) contained the LAA chicken-wing morphologies since they are particularly different from the remaining LAA shapes.

2.2   Methodological Pipeline

2.3   Data Pre-processing

From the CT images, the LA geometry was obtained using the (semi-automatic) region-growing segmentation tools available in Slicer 4.10.112[1]. The LA binary masks obtained were introduced to a Marching Cubes algorithm to generate 3D surface mesh reconstructions. Subsequently, a data pre-processing pipeline was implemented for LA, LAA and PV morphological parameter extraction. The pre-processing pipeline can be divided into two blocks: a method for automatic ostium detection and LAA clipping; and an algorithm for PV ostium detection (see Fig. 1).

For an automatic extraction of LA morphological parameters, the LAA had to be separated or clipped from the main LA structure. Thus, the first block aims at obtaining an automatic LAA clipping through the orifice of the LAA, otherwise known as ostium. First, the extraction of the LA centreline was performed with the VMTK[2] library in Python, from the definition of two seed points: the

---

[1]   https://www.slicer.org/.
[2]   https://www.vmtk.org/.

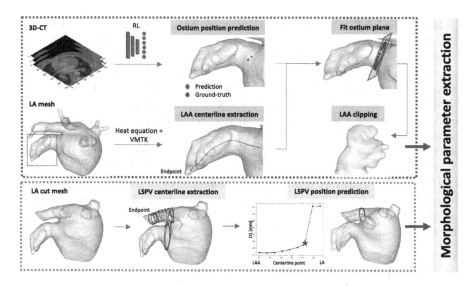

Fig. 1. Pipeline for the automatic extraction of left atrial (LA) morphological parameters. The purple and green blocks represent left atrial appendage (LAA) and left superior pulmonary vein (LSPV) pre-processing steps, respectively. CT: computed tomography images. VMTK: Vascular Modelling Toolkit. (Color figure online)

geometrical center of the LA mesh and the tip point of the LAA. Similar to [10], the furthermost or tip point of the LAA was automatically detected by applying a steady-state heat transfer equation on the base of the LAA.

A reinforcement learning (RL) algorithm, trained with CT images and expert annotations of the ostium position, was implemented for an image-based automatic ostium detection. The landmark detection algorithm is a Deep Q-Network [2] modified for cardiac applications. The model was trained with 158 CT images provided from Hospital Haut-Lévêque (Bordeaux, France) and Ringshospitalen (Copenhagen, Denmark) hospitals. The expert-based training labels were extracted from the VIDAA platform [1]. For training and inference, the CT images were resampled to $256 \times 256 \times 305$ pixels. The model was evaluated with 16 testing images, the mean distance error between ostium label and prediction was 2.92 ± 1.34 mm. The trained model and pre-processing method can be found in https://github.com/marcos-mc/RL_landmark_detection_for_cardiac_applications.

The inferred voxel coordinate prediction was transformed to a world coordinate from which the closest LA centreline point was selected as the final ostium position. Once the LAA centreline and ostium position were extracted, the ostium orientation was defined as the unit vector computed from two centreline points, located 5 points before and after the ostium point. With the ostium position point coordinate and the orientation vector as the normal, the ostium plane was defined.

To clip the LAA in a consistent manner, we adopted the definition of Walker et al. [17], defining the ostium as the narrowest part of the LAA neck. Thus, for the ostium plane estimation, similar to the work of Leventic et al. [12], random perturbations were added to the estimated normal vector and ostium position to select the plane that clips the LAA with the smallest cross-sectional area.

Similarly, the second block aimed at the detection of the PV ostium of the LSPV. For this, the PV was automatically detected with Paraview[3], after which the endpoint of the PV was extracted by estimating the center of this region. Centrelines of each PV were then automatically computed from the LA centre to each estimated PV endpoint. Inspired by Leventic et al. [12], the ostium of the PV was automatically estimated by defining a set of contours along the PV centreline. Then, for each contour the maximal ostium diameter was computed, and the centreline point prior to the highest increase in contour diameter was defined as the PV ostium point (Fig. 1, green block).

2.4 Morphological Parameter Extraction

A set of morphological parameters were automatically computed once the LAA structure was clipped. These were selected to characterise the size, shape and complexity of the LA and LAA, including: LA area/volume, maximal/minimal ostium diameter, the height/depth/length of the LAA, ostium perimeter, area and mean diameter and the LA centreline length. Furthermore, the bending angle of the LAA, the ostium irregularity and the LAA tortuosity were also extracted. In addition, to detect multi-lobular surfaces, similar to Alenyà et al. [3], a curvature metric and a pseudo-gyrification index (i.e., convex hull area divided by the LAA area; more irregular shapes would have smaller values) was extracted from the surface of LAA meshes.

Similar to [13], the angle between the LAA and LSPV main directions was computed. Higher values of this angle would indicate increased bending of the LAA, as the LAA geometrical center moves away from the LSPV. Furthermore, if the direction of the LSPV varies from the direction of the LAA from a frontal view, what has been defined as LPSV/LAA alignment, this angle would increase. Therefore, we hypothesized that this angle could provide useful information about the LSPV/LAA alignment and the LAA morphology.

As in-silico haemodynamic parameters, the total number of particles and particle age were provided. The total number of particles referred to the number of particles found in the LAA after three cardiac beat fluid simulation, while the particle age referred to the average time spent of particles in the LA.

2.5 Unsupervised MKL Analysis

An unsupervised formulation of MKL was used to obtain a low-dimensional output space integrating LA morphological and haemodynamic parameters. The

---

[3] https://www.paraview.org/.

MKL algorithm positions subjects in the output space based on their input feature similarity, quantified by Gaussian kernel affinity matrices which are feature-specific, resulting in an easily-interpretable latent space. For a detailed description of the algorithm, the interested reader is referred to [16]. The input features of the algorithm were the morphological and haemodynamic parameters described above (total of 19). For the affinity matrices, the Euclidean distance was used as a metric. Once the latent space was obtained, a K-Means algorithm was applied as a unsupervised clustering technique to group subjects based on phenotypical similarities related to LA/LAA/PV morphology and haemodynamics.

To validate the phenogrouping, the clustering obtained on the MKL output space was compared to that obtained on the principal component analysis (PCA) space.

3 Results

3.1 Latent Space Exploration

Population-Wise Distribution. The MKL output space combining morphological and haemodynamic features is depicted in Fig. 2(a)-(c). As can be easily observed in Fig. 2(a), there are two regions with higher thrombus (left) and higher control (right) density.

Feature Regression. As previously observed in the probability distribution plots, patients located at the left-most region in the plots of Fig. 2 show a trend of increased thrombus formation. Focusing on features that characterize each region, we observed that the maximal ostium diameter (Fig. 2(g)) and the total number of particles (Fig. 2(f)) exhibit a trend with increasing values as we move towards the left-most output space region. Regarding the LAA-LSPV orientation angle (Fig. 2(e)), the opposite trend is observed, where the angle increases as we shift towards the right-most or control group region. Interestingly, the LSPV/LAA alignment & shape classification labels (Fig. 2(d)) also exhibit a trend in this output space where chicken-wing (yellow) and misaligned (purple and blue) morphologies are increasingly found as we move towards the control group region. In general, patients located at the left-most region, showed increased maximal ostium diameter, total number of particles and proportion of LSPV/LAA aligned morphology and decreased orientation angle.

3.2 Cluster Analysis

Clustering was performed on the first three dimensions of the output space with three clusters as the optimal number of clusters revealed by the silhouette score. The MKL statistical cluster analysis results are shown in Table 1. As observed, the subjects revealed significant inter-cluster differences in the proportion of thrombus cases (61% in Cluster 1 vs 37% and 29% in Cluster 2

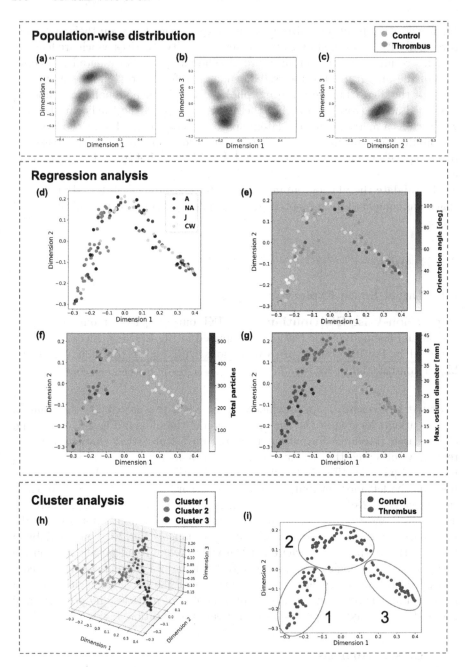

Fig. 2. Physiological interpretation of the MKL output space for the first dimensions of the projected data. (a)-(c): probability density function for control and thrombus sample populations of different pairs of dimensions, higher color intensity indicates higher density. (d)-(g): feature-based regression analysis for the first two dimensions of projected data, (h)-(i): cluster distribution in the first three (h) and first two (i) dimensions. A, complete; NA, moderate; J, no LAA-LSPV misalignment; CW, chicken-wing.

and 3, respectively). Furthermore, the MKL input features that revealed statistical significance were: maximal and minimal ostium diameter, LAA height and depth, LAA tortuosity, ostium perimeter, area and mean diameter, LAA bending angle, ostium irregularity, LA and LAA centreline length, curvature, total particles and orientation angle. Cluster 1 was characterized by increased ostium diameter, height and depth, LAA and LA and LAA centreline length and a decreased orientation angle. A significantly higher curvature index and

Table 1. Morphological and haemodynamic characteristics of subjects by cluster for MKL and PCA methods. Results are given as the mean for continuous variables and as a proportion (%) for categorical variables. N; number of subjects. * Features with statistically significant inter-cluster differences (for both methods). In bold, significant values of cluster with highest proportion of thrombus.

	MKL			PCA	
	Cluster 1	Cluster 2	Cluster 3	Cluster 1	Cluster 2
N	38	49	34	65	56
Thrombus prop. (%) *	**60.5**	36.7	29.4	**51.8**	33.8
Max. ostium diameter (mm) *	**34.14**	27.55	21.85	**32.53**	24.13
Min. ostium diameter (mm) *	**24.58**	20.13	15.32	**23.42**	17.38
LAA height (mm) *	**21.59**	17.55	14.80	**21.28**	15.18
LAA depth (mm)*	**39.78**	36.27	31.58	**39.54**	33.05
LAA tortuosity *	**0.43**	0.39	0.38	**0.43**	0.38
LA area (mm^2)	22403	22663	23304	23486	22137
LA volume (mm^3)	169178	170798	181698	165774	182148
Ostium perimeter(mm) *	**94.74**	76.09	62.14	**90.08**	66.03
Ostium area (mm^2) *	**673.5**	435.5	265.9	**611.6**	334.1
Ostium mean diameter (mm) *	**29.36**	23.84	18.58	**28.00**	20.75
LAA bending angle (deg) *	**118.10**	108.64	107.11	**117.38**	105.85
Ostium irregularity	0.300	0.290	0.320	0.032	0.028
LA centreline length (mm) *	**192.20**	171.70	166.15	**191.2**	164.0
LAA centreline length (mm) *	**47.82**	42.15	36.60	**47.00**	38.39
LAA curvature *	**0.058**	0.068	0.079	**0.058**	0.076
LAA GI	0.95	0.94	0.95	0.97	0.98
Particle age	1.32	1.29	1.31	1.30	1.31
Total nº particles *	**240.50**	173.70	111.6	**229.9**	131.9
Orientation angle (deg) *	**33.06**	40.75	46.70	**30.52**	48.19
*LAA shape/orientation (%) *￼*					
(i) A (n = 15)	**7.9**	16.3	11.7	**8.9**	15.4
(ii) NA (n = 28)	**15.8**	22.4	32.4	**14.3**	30.8
(iii) J (n = 54)	**68.4**	38.8	26.5	**71.4**	21.5
(iv) CW (n = 24)	**7.9**	22.4	29.4	**5.4**	32.3

smaller bending angle was found in Cluster 3 compared to Cluster 1, likely due to the increased proportion of chicken-wing morphologies. As for haemodynamic parameters, only the total number of particles exhibited statistical significance, with increased values shown in the first cluster.

Furthermore, significant inter-cluster differences were also found in the alignment/shape classification labels (Table 1). Cluster 1 exhibited a markedly increased proportion of completely aligned morphologies, as seen in Fig. 2(d). In parallel, Cluster 2 showed the highest proportion of completely misaligned cases, whilst Cluster 3 showed the highest proportion of chicken-wing and slightly misaligned cases.

3.3   Validation of Cluster Analysis Results

The clustering results obtained with the MKL dimensionality reduction method were validated through the implementation of PCA. For this, the first 6 principal components (accounting for 80% of the variance) were selected to perform cluster analysis. Similar to the methodology followed for MKL space analysis, a silhouette plot and K-Means algorithm were employed. The optimal number of clusters for the first 6 dimensions resulted in two clusters. In Table 1 the PCA statistical cluster analysis results are additionally shown.

Through a comparison of MKL and PCA cluster results, we observe that the cluster with the highest proportion of thrombus (Cluster 1 for both methods) shows similar feature value trends with respect to other clusters. In fact, the same trend is observed for all features with statistical significant differences (highlighted in bold).

4   Discussion and Conclusions

In this work we aimed to explore a combination of LAA morphological and particle-flow based indices through unsupervised ML techniques. With the low-dimensional latent space extracted, probability distributions of thrombus and non-thrombus populations were observed jointly with feature-wise trends. As observed in Fig. 2, morphological features such as the maximal ostium diameter exhibited a decreasing trend from higher to lower proportion of thrombus formation, in agreement with literature [9,10]. Moreover, we found a relation of LAA-LSPV alignment with the risk of thrombus formation, with less risk associated to the groups with more misalignment. The results suggests that quantifying the configuration of LSPV with respect to LAA could provide useful information related to the risk of thrombus formation and further work should be performed in this area.

Through the examination of haemodynamic indices, a higher number of particles in the LAA were observed in subjects distributed in the left-hand side of the MKL plots, coinciding with the assumption that aligned LSPV and LAA structures lead to higher blood flow entrance in the LAA. However, whether this would lead to increased blood stagnation and risk of thrombus formation cannot be concluded. As reported in [13], the configuration of other PV such as the right superior pulmonary vein (RSPV) could also play an important role in the stagnation process.

The MKL cluster analysis provided three groups with similar morphological and haemodynamic characteristics and distinct thrombus proportion (Table 1). Cluster 1 was associated with the highest proportion of thrombus formation and was characterized by increased maximal and minimal ostium diameter, height, depth, tortuosity and LAA centreline length. This coincides with previous literature where LAA size and complexity have been related to an increased risk of thrombus formation [9]. With respect to qualitative-based labels, we observe that Cluster 1 has a predominant proportion of J type or completely aligned morphology. In fact, shifting from Cluster 1 to 3 we observed that the proportion of J morphologies significantly decreased whilst chicken-wing morphologies increased. This can be linked with previous literature suggesting that chicken-wing morphologies are associated with lower risk of thrombus [7]. Interestingly, Cluster 2 showed the highest proportion of misaligned cases, although completely aligned morphologies were still predominant. On the other hand, Cluster 3 showed a clear superiority of chicken-wing and misaligned morphologies. Furthermore, a posterior PCA cluster analysis revealed similar feature trends related to the proportion of thrombus formation (Table 1).

Based on our analysis we conclude that the configuration of other PVs should be extracted for a more comprehensive assessment of blood flow stagnation processes and that further work is required to better quantify the distribution of PVs. In addition, further validation should be performed on the proposed automatic parameter extraction method as the accuracy of the feature values will have an effect on the output low-dimensional space. Finally, the exploration and extraction of a low-dimensional latent space based on these parameters enabled a good starting point for the future generation of unsupervised thrombus prediction models.

Acknowledgements. This project has received funding from the European Union's Horizon 2020 research and innovation programme under grant agreement No 101016496 (SimCardioTest).

References

1. Aguado, A.M., et al.: In silico optimization of left atrial appendage occluder implantation using interactive and modeling tools. Front. Physiol., 237 (2019)
2. Alansary, A., et al.: Evaluating reinforcement learning agents for anatomical landmark detection. Med. Image Anal. **53**, 156–164 (2019)
3. Alenyà, M., et al.: Computational pipeline for the generation and validation of patient-specific mechanical models of brain development. Brain Multiphys. **3**, 100045 (2022)
4. Ammash, N., et al.: Left atrial blood stasis and von Willebrand factor-adamts13 homeostasis in atrial fibrillation. Arterioscler. Thromb. Vasc. Biol. **31**(11), 2760–2766 (2011)

5. Beigel, R., Wunderlich, N.C., Ho, S.Y., Arsanjani, R., Siegel, R.J.: The left atrial appendage: anatomy, function, and noninvasive evaluation. JACC: Cardiovasc. Imaging **7**(12), 1251–1265 (2014)

6. Cresti, A., et al.: Prevalence of extra-appendage thrombosis in non-valvular atrial fibrillation and atrial flutter in patients undergoing cardioversion: a large Transoesophageal echo study. EuroIntervention **15**(3), e225–e230 (2019)

7. Di Biase, L., et al.: Does the left atrial appendage morphology correlate with the risk of stroke in patients with atrial fibrillation? Results from a multicenter study. J. Am. Coll. Cardiol. **60**(6), 531–538 (2012)

8. Fang, R., Li, Y., Zhang, Y., Chen, Q., Liu, Q., Li, Z.: Impact of left atrial appendage location on risk of thrombus formation in patients with atrial fibrillation. Biomech. Model. Mechanobiol. **20**(4), 1431–1443 (2021). https://doi.org/10.1007/s10237-021-01454-4

9. García-Isla, G., et al.: Sensitivity analysis of geometrical parameters to study Haemodynamics and thrombus formation in the left atrial appendage. Int. J. Numer. Methods Biomed. Eng. **34**(8), e3100 (2018)

10. Genua, I., et al.: Centreline-based shape descriptors of the left atrial appendage in relation with thrombus formation. In: Pop, M., et al. (eds.) STACOM 2018. LNCS, vol. 11395, pp. 200–208. Springer, Cham (2019). https://doi.org/10.1007/978-3-030-12029-0_22

11. Harrison, J., Lorenzi, M., Legghe, B., Iriart, X., Cochet, H., Sermesant, M.: Phase-independent latent representation for cardiac shape analysis. In: de Bruijne, M., et al. (eds.) MICCAI 2021. LNCS, vol. 12906, pp. 537–546. Springer, Cham (2021). https://doi.org/10.1007/978-3-030-87231-1_52

12. Leventić, H., et al.: Left atrial appendage segmentation from 3D CCTA images for Occluder placement procedure. Comput. Biol. Med. **104**, 163–174 (2019)

13. Mill, J., et al.: In-Silico analysis of the influence of pulmonary vein configuration on left atrial Haemodynamics and thrombus formation in a large cohort. In: Ennis, D.B., Perotti, L.E., Wang, V.Y. (eds.) FIMH 2021. LNCS, vol. 12738, pp. 605–616. Springer, Cham (2021). https://doi.org/10.1007/978-3-030-78710-3_58

14. Nedios, S., et al.: Left atrial appendage morphology and thromboembolic risk after catheter ablation for atrial fibrillation. Heart Rhythm **11**(12), 2239–2246 (2014)

15. Pons, M.I., et al.: Joint analysis of morphological parameters and in silico Haemodynamics of the left atrial appendage for thrombogenic risk assessment. J. Interv. Cardiol. **2022**, 9125224 (2022)

16. Sanchez-Martinez, S., Duchateau, N., Erdei, T., Fraser, A.G., Bijnens, B.H., Piella, G.: Characterization of myocardial motion patterns by unsupervised multiple kernel learning. Med. Image Anal. **35**, 70–82 (2017)

17. Walker, D.T., Humphries, J.A., Phillips, K.P.: Anatomical analysis of the left atrial appendage using segmented, three-dimensional cardiac CT: a comparison of patients with paroxysmal and persistent forms of atrial fibrillation. J. Interv. Card. Electrophysiol. **34**(2), 173–179 (2012). https://doi.org/10.1007/s10840-011-9638-1

18. Watson, T., Shantsila, E., Lip, G.Y.: Mechanisms of Thrombogenesis in atrial fibrillation: Virchow's triad revisited. Lancet **373**(9658), 155–166 (2009)

19. Yaghi, S., et al.: Left atrial appendage morphology improves prediction of stagnant flow and stroke risk in atrial fibrillation. Circ. Arrhythm. Electrophysiol. **13**(2), e008074 (2020)

Geometrical Deep Learning for the Estimation of Residence Time in the Left Atria

Daniel Cañadas Gómez, Xabier Morales Ferez[✉], and Oscar Camara Rey

Physense, Department of Information and Communication Technologies,
Universitat Pompeu Fabra, Barcelona, Spain
xabier.morales@upf.edu

Abstract. Atrial fibrillation (AF) is the most common clinically significant arrhythmia, which can lead to clot formation in the left atrial appendage (LAA). Computational fluid dynamics simulations can provide advanced hemodynamic descriptors to help in the stratification of patient thrombogenic risk. For instance, the residence time measures the degree of fluid entrapment inside the domain of interest, which can be related to the risk of thrombus formation. However, classic fluid modeling approaches are notoriously time-consuming and resource intensive, preventing their use in large patient cohorts. Therefore, in this study, we leveraged deep learning (DL) methods to predict the residence time, solely based on the patient-specific left atrial geometry at inference time. To this end, we used graph neural networks since they are well-suited to non-structured data such as meshes, in which medical data and simulation results are often best represented. Two slightly different neural network architectures were trained on a cohort of 106 patients, with velocity fields derived from computational fluid dynamics simulations as training data. The average RT values were accurately predicted, with a mean normalized root mean squared error of 0.08, in less than 5 s, demonstrating the potential of DL-based surrogates of fluid simulation outcomes to save computational time and resources without compromising result accuracy.

Keywords: Neural network · Fluid simulations · Atrial fibrillation ·
Left atrial appendage · Thrombus formation

1 Introduction

Atrial fibrillation (AF) is the most common form of arrhythmia. It is estimated that in Europe alone, some 17.9 million people will suffer from this condition by 2060 [1]. The higher mortality rate observed in AF patients is often related to major system thromboembolism [2]. Notably, up to 99% of blood clots associated with AF thromboembolism are said to arise from the left atrial appendage (LAA), a small tubular-shaped sac located in the wall of the left atrium (LA).

O. Camara et al. (Eds.): STACOM 2022, LNCS 13593, pp. 211–220, 2022.
https://doi.org/10.1007/978-3-031-23443-9_20

Unfortunately, the limited spatiotemporal resolution of current imaging techniques is often ill-suited to the complex 4D nature of LA hemodynamics, key to analyzing the risk of thrombosis. Computational fluid dynamics (CFD), on the other hand, can derive advanced hemodynamic parameters such as the endothelial cell activation potential (ECAP) or the residence time (RT) that can provide an estimate of this risk. The latter is particularly interesting since it quantifies the degree of fluid entrapment in a given domain, and can provide information on flow compartmentalization and re-circulation. However, similarly to CFD simulations, the computation of RT involves solving partial differential equations (PDE), which is computationally expensive.

As such, there is growing interest in the development of CFD deep learning (DL) surrogates, which once trained can be run orders of magnitude faster and with fewer computational resources to derive such advanced hemodynamic indices. However, the majority of current DL models are most suited to regularly structured data such as images. Anatomical data and simulation results, on the other hand, are often most efficiently represented through graphs or meshes; due to their irregular neighborhood structure, the proper definition of operators such as convolution is nontrivial [6]. Costly transformations of mesh data onto Cartesian grids can be employed but they often lead to the distortion or loss of information [9]. Aiming to circumvent this issue, a new set of methods have emerged, known as geometric DL (GDL), which are capable of handling irregularly structured data. Such models have already been applied as fast alternatives to fluid simulations. For instance, Suk et al. [7] used gem-equivariant neural networks to predict wall shear stress (WSS) in coronary artery surface meshes. Morales et al. [9] also used GDL models to predict the endothelial cell activation potential (ECAP) inside the LA without requiring fluid simulations at inference time.

Therefore, in this work, we developed a DL surrogate for the rapid and efficient estimation of the RT for the stratification of blood clot formation risk inside the LA. Two distinct GDL architectures (PointNet and U-Net) were trained on a large dataset of patient-specific LA geometries and corresponding CFD simulations. At inference time, the developed models estimate RT in real-time, orders of magnitude faster than conventional finite-element solvers.

2   Methods

2.1   Dataset

A dataset consisting of 106 LA volumetric meshes was used, created from the semi-automatic segmentation of pre-intervention cardiac computed tomography (CT) scans of AF patients that underwent an LAA occlusion (LAAO) procedure. Imaging data were collected in Hospital Haut-Lévêque (Bordeaux, France), after approval by the institutional Ethics Committee and informed consent from patients. The LA meshes contained between 3 and 7×10^5 elements, depending on LA geometrical complexity. CFD simulations were run for 3 cardiac beats (84 time-steps each) on the LA volumetric meshes, creating 4D velocity fields simulating LA blood flow patterns. The interested reader can find more details on the CFD simulations in [9].

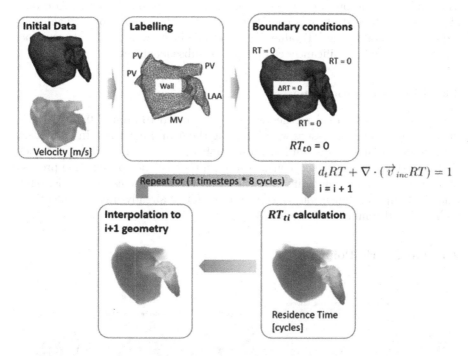

Fig. 1. Computational pipeline to estimate the residence time (RT) vector fields from computational fluid velocities (v_{inc}), for the training of geometrical deep learning architectures. PV: pulmonary veins. LAA: left atrial appendage. MV: mitral valve.

2.2   Generation of Residence Time Fields for Training

The developed computational pipeline to generate the residence time fields for training the GDL architectures is illustrated in Fig. 4. From the CFD simulations, the following advection equation was solved to obtain the RT scalar field:

$$d_t RT + \nabla(v_{inc}RT) = 1, \tag{1}$$

where v_{inc} is the velocity field obtained from the CFD solver. Eq. 1 was solved with FiPy[1], a finite-volume PDE solver toolbox for Python. The two initial beats were intended to stabilize the simulation, so only the velocity data from the last beat was considered for model training. As boundary conditions, the RT was set to 0 for both inlets and outlets of the LA (pulmonary veins and mitral valve,

---

[1] https://www.ctcms.nist.gov/fipy.

respectively), as in [5]. The divergence of the RT in the LA walls was similarly set to 0, as well as RT values everywhere as initial conditions.

The residence time values computed for a given time step were interpolated as boundary conditions for the next one, due to the motion of the LA. Additionally, a convergence study was completed to assess the number of cardiac cycles required to stabilize the RT value. Eight cardiac beats were finally chosen since RT values did not significantly change in the subsequent beats.

2.3   Mesh Pre-processing

All input LA geometrical meshes were downsampled to around 0.3×10^5 elements to avoid memory issues while training the neural networks. In addition, a surface interpolation based on k-nearest neighbors was used to project volumetric RT fields onto the surface, considerably simplifying the optimization process. Moreover, we focused on estimating the mean RT over the cardiac cycle, omitting the temporal dimension, since the mean RT is already useful to identify a high risk of thrombus formation [10].

2.4   Geometric PointNet

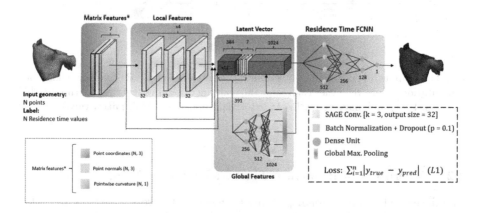

Fig. 2. Architecture of the Geometric PointNet. It consists of two main blocks: a local feature extractor and a global feature extractor. The first one uses SAGE convolutions [14] to extract a combination of local vertex features. These are then concatenated and fed to multilayer perceptron before extracting global features through max pooling. Finally, both sets of features are grouped before predicting the RT for each surface vertex. Numbers in each layer indicate the number of features.

The first evaluated network architecture is shown in Fig. 3, and was based on the approach developed by Morales et al. [9], who employed a graph convolutional neural network (GCNN) based on the popular PointNet network [12], using spline convolution operators to predict advanced hemodynamic parameters on the left atrial appendage.

The mesh data was considered an un-directed graph, $G = <v, \epsilon, X>$, composed of N vertices belonging to the set of vertices, v. N can take a different value for each of the analyzed meshes, making GCNN attractive to work with unstructured meshes in which it is complex to define correspondences between subjects. The edges in ϵ connect pairs of vertices from v. The trainable graph convolutional filters are then applied to a set of point-wise features X.

The un-directed graph was used to extract and concatenate an initial set of input features, including the vertex coordinates, the vertex normals, and the point-wise curvature. The local feature extractor was defined as a sequence of 12 SAGE convolutional layers [14] with a hidden size equal to 32, followed by a batch normalization layer and a dropout with probability p = 0.1. SAGE convolutional layers process each feature vector, h, belonging to each of the vertices $j \in v$ and aggregate the information of the j's immediate neighborhood, including all the neighbors of vertex j that are directly connected by an edge $e \in \epsilon$.

The concatenation of all the convolutional layer outputs from the local feature extractor, with the input matrix of features, will then be passed to the global feature extractor. It was defined as a multilayer perceptron (depth = 3), with shared weights and max-pooling. Lastly, the concatenation of both extractor outputs was processed once again by a fully connected layer (depth = 4), with decreasing hidden sizes of 512, 256, and 128, producing a single RT value in the last layer for each vertex j.

2.5 Mesh U-Net

A second GDL methodology was evaluated by following the approach developed by Suk et al. [7], who proposed a U-Net-based architecture to predict WSS in coronary arteries. Unlike the Geometric PointNet approach, this model introduces pooling operations that allow extracting multi-scale features.

Some of the input features included the edge connectivity, the coordinates of the vertices of the mesh, the surface normals (in the form of unit vectors), and the combined geodesic distance to the inlets. The latter gives the model a notion of the direction of the blood flow inside the LA, thus making the model rotation-equivariant. These are decomposed into factors of irreducible representations of two-dimensional rotation group (SO(2)) [8].

The input vector features for each of the vertices are then passed through the first of four pooling blocks composing the mesh CNN architecture. Each of these pooling blocks is composed of a few residual blocks and a pooling layer. The residual blocks are in the same way divided into two graph SAGE convolutions [14] and a skip connection between them. Anisotropic gauge-equivariant kernels were introduced [8], which carry relative orientation information for the different features through the mesh edges (message passing). The general convolution scheme can then be defined as follows:

$$((K^1, K^2(\cdot, \cdot)) * f)p := fp \cdot K^1 + \sum_{q \in N(p)} \rho(p, q) \cdot fq \cdot K^2(p, q), \qquad (2)$$

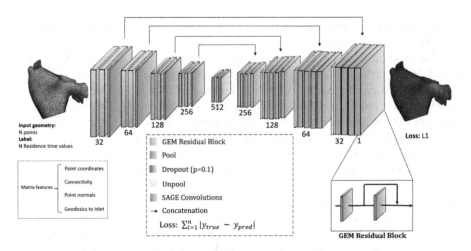

Fig. 3. Layout of the Mesh U-Net. It follows a typical U-shaped structure, where the input features are passed through an encoder, which downsamples the data following a sequence of residual layers with convolutions and poolings. The lower dimensional vector is again upsampled in the decoder branch, which receives skip connections from same-level encoder data. This produces an output of size one for each of the vectors of the mesh.

where K^1 and K^2 are trainable kernels, and weigh both the features of $p \in \upsilon$ and its neighbours q of size c, as detailed in [8]. After all, the residual blocks are applied in a single pooling block, then a pooling layer, defined as clustering, is applied to downsample the data, thus enlarging the model receptor field.

2.6 Hyperparameter Tuning

A grid-search approach was employed to optimize the values of the hyperparameters. All possible combinations were tested using the Adam optimizer, and the L_1 loss function. A train-validation-test split of 80%, 5%, and 15% was used in both models. The Geometric PointNet and Mesh U-Net models were trained for 650 and 1300 epochs, with a batch size of 16 and 2 and a learning rate $lr = 0.0001$ and $lr = 0.01$, respectively. For both models, a dropout with $p = 0.1$ was used.

2.7 Model Evaluation

The ability of the models to generalize to unseen samples was evaluated. To quantify the similarity between the RT from the CFD simulations (ground truth) and the GDL architectures (prediction), the normalized root mean squared error (NRMSE) was computed. In addition, the maximum difference, (Δ_{max}), defined as the maximum of the subtraction between ground truth (GT) and predictions, was also used. The mean and median of the previous measures were calculated over all the subjects in the test set (15% of the total data). The absolute difference between GT and predictions was also calculated. Finally, a threshold equal to 2 of the mean RT was chosen, following [5], to evaluate the potential of the trained neural networks to perform a correct LA thromboembolic risk assessment. The sensitivity, specificity, and accuracy of the analyzed models were then calculated.

3   Results

The CFD simulation results were successfully processed in the analyzed 106 LA geometries to compute the RT maps that will serve for the training of the evaluated GDL methods. Figure 4 (left column, ground truth) shows some examples of the CFD-derived RT maps.

Inference of the mean RT maps during tool less than 5 s for both of the deep learning approaches, orders of magnitude faster than conventional PDE solvers (>24 h). Some examples of mean RT prediction of both the assessed GDL architectures are shown in Fig. 4. The NRMSE compared to the CRT-derived mean RT maps was 0.08 for both methods, and a mean maximum difference (Δ_{max}) of 5 and 3.38 for the Geometric PointNet and Mesh U-Net, respectively.

Regarding the sensitivity, specificity, and accuracy to identify regions with a high risk of thrombus formation (mean RT > 2 cycles), a similar good sensitivity was obtained for the Geometric PointNet and Mesh U-Net architectures (0.820 and 0.824, respectively), while a slightly higher specificity (0.984) and accuracy (0.966) were achieved by the Mesh U-Net, comparing to the Geometric PointNet (0.956 and 0.939, respectively).

4   Discussion

The present study has demonstrated the potential of GNNs to estimate surrogates of haemodynamic metrics for thrombus formation in the LA, such as the mean residence time, in real-time at inference time. Three main steps were required to train the two evaluated geometrical deep learning architectures: the creation of the mean RT dataset from the initial simulated velocity fields; the pre-processing of the resulting LA meshes with RT maps; and the training of the geometric DL models.

Regarding the creation of the initial RT dataset, it could be observed how regions with high mean RT values were mainly found at the tip (deepest part) of the LAA (see Fig. 4, left column), with lower values mostly located outside the LAA cavity. This strongly coincides with the RT maps estimated in García-Villalba et al. [5], and with the fact that more than 99% of AF-related clots are formed inside the LAA [2,11].

The two GDL approaches correctly identified the LA regions with the highest mean RV values networks, mainly the LAA, as can be seen in Fig. 4. However, the Geometric PointNet method provided mean RT maps with values closer to the ground-truth ones (from CFD simulations) but with slightly larger regions.

On the contrary, the Mesh U-Net better matched the regions with high RT values (e.g., see Case 4 and Case 7 in the figure). The same conclusion can be drawn from the quantitative analysis of high risk region detection, with both GDL techniques providing similar sensitivity, but with better specificity, and thus accuracy, for the Mesh U-Net, which is able to better predict true negative zones.

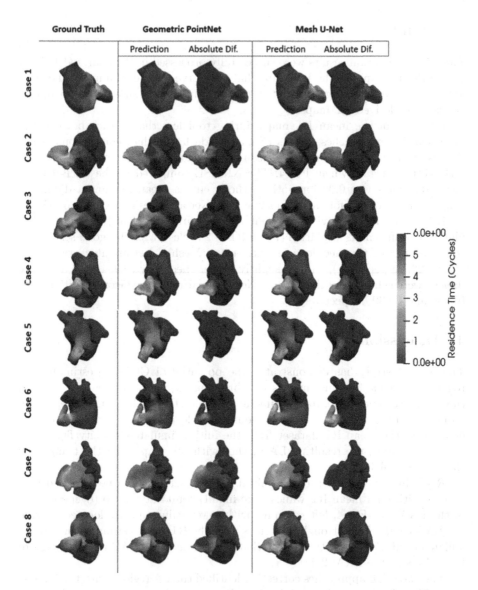

Fig. 4. Mean residence time (RT) maps predicted by the Geometric PointNet and Mesh-UNet geometrical deep learning models (GDL) in eight test left atrial geometries (out of the 106 processed ones). Left: ground-truth mean RT maps directly processed from the computational fluid dynamics (CFD) simulations. The mean RT maps predicted by the GDL techniques are shown next to the absolute difference with the CFD-derived ones.

The performance differences between both models could be due to architectural features of the Mesh U-Net that makes it more sensitive to geometrical features associated with certain RT values. Firstly, the pooling layers increase its receptive field, while countering in the same way the mesh resolution effects on the convolutions. Also, as mentioned above, vertex features in the Mesh U-Net carry orientation information, and are rotation and translation equivariant, having also notion of the flow direction due to the geodesic distance to the input feature.

5   Conclusion

In this study, we have successfully implemented geometrical deep learning networks for the prediction of hemodynamic indices from CFD simulations on patient-specific LA geometries. The obtained preliminary results prove that GDL architectures can properly extract features related to the morphology of LA geometries, and adequately relate them with the corresponding RT values. The mean RT predictions, when translated to the quantification of the risk of thrombus formation in the LAA showed promising results, with the best accuracy of 96.6%, a sensitivity of 82.4%, and a specificity of 98.4%.

Further research should focus on improving the results of these networks for a more accurate prediction of the RT patterns and better classification of high-risk LA geometries. Potential future work includes considering the temporal dimension, obtaining more data, and using models that are subject to physical constraint, approaching these techniques to their use in clinical routine. However, research is needed to accelerate the generation of the required initial velocity vector field (now still obtained from computationally costly CFD simulations), which could be provided by novel imaging sequences such as 4D flow magnetic resonance imaging.

Funding Information. This work was supported by the Agency for Management of University and Research Grants of the Generalitat de Catalunya under the Grants for the Contracting of New Research Staff Programme - FI (2020 FI_B 00608) and the Retos Investigación project (RTI2018-101193-B-I00). Additionally, this work was supported by the H2020 EU SimCardioTest project (Digital transformation in Health and Care SC1-DTH-06-2020; grant agreement No. 101016496).

Conflict of Interest Statement. The authors declare that the research was conducted in the absence of any commercial or financial relationships that could be construed as a potential conflict of interest.

References

1. Lippi, G., Sanchis-Gomar, F., Cervellin, G.: Global epidemiology of atrial fibrillation: an increasing epidemic and public health challenge. Int. J. Stroke **16**(2), 217–221 (2021)
2. Freedman, B., Potpara, T.S., Lip, G.Y.: Stroke prevention in atrial fibrillation. Lancet **388**(10046), 806–17 (2016)

3. Duncan, P.W., Zorowitz, R., Bates, B., Choi, J.Y., Glasb, J.J.: Management of adult stroke rehabilitation care. Stroke **36**(9), e100–e143 (2005)
4. Blackshear, J.L., Odell, J.A.: Appendage obliteration to reduce stroke in cardiac surgical patients with atrial fibrillation. Ann. Thorac. Surg. **61**, 755–759 (1996)
5. García-Villalba, M., Rossini, L., Gonzalo, A., et al.: Demonstration of patient-specific simulations to assess left atrial appendage Thrombogenesis risk. Front. Physiol. **12**, 596596 (2021)
6. Fey, M., Lenssen, J.E., Weichert, F., Müller, H.: SplineCNN: fast geometric deep learning with continuous B-Spline kernels. In: IEEE/CVF Conference on Computer Vision and Pattern Recognition, pp. 869–87 (2018)
7. Suk, J., Haan, P., Lippe, P., Brune, C., Wolterink, J.M.: Mesh convolutional neural networks for wall shear stress estimation in 3D artery models. In: Puyol Antón, E., et al. (eds.) STACOM 2021. LNCS, vol. 13131, pp. 93–102. Springer, Cham (2022). https://doi.org/10.1007/978-3-030-93722-5_11
8. de Haan, P., Weiler, M., Cohen, T., Welling, M.: Gauge equivariant mesh CNNs: anisotropic convolutions on geometric graphs. In: ICLR (2021)
9. Ferez, X.M., et al.: Deep learning framework for real-time estimation of in-silico thrombotic risk indices in the left atrial appendage. Front. Physiol. **12**, 694945 (2021)
10. Martinez-Legazpi, P., Rossini, L., Perez del Villar, C., Benito, Y., Devesa-Cordero, C., Yotti, R.: Stasis mapping using ultrasound: a prospective study in acute myocardial infarction. JACC Cardiovasc. Imaging **11**, 514–515 (2017)
11. Oladiran, O., Nwosu, I.: Stroke risk stratification in atrial fibrillation: a review of common risk factors. J. Community Hosp. Intern. Med. Perspect. **9**(2), 113–120 (2019)
12. Qi, C., Su, H., Mo, K., Guibas, L., Su, H.: PointNet: deep learning on point sets for 3D classification and segmentation. In: CVPR (2017)
13. Mill, J., et al.: Patient-specific flow simulation analysis to predict device-related thrombosis in left atrial appendage Occluders. REC Interv. Cardiol. **3**(4), 278–85 (2021)
14. Hamilton, W.L., Ying, R., Leskovec, J.: Inductive representation learning on large graphs (2017). CoRR, abs/1706.02216

Explainable Electrocardiogram Analysis with Wave Decomposition: Application to Myocardial Infarction Detection

Yingyu Yang[1]([⊠]), Marie Rocher[2], Pamela Moceri[2], and Maxime Sermesant[1]

[1] Université Côte d'Azur, Inria Epione Team, Sophia Antipolis, France
`yingyu.yang@inria.fr`
[2] UR2CA, Université Côte d'Azur, Faculté de Médecine, Nice, France

Abstract. Automatic analysis of electrocardiograms with adequate explain-ability is a challenging task. Many deep learning based methods have been proposed for automatic classification of electrocardiograms. However, very few of them provide detailed explainable classification evidence. In our study, we explore explainable ECG classification through explicit decomposition of single-beat (median-beat) ECG signal. In particular, every single-beat ECG sample is decomposed into five subwaves and each subwave is parameterised by a Frequency Modulated Moebius. Those parameters have explicit meanings for ECG interpretation. In stead of solving the optimisation problem iteratively which is time-consuming, we make use of an Cascaded CNN network to estimate the parameters for each single-beat ECG signal. Our preliminary results show that with appropriate position regularisation strategy, our neural network is able to estimate the subwave for P, Q, R, S, T events and maintain a good reconstruction accuracy (with R2 score 0.94 on test dataset of PTB-XL) in a unsupervised manner. Using the estimated parameters, we achieve very good classification and generalisation performance on myocardial infarction detection on four different datasets. The features of high importance are in accordance with clinical interpretations.

Keywords: ECG analysis · Reconstruction · Explainable ML · Myocardial infarction classification

1 Introduction

Myocardial infarction (MI) is a kind of pathology where myocardial cells are dead due to the prolonged lack of oxygen (ischaemia). A patient is diagnosed as MI if he/she has elevated cardiac troponin values and falls into at least one of the following conditions: symptoms of myocardial ischaemia, new changes of ST-segment/T-wave in electrocardiogram (ECG), development of pathological Q waves in ECG, abnormal myocardium motion and presence of coronary thrombus [22]. Among all the diagnostic approaches, ECG is very easy and fast to perform on patients, even with non-experts. Whereas, to diagnose MI patients with ECG is a challenging task. For example, one study shows that experienced cardiologists only identified 82% of the real ST-segment elevated MI patients [17].

© The Author(s), under exclusive license to Springer Nature Switzerland AG 2022
O. Camara et al. (Eds.): STACOM 2022, LNCS 13593, pp. 221–232, 2022.
https://doi.org/10.1007/978-3-031-23443-9_21

Computer-assisted ECG analysis could help cardiologists, non-experts better interpret ECG recordings for MI detection.

There exist many research works on automatic MI detection using ECG signals and they can be categorised into feature-based methods and neural network based methods. The feature-based methods usually contains three stages: ECG delineation (segmentation), feature extraction and classification. Different kinds of features were explored: morphological features (such as ST-elevation value, QRS duration, T wave amplitude, Q wave amplitude etc.) [4,10,14], wavelet transform related features (coefficients) [5,11,19], empirical mode decomposition features [1] and so on. Compared with other features, morphological features are explainable but very sensitive to ECG delineation results. ECG delineation methods [9,18], usually depend on annotations for all the events (P,Q,R,S,T) to train models. They are constrained by the size and type of pathology in the annotated dataset. The neural network based methods for MI detection have overwhelmed in recent years. For example, single-beat ECG signals are directly classified using a 1D convolutional neural network (CNN) [2] or are transformed into 12-lead 2D image for a 2D CNN [7]. For detailed review of MI detection methods, interested readers are invited to read reviews [3,24].

In our work, we tackle this MI detection problem following the feature-based approach by ECG parameterisation. Actually, ECG parameterisation/modelling is not a new idea. For example, Liu et al. proposed to fit a 20th order polynomial function to a given ECG signal and used the fitted coefficients as ECG features [12]. Second-order ODEs were applied to model the 12-lead ECGs and the estimated time-varying coefficients were used as features [25]. They both achieved good accuracy for MI detection but the features do not have an explainable meaning for clinicians. In our case, we adapt the explainable Frequency Modulated Moebius model as our parameterisation model [20] for single-beat ECG.

The contribution of our work lies in two folds.

- We propose a time-efficient and automatic pipeline for ECG decomposition and reconstruction by passing through a deep learning model: Cascaded FMMnet, which is capable to reconstruct single beat signal with high quality. The training is unsupervised and make it accessible for all kinds of ECG datasets, with or without annotations. The estimated parameters have explainable meanings for each subwave, such as the amplitude, the position etc.
- We present our preliminary results of using the estimated parameters as features to classify normal and myocardial infarction patients. The important features identified are in accordance with the clinical interpretation indexes, such as T wave and Q wave change.

2   Methods

The pipeline of our explainable ECG analysis consists of three stages (Fig. 1(a)). First, the 12-lead ECG recordings are filtered and segmented to obtain single-beat median ECG signals. Second, each lead-wise median signal is passed through an encoder network (Cascaded FMMnet) for subwave decomposition. The encoder network outputs 21 estimated parameters that are crucial for signal reconstruction and interpretation. Third, for each sample (12-lead median ECG signal), myocardial infarction classification is conducted based on the $264(21 \times 12 + 12)$ estimated parameters, where 12 additional parameters comes from estimated ST-segment voltage value. To explain the classification result, we use the additive weighted features for linear classification models and SHAP value [15] for non-linear classification models. Decomposed waveform and feature importance from classifier together give visual and quantitative explainability of our prediction result.

2.1 Data Preprocessing

The preprocessing includes 5 steps: resampling, filtering, R-peak detection, ECG segmentation and median signal generation.

The original 12-lead ECG recording is resampled 500 Hz if its original sampling rate is 500 Hz. A butterworth high-pass filter with cutoff frequency at 0.5 Hz is then applied to remove baseline wander. The R-peaks of Lead II are automatically detected and used as reference for all the other leads. For every lead ECG, each single beat segment is set from 35% heart beat duration (s) before the R-peak to 50% heart beat duration (s) after the R-peak. One 1.2-second median beat signal is calculated by aligning the R-peaks of all the single beats (at 0.5-second position) and padded by neighbouring values at the two ends if the medial signal is shorter than 1.2 s. Neurokit2 package [16] is used for filtering, R-peak detection and single beat segment calculation.

2.2 Cascaded FMMnet

In order to reinforce explain-ability in automatic ECG analysis, we utilise the decomposition model proposed by [20]. The idea is to approximate the single-beat ECG signal by composition of five subwaves (P,Q,R,S,T), each of which is parameterised by a Frequency Modulated Moebius.

Assuming $X(t_i)$, $t_i \in [0, 2\pi]$, the original signal of a single-beat ECG record, could be decomposed into five subwaves W_s, $s \in \{P, Q, R, S, T\}$. Each subwave is described by a four-dimensional parameter $p_s = \{A_s, \alpha_s, \beta_s, \omega_s\}$ respectively,

$$W_s(t, p_s) = A_s \cos(\beta_s + 2\arctan(\omega_s \tan(\frac{t - \alpha_s}{2}))) \qquad (1)$$

where A, α, β, ω control the absolute amplitude, the position, the skewness and the kurtosis of the waveform (see Fig. 1 (c, d)). The approximation of original signal $\hat{X}(t)$ is defined as the addition of the five subwaves W_s and an additional baseline parameter M,

$$\hat{X}(t, \theta, M) = M + \sum_{s \in \{P,Q,R,S,T\}} W_s(t, p_s) \qquad (2)$$

where $\theta = (M, p_S, p_Q, p_R, p_S, p_T)$ and they verify the following ranges

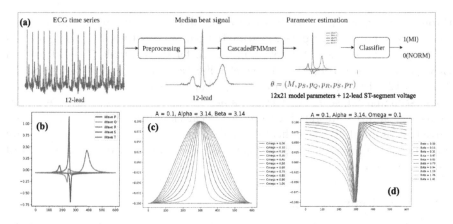

Fig. 1. (a) Pipeline of our proposed explainable ECG classification. (b) An example of ECG decomposition. (c)ω controls the kurtosis of the wave signal. (d) β controls the skewness of the wave signal.

- 1. $M \in \mathcal{R}$
- 2. $p_s \in \mathcal{R}^+ \times [0, 2\pi] \times [0, 2\pi] \times [0, 1], s \in \{P, Q, R, S, T\}$
- 3. $\alpha_P \leq \alpha_Q \leq \alpha_R \leq \alpha_S \leq \alpha_T$

The aim of decomposition is to estimate the optimal 21 parameters $\hat{\theta}$ that best fit $\hat{\theta} = argmin_\theta \sum_{i=1}^n [X(t_i) - \hat{X}(t_i)]^2$.

Instead of using the computationally intensive iterative optimisation [20], we estimate the 21 parameters through a data-driven deep learning model: Cascaded FMMnet. The network consists of 5 identical cascaded sub-network, each of which is responsible for estimating 5 parameters $(M_i, A_i, \alpha_i, \beta_i, \omega_i)$ of one sub-wave S_i, where $S_i(t) = M_i + A_i \cos(\beta_i + 2 \arctan(\omega_i \tan(\frac{t-\alpha_i}{2})))$. Assuming the original signal $X(t)$, the input of the ith subnet $X_i(t)$ is the residual of the original signal subtracting former subwaves, i.e. $X_i(t) = X_{i-1}(t) - S_{i-1}(t)$, where $i \in [1, 5], X_0(t) = X(t), S_0(t) = 0$.

The encoder block in our network comprises 2 stacks of causal convolution with down-sampled skip-connection, 1 max-pooling layer and 2 linear layers. It takes an input of 1×600 dimension and outputs a 21-dimension vector which is the estimation of the parameters. The input median ECG signal is resized to be within the range of $[-1, 1]$ and the last linear layer has a Sigmoid activation for $(A_i, \alpha_i, \beta_i, \omega_i)$ and a Tanh activation for M_i parameter. The final M is the sum of all the $M_i, i \in [1, 5]$. The final estimation of θ are obtained by multiplying M, A_i with the resize factor and by multiplying α_i, β_i with 2π.

We penalise the network by minimising the mean square error of reconstructed signal $S_i(t)$ and the input signal $X_i(t)$. In order to force each subnet to capture a fixed subwave, a regulariser called prior loss is added in the loss function. We randomly chose 100 median ECG samples and estimated the 21 hidden parameters (5 hidden waves) using the FMM R package [21]. The mean μ_α and variance σ_α^2 of parameter α are computed and are used to constrain the

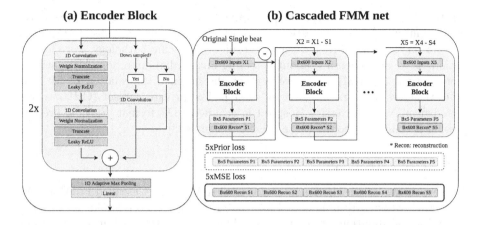

Fig. 2. The detailed architecture of Cascaded FMM net.

subwaves' position. We let the Cascaded FMMnet estimate the T,R,S,P,Q subwaves sequentially by regularising the position α to be close to its corresponding pre-calculated distribution. The total loss function is

$$loss = loss_{mse} + \gamma loss_{prior} \tag{3}$$

$$= \sum_{i=1}^{5} |X_i(t) - S_i(t)|^2 + \gamma \sum_{i=1}^{5} \frac{(\alpha_i - \mu_\alpha)^2}{\sigma_\alpha^2} \tag{4}$$

where γ controls the balance of signal fitting and parameter distribution.

In order to ease the estimation for all leads, we assume that the P,Q,R,S,T are sequentially positioned in all single beat ECG, which may be different with the conventional names of ECG wave peaks in some leads. For example, in Fig. 6(a), the decomposed Q wave represents the conventional R wave for lead V1.

3   Experiments and Results

3.1   Datasets

Table 1. The detailed information of the 4 datasets used in our study. AMI/IMI/LMI refer to anterior/inferior/lateral myocardial infarction.

Dataset	NORM	MI	AMI	IMI	LMI	Frequency (Hz)	Folds
PTB-XL [23]	7185	2955	1937	1447	70	500	10
PTB [6]	80	368	181	175	130	1000	5
CPSC(+Extra) [13]	922	370	–	–	–	500	5
CHU	9	12	–	–	–	Paper scan	-

We included three public ECG datasets and one private dataset in our study. All the four datasets contain standard 12-lead ECG recordings and are (re)sampled

500 Hz. The PTB-XL dataset [23] contains 21837 12-lead ECG recordings and covers multiple ECG diagnostics and morphologies. Our Cascaded FMMnet is trained on this big dataset without considering specific pathology. For classification, as we are concerned about detecting myocardial infarction (MI) patients from normal (NORM) cases, we present the detailed information of MI and NORM across different datasets in Table 1. The private dataset (CHU) is collected from Nice University Hospital in scanned PDF format. Specific preprocessing is conducted to digitalize the ECG signals [8].

3.2 Reconstruction

Experiment. The PTB-XL dataset is used to train the Cascaded FMMnet. As it is provided with 10 pre-defined folds, we randomly choose 8 folds as training set, 1 fold as validation set and the rest fold as test set. 100 random samples from training set are picked to compute the mean and variance used in regulariser (Eq. 4). The encoder network is implemented in Pytorch and is trained with batch size of 192, learning rate of 0.0001. γ is initialised from 1 and it is updated to $\gamma = 0.1\gamma$ if $loss_{mse} \leq \gamma loss_{prior}$.

Results. We evaluated the reconstruction performance by mean absolute error ($MAE = \frac{1}{600} \sum_{i=1}^{600} |X(t_i) - \hat{X}(t_i)|$) and R2 score. The proposed Cascaded FMMnet demonstrated very good reconstruction results and generalised well on three unseen datasets from different centers. First, the FMMnet was trained on 80% of the whole PTB-XL dataset and it demonstrated similar reconstruction result on the train/validation/test set of PTB-XL: they all presented a mean MAE of $0.016\,mV$ and a mean R2 score of 0.94. Second, all the four datasets showed consistent reconstruction error on NORM/MI patients (Fig. 3(a–b)) and lead-wisely (Fig. 3(c–d)). Our Cascaded FMMnet takes 0.09 s/12 leads on a Dell laptop (Intel© Core™ i7-8650U CPU @ 1.90 GHz × 4) while the original FMM optimisation [21] takes more than 10s/1 lead on the same machine. We show an example of 12-lead ECG decomposition in Fig. 6(a).

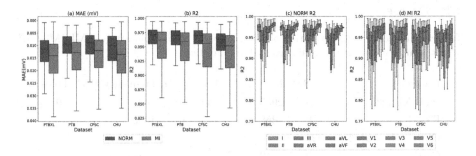

Fig. 3. (a–b) The reconstruction metrics of different evaluation datasets on HC/MI separately. (c–d) Lead-wise R2 score of signal reconstruction of NORM patients and MI patients.

3.3 Classification

Experiment. For each 12-lead ECG sample (median beat), we obtain a 252-dimension feature vector from the Cascaded FMMnet. In addition, we include ST-segment voltage feature for each lead to form a 264-dimension vector. The ST-segment is identified as the short flat platform between peak S and peak T from each lead with the help of positional parameter α_T and α_S.

Explainable Classification: we explored two approaches to provide explainable classification. The first method begins with a partial least squares regression that projects the 264-dimension vector to 3-dimension. A support vector machine (SVM) with linear kernel is applied then to find the best hyperplane that separates the MI/NORM patients. The additive nature of weighted features help to explain the classification results. The second method is to use SHAP value to explain a Logistic regression (LR) model for MI/NORM classification. We trained separate classifiers on PTB-XL, PTB and CPSC(+extra) datasets using 10-fold, 5-fold, 5-fold cross-validation respectively. Two more specific classifiers for detecting AMI/IMI (vs. non AMI/IMI) were established on PTB-XL using 10-fold cross-validation.

Generalisable Classification: we tested the classifiers trained on PTB-XL to other three datasets: PTB, CPSC(+extra) and our private dataset (CHU) to evaluate the generalisation of our proposed classification pipeline.

Table 2. Classification results on different datasets using Cascaded FMM net.

Model	Dataset (evaluation)	Dataset (train)	Class	CV	AUROC	Accuracy	Sensitivity	Specificity
VGG [7]	PTB-XL	PTB-XL	MI	10-fold	1.00	0.97	0.96	0.98
Ours (LR)	PTB-XL	PTB-XL	MI	10-fold	0.99	0.96	0.93	0.96
Ours (SVM)	PTB-XL	PTB-XL	MI	10-fold	0.98	0.94	0.92	0.95
Ours (LR)	PTB-XL	PTB-XL	AMI	10-fold	0.98	0.93	0.92	0.93
Ours (LR)	PTB-XL	PTB-XL	IMI	10-fold	0.97	0.91	0.92	0.91
VGG [7]	PTB	PTB	MI	5-fold	0.98	0.96	0.97	0.91
Ours (LR)	PTB	PTB	MI	5-fold	0.95	0.91	0.92	0.88
Ours (SVM)	PTB	PTB	MI	5-fold	0.95	0.90	0.92	0.81
Ours (LR)	PTB	PTB-XL	MI	–	0.95	0.84	0.83	0.99
Ours (SVM)	PTB	PTB-XL	MI	–	0.94	0.82	0.79	0.99
Ours (LR)	CPSC	CPSC	MI	5-fold	0.97	0.93	0.88	0.96
Ours (SVM)	CPSC	CPSC	MI	5-fold	0.97	0.92	0.88	0.93
Ours (LR)	CPSC	PTB-XL	MI	–	0.97	0.94	0.86	0.97
Ours (SVM)	CPSC	PTB-XL	MI	–	0.95	0.91	0.83	0.95
Ours (LR)	Private	PTB-XL	MI	–	0.80	0.76	0.92	0.56
Ours (SVM)	Private	PTB-XL	MI	–	0.77	0.71	0.83	0.56

Results. We present the detailed classification evaluation in Table 2. First, it can be observed that using our classification pipeline, we are able to obtain satisfactory classification performance on different datasets (PTB-XL/PTB/CPSC)

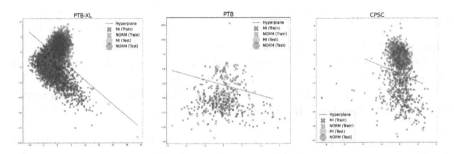

Fig. 4. The classification boundary (hyperplane) of trained linear SVM classifiers and data points (3-dim) projected on one of the 2D plane orthogonal to the corresponding hyperplane.

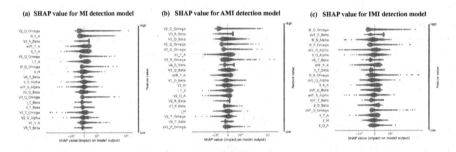

Fig. 5. Explanations of feature importance for Myocardial Infarction (MI), Anterior Myocardial Infarction (AMI) and Inferior Myocardial Infarction (IMI) classification respectively using SHAP value on models trained on PTB-XL dataset. Higher shap value helps to augment the chances of detecting positive classes, in our cases, the MI/AMI/IMI classes.

compared with other methods. In Fig. 4, we can observe that a linear classifier (SVM with linear kernel) is already capable to obtain good separation of MI/NORM patients on both training and test data for PTB-XL, PTB and CPSC datasets respectively. Using SHAP values, our models (Logistic regression classifiers) are capable to identify important infarction related features such as T wave amplitude change (T_A), T inversion (T_β) etc. They also distinguish the influenced leads for infarction. For example, as shown in Fig. 5, it distinguishes the V1,V2,V3 for AMI (Fig. 5b), the II,III,avF for IMI (Fig. 5c). Since the MI label combines MI of different localisation, the important features for MI/NORM classification are spread widely across different leads (Fig. 5a).

Second, models trained on PTB-XL generalise well on other datasets, without to much drop of accuracy. However, we observe a drop of specificity on our small private dataset. Since the signals are extracted from scanned ECG papers, the domain gap could be enlarged. In addition, the manner of annotation may be different. The label of NORM/MI in our private dataset is the final diagnosis decision. It's made by an experienced cardiologist after an overall examination of

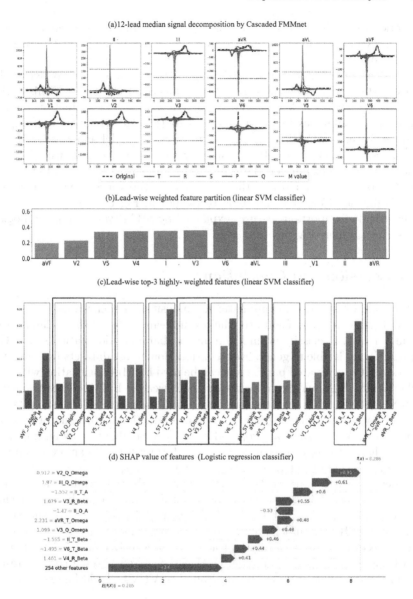

(a)12-lead median signal decomposition by Cascaded FMMnet

(b)Lead-wise weighted feature partition (linear SVM classifier)

(c)Lead-wise top-3 highly- weighted features (linear SVM classifier)

(d) SHAP value of features (Logistic regression classifier)

Fig. 6. An example from PTB-XL test dataset classified as MI by the linear SVM classifier/Logistic regression classifier. According to the PTB-XL description, the patient is diagnosed with old anteroseptal infarction (tiny R waves present in V2,V3), anterolateral ischaemia (inverted T waves and depressed ST-segments in I, avL, flat T waves in V5,V6) and inferior infarction (flat T wave in II). (a) Decomposed 12-lead median ECG. (b)(c) Explainability provided by linear SVM model. Red rectangles mark the T-ST change related leads (acute infarction/ischaemia related) and blue rectangles mark the R wave change related leads (old infarction related and are represented here by our parameters prefixed by Q). (d) Explainability provided by Logistic regression model by SHAP value. (Color figure online)

the patient, including echocardiography, ECG and sometimes coronary angiography. This may be different from the direct diagnosis based on ECG alone from PTB-XL dataset.

We show an example from PTB-XL test data that was correctly detected as MI (MI/NORM classifier) in Fig. 6. The influenced ECG leads of this patient are I, II, avL, V2, V3, V5 and V6. First, our Cascaded FMMnet successfully identified the subwaves for most of the leads (Fig. 6a). Second, we group the weighted features lead-wisely using the linear SVM model and observe that all leads helped to identify this MI patient (they are all positive as shown in Fig. 6b). In addition, when examining the top-3 highest weighted features of every lead (Fig. 6c), we find that most of the important features are in accordance with doctor's diagnosis (see red and blue rectangles in Fig. 6c). We also observe a similar trend of important features explained by SHAP value from the Logistic regression classifier (Fig. 6d).

4   Discussion and Conclusion

In this work, we propose an automatic decomposition model, Cascaded FMMnet for ECG analysis which facilitates the downstream tasks, such as classification and in our case, to classify normal and myocardial infarction patients. The Cascaded FMMnet is able to generalise reconstruction across datasets from different centers and the estimated parameters can be used for MI classification with good explainability. It should be noticed that the Cascaded FMMnet was trained on PTB-XL dataset, which contains not only healthy samples but also samples with different pathologies (myocardial infarction, conduction disturbance and hypertrophy etc.). We believe that proposed network is capable to play an important role in other pathology classification. In the future, we will explore to improve the decomposition of different ECG leads with more specification. Our current Cascaded FMMnet assumes P,Q,R,S,T present in all leads while in some cases, some subwaves can be absent. For example, in Fig. 6a, we can observe the small S wave (in green) does not represent a meaningful peak and it's superposed with the large R wave in lead avF, V1-4. This phenomenon is supported by the superior reconstruction accuracy on leads close to lead II(I, II, V5, V6) than the other leads (Fig. 3c–d). The original FMM paper [20] also presented their results on lead II only. Some specific care for leads like V1-V4 should be explored.

Acknowledgements. This work has been supported by the French government through the National Research Agency (ANR) Investments in the Future with 3IA Côte d'Azur (ANR-19-P3IA-0002) and by Inria PhD funding.

References

1. Acharya, U.R., et al.: Automated characterization and classification of coronary artery disease and myocardial infarction by decomposition of ECG signals: a comparative study. Inf. Sci. **377**, 17–29 (2017)

2. Acharya, U.R., Fujita, H., Oh, S.L., Hagiwara, Y., Tan, J.H., Adam, M.: Application of deep convolutional neural network for automated detection of myocardial infarction using ECG signals. Inf. Sci. **415**, 190–198 (2017)

3. Ansari, S., et al.: A review of automated methods for detection of myocardial ischemia and infarction using electrocardiogram and electronic health records. IEEE Rev. Biomed. Eng. **10**, 264–298 (2017)

4. Arif, M., Malagore, I.A., Afsar, F.A.: Detection and localization of myocardial infarction using K-nearest neighbor classifier. J. Med. Syst. **36**(1), 279–289 (2012). https://doi.org/10.1007/s10916-010-9474-3

5. Bhaskar, N.A.: Performance analysis of support vector machine and neural networks in detection of myocardial infarction. Procedia Comput. Sci. **46**, 20–30 (2015)

6. Bousseljot, R., Kreiseler, D., Schnabel, A.: Nutzung der EKG-signaldatenbank CARDIODAT der PTB über das internet (1995)

7. Fang, R., Lu, C.C., Chuang, C.T., Chang, W.H.: A visually interpretable detection method combines 3-D ECG with a multi-VGG neural network for myocardial infarction identification. Comput. Methods Programs Biomed. **219**, 106762 (2022)

8. Fortune, J.D., Coppa, N.E., Haq, K.T., Patel, H., Tereshchenko, L.G.: Digitizing ECG image: a new method and open-source software code. Comput. Methods Programs Biomed. 106890 (2022)

9. Graja, S., Boucher, J.M.: Hidden Markov tree model applied to ECG delineation. IEEE Trans. Instrum. Meas. **54**(6), 2163–2168 (2005)

10. Jaleel, A., Tafreshi, R., Tafreshi, L.: An expert system for differential diagnosis of myocardial infarction. J. Dyn. Syst. Meas. Control **138**(11), 111012 (2016)

11. Jayachandran, E.S., Joseph, K.P., Acharya, U.R.: Analysis of myocardial infarction using discrete wavelet transform. J. Med. Syst. **34**(6), 985–992 (2010). https://doi.org/10.1007/s10916-009-9314-5

12. Liu, B., et al.: A novel electrocardiogram parameterization algorithm and its application in myocardial infarction detection. Comput. Biol. Med. **61**, 178–184 (2015)

13. Liu, F., et al.: An open access database for evaluating the algorithms of electrocardiogram rhythm and morphology abnormality detection. J. Med. Imaging Health Infor. **8**(7), 1368–1373 (2018)

14. Lu, H., Ong, K., Chia, P.: An automated ECG classification system based on a neuro-fuzzy system. In: Computers in Cardiology 2000, vol. 27 (Cat. 00CH37163), pp. 387–390. IEEE (2000)

15. Lundberg, S.M., Lee, S.I.: A unified approach to interpreting model predictions. In: Guyon, I., et al. (eds.) Advances in Neural Information Processing Systems, vol. 30. Curran Associates, Inc. (2017)

16. Makowski, D., et al.: NeuroKit2: a Python toolbox for neurophysiological signal processing. Behav. Res. Methods **53**(4), 1689–1696 (2021). https://doi.org/10.3758/s13428-020-01516-y

17. Mixon, T.A., et al.: Retrospective description and analysis of consecutive catheterization laboratory ST-segment elevation myocardial infarction activations with proposal, rationale, and use of a new classification scheme. Circ. Cardiovasc. Qual. Outcomes **5**(1), 62–69 (2012)

18. Peimankar, A., Puthusserypady, S.: Dens-ECG: a deep learning approach for ECG signal delineation. Expert Syst. Appl. **165**, 113911 (2021)

19. Pereira, H., Daimiwal, N.: Analysis of features for myocardial infarction and healthy patients based on wavelet. In: 2016 Conference on Advances in Signal Processing (CASP), pp. 164–169. IEEE (2016)

20. Rueda, C., Larriba, Y., Lamela, A.: The hidden waves in the ECG uncovered revealing a sound automated interpretation method. Sci. Rep. **11**(1), 1–11 (2021)
21. Rueda, C., Larriba, Y., Peddada, S.D.: Frequency modulated möbius model accurately predicts rhythmic signals in biological and physical sciences. Sci. Rep. **9**(1), 1–10 (2019)
22. Thygesen, K., et al.: Fourth universal definition of myocardial infarction (2018). Eur. Heart J. **40**(3), 237–269 (2018). https://doi.org/10.1093/eurheartj/ehy462
23. Wagner, P., et al.: PTB-XL, a large publicly available electrocardiography dataset. Sci. Data **7**(1), 1–15 (2020)
24. Xiong, P., Lee, S.M.Y., Chan, G.: Deep learning for detecting and locating myocardial infarction by electrocardiogram: a literature review. Front. Cardiovasc. Med. **9**, 860032 (2022)
25. Zewdie, G., Xiong, M.: Fully automated myocardial infarction classification using ordinary differential equations. arXiv preprint arXiv:1410.6984 (2014)

A Systematic Study of Race and Sex Bias in CNN-Based Cardiac MR Segmentation

Tiarna Lee[1]([✉]), Esther Puyol-Antón[1], Bram Ruijsink[1,2], Miaojing Shi[3], and Andrew P. King[1]

[1] School of Biomedical Engineering and Imaging Sciences, King's College London, London, UK
tiarna.lee@kcl.ac.uk
[2] Guy's and St Thomas' Hospital, London, UK
[3] Department of Informatics, King's College London, London, UK

Abstract. In computer vision there has been significant research interest in assessing potential demographic bias in deep learning models. One of the main causes of such bias is imbalance in the training data. In medical imaging, where the potential impact of bias is arguably much greater, there has been less interest. In medical imaging pipelines, segmentation of structures of interest plays an important role in estimating clinical biomarkers that are subsequently used to inform patient management. Convolutional neural networks (CNNs) are starting to be used to automate this process. We present the first systematic study of the impact of training set imbalance on race and sex bias in CNN-based segmentation. We focus on segmentation of the structures of the heart from short axis cine cardiac magnetic resonance images, and train multiple CNN segmentation models with different levels of race/sex imbalance. We find no significant bias in the sex experiment but significant bias in two separate race experiments, highlighting the need to consider adequate representation of different demographic groups in health datasets.

Keywords: Segmentation · Fairness · CNN · Cardiac MRI

1 Introduction

In the field of healthcare, artificial intelligence (AI) is increasingly being used to automate decision-making processes by aiding the diagnosis and analysis of medical images, producing performances that are equal to, or better than, that of clinicians [12]. Although the use of AI has improved performances for a broad range of tasks, biases found in the wider world have also been found in these AI models. For example, commercial gender classification models were found to perform better on lighter-skinned male faces than on darker-skinned female faces [2]. This difference was attributed to the lack of representation of women

---

M. Shi and A. P. King—Joint last authors.

Supplementary Information The online version contains supplementary material available at https://doi.org/10.1007/978-3-031-23443-9_22.

and non-white faces in the publicly available datasets that the models were trained on.

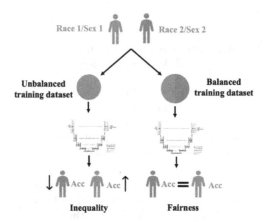

Fig. 1. An illustration showing that biases occur in medical image segmentation models as a result of the imbalances of protected attributes in the training datasets.

In the past few years, research has highlighted the impact of data imbalance on medical imaging tasks. For example, [1] investigated the impact of demographic imbalance on skin lesion classification models, whilst [7] performed a systematic study of sex bias for X-ray classification tasks. Furthermore, [10,11] demonstrated the existence of racial bias in AI models for segmenting cardiac magnetic resonance (CMR) images. These biases are caused by a combination of training data imbalance and distributional differences between the images acquired from different demographic groups. For example, in X-ray imaging, breast tissue can cause perceptible differences in image characteristics between males and females [4]. Other recent work has suggested that information about protected attribute status can be predicted from a range of medical imaging modalities [3,5], suggesting that these distributional differences are widespread and there is a high potential for bias in AI models applied to medical imaging.

In this work, we perform a systematic study of how training set imbalances in the numbers of subjects from protected groups, such as race and sex, affect the performance of an AI-based segmentation model (see Fig. 1). We use a dataset of CMR images to design three experiments: the first studies imbalances in sex by using male and female subjects, the second studies imbalances in race by using white and black subjects, and the third also studies imbalances in race using white and Asian subjects. For each of these experiments, we systematically vary the level of demographic imbalance and measure the performance of the resulting AI models for different protected groups. Our key contributions are:

1. We significantly extend the preliminary work of [10,11] (which only highlighted the presence of race bias) to systematically analyse the impact of sex and race imbalance on CMR segmentation model bias.

2. We assess the bias in terms of conventional segmentation performance metrics as well as derived clinical measures of cardiac function.
3. We perform an intersectional study of AI-based segmentation bias, analysing bias effects on groups such as black females, white males, etc.

2 Materials

In this study we use CMR images from the UK Biobank [9]. The images were acquired at four centres across the UK and all centres used the same CMR acquisition protocol. The dataset consists of end diastolic (ED) and end systolic (ES) cine short-axis images from 1,761 subjects. These were a random subset of subjects with no cardiovascular disease or cardiovascular risk factors. We limited our dataset to these subjects to minimise the impact of other potential sources of variation on our analysis (i.e. apart from sex and race). The demographic data for these subjects were also gathered from the UK Biobank database and can be seen in Table 1 for the subjects used in this study. It is important to note that although the UK Biobank uses the term 'gender' in their records, in practice the term is more similar to biological 'sex' [13].

For each subject, ground truth segmentations for the left ventricular blood pool (LVBP), left ventricular myocardium (LVM) and right ventricular blood pool (RVBP) were obtained for the ED and ES images via manual segmentation of the LV endocardial and epicardial borders and the RV endocardial border using cvi42 (version 5.1.1, Circle Cardiovascular Imaging Inc., Calgary, Alberta, Canada). Each ground truth image was annotated by one expert from a panel of ten who were briefed using the same guidelines. Each expert received a random sample of images from subjects with different sexes and races. The experts were not provided with demographic information about the subjects such as their race or sex.

Table 1. Clinical characteristics of subjects used in the experiments. Average values are presented for each characteristic with standard deviations given in brackets. Statistically significant differences, determined using a two-tailed Student's t-test, are indicated with an asterisk * ($p < 0.05$).

Health measure	Overall	Male	Female	White	Black	Asian
#Subjects	1761	889	872	1220	238	303
Age (years)	63.3 (7.9)	63.7 (8.1)	62.9 (7.7)	64.8 (7.6) *	58.9 (7.0) *	60.6 (8.0) *
Weight (kg)	76.5 (14.7)	83.0 (12.9) *	69.9 (13.3) *	76.8 (14.5)	81.5 (15.9) *	71.3 (12.9) *
Standing height (cm)	169.0 (9.2)	175.3 (6.5) *	162.5 (6.8) *	169.8 (9.2) *	168.7 (9.3)	165.9 (8.8) *
Body Mass Index (kg)	26.7 (4.3)	27.0 (3.6)	26.5 (4.8)	26.6 (4.2)	28.6 (5.0) *	25.8 (3.6) *

2.1 Experimental Setup

To investigate how imbalances in the training datasets affect segmentation performance for different protected groups, we designed three experiments using datasets with varying levels of imbalance in both sex and race. For each of the three experiments, we created five training datasets in which the proportions of the subjects varied according to the protected attribute being investigated. For example, when investigating the effect of varying proportions of males and females in the dataset (Experiment 1), the proportion of males and females in Dataset 1 is 0%/100%, the proportion in Dataset 2 is 25%/75% and so on. We controlled for race in this sex experiment, *i.e.* for each of the five datasets 50% of the subjects were black and 50% were white. To investigate the effect of racial imbalance (Experiment 2), the same method as above was applied but here, the five datasets had varying proportions of black and white subjects. Each of the datasets was 50% male and 50% female. For both of these experiments, the same test data were used which contained 50% black and 50% white subjects, and 50% male and 50% female subjects. Experiment 3 also investigated racial bias and was performed using the same method as Experiment 2 but used white and Asian subjects, whilst controlling for sex. The test set for this experiment was comprised of 50% Asian and 50% white subjects, and 50% male and 50% female subjects. For each of the experiments, there were 176 subjects in the training set and 84 in the test set.

3 Methods

To assess segmentation performance, the nnU-Net segmentation network [6] was used to segment the LVBP, LVM and RVBP from the subjects' ED and ES images. The loss function $\mathcal{L}$ was a combination of the cross entropy loss $\mathcal{L}_{ce}$ and the Dice loss $\mathcal{L}_{dice}$,

$$\mathcal{L} = \mathcal{L}_{ce} + \mathcal{L}_{dice}. \tag{1}$$

Specifically, $\mathcal{L}_{dice}$ is implemented as

$$\mathcal{L}_{dice} = -\frac{2}{|K|} \sum_{k \in K} \frac{\sum_{i \in I} u_i^k v_i^k}{\sum_{i \in I} u_i^k + \sum_{i \in I} v_i^k}, \tag{2}$$

where K is the number of classes, I is the number of pixels in the training image, u_i is a vector of softmaxed predicted class probabilities at the i-th pixel, and v_i is the corresponding one hot encoding of the ground truth class at this pixel.

All models were trained on an NVIDIA RTX A6000. The models were optimised using stochastic gradient descent with a 'poly' learning rate schedule, where the initial learning rate was 0.01 and the Nesterov momentum was 0.99. A batch size of 16 was used and the models were trained for 500 epochs. During training, data augmentation was applied to the images including mirroring, elastic deformation, gamma augmentation, rotation and scaling. The nnU-Net was trained using five-fold cross validation on the training set and the resulting five models were used as an ensemble when applied to the test set. Connected component analysis was applied to the predicted segmentations, with only the largest component retained for each class. The softmax probabilities of the five

Table 2. Overall median DSC for each of the three experiments broken down by the protected attributes used in the experiments. In each experiment we report the results for the respective protected groups and the whole test set. The train percentage signifies the percentages of protected groups used in training in the three experiments. The first and second percentage values correspond with the order of the protected groups in the three experiments, *i.e.* in Experiment 1, 0%/100% corresponds with 0% female, 100% male; in Experiment 2, 0%/100% corresponds with 0% black, 100% white, etc. Best results shown in bold.

Train	Experiment 1			Experiment 2			Experiment 3		
percentage	Female	Male	All	Black	White	All	Asian	White	All
0%/100%	0.955	0.954	0.955	0.924	0.953	0.940	0.912	0.944	0.930
25%/75%	0.957	**0.963**	0.959	0.956	**0.954**	0.950	0.955	0.944	0.949
50%/50%	0.957	0.959	0.958	0.964	0.952	0.955	0.972	**0.944**	0.954
75%/25%	0.958	0.962	**0.961**	0.969	0.951	**0.960**	0.974	0.936	**0.957**
100%/0%	**0.959**	0.960	0.960	**0.970**	0.928	0.953	**0.974**	0.913	0.943

models were averaged when applying the ensemble to the test set to produce the final predicted segmentations.

3.1   Model Evaluation

Model performance was assessed using the *Dice similarity coefficient* (DSC) which measures the spatial overlap between two sets. For a ground truth segmentation A and its corresponding prediction B, the DSC is given by $DSC = \frac{2|A \cap B|}{|A|+|B|}$.

Clinical measures of cardiac function were also calculated for each of the experiments. The end-diastolic volume (EDV) and end-systolic volume (ESV) were first calculated and used to find the ejection fraction (EF) given by $EF = \frac{EDV-ESV}{EDV}$.

4   Results

The overall median DSC scores for the protected group test sets in each of the three experiments are provided in Table 2. We also show a visual representation of the three experimental results in Fig. 2 and performance broken down by region in Fig. 3. Intersectional analysis of the DSC for each of the experiments can be found in Table 3 and analysis of clinical measures can be found in Fig. 4.

4.1   Experiment 1: Male *vs.* Female

The results from Experiment 1 investigating the effect of the imbalance of sex can be seen in Fig. 2a. The performance was reasonably consistent across the five different levels of imbalance. There were no significant differences in overall median DSC for the male and female test sets. However, the intersectional analysis in

Table 3 shows that there were significant differences in performance between black and white females, and black and white males, with the black participants achieving significantly higher DSC scores. The clinical measures reported in Fig. 4 also show no statistically significant differences, with the exception that RVESV shows a consistent under-estimation for female subjects.

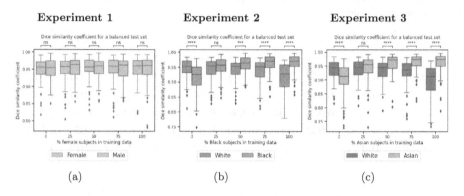

Fig. 2. Overall median DSC for the three experiments. Statistical significance was found using a Mann-Whitney U test and is denoted by **** ($p \leq 0.0001$), *** ($0.001 < p \leq 0.0001$), ** ($0.01 < p \leq 0.001$), * ($0.01 < p \leq 0.05$), ns ($0.05 \leq p$).

4.2   Experiment 2: White vs. Black

Table 2 shows that the highest overall median DSC was achieved when the dataset was comprised of 75% black subjects and 25% white subjects. The results from Table 2 and Fig. 2b show that as the proportion of the protected group increases (e.g. 0% black subjects to 100% black subjects), the DSC for this group also increases.

Accuracy parity (i.e. approximately equal accuracy between groups) was achieved when the proportion of white subjects was 75% and the proportion of black subjects was 25%. By increasing the proportion of black subjects from 0% to 25% to achieve accuracy parity, the median DSC for the minority race increased significantly (from 0.924 to 0.956, p<0.0001) and increased slightly from 0.953 to 0.954 for the white subjects. This resulted in a median DSC which was significantly higher overall (0.950 compared to 0.940, p<0.0001). Further increasing the proportion of black subjects to 50% increased the DSC for the black subjects to 0.964.

Interestingly, the results for the clinical measures (Fig. 4) show no clear bias. The errors observed, shown by the lengths of the whiskers in the box plots, are consistent across training sets. However, there does appear to be an over-estimation of LVESV for both white and black subjects when black subjects are underrepresented in the training set, and an underestimation when white subjects are underrepresented.

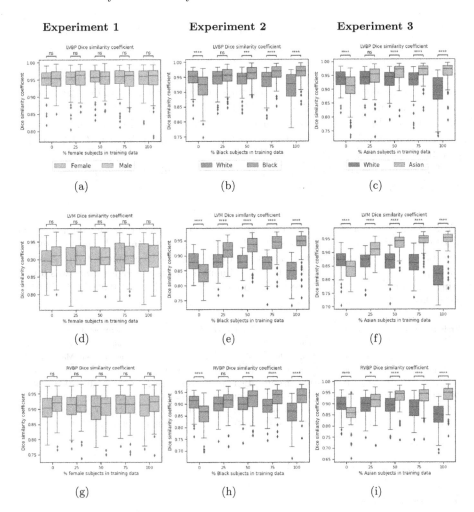

Fig. 3. Comparison of DSC for the left ventricular blood pool (LVBP) (a–c), left ventricular myocardium (LVM) (d–f), and right ventricular blood pool (RVBP) (g–i) for the three experiments. Statistical significance was found using a Mann-Whitney U test and is denoted by **** ($p \leq 0.0001$), *** ($0.001 < p \leq 0.0001$), ** ($0.01 < p \leq 0.001$), * ($0.01 < p \leq 0.05$), ns ($0.05 \leq p$).

4.3   Experiment 3: White *vs.* Asian

Table 2 shows that the highest overall median DSC for Experiment 3 was achieved when the dataset was comprised of 75% Asian subjects and 25% white subjects. Figure 2c shows a similar trend to the results from Experiment 2. However, in this experiment, with a training dataset that is 100% Asian, larger differences can be observed between the white and Asian subjects than were observed for the white and black subjects. For both race experiments, the largest differences in DSC scores between the groups were found in the LVM (Fig. 3e and Fig. 3f).

Accuracy parity was also achieved here with a split of 75% white subjects and 25% Asian patients. Increasing the proportion of Asian subjects from 0% to 25% increased the DSC for Asian subjects from 0.912 to 0.955 (p<0.0001) and the overall DSC from 0.930 to 0.949 (p<0.0001) but did not increase the DSC for the white subjects. Further increasing the proportion of Asian subjects to 50% significantly increased the DSC for the Asian subjects to 0.972 (0.01< p<0.001) and the overall DSC to 0.954.

The results for the clinical measures (Fig. 4) for this experiment show similar findings to those for Experiment 2.

Table 3. Intersectional analysis of median DSC scores broken down by the protected attributes used in the experiments. The train percentages signify the proportions of each protected group used in training, i.e. female/male for Experiment 1, black/white for Experiment 2 and Asian/white for Experiment 3. Statistical significance was found using a Mann-Whitney U test and is denoted by *** $(0.001 < p \leq 0.0001)$, ** $(0.01 < p \leq 0.001)$, * $(0.01 < p \leq 0.05)$.

Train percentage	Experiment 1				Experiment 2				Experiment 3			
	Female		Male		Black		White		Asian		White	
	White	Black	White	Black	Male	Female	Male	Female	Male	Female	Male	Female
0%/100%	0.950	0.963**	0.948	0.963**	0.922	0.925	0.948	0.954	0.919	0.909	0.945	0.944
25%/75%	0.952	0.962*	0.955	0.969*	0.951	0.959	0.956	0.952	0.952	0.960	0.944	0.944
50%/50%	0.951	0.968***	0.957	0.962	0.961	0.967	0.945	0.953	0.970	0.973	0.945	0.944
75%/25%	0.952	0.970**	0.951	0.965*	0.964	0.971	0.952	0.949	0.970	0.975	0.934	0.937
100%/0%	0.954	0.969**	0.950	0.971**	0.971	0.969	0.936	0.913	0.965	0.975	0.901	0.923

5   Discussion

To the best of our knowledge, this work has presented the first systematic study of how imbalances in training datasets affect AI-based CMR segmentation performance. Our results show that a significant bias towards the majority group can be seen when the datasets are racially imbalanced. We did not find any significant sex bias.

As discussed in Sect. 4, the median DSC for both the black and Asian subjects increased significantly when their representation in the dataset was increased from 0% to 25%. It could be argued that a model trained with this dataset is more fair than a model trained with 100% white subjects as the DSC for the minority subjects increased while the DSC for the white subjects remained approximately the same. Increasing the proportion of the minority races to 50% further increased the overall DSC scores, suggesting that a more diverse training set will produce a model with a better segmentation performance. In clinical applications, cardiac segmentations are used to obtain important measures of

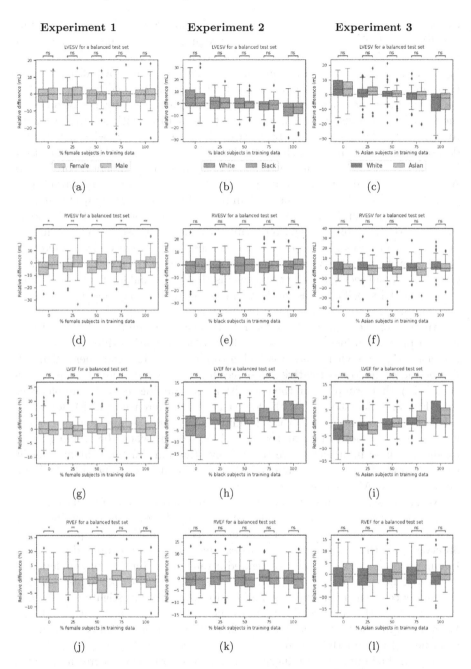

Fig. 4. Relative difference in LV and RV end-systolic volume (ESV) and ejection fraction (EF) for the three experiments. The differences were calculated by subtracting ground truth values from predicted values e.g. $LVESV_{difference} = LVESV_{predicted} - LVESV_{ground\ truth}$. Statistical significance was found using a Mann-Whitney U test and is denoted by ** $(0.01 < p \leq 0.001)$, * $(0.01 < p \leq 0.05)$, ns $(0.05 \leq p)$.

cardiac function such as left- and right-ventricular ES volume and ED volume (see Fig. 4), which are used for diagnosis, prognosis and treatment planning for patients [8]. Given that the prevalence of cardiovascular diseases is known to be higher in minority races, producing models which are fairer overall is essential to improve patient management and reduce healthcare inequalities.

However, we note that if the tasks of segmenting CMR images of different races had equal difficulty, we would expect the dataset with a 50%/50% split to have equal DSC scores for both races. In Experiment 2, with a balanced dataset, the black subjects achieved a DSC score which was higher than that for white subjects. The same pattern can be seen for Experiment 3 considering Asian subjects. This suggests that there may be more 'within-group' variation in the anatomies of the white subjects than in the black or Asian subjects, making segmenting white subjects' hearts an inherently harder problem for the AI model to solve. Applying bias mitigation techniques such as those introduced in [10] may produce a model which does not generalise well to anatomical differences between protected groups which may carry important diagnostic information. Given that a subject's self-identified race is (likely) available at inference time, using a model specific to the subject's protected group may prove to be more fair than using a single model for all subjects.

Analysis of the clinical measures (Fig. 4) revealed some interesting and perhaps surprising findings. There was no clear trend in relative differences in clinical measures, even when there was such a trend in DSC in Experiments 2 and 3. Furthermore, the lowest relative differences for a protected group did not always occur when that group was in the majority in the training set. For example, consider Fig. 4h which shows that the lowest relative difference in LVEF for white subjects does not occur when the training set is 100% white, but rather the lowest errors for both black and white subjects occur when the training set is balanced. These findings require further investigation, but add further weight to our argument that considering training set (im)balance is crucial for training fair and robust AI-based segmentation models.

Nevertheless, in terms of spatial overlap measures, our results have shown clear racial biases in segmentation performance. However, it is possible that race is not the underlying factor that explains the differences in performance observed, and the differences could instead be caused by a confounding factor. In our previous work [11] we investigated a range of potential confounders and discovered none that could explain the bias. We remain open-minded about the existence of such confounders, and future work will investigate the effect of imbalances of other protected attributes such as socioeconomic status and age. We also emphasise that racial bias in CMR segmentation is a cause for concern regardless of its underlying explanation.

Overall, including more minorities increased overall DSC scores and scores for individual minority groups. Therefore, we advocate for the inclusion of more minorities in public medical datasets and the transparent reporting of performance of AI models separately for all protected groups. Although varying the proportion of females in the dataset did not result in significant differences in segmentation performance, it is essential that females have equal representation in publicly available medical datasets as previous work discussed in Sect. 1 has found that AI-based diagnosis methods frequently underperform for females. This work has highlighted the need for the fair representation of minority groups in medical imaging datasets. To the best of our knowledge, this work is the first to systematically investigate the effect of dataset imbalances on segmentation accuracy.

Acknowledgements. This work was supported by the Engineering & Physical Sciences Research Council Doctoral Training Partnership (EPSRC DTP) grant EP/T517963/1. This research has been conducted using the UK Biobank Resource under Application Number 17806.

References

1. Abbasi-Sureshjani, S., Raumanns, R., Michels, B.E.J., Schouten, G., Cheplygina, V.: Risk of training diagnostic algorithms on data with demographic bias. In: Cardoso, J., et al. (eds.) IMIMIC/MIL3ID/LABELS -2020. LNCS, vol. 12446, pp. 183–192. Springer, Cham (2020). https://doi.org/10.1007/978-3-030-61166-8_20
2. Buolamwini, J., Gebru, T.: Gender shades: intersectional accuracy disparities in commercial gender classification. In: Proc MLR, vol. 81, pp. 1–15 (2018). https://doi.org/10.2147/OTT.S126905
3. Coyner, A.S., et al.: Not color blind: AI predicts racial identity from black and white retinal vessel segmentations. arXiv preprint arXiv:2109.13845 (2021)
4. Ganz, M., Holm, S.H., Feragen, A.: Assessing bias in medical AI. In: Workshop on Interpretable ML in Healthcare at International Conference on Machine Learning (ICML) (2021)
5. Gichoya, J.W., et al.: AI recognition of patient race in medical imaging: a modelling study. Lancet Digit. Health **4**(6), e406–e414 (2022). https://doi.org/10.1016/S2589-7500(22)00063-2
6. Isensee, F., et al.: nnU-Net: a self-configuring method for deep learning-based biomedical image segmentation. Nat. Methods **18**(2), 203–211 (2021). https://doi.org/10.1038/s41592-020-01008-z
7. Larrazabal, A.J., et al.: Gender imbalance in medical imaging datasets produces biased classifiers for computer-aided diagnosis. Proc. Natl. Acad. Sci. **117**(23), 12592–12594 (2020). https://doi.org/10.1073/pnas.1919012117
8. Chen, C., et al.: Deep learning for cardiac image segmentation: a review. Front. Cardiovasc. Med. **7**, 25 (2020). https://doi.org/10.3389/fcvm.2020.00025
9. Petersen, S.E., et al.: UK Biobank's cardiovascular magnetic resonance protocol. J. Cardiovasc. Magn. Reson. **18**(1), 1–7 (2015). https://doi.org/10.1186/s12968-016-0227-4
10. Puyol-Antón, E., et al.: Fairness in cardiac magnetic resonance imaging: assessing sex and racial bias in deep learning-based segmentation. Front. Cardiovasc. Med. 664 (2022). https://doi.org/10.3389/fcvm.2022.859310
11. Puyol-Antón, E., et al.: Fairness in cardiac MR image analysis: an investigation of bias due to data imbalance in deep learning based segmentation. In: de Bruijne, M., et al. (eds.) MICCAI 2021. LNCS, vol. 12903, pp. 413–423. Springer, Cham (2021). https://doi.org/10.1007/978-3-030-87199-4_39

12. Shen, J., et al.: Artificial intelligence versus clinicians in disease diagnosis: systematic review. JMIR Med. Inform. **7**(3), e10010 (2019). https://doi.org/10.2196/10010
13. Tannenbaum, C., et al.: Sex and gender analysis improves science and engineering. Nature **575**(7781), 137–146 (2019). https://doi.org/10.1038/s41586-019-1657-6

Mesh U-Nets for 3D Cardiac Deformation Modeling

Marcel Beetz[1(✉)], Jorge Corral Acero[1], Abhirup Banerjee[1,2], Ingo Eitel[3,4,5], Ernesto Zacur[1], Torben Lange[6,7], Thomas Stiermaier[3,4,5], Ruben Evertz[6,7], Sören J. Backhaus[6,7], Holger Thiele[8,9], Alfonso Bueno-Orovio[10], Pablo Lamata[11], Andreas Schuster[6,7], and Vicente Grau[1]

[1] Department of Engineering Science, Institute of Biomedical Engineering, University of Oxford, Oxford, UK
marcel.beetz@eng.ox.ac.uk
[2] Division of Cardiovascular Medicine, Radcliffe Department of Medicine, University of Oxford, Oxford, UK
[3] University Heart Center Lübeck, Medical Clinic II, Cardiology, Angiology, and Intensive Care Medicine, Lübeck, Germany
[4] University Hospital Schleswig-Holstein, Lübeck, Germany
[5] German Centre for Cardiovascular Research, Partner Site Lübeck, Lübeck, Germany
[6] Department of Cardiology and Pneumology, University Medical Center Göttingen, Georg-August University, Göttingen, Germany
[7] German Centre for Cardiovascular Research, Partner Site Göttingen, Göttingen, Germany
[8] Department of Internal Medicine/Cardiology, Heart Center Leipzig at University of Leipzig, Leipzig, Germany
[9] Leipzig Heart Institute, Leipzig, Germany
[10] Department of Computer Science, University of Oxford, Oxford, UK
[11] Department of Biomedical Engineering, King's College London, London, UK

Abstract. During a cardiac cycle, the heart anatomy undergoes a series of complex 3D deformations, which can be analyzed to diagnose various cardiovascular pathologies including myocardial infarction. While volume-based metrics such as ejection fraction are commonly used in clinical practice to assess these deformations globally, they only provide limited information about localized changes in the 3D cardiac structures. The objective of this work is to develop a novel geometric deep learning approach to capture the mechanical deformation of complete 3D ventricular shapes, offering potential to discover new image-based biomarkers for cardiac disease diagnosis. To this end, we propose the *mesh U-Net*, which combines mesh-based convolution and pooling operations with U-Net-inspired skip connections in a hierarchical step-wise encoder-decoder architecture, in order to enable accurate and efficient learning directly on 3D anatomical meshes. The proposed network is trained to model both cardiac contraction and relaxation, that is, to predict the 3D cardiac anatomy at the end-systolic phase of the cardiac cycle based on the corresponding anatomy at end-diastole and vice versa. We evaluate our

© The Author(s), under exclusive license to Springer Nature Switzerland AG 2022
O. Camara et al. (Eds.): STACOM 2022, LNCS 13593, pp. 245–257, 2022.
https://doi.org/10.1007/978-3-031-23443-9_23

method on a multi-center cardiac magnetic resonance imaging (MRI) dataset of 1021 patients with acute myocardial infarction. We find mean surface distances between the predicted and gold standard anatomical meshes close to the pixel resolution of the underlying images and high similarity in multiple commonly used clinical metrics for both prediction directions. In addition, we show that the mesh U-Net compares favorably to a 3D U-Net benchmark by using 66% fewer network parameters and drastically smaller data sizes, while at the same time improving predictive performance by 14%. We also observe that the mesh U-Net is able to capture subpopulation-specific differences in mechanical deformation patterns between patients with different myocardial infarction types and clinical outcomes.

Keywords: Geometric deep learning · 3D heart contraction · Cardiac mechanics · Acute myocardial infarction · Cardiac MRI · Spectral graph convolutions · Mesh sampling

1 Introduction

The human heart undergoes a series of complex 3D deformations during a normal heartbeat that are crucial for its correct operation. Accordingly, cardiac contraction and relaxation patterns are routinely analyzed in clinical practice to detect, diagnose, and treat a variety of cardiac diseases, including acute myocardial infarction. However, current clinical decision-making is usually based on relatively coarse global metrics of cardiac function, such as ejection fraction, that neglect more localized and intricate cardiac shape changes and therefore limit the clinical accuracy. Hence, many previous works have studied temporal deformations of the heart in a more holistic manner and in a variety of different settings using generative adversarial networks [24], graph neural networks [9,21–23], Siamese recurrent neural networks [26], principal component analysis [7], manifold learning [11], probabilistic encoder-decoder models [16], time-series autoencoders [5], point cloud autoencoders [1–4], or systems of differential equations [14,17,19].

However, these approaches are either limited to synthetic anatomy models with only a limited number of different cardiac shapes [9], lack validation on pathological datasets [4,21,22,24], use low-dimensional representations of the cardiac anatomy [5,21,22,24], consider cardiac contraction but not relaxation [21,22,24], use inefficient and memory-intensive voxelgrids to represent cardiac anatomical surfaces [16,26], or require special post-processing steps for common follow-up tasks due to a lack of explicit connectivity information in point clouds [1–4]. In this work, our aim is to develop a novel data-driven approach that is able to accurately capture and predict the complex deformations of pathological 3D cardiac shapes between the end-diastolic (ED) and the end-systolic (ES) phases of the cardiac cycle in an effective and efficient manner. To this end, we propose the *mesh U-Net*, a new geometric deep learning architecture, which combines mesh-based convolution and pooling operations with the hierarchical encoder-decoder structure of the U-Net to enable the direct and effective processing of high-resolution 3D anatomical surface meshes. We find that the mesh U-Net can successfully model and differentiate complex cardiac mechanics patterns on an individual, subpopulation, as well as the whole-population level, allowing for improved understanding of different myocardial infarction (MI) types and more interpretable risk identification of major adverse cardiac events based on novel image-based biomarkers.

2   Methods

2.1   Dataset and Preprocessing

The dataset used in this work consists of cine MRI acquisitions from 1021 patients with MI. The MI event occurred at a median of 3 days prior to imaging and was classified as ST-elevation myocardial infarction (STEMI) or non-ST-elevation myocardial infarction (non-STEMI). After a 12-months post-infarct follow-up period, 74 patients suffered a major adverse cardiac event (MACE) defined as either reinfarction, new congestive heart failure, or all-cause death. All images were acquired as part of two multi-center studies with a mean pixel resolution of 1.36 mm (range: [1.16, 2.08] mm, SD: 0.21 mm). Both the studies and acquisition protocols have previously been described in greater detail in [12,29]. We select the short-axis slices at both the ED and ES phases of the cardiac cycle of each patient and reconstruct the corresponding 3D mesh representations of the left ventricular (LV) myocardial anatomy using the multi-step process outlined in [7,8,18]. We then create a mean template mesh by averaging each of the vertex coordinates across the dataset and represent the anatomy of each patient as its vertex-wise deformation with respect to this mean mesh for network training. Finally, we apply standardization to all vertex coordinate values of both the input and output meshes of the network.

2.2   Network Design

The architecture of the proposed mesh U-Net follows a symmetric encoder-decoder design with 3D meshes as both network inputs and outputs (Fig. 1).

The encoder consists of four hierarchical mesh down-sampling steps creating successively coarser representations of the input mesh with each lower network level, while the decoder employs four corresponding up-sampling operations for symmetric step-wise increases in mesh resolution. Spectral graph convolutions are then used at each of the four resolution levels to enable multi-scale feature learning in both local and global contexts across a variety of cardiac shapes. Similar to the original U-Net architecture [28], we increase the number of feature maps of the graph convolution operations at deeper network levels as the mesh resolution decreases to facilitate the learning of more complex features. We also incorporate skip connections between the corresponding levels of the encoder

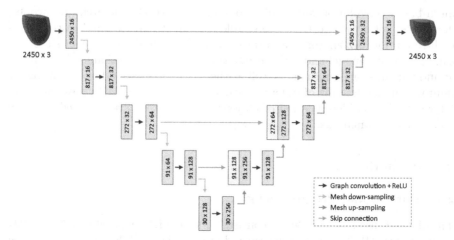

Fig. 1. Network architecture of the proposed mesh U-Net. A symmetric encoder-decoder structure with hierarchical mesh sampling operations and spectral graph convolutions enables efficient multi-scale feature learning directly on mesh data. U-Net inspired skip connections facilitate information flow between the encoder and decoder branches, while the number of feature maps steadily increases in lower network levels to capture more complex patterns. Surface meshes with 2450 vertices are used as network inputs and outputs to represent the cardiac anatomy at either ED or ES.

and decoder branches to allow for efficient information flow between earlier and later layers of the network. Both down-sampling and up-sampling layers are implemented using the quadric error minimization criterion [27] with three-fold reductions and increases in mesh resolution respectively at each network level. All graph convolutions utilize the Chebyshev polynomial approximation [10,13] of order 5 and are followed by rectified linear units (ReLU) as activation functions.

2.3   Implementation and Training

The deep learning code base is implemented using the PyTorch [25] framework. We train our models using the Adam optimizer [15] with a learning rate of 0.007, a batch size of 16, and a weight decay of 0.0006 for 250 epochs on a CPU with 8 GB memory. We select the L1 distance between the corresponding vertices in the predicted and gold standard meshes as our loss function and use a random train/validation/test dataset split of 80%/5%/15% for the prediction experiments in Sect. 3.1 and 3.2. The test dataset size in the subpopulation-specific experiments (Sect. 3.3) is set to the same number as the respective left-out subpopulation (i.e. n = 305 for STEMI/non-STEMI; n = 74 for non-MACE/MACE), while the remaining data is split into 95% training and 5% validation data. In all experiments, we select the model weights of the training checkpoint with the best performance on the validation dataset as our final model for evaluation on the test dataset.

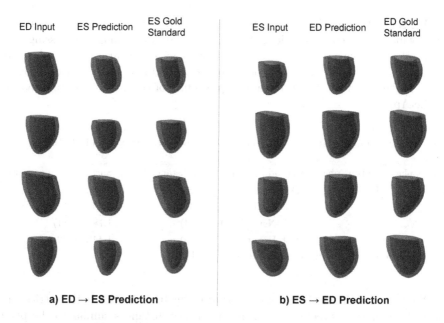

Fig. 2. ES prediction (a) and ED prediction (b) results of the mesh U-Net for four sample patients (in rows) of the unseen test dataset.

3   Experiments and Results

3.1   Prediction Quality

We first want to assess the mesh U-Net's ability to correctly predict cardiac shapes at ES from ED inputs and vice versa. To this end, we train two separate networks, one for each prediction direction, to analyze both cardiac contraction and relaxation. We then pass the input ED/ES cardiac meshes from the unseen test dataset through the pertinent trained network and compare the prediction outputs with the corresponding test dataset meshes at ES/ED, which we assume to be our gold standard. The qualitative results for four sample cases are displayed in Fig. 2-a and Fig. 2-b for the ES and ED prediction tasks, respectively.

We observe a good alignment between the predicted and gold standard meshes on both a local and global level and for a variety of different shapes, in both the ED and ES prediction tasks.

Next, we quantify the mesh U-Net's prediction performance by calculating both the mean surface distance (MSD) (Eq. 1) and the Hausdorff distance between each predicted mesh X and their corresponding gold standard mesh Y of the test dataset.

$$MSD(X,Y) = \frac{1}{2}\left(\frac{1}{|X|} \sum_{x \in X} d(x,Y) + \frac{1}{|Y|} \sum_{y \in Y} d(y,X) \right) \qquad (1)$$

We select the 3D U-Net [6], whose voxelgrid-based design components have been widely utilized in previous works for 3D medical image processing and anatomical surface modelling, as a benchmark to evaluate our method and report the distance values obtained by both approaches and in both prediction tasks in Table 1. Further details on the training procedure of the 3D U-Net are provided in the Appendix.

Table 1. Quantitative prediction results of the mesh U-Net and 3D U-Net.

Input phase	Output phase	Method	Hausdorff distance (mm)	Surface distance (mm)
ED	ES	3D U-Net	8.02 (±3.26)	1.85 (±0.62)
		Mesh U-Net	7.06 (±2.31)	1.69 (±0.57)
ES	ED	3D U-Net	7.94 (±3.74)	1.78 (±0.81)
		Mesh U-Net	6.27 (±1.85)	1.44 (±0.39)

Values represent mean (± standard deviation (SD)).

We find that the mesh U-Net outperforms the 3D U-Net for both metrics and in both prediction directions, with mean surface distances similar to the pixel resolution of the underlying image acquisitions.

In addition to the prediction accuracy, we also compare the mesh U-Net and 3D U-Net architectures in terms of their data formats and number of network parameters (Table 2).

Table 2. Technical comparison of mesh U-Net and 3D U-Net.

Method	Data type	Data instance size	Network parameters
3D U-Net	Voxelgrid	$\sim 2.1 \times 10^6$ ($128 \times 128 \times 128$)	$\sim 2.0 \times 10^6$
Mesh U-Net	Mesh	$\sim 7.4 \times 10^3$ (2450×3)*	$\sim 6.8 \times 10^5$

*Vertex connectivity is the same for each mesh in the dataset.

We observe that the mesh U-Net only needs about 34% as many parameters as the 3D U-Net, while its usage of meshes to represent anatomical surface data allows for a drastic reduction in the required data sizes.

3.2 Clinical Evaluation

Next, we evaluate the mesh U-Net's ability to accurately capture the mechanical deformation of the heart from a clinical perspective. To this end, we first obtain the ES/ED predictions for all ED/ES input meshes of the test dataset from the respective networks. Then, we use these predicted mesh populations as well as the corresponding test set mesh populations, which we consider as our gold standard, to calculate multiple image-based biomarkers commonly used in clinical practice. We consider both cardiac anatomy metrics focusing on the heart at either ED or

Table 3. Clinical anatomy and function metrics results achieved by the mesh U-Net.

Input phase	Predicted phase	Clinical metric	Gold standard	Prediction
ED	ES	LV ES volume (ml)	80 ($\pm$31)	78 ($\pm$27)
		LV myocardial mass (g)	121 ($\pm$33)	120 ($\pm$27)
		LV stroke volume (ml)	68 ($\pm$21)	71 ($\pm$18)
		LV ejection fraction (%)	47 ($\pm$11)	47 ($\pm$14)
ES	ED	LV ED volume (ml)	148 ($\pm$37)	149 ($\pm$36)
		LV myocardial mass (g)	113 ($\pm$29)	115 ($\pm$24)
		LV stroke volume (ml)	73 ($\pm$22)	74 ($\pm$18)
		LV ejection fraction (%)	50 ($\pm$12)	51 ($\pm$11)

Values represent mean ($\pm$ SD) in all cases.

ES (LV volume, LV mass) and cardiac function metrics which assess the temporal changes between cardiac shapes at ED and at ES (LV stroke volume, LV ejection fraction). The results for both prediction directions are reported in Table 3.

We find highly similar mean and standard deviation values between the gold standard and predicted mesh populations across both cardiac anatomy and function metrics and for both prediction directions.

3.3 Subpopulation-Specific Deformations

In addition to predicting mechanical shape changes for the whole population in the dataset, we also want to investigate whether the mesh U-Net can learn 3D deformation patterns that are specific to certain subpopulations. To this end, we first partition the original dataset into two subpopulations based on STEMI vs non-STEMI and also separately for non-MACE vs MACE. We then train a mesh U-Net on the data from one subpopulation (STEMI and non-MACE respectively) and evaluate its performance on both the unseen test dataset from the same subpopulation, as well as the meshes from the other subpopulation (non-STEMI and MACE respectively) that are not used during training. In this way, we are able to analyze possible differences in the distributions of surface distance values for both the matching and non-matching test datasets. We show the results for both outcomes and for both prediction directions in Fig. 3.

We find statistically significant differences in terms of Kolmogorov-Smirnov (KS) test (p<0.005 in all experiments) between the matching and non-matching performance distributions in all four experiments, with lower prediction errors for the matching datasets. This indicates that the mesh U-Net has learnt subpopulation-specific patterns, which cause the performance to deteriorate when applying the network to the other subpopulation whose unique patterns were not seen during training. While the findings displayed in Fig. 3 are based on surface distances, we observe similar results using Hausdorff distances as the comparative metric, which can be found in the Appendix.

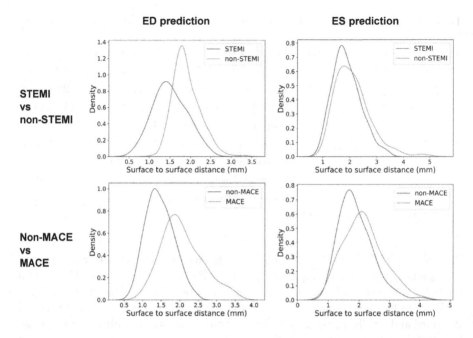

Fig. 3. Surface distance distributions achieved by mesh U-Nets trained on one subpopulation and evaluated on unseen test dataset of the same subpopulation (blue color) and the complementary subpopulation (orange color). Plots are shown for both ED and ES prediction tasks (columns) and both STEMI and MACE population splits (rows). p-values for KS-test: <0.0001 (ED prediction for both STEMI and MACE); <0.001 (ES prediction for MACE); <0.005 (ES prediction for STEMI). (Color figure online)

4   Discussion and Conclusion

In this work, we have presented the mesh U-Net, a novel geometric deep learning method, for efficient deformation modelling of 3D cardiac anatomy shapes. Its architecture relies on recent advances in mesh-specific convolution and pooling operations, giving it the ability to directly and effectively process anatomical surface mesh data and capture important features at multiple different scales in a hierarchical manner. Similar to the original U-Net architecture, skip connections enable an easy information exchange across different parts of the network, while the step-wise sampling operations along the symmetric encoder and decoder branches allow for smooth transitions between different scales. The mesh U-Net outperforms a 3D U-Net benchmark by a considerable margin in both the ED and ES prediction tasks, showing the high efficacy of the chosen architectural design in different bitemporal settings. The results are consistent across both the surface distance and Hausdorff distance metrics, demonstrating the high average prediction accuracy as well as an improved robustness to outliers and artifacts. We also visually observe the good alignment between the predicted and gold standard cardiac shapes on both a local and global level further confirming its good performance. Prediction errors are generally the largest at the apical and basal areas of the heart, which we suspect to be at least partially caused by the limited amount of spatial information available in those areas in the original MRI acquisition and the 3D reconstruction process.

The mesh U-Net achieves high prediction accuracy with 66% fewer parameters than the 3D U-Net, while using meshes to store the same anatomical surface information with 99.6% smaller data instance sizes than the corresponding voxelgrid representations. This shows the high efficiency of the design, enabling easy training on a CPU instead of a GPU and the usage of fewer memory resources with higher resolution meshes without being limited by any voxel resolution, as is the case for voxelgrids. Furthermore, the mesh U-Net's predictions obtain highly realistic values in both cardiac anatomy and function metrics for both prediction directions with typical morphological differences between the ED and ES phases (e.g. thicker myocardium at ES, larger heart size at ED) correctly captured. This provides further evidence of a good population-wide performance from a clinical perspective and indicates that accurate deformation patterns in line with clinical practice are captured by the network for both cardiac contraction and relaxation. This is important for the accuracy of any computer-based model as well as its clinical acceptance. While we did not explicitly include any mechanical properties or patient metadata in this study, we hypothesize that the network was able to implicitly learn many pertinent features from the large population of heart shapes in order to still achieve the observed good prediction accuracy.

Finally, we find that the mesh U-Net is capable of learning deformation patterns specific to different pathological subpopulations. This shows that diseases affect the mechanical deformation of the heart in different ways, and allows for a better understanding and potential stratification of myocardial infarction types (such as STEMI and non-STEMI) based on differences between the predicted and true 3D shapes. The results are also plausible from a biomechanical perspective as mechanical properties within the left ventricular tissue are typically shared by patients with similar pathophysiological conditions and differ between patients with separate pathologies. In addition, the non-MACE vs MACE comparison demonstrates that deviations in prediction, which represents "healthy" cardiac function, against actual function measured from images can help identify subjects at risk. This approach is also highly interpretable, as its usage of high-resolution 3D meshes for both predicted and true cardiac anatomies allows an easy visualization of global and local differences between actually observed and expected healthy hearts, which is crucial for an application in clinical practice and the discovery of novel image-based biomarkers.

Acknowledgments. The authors express no conflict of interest. The work of MB is supported by the Stiftung der Deutschen Wirtschaft (Foundation of German Business). AB is a Royal Society University Research Fellow and is supported by the Royal Society (Grant No. URF\R1\221314). The work of AB and VG is supported by the British Heart Foundation (BHF) Project under Grant PG/20/21/35082. The work of VG is supported by the CompBioMed 2 Centre of Excellence in Computational Biomedicine (European Commission Horizon 2020 research and innovation programme, grant agreement No. 823712). The work of JCA is supported by the EU's Horizon 2020 research and innovation program under the Marie Sklodowska-Curie (g.a. 764738) and the EPSRC Impact Acceleration Account (D4D00010 DF48.01), funded by UK Research and Innovation. ABO holds a BHF Intermediate Basic Science Research Fellowship (FS/17/22/32644). The work is also supported by the German Center for Cardiovascular Research, the British Heart Foundation (PG/16/75/32383), and the Wellcome Trust (209450/Z/17).

A   3D U-Net

In this section, we describe in greater detail the training and validation procedure of the 3D U-Net [6] used for both the end-diastolic and end-systolic prediction tasks. First, we convert the mesh representations of the cardiac anatomies of the whole dataset into voxelgrids to allow for the same dataset to be used for both the 3D U-Net and mesh U-Net evaluation. We achieve this by voxelizing the 3D meshes and placing them in the center of $128 \times 128 \times 128$ voxelgrids where each voxel is encoded as either background (value: "0") or left ventricular myocardium (value: "1"). We then select a 3D U-Net architecture and train it using binary cross entropy as a loss function. Next, we pass the unseen test data through the trained 3D U-Net and convert the resulting predictions and corresponding gold standard anatomies from voxelgrid to 3D surface mesh representations with the marching cubes algorithm [20]. Finally, we use the obtained mesh representations to calculate both surface distances and Hausdorff distances between the meshes predicted by the 3D U-Net and the respective gold standard meshes.

B   Subpopulation-Specific Deformations

We display the results of the subpopulation-specific training experiments using the Hausdorff distance as a quantitative metric in Fig. 4.

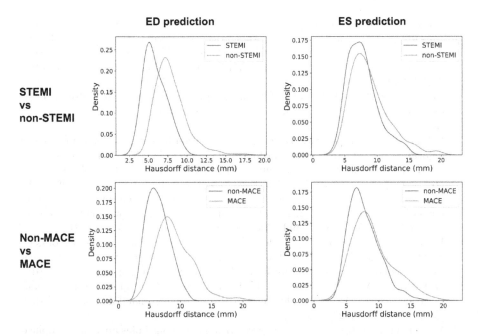

Fig. 4. Hausdorff distance distributions achieved by mesh U-Nets trained on one subpopulation and evaluated on unseen test dataset of the same subpopulation (blue color) and the complementary subpopulation (orange color). Plots are shown for both ED and ES prediction tasks (columns) and both STEMI and MACE population splits (rows). p-values for KS-test: <0.0001 (ED prediction for both STEMI and MACE); <0.001 (ES prediction for STEMI); <0.005 (ES prediction for MACE). (Color figure online)

References

1. Beetz, M., Banerjee, A., Grau, V.: Generating subpopulation-specific biventricular anatomy models using conditional point cloud variational autoencoders. In: Puyol Antón, E., et al. (eds.) STACOM 2021. LNCS, vol. 13131, pp. 75–83. Springer, Cham (2022). https://doi.org/10.1007/978-3-030-93722-5_9

2. Beetz, M., Banerjee, A., Grau, V.: Multi-domain variational autoencoders for combined modeling of MRI-based biventricular anatomy and ECG-based cardiac electrophysiology. Front. Physiol., 991 (2022)

3. Beetz, M., Banerjee, A., Sang, Y., Grau, V.: Combined generation of electrocardiogram and cardiac anatomy models using multi-modal variational autoencoders. In: 2022 IEEE 19th International Symposium on Biomedical Imaging (ISBI), pp. 1–4 (2022)

4. Beetz, M., Ossenberg-Engels, J., Banerjee, A., Grau, V.: Predicting 3D cardiac deformations with point cloud autoencoders. In: Puyol Antón, E., et al. (eds.) STACOM 2021. LNCS, vol. 13131, pp. 219–228. Springer, Cham (2022). https://doi.org/10.1007/978-3-030-93722-5_24

5. Bello, G.A., et al.: Deep-learning cardiac motion analysis for human survival prediction. Nat. Mach. Intell. **1**(2), 95–104 (2019)

6. Çiçek, Ö., Abdulkadir, A., Lienkamp, S.S., Brox, T., Ronneberger, O.: 3D U-Net: learning dense volumetric segmentation from sparse annotation. In: Ourselin, S., Joskowicz, L., Sabuncu, M.R., Unal, G., Wells, W. (eds.) MICCAI 2016. LNCS, vol. 9901, pp. 424–432. Springer, Cham (2016). https://doi.org/10.1007/978-3-319-46723-8_49

7. Corral Acero, J., et al.: Understanding and improving risk assessment after myocardial infarction using automated left ventricular shape analysis. JACC: Cardiovasc. Imaging (2022)

8. Corral Acero, J., et al.: SMOD - data augmentation based on statistical models of deformation to enhance segmentation in 2D cine cardiac MRI. In: Coudière, Y., Ozenne, V., Vigmond, E., Zemzemi, N. (eds.) FIMH 2019. LNCS, vol. 11504, pp. 361–369. Springer, Cham (2019). https://doi.org/10.1007/978-3-030-21949-9_39

9. Dalton, D., Lazarus, A., Rabbani, A., Gao, H., Husmeier, D.: Graph neural network emulation of cardiac mechanics. In: Proceedings of the 3rd International Conference on Statistics: Theory and Applications (ICSTA 2021), pp. 127-1-8 (2021)

10. Defferrard, M., Bresson, X., Vandergheynst, P.: Convolutional neural networks on graphs with fast localized spectral filtering. In: Proceedings of the 30th International Conference on Neural Information Processing Systems, pp. 3844–3852 (2016)

11. Di Folco, M., Moceri, P., Clarysse, P., Duchateau, N.: Characterizing interactions between cardiac shape and deformation by non-linear manifold learning. Med. Image Anal. **75**, 102278 (2022)

12. Eitel, I., et al.: Intracoronary compared with intravenous bolus abciximab application during primary percutaneous coronary intervention in ST-segment elevation myocardial infarction: cardiac magnetic resonance substudy of the AIDA STEMI trial. J. Am. Coll. Cardiol. **61**(13), 1447–1454 (2013)

13. Hammond, D.K., Vandergheynst, P., Gribonval, R.: Wavelets on graphs via spectral graph theory. Appl. Comput. Harm. Anal. **30**(2), 129–150 (2011)

14. Hong, B.D., Moulton, M.J., Secomb, T.W.: Modeling left ventricular dynamics with characteristic deformation modes. Biomech. Model. Mechanobiol. **18**(6), 1683–1696 (2019). https://doi.org/10.1007/s10237-019-01168-8

15. Kingma, D.P., Ba, J.: Adam: a method for stochastic optimization. arXiv preprint arXiv:1412.6980 (2014)

16. Krebs, J., Mansi, T., Ayache, N., Delingette, H.: Probabilistic motion modeling from medical image sequences: application to cardiac cine-MRI. In: Pop, M., et al. (eds.) STACOM 2019. LNCS, vol. 12009, pp. 176–185. Springer, Cham (2020). https://doi.org/10.1007/978-3-030-39074-7_19

17. Krishnamurthy, A., et al.: Patient-specific models of cardiac biomechanics. J. Comput. Phys. **244**, 4–21 (2013)

18. Lamata, P., et al.: An automatic service for the personalization of ventricular cardiac meshes. J. Roy. Soc. Interface **11**(91), 20131023 (2014)

19. Lopez-Perez, A., Sebastian, R., Ferrero, J.M.: Three-dimensional cardiac computational modelling: methods, features and applications. Biomed. Eng. Online **14**(1), 1–31 (2015)

20. Lorensen, W.E., Cline, H.E.: Marching cubes: a high resolution 3D surface construction algorithm. ACM SIGGRAPH Comput. Graph. **21**(4), 163–169 (1987)

21. Lu, P., Bai, W., Rueckert, D., Noble, J.A.: Modelling cardiac motion via spatiotemporal graph convolutional networks to boost the diagnosis of heart conditions. In: Puyol Anton, E., et al. (eds.) STACOM 2020. LNCS, vol. 12592, pp. 56–65. Springer, Cham (2021). https://doi.org/10.1007/978-3-030-68107-4_6

22. Lu, P., Bai, W., Rueckert, D., Noble, J.A.: Multiscale graph convolutional networks for cardiac motion analysis. In: Ennis, D.B., Perotti, L.E., Wang, V.Y. (eds.) FIMH 2021. LNCS, vol. 12738, pp. 264–272. Springer, Cham (2021). https://doi.org/10. 1007/978-3-030-78710-3_26

23. Meister, F., et al.: Graph convolutional regression of cardiac depolarization from sparse endocardial maps. In: Puyol Anton, E., et al. (eds.) STACOM 2020. LNCS, vol. 12592, pp. 23–34. Springer, Cham (2021). https://doi.org/10.1007/978-3-030-68107-4_3

24. Ossenberg-Engels, J., Grau, V.: Conditional generative adversarial networks for the prediction of cardiac contraction from individual frames. In: International Workshop on Statistical Atlases and Computational Models of the Heart, pp. 109–118 (2019)

25. Paszke, A., et al.: PyTorch: an imperative style, high-performance deep learning library. In: Proceedings of the 33rd International Conference on Neural Information Processing Systems, pp. 8026–8037 (2019)

26. Qin, C., et al.: Joint learning of motion estimation and segmentation for cardiac MR image sequences. In: Frangi, A.F., Schnabel, J.A., Davatzikos, C., Alberola-López, C., Fichtinger, G. (eds.) MICCAI 2018. LNCS, vol. 11071, pp. 472–480. Springer, Cham (2018). https://doi.org/10.1007/978-3-030-00934-2_53

27. Ranjan, A., Bolkart, T., Sanyal, S., Black, M.J.: Generating 3D faces using convolutional mesh autoencoders. In: Proceedings of the European Conference on Computer Vision (ECCV), pp. 704–720 (2018)

28. Ronneberger, O., Fischer, P., Brox, T.: U-Net: convolutional networks for biomedical image segmentation. In: Navab, N., Hornegger, J., Wells, W.M., Frangi, A.F. (eds.) MICCAI 2015. LNCS, vol. 9351, pp. 234–241. Springer, Cham (2015). https://doi.org/10.1007/978-3-319-24574-4_28

29. Thiele, H., et al.: Effect of aspiration thrombectomy on microvascular obstruction in NSTEMI patients: the TATORT-NSTEMI trial. J. Am. Coll. Cardiol. **64**(11), 1117–1124 (2014)

Skeletal Model-Based Analysis of the Tricuspid Valve in Hypoplastic Left Heart Syndrome

Jared Vicory[1]([✉]), Christian Herz[2], Ye Han[1], David Allemang[1], Maura Flynn[2], Alana Cianciulli[2], Hannah H. Nam[2], Patricia Sabin[2], Andras Lasso[3], Matthew A. Jolley[2,4], and Beatriz Paniagua[1]

[1] Kitware Inc, North Carolina, USA
jared.vicory@kitware.com
[2] Department of Anesthesia and Critical Care Medicine,
Children's Hospital of Philadelphia, Philadelphia, PA 02115, USA
[3] Queen's University, Ontario, Canada
[4] Division of Pediatric Cardiology, Children's Hospital of Philadelphia,
Philadelphia, PA 02115, USA

Abstract. Hypoplastic left heart syndrome is a congenital heart disease characterized by incomplete development of the left heart. Affected children undergo a series of operations which result in the tricuspid valve becoming the only functional atrioventricular valve. Many of these patients go on to develop complications associated with heart failure and death such as tricuspid valve regurgitation. Predicting which patients will develop regurgitation as well as planning for corrective procedures could be greatly enhanced through better understanding of the relationship between geometry and function of the tricuspid valve. Traditional analysis has relied on simple, global anatomical measures which often can not capture localized structural changes. Recently, statistical shape modeling has proven to be useful for analyzing the geometry of the tricuspid valve. We propose to use skeletal representations (s-reps) for modeling the leaflets of the tricuspid valve in these patients. S-reps are a more feature-rich representation than traditional boundary-based models and have been shown to have advantages for statistical analysis. Unfortunately, it is more difficult to fit s-reps to many geometries which limits the application of their powerful analysis techniques. We propose an extension to previous s-rep fitting approaches which yields improved models for difficult to fit objects such as the leaflets of the tricuspid valve. We incorporate application-specific anatomical landmarks and population information to improve correspondence. We use several traditional shape analysis techniques to compare the efficiency of s-reps with boundary representations created using SPHARM-PDM. We observe that principal component analysis produces a more compact shape space using s-reps, needing fewer modes to represent 90% of the population variation, while distance-weighted discrimination shows that s-reps provide more significant classification results between valves with less regurgitation and those with more. These results demonstrate the power of using s-reps for relating structure and function of the tricuspid valve.

© The Author(s), under exclusive license to Springer Nature Switzerland AG 2022
O. Camara et al. (Eds.): STACOM 2022, LNCS 13593, pp. 258–268, 2022.
https://doi.org/10.1007/978-3-031-23443-9_24

Keywords: Statistical shape modeling · Statistical shape analysis · Skeletal models · Hypoplastic left heart syndrome · Cardiac imaging · 3D echocardiography

1   Introduction

Hypoplastic left heart syndrome (HLHS) is a form of congenital heart disease characterized by incomplete development of the left heart, affecting over 1,000 infants in the US per year and proving uniformly fatal without surgical intervention [5]. Surgery allows the children to survive, but many will go on to develop tricuspid regurgitation (TR). Because the tricuspid valve (TV) is the only remaining functional atrioventricular valve, this condition is highly associated with heart failure, requiring surgical intervention in as many as 30% of HLHS patients [7,11].

The exact relationship between TV structure and TR is not known. Both imaging and surgical inspection are used to examine the TV but are not without limitations [22]. Metrics such as annular area [3,14], septolateral diameter [3,4, 14], bending angle [12], anterior leaflet prolapse [4], and anterior papillary muscle location [3,14] computed from 3D echocardiograms have been used to detect the presence of TR. While some of these measures are semi-local, none of them can capture more localized or subtle differences in tricuspid valve geometry.

Previous work [24] has shown that boundary-based shape models of the TV can be used as a basis for analyzing a population of valves and performing tasks such as discriminating between valves with lower and higher amounts of TR. However, there are inherent limitations to representing objects using purely boundary information. In this work, we instead explore the use of discrete skeletal representations (s-reps) for modeling the TV.

S-reps have the benefit of explicitly modeling the entire object, including its interior, unlike pure boundary models such as those produced by methods like spherical harmonic point distribution model (SPHARM-PDM) [21]. S-reps have been shown to be a powerful representation in a variety of tasks, including segmentation [23], classification [9] and hypothesis testing [19].

These advantages are tempered by the fact that it is often more difficult to fit s-reps to objects than it is simpler boundary-based models. We introduce changes to the traditional s-rep fitting pipeline to make it easier to create s-reps for objects like TV leaflets which cause difficulty for existing approaches. We produce s-reps which have improved correspondence across the population compared with previous s-rep fitting techniques. We then compare the power of s-reps and s-rep-based analysis of the TV with previous work using boundary based models and show that it provides advantages for both general population analysis and discrimination.

2   Materials

2.1   Subjects

Acquisition of transthoracic 3DE images of the TV is part of the standard clinical echo lab protocol for HLHS at Children's Hospital of Philadelphia (CHOP). Patients were retrospectively identified who had HLHS with a Fontan circulation and in whom 3DE of the TV had been previously performed. Exclusion criteria included presence of significant artifacts and inability to segment the TV. This study was approved by an Institutional Review Board. We utilized 100 3DE scans with age range 2 to 30 years (mean 10.36 years). Images were acquired using sector narrowed Full Volume or 3D Zoom mode with a wide field of view. Electrocardiogram-gated acquisitions were obtained when patient cooperation allowed. Transthoracic probes (X7 or X5) were used with the Philips IE33 and EPIQ 7 ultrasound systems.

3   Methods

3.1   Image Segmentation and Model Creation

Images were exported to DICOM via Philips QLAB and imported into 3D Slicer [1] using the SlicerHeart [2,3] Philips DICOM converter. A single mid-systolic frame was chosen for modeling the TV. TV segmentation was performed using the 3D Slicer Segmentation module (Fig. 1).

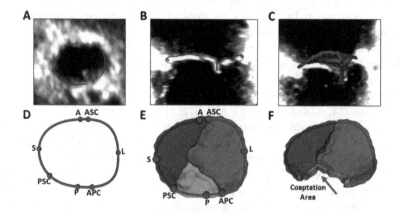

Fig. 1. A, B and C. Segmentation of TV from ultrasound; D and E. Atrial views of the annulus (D) and TV (E) with landmarks. Of particular note in this work are the commissures: ASC (anterior/septal commisure), APC (anterior/posterior commisure), and PSC (posterior/septal commisure). F. Cutaway showing the coaptation surface where leaflets meet.)

3.2   Skeletal Representations

In previous work [24], corresponding boundary point distribution models
(PDMs) of each leaflet in the tricuspid valve were created using SPHARM-
PDM [21] with correspondence-correcting post-processing [24]. While this process produces high quality models suitable for statistical analysis, there are inherent limitations to representing objects using purely boundary information. This
is particularly true for TV leaflets which are very thin and sheet-like as the relationship between points which are close in space but on opposite sides of the
leaflet can be just as if not more important than those of nearby points on the
same side. For this reason we choose to use s-reps to model the leaflets of the
TV in this work. As seen in Fig. 2, an s-rep consists of a grid of points on the
skeleton and a set of vectors emanating from the skeleton to the boundary called
spokes.

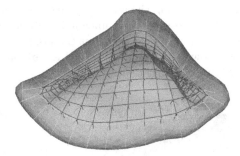

Fig. 2. An s-rep fit to a tricuspid valve anterior leaflet. The green lines are the mesh of
the object's skeleton. The cyan spokes point to the top side of the object, the magenta
to the bottom, and the yellow to the crest. (Color figure online)

The general process for fitting an s-rep to an object consists of several
steps [15]. Because directly computing a medial or skeletal model of an arbitrary object is a non-trivial process, the objects are first deformed to a simpler
object such as an ellipsoid. This deformation is currently done via mean curvature flow. From there an s-rep of the ellipsoid (the skeleton of an ellipsoid is
actually medial) can be directly computed. Finally, thin-plate splines (TPS) are
used to interpolate a space-filling reverse deformation from the ellipsoid back
to the original object. Finally, a refinement step that corrects any deficiencies
in the s-rep fit caused by TPS warping by ensuring that the spokes touch the
original object boundary and are approximately orthogonal to it is done.

This process works well for objects which are mostly ellipsoidal in shape
and have three dimensions with obviously different principal radii. However, for
objects such as TV leaflets where these assumptions do not hold, there can
be substantial differences in how ellipsoids are mapped to the leaflets across
a population. This will in turn yield s-reps which are not in correspondence
because the same anatomical regions, such as the coaptation surfaces, will be

mapped to very different parts of the s-rep. S-reps generated using this approach are thus not suitable for statistical analysis. In this work we instead leverage SPHARM-PDM models previously fit to the leaflet boundaries. Because these fits are based on a decomposition of the object's boundary using spherical harmonics, by choosing to reconstruct the object using only first degree polynomials we obtain an ellipsoid that best fits the object and is in correspondence with the original SPHARM-PDM mesh. We can then proceed as before, computing an s-rep directly from this ellipsoid and using a TPS warp based on the existing correspondence. Because correspondence of these models across the entire population was already established, this process yields a set of s-reps which share this correspondence.

The final problem with fitting s-reps to TV leaflets is the choice of which dimensions of the leaflets should correspond to the axes of the ellipsoid. We choose to let the commisure-to-commisure axis of the leaflet be the major axis of the ellipsoid. The second axis of the ellipsoid is chosen to be the axis between the centerpoint of the leaflet along the annulus to its point closest to the center of the valve. Note that, in cases with significant billow or tenting, this center point is not precisely the center of the valve but the point along the ridge of the valve either above or below the annuluar plane. Figure 3 shows an example of these axes on a leaflet with tenting.

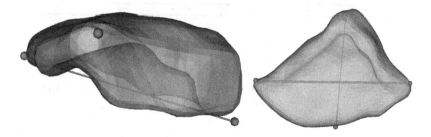

Fig. 3. The leaflet axes used to map the first two ellipsoid axes. On the left, the line between the midpoint along the annulus and lowest part of the valve maps to the next longest ellipsoid axis. On the right, the left-to-right axis between the two commisure points maps to the long ellipsoid axis.

3.3   Analysis of s-reps

Discrete s-reps are represented as a series of tuples: a tuple of points representing samples on the object skeleton, a tuple of unit vectors representing spokes pointing from each skeletal point toward the object boundary, and a tuple of distances representing the length of these vectors. Because this representation is highly non-Euclidean, traditional statistical analysis methods which make Euclidean assumptions cannot be applied directly. Techniques such as principal geodesic analysis (PGA) were designed to act as analogues to traditional Euclidean methods on general non-Euclidean spaces. These techniques have been previously applied to s-reps, but because we know what manifolds s-rep features live on we can use a more customized approach. We use a method called Principal Nested Spheres (PNS) [10] which has been shown highly effective in Euclideanizing the s-rep representation [16].

PNS works by performing a principal component analysis (PCA)-like analysis, but operating on a data which lives on a sphere rather than on Euclidean data. It starts with data on an d dimensional sphere and finds the $d-1$ dimensional subsphere on that sphere that minimizes the projected distances of each data point to that subsphere. This subsphere can be represented by an axis and distance which we call a polar system. The data is then projected to this subsphere and the process repeats until a 0 dimensional datapoint, the mean, is found. In each step, the projected distance is a Euclideanized feature. An example is shown in Fig. 4. Because many of the non-Euclidean features of s-reps live on spheres, this technique can be used to analyze them.

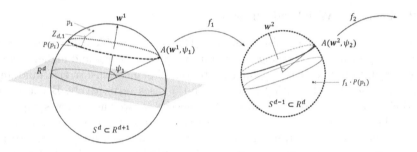

Fig. 4. An example PNS projection step. The point p_1 on a d-dimensional sphere is projected onto the $d-1$-dimensional subsphere defined by the polar system $A(w^1, \psi_1)$. $Z_{d,1}$ is the Euclideanized feature resulting from this projection. This process is repeated until projecting onto a 0-dimensional subsphere.

To create a fully Euclideanized s-rep, we must perform several separate PNS analyses: one high-dimensional PNS for the tuple of skeletal points (n skeletal points live on the sphere $\mathbf{S}^{3n-4}$ and a two-dimensional PNS for each individual unit spoke direction which live on $\mathbf{S}^2$. The Euclideanized features from these analyses are combined with appropriately normalized spoke lengths to create a matrix of Euclidean features suitable for being used with traditional Euclidean statistical analysis methods. Full details of this process can be found in [16].

3.4 Normalization

The question of how to normalize valve scale in group of children of varying size is an open one that requires further study. A traditional way to normalize shapes is using gross geometric scale using Procrustes analysis. In previous work on heart valves, body surface area (BSA) has been shown to be a meaningful basis for rescaling [20]. In this work we compare results using these two approaches but we do not directly address the question of which is the better approach or if there is another approach which provides better results.

4   Evaluation

4.1   Principal Component Analysis

We use principal component analysis (PCA) on the composite matrix described in Sect. 3.3 to find a mean and modes of variation of our population of s-reps. One major use of PCA is for dimensionality reduction by finding a lower-dimensional space which captures most of the variation in the population. Compared to results using a traditional PCA approach on the SPHARM-PDM models, s-reps provide a more compact representation, being able to capture 90% of the population variation in 11 modes rather than 16.

4.2   Distance Weighted Discrimination

We use distance-weighted discrimination (DWD) [13], an SVM-like binary classification algorithm, for learning to separate valves with lower and higher amounts of regurgitation. Our data is grouped into four classifications: trivial, mild, moderate, and severe TR, with 19, 46, 30, and 5 cases respectively. Because these gradings are qualitative and somewhat subjective, as well as the low sample sizes particularly in the trivial and severe classes, we group the trivial and mild cases into one class and moderate and severe cases into another for training DWD.

In Fig. 5, we compare DWD results on both SPHARM-PDM and s-rep models using both normalization strategies discussed in Sect. 3.4. These plots are generated by projecting all cases onto the computed separating direction and using kernel density estimation to fit distributions to each class. Though separation results are generally worse using BSA-based normalization than standard Procrustes analysis, we can see that there tends to be better separation, particularly between the largest mild and moderate groups, using s-reps.

To further validate this result we use 10 rounds of 5-fold cross-validation to assess the accuracy of classifiers trained on both models and using both normalization methods. The performance on the trivial and severe groups, the smallest and most extreme, is similar across all combinations. For mild vs moderate, while there is still significant overlap between the two groups, s-reps show generally better separation, correctly classifying approximately 65% of moderates and 90% of mild cases using Procrustes normalization and 55% of moderates and 85% of milds using BSA normalization. This compares to approximately 50%/95% and 45%/90% for SPHARM. Interestingly, s-reps show significantly better results for the moderate group but sometimes slightly worse results for mild cases.

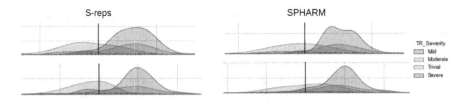

Fig. 5. Comparison of DWD projection plots between s-reps (left) and SPHARM-PDM (right) using both Procrustes normalization (top) and BSA-based normalization (bottom). Generally, Procrustes normalization performed better than BSA and s-reps performed better than SPHARM.

5 Discussion

In this work we propose a method for creating discrete skeletal representations for performing modeling and analysis of the individual leaflets of the tricuspid valve. We introduce improvements to previous s-rep fitting approaches to leverage the ease of creating population-level correspondences via methods which are less computationally intensive methods than s-rep fitting but which in turn produce less powerful models. This approach leads to models with improved suitability for statistical analysis. We employ several standard analysis techniques and demonstrate that s-reps provide a richer geometric representation than previous work using boundary-based PDMs. In addition, S-reps seem particularly well-suited to representing valve leaflets as these thin structures lend themselves to being represented by skeletal models with a defining medial surface, unlike boundary models. In contrast, for modeling of more globular shapes, such as the right ventricle, SPHARM may have benefits as the boundary may be more relevent than the internal volume in that setting.

The methods presented in this work are distributed as part of Slicer-SALT [25], an open-source, free, comprehensive software that will allow biomedical scientists to precisely locate shape changes in their imaging studies, and SlicerHeart [2,3], a 3DSlicer extension containing tools for cardiac image import (3D/4D ultrasound, CT, MRI), quantification, and implant placement planning and assessment.

The time needed for initial SPHARM-PDM and s-rep mapping are relatively low (5 min per case on an AMD Ryzen 3950X in a mostly single-threaded implementation), however, the current implementation of s-rep refinement can be quite slow (30 min–1 h, depending on how much refinement is needed) and may require much manual parameter tuning. We are currently exploring deep learning-based approaches for generating s-reps of objects to reduce required computation time as well as lower the need for manual adjustment for each new setting.

As discussed in Sect. 3.4, the question of how to normalize TV leaflets in a population of children of varying size is an open one. We continue to investigate additional normalization schemes, including normalizing by age-adjusted average

BSA to attempt to separate normal size increases due to aging from disease-related changes.

Clinically, shape methods such as s-reps could be used in multiple ways. They could be used to augment rapid segmentation of valves via inclusion into atlas based [17] or machine learning [8] approaches to create valve models while reducing time-consuming manual work. Shape analysis can be used to characterize the population and compare any individual valve in the context of that population. In association with clinical outcome data such as degree of regurgitation or other patient outcome, such information could allow for risk stratification and inform development of strategies for image-derived patient specific repair [6].

A future area that we wish to explore is in the use of skeletal models as a method of extraction of shell models for the finite element modeling (FEM) of heart valves. We believe that s-reps provide a powerful basis for an open-source application for more easily creating these models compared to previous medial approaches [18]. The structure of s-reps could also lend themselves well to create both skeletal and volumetric meshes for FEM. In addition, the inherent parameterization of valve leaflets may be beneficial in solving inverse problems in FEM such as those to derive leaflet material properties from images themselves.

Acknowledgements. Research reported in this publication was supported by NIH NHLBI award R01HL153166 as well as NIBIB R01EB021391 and R56EB021391. This work was also supported by a CHOP Frontier program (Pediatric Valve Center), as well as the Cora Topolewski fund at CHOP. The content is solely the responsibility of the authors and does not necessarily represent the official views of the National Institutes of Health.

References

1. Fedorov, A., et al.: 3D slicer as an image computing platform for the quantitative imaging network. Magn. Reson. Imaging **30**(9), 1323–1341 (2012). https://doi.org/10.1016/J.MRI.2012.05.001
2. Scanlan, A.B., et al.: Comparison of 3D echocardiogram-derived 3D printed valve models to molded models for simulated repair of pediatric atrioventricular valves. Pediatr. Cardiol. **39**(3), 538–547 (2018). https://doi.org/10.1007/S00246-017-1785-4
3. Nguyen, A.V., et al.: Dynamic three-dimensional geometry of the tricuspid valve annulus in hypoplastic left heart syndrome with a fontan circulation. J. Am. Soc. Echocardiogr. **32**(5), 655–666 (2019). https://doi.org/10.1016/J.ECHO.2019.01.002. Official Publication of the American Society of Echocardiography
4. Bautista-Hernandez, V., et al.: Mechanisms of tricuspid regurgitation in patients with hypoplastic left heart syndrome undergoing tricuspid valvuloplasty. J. Thorac. Cardiovasc. Surg. **148**(3), 832–840 (2014)
5. Gordon, B.M., Rodriguez, S., Lee, M., Chang, R.K.: Decreasing number of deaths of infants with hypoplastic left heart syndrome. J. Pediatr. **153**(3), 354–358 (2008). https://doi.org/10.1016/J.JPEDS.2008.03.009
6. Bouma, W., et al.: Preoperative three-dimensional valve analysis predicts recurrent ischemic mitral regurgitation after mitral annuloplasty. Ann. Thorac. Surg. **101**(2), 567–575 (2016)

7. Barber, G., et al.: The significance of tricuspid regurgitation in hypoplastic left-heart syndrome. Am. Heart J. **116**(6 Pt 1), 1563–1567 (1988). https://doi.org/10.1016/0002-8703(88)90744-2
8. Herz, C., et al.: Segmentation of tricuspid valve leaflets from transthoracic 3D echocardiograms of children with hypoplastic left heart syndrome using deep learning. Front. Cardiovasc. Med., 1839 (2021)
9. Hong, J., Vicory, J., Schulz, J., Styner, M., Marron, J.S., Pizer, S.M.: Non-euclidean classification of medically imaged objects via s-reps. Med. Image Anal. **31**, 37–45 (2016)
10. Jung, S., Dryden, I.L., Marron, J.S.: Analysis of principal nested spheres. Biometrika **99**(3), 551–568 (2012). https://doi.org/10.1093/biomet/ass022
11. King, G., et al.: Atrioventricular valve failure in Fontan palliation. J. Am. Coll. Cardiol. **73**(7), 810–822 (2019)
12. Kutty, S., et al.: Tricuspid regurgitation in hypoplastic left heart syndrome: mechanistic insights from 3-dimensional echocardiography and relationship with outcomes. Circ.: Cardiovasc. Imaging **7**(5), 765–772 (2014)
13. Marron, J.S., Hill, C., Todd, M.: Distance Weighted Discrimination
14. Nii, M., Guerra, V., Roman, K.S., Macgowan, C.K., Smallhorn, J.F.: Three-dimensional tricuspid annular function provides insight into the mechanisms of tricuspid valve regurgitation in classic hypoplastic left heart syndrome. J. Am. Soc. Echocardiogr. **19**(4), 391–402 (2006)
15. Pizer, S.M., et al.: Object shape representation via skeletal models (s-reps) and statistical analysis. In: Riemannian Geometric Statistics in Medical Image Analysis, pp. 233–271. Elsevier (2020)
16. Pizer, S.M., et al.: Nested sphere statistics of skeletal models. In: Breuß, M., Bruckstein, A., Maragos, P. (eds.) Innovations for Shape Analysis, pp. 93–115. Springer, Heidelberg (2013). https://doi.org/10.1007/978-3-642-34141-0_5
17. Pouch, A.M., et al.: Image segmentation and modeling of the pediatric tricuspid valve in hypoplastic left heart syndrome. In: Pop, M., Wright, G.A. (eds.) FIMH 2017. LNCS, vol. 10263, pp. 95–105. Springer, Cham (2017). https://doi.org/10.1007/978-3-319-59448-4_10
18. Rego, B.V., Pouch, A.M., Gorman, J.H., Gorman, R.C., Sacks, M.S.: Patient-specific quantification of normal and bicuspid aortic valve leaflet deformations from clinically derived images. Ann. Biomed. Eng. **50**(1), 1–15 (2022). https://doi.org/10.1007/s10439-021-02882-0
19. Schulz, J., Pizer, S.M., Marron, J., Godtliebsen, F.: Non-linear hypothesis testing of geometric object properties of shapes applied to hippocampi. J. Math. Imaging Vis. **54**(1), 15–34 (2016)
20. Sluysmans, T., Colan, S.D.: Theoretical and empirical derivation of cardiovascular allometric relationships in children. J. Appl. Physiol. **99**(2), 445–457 (2005)
21. Styner, M., et al.: Framework for the statistical shape analysis of brain structures using SPHARM-PDM. Insight J. **1071**, 242 (2006)
22. Bharucha, T., Khan, R., Mertens, L., Friedberg, M.K.: Right ventricular mechanical dyssynchrony and asymmetric contraction in hypoplastic heart syndrome are associated with tricuspid regurgitation. J. Am. Soc. Echocardiogr. **26**(10), 1214–1220 (2013). https://doi.org/10.1016/J.ECHO.2013.06.015. Official publication of the American Society of Echocardiography
23. Vicory, J., Foskey, M., Fenster, A., Ward, A., Pizer, S.M.: Prostate segmentation from 3dus using regional texture classification and shape differences. In: Proceedings of the Shape-Symposium on Statistics Shape Models Application, p. 24. Citeseer (2014)

24. Vicory, J., et al.: Statistical shape analysis of the tricuspid valve in hypoplastic left heart syndrome. In: Puyol Antón, E., et al. (eds.) STACOM 2021. LNCS, vol. 13131, pp. 132–140. Springer, Cham (2022). https://doi.org/10.1007/978-3-030-93722-5_15

25. Vicory, J.: SlicerSALT: shape analysis toolbox. In: Reuter, M., Wachinger, C., Lombaert, H., Paniagua, B., Lüthi, M., Egger, B. (eds.) ShapeMI 2018. LNCS, vol. 11167, pp. 65–72. Springer, Cham (2018). https://doi.org/10.1007/978-3-030-04747-4_6

Simplifying Disease Staging Models into a Single Anatomical Axis - A Case Study of Aortic Coarctation In-utero

Uxio Hermida[1(✉)], Milou P.M. van Poppel[2], David Stojanovski[1],
David F.A. Lloyd[3,4], Johannes K. Steinweg[2], Trisha V. Vigneswaran[4,5],
John M. Simpson[4,5], Reza Razavi[2,4], Adelaide De Vecchi[1],
Kuberan Pushparajah[2,4], and Pablo Lamata[1]

[1] Department of Biomedical Engineering, School of Biomedical Engineering and
Imaging Sciences, King's College London, St Thomas' Hospital, London, UK
`uxio.hermida_nunez@kcl.ac.uk`
[2] Department of Cardiovascular Imaging, School of Biomedical Engineering
and Imaging Sciences, King's College London, St Thomas' Hospital,
London, UK
[3] Department of Perinatal Imaging, School of Biomedical Engineering and Imaging
Sciences, King's College London, St Thomas' Hospital, London, UK
[4] Department of Congenital Heart Disease, Evelina London Children's Hospital,
London, UK
[5] Harris Birthright Centre, Fetal Medicine Research Institute, King's College
Hospital, London, UK

Abstract. Statistical shape modelling and classification methods are
used to study characteristic disease phenotypes, to derive novel shape
biomarkers, and to extract insights into disease mechanisms. Linear classification models are commonly chosen due to their ability to provide a
single score, as well as easy-to-interpret characteristic shapes. In disease
staging models, a multi-class problem is generally set. Then, one would
expect that the set of linear models comparing any pair of classes will
lead to the same unique anatomical axis capturing characteristic shapes
for each group in the disease spectrum, and a unique mechanistic interpretation. In this work, we aim to explore the validity of this assumption
and assess the confidence in simplifying both mechanistic interpretations
and clinical classification performance into a single axis in disease staging models. To do so, we used a statistical shape model of fetal great
arteries in cases with suspected coarctation of the aorta. Data included
control, false positive and confirmed coarctation cases, representative of
three categories of developmental impairment from the disease spectrum.
Principal component analysis combined with a fisher linear discriminant
analysis was used to explore phenotypes associated with each group and
classification performance. A combination of classification overfitting,
a co-linearity index between axes, and the three-dimensional extreme
phenotypes provided useful information for simplification into a single
anatomical axis. Careful consideration should be taken in disease progression studies where either overfitting or co-linearity are compromised,

© The Author(s), under exclusive license to Springer Nature Switzerland AG 2022
O. Camara et al. (Eds.): STACOM 2022, LNCS 13593, pp. 269–279, 2022.
https://doi.org/10.1007/978-3-031-23443-9_25

as the simplification with a single anatomical axis might lead to the inference of misleading mechanisms associated with disease.

Keywords: Computational anatomy · Remodelling · Clinical biomarker · Discriminant analysis

1   Introduction

Among digital twin technologies [3], statistical shape modelling (SSM) has been widely used to derive novel shape biomarkers, explore characteristic disease phenotypes and understand pathophysiological mechanisms [2,6,8,10,14,18]. Most SSM pipelines rely on a combination of dimensionality reduction and classification methods. Principal Component Analysis (PCA) is the most used dimensionality reduction method and allows the exploration of shape changes with respect to a population average. After PCA, shape information is decomposed into several modes of anatomical variation (i.e., PCA modes), and the task is to find the anatomical direction that best discriminates between two clinical groups. Linear classification models are commonly chosen as they allow simplification of shape changes associated with disease into a single axis with the aim of having: 1. Single classification score; 2. Easy-to-interpret shape remodelling patterns.

In disease staging SSM studies, a partition of the population into C separate classes may be conceptualised, each corresponding to a different point along the disease spectrum. Such partitioning using a pairwise discriminative approach results in $C(C-1)/2$ classifiers and an associated set of linear models. It would then be expected that there exists a unique anatomical axis of disease staging and mechanistic interpretation where all pairwise linear models converge.

In this work, we aim to explore the validity of this assumption and assess the confidence in simplifying both mechanistic interpretations and clinical classification performance into a single axis in disease staging SSM studies. To do so, we used a statistical shape model of fetal great arteries in cases with suspected coarctation of the aorta (CoA), including data from three groups representative of three categories of developmental impairment from the spectrum of CoA: control, false positive (FP) and confirmed CoA cases.

2   Methods

2.1   Clinical Background and Dataset

CoA is one of the most common congenital heart defects [15]. Current prenatal diagnosis of neonatal or duct-dependant CoA using standard echocardiographic assessment remains challenging, with high FP diagnosis rates up to 80% [4,13,19]. Moreover, most ultrasound metrics have been reported to have relatively poor generalizability [4] and there is a lack of understanding of the disease spectrum. Recently, the utility of fetal three-dimensional (3D) cardiac magnetic resonance imaging (CMR) for antenatal diagnosis of CoA [11] was explored, showing evidence of anatomical and functional differences between control, FP and CoA cases. This work was further expanded using a detailed SSM pipeline, providing an accurate and robust shape biomarker score and mechanistic insights about CoA in-utero [7].

In the present study, we used a cohort of 53 control cases (31.0 ± 1.6 weeks) and 108 cases with suspected CoA, including 65 FP (32.2 ± 1.7 weeks) and 43 confirmed CoA cases (32.1 ± 1.7 weeks). Each group represents a category of developmental impairment along the disease spectrum. Note that such spectrum does not represent longitudinal changes during fetal gestation as no longitudinal data was available for study, but rather discrete categories capturing different levels of anatomical remodelling as a consequence of their developmental impairment.

Multiple T2-weighted 'black-blood' standard single-shot fast spin echo sequences were acquired with a repetition time (TR), 20,000 ms; echo time (TE), 80 ms; flip angle, 90°; voxel size, 1.25×1.25 mm; slice thickness, 2.5 mm; sensitivity encoding (SENSE) factor, 2; partial Fourier factor, 0.547; slice duration, 546 ms. 2D data was processed with a motion-corrected slice-volume registration method [12], resulting in high-resolution 3D volumes (0.55–0.75 mm isotropic). Segmentations of the fetal heart were done semi-automatically, with manual refinements by a clinician with fetal CMR experience. Cardiac outcomes were collected using hospital records as described in [12]. Informed consent was obtained for all patients included in the study as part of the Intelligent Fetal Imaging and Diagnosis (iFIND) Project (Research Ethics Committee (REC) no:14/LO/1806) or Quantification of fetal growth and development using MRI (REC no: 07/H0707/105).

2.2 Shape Encoding

3D surfaces of the fetal great arteries can be extracted from semi-automatic fetal CMR segmentations. However, such models are prone to surface irregularities. Therefore, to mitigate these effects, we used centerline points and radius to encode shape for the SSM. To ensure topological consistency between cases, centerlines for each segment of interest (i.e. ascending aorta (AAo), descending aorta (DAo) and arterial duct (AD)) were extracted using 4 anatomical landmarks: 1. DAo at the level of the diaphragm; 2. AD at pulmonary artery bifurcation; 3. Medial point of the AD; 4. AAo at right pulmonary artery. Landmarks were manually placed by a single observer for all cases. After extraction, centerlines were merged and aligned to a common reference space. All steps were implemented using open-source Python toolkits, such as the Visualization Toolkit (VTK) [17] or the Vascular Modeling Toolkit (VMTK) [1]. For more details, the reader is referred to [7] (Fig. 1).

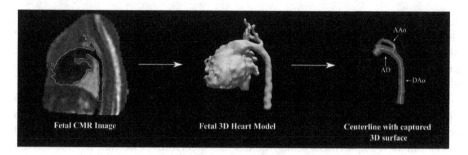

Fig. 1. From a 'black-blood' motion-corrected three-dimensional (3D) cardiovascular magnetic resonance image (CMR) of the fetal thorax, semi-automatic segmentation of the cardiac anatomy was performed. A 3D surface model is then extracted and used, with a set of anatomical landmarks, to extract the centerlines of the fetal arteries and their captured 3D surface. AAo: Ascending aorta; AD: Arterial duct; DAo: Descending aorta.

2.3   Dimensionality Reduction

After centerline extraction, shape information is encoded with a common vector space $R^{N(3+1)}$ (i.e., the coordinates of each point along the centerline plus the captured radius) where N is the number of centerline points. Centerlines for each segment were resampled to a user-defined set of points: 25 points for AAo and AD; 51 for DAo. A total of $N = 101$ points were then used to construct a feature vector x_i of length $4N$ for each case i. PCA was then used to reduce the high dimensionality of the problem and build the SSM. Note that all features were standardized to avoid biasing the SSM due to feature range differences [5]:

$$x_i' = \frac{(x_i - \mu)}{\sigma} \tag{1}$$

where μ and σ are the per-feature mean and standard deviation (SD) vectors. Anatomical modes of variation ϕ (eigenvectors) and their respective variances (eigenvalues) are obtained by singular value decomposition of the covariance matrix built using the standardized feature vector x_i' and the population average. PCA modes are ranked by the amount of variance explained. Therefore, the first modes represent the most common shape features in the cohort.

2.4   Phenotype Exploration

We used a Fisher Linear Discriminant Analysis (LDA) to explore characteristic phenotypes associated with each group in our population [18]. For the multi-class analysis, we used a pairwise classification approach, resulting in 3 linear models: 1. Controls (1) against FP (2) ($LDA_{1,2}$); 2. FP (2) against confirmed CoA (3) ($LDA_{2,3}$); 3. Controls (1) against confirmed CoA (3) ($LDA_{1,3}$). Each anatomy can be described with a case-specific shape score (i.e., Z-score) along each LDA axis. Therefore, the classification performance of each linear model can be assessed. The linear nature of each LDA axis allows the generation of extreme 3D phenotypes along their direction (i.e., ± 3 standard deviations (SD) from the average population shape). 3D extreme phenotypes represent an idealized or simplified representation of the phenotypes associated with each group. Therefore, their visual inspection allows the interpretation of potential mechanisms associated with shape changes in CoA. All three LDA models were built using the 10 first PCA modes.

2.5 The Uniqueness of a Disease Staging Axis

Once a set of pairwise linear models are built, the task is to explore the uniqueness of the disease spectrum axis, capturing the different levels of developmental impairment, and test whether such an axis can confidently provide both accurate classification performance and mechanistic insights. First, the area under the receiver operating characteristic (AUC) in resubstitution (RS) and leave-one-out cross-validation (LOOCV) was used to assess the classification performance of each LDA axis, and their difference (RS - LOOCV) defined their overfitting. Then, to further inform about the similarity between axes, we introduce a co-linearity index (CI) metric. The CI between two LDA axes 1 and 2 was computed as:

$$CI_{1|2} = |\vec{w_1} \cdot \vec{w_2}| \tag{2}$$

where $\vec{w_1}$ and $\vec{w_2}$ are the two unit vectors corresponding to each LDA axis compared. Therefore, a CI of 1.0 is the best CI and 0.0 is the worst. Three CI values were computed: $CI_{1,2|2,3}$ (between $LDA_{1,2}$ and $LDA_{2,3}$); $CI_{1,3|2,3}$ (between $LDA_{1,3}$ and $LDA_{2,3}$); $CI_{1,2|1,3}$ (between $LDA_{1,2}$ and $LDA_{1,3}$).

We further assessed the impact of a limited sample size on the robustness of the findings provided by both the AUC in RS and LOOCV, CI between axes, and 3D extreme phenotypes. A set of 9 different SSM were built with random subsets of increasing size, from 20% to 100%. We used a stratified random approach to ensure keeping constant the proportion of cases from each class in each subset. Each model was run 10 times with different random subsets to remove potential bias in the results due to subset selection. For each subset LDAs, PCA modes capturing at least 85% of the shape variability in each subset were included.

3 Results

The SSM with the full cohort resulted in the first 10 PCA modes capturing 85% of the shape variability in the population. Figure 2 shows the extreme phenotypes captured by each LDA axis. Table 1 shows the multi-class classification performance of each LDA model built with 10 PCA modes. All linear models perfectly classified control (1) and CoA cases (3) (AUC = 1.0). Excellent classification accuracy between FP (2) and CoA (3) cases was obtained with both $LDA_{2,3}$ and $LDA_{1,3}$ axes. Moreover, both axes were able to distinguish between control (1) and FP (2) cases with reasonable accuracy (AUC> 0.84).

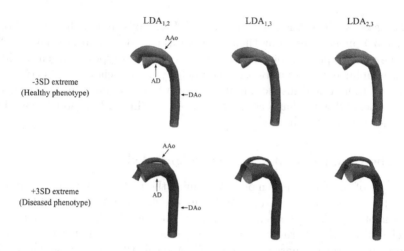

Fig. 2. Extreme phenotypes captured by each linear discriminant axis (LDA) comparing the three different groups along the disease spectrum: (1) Controls; (2) False positives; (3) Confirmed CoA. SD: Standard deviation; AAo: Ascending aorta; AD: Arterial duct; DAo: Descending aorta. SD: Standard deviation.

The best CI was obtained between $LDA_{1,3}$ and $LDA_{2,3}$ axes ($CI_{1,3|2,3} = 0.96$), capturing differences between control (1) and CoA (3), and FP (2) and CoA (3) respectively. The worst CI was obtained between $LDA_{1,2}$ and $LDA_{2,3}$ axes ($CI_{1,2|2,3} = 0.82$), where each LDA axis captures shape differences ending and starting from FP cases respectively. A $CI_{1,2|1,3} = 0.91$ was obtained between $LDA_{1,2}$ and $LDA_{1,3}$ axes.

Table 1. Classification performance (AUC) of each LDA axis

Classes	$LDA_{1,2}$	$LDA_{2,3}$	$LDA_{1,3}$
Control (1) - FP (2)	0.96	0.84	0.88
FP (2) - CoA (3)	0.80	0.94	0.92
Control (1) - CoA (3)	1.00	1.00	1.00

Figure 3 shows the sensitivity of the classification performance and the CI to a limited sample size. Classification performance is shown for the two classes used to build each LDA axis. Excellent classification was obtained with all models, even when using 50% of the initial population. Decreasing the number of cases below 50% was reflected as an increase in overfitting (see the gap between RS and LOOCV performance in Fig. 3 panels A, B and C). The smaller the number of cases used, the higher the variability in classification performance, as each model depends on the specific subset of the population used.

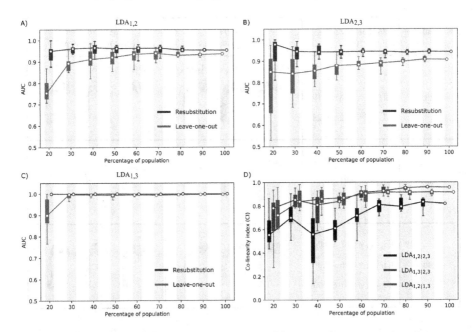

Fig. 3. Boxplots showing the classification performance depending on the percentage of the population used for analysis for A) LDA between control cases (1) and FP cases (2); B) LDA between FP cases (2) and confirmed CoA cases (3); C) LDA between control cases (1) and CoA cases (3); D) Boxplots showing the CI metric for each pair of LDA axes depending on the percentage of the population used.

The relationship between CI and extreme phenotypes captured by each LDA with each subset of the population is shown in Fig. 4. Each shape represents the median from the 10 different LDA models built. As expected, the larger the population subset, the more similar the extreme phenotypes to the 100% population models. Nevertheless, a low CI between axes ($CI \approx 0.5$) was still associated with the same mechanistic interpretations of the extreme shape models: in addition to the relative size between AD and aortic isthmus, all diseased phenotypes capture a proximal displacement of the aortic isthmus inserting vertically in the AD, while the healthy phenotypes capture a more aneurysmal AD with lateral insertion into the aortic isthmus and descending aorta. Although with a reduced sample size chances of clear differences compared to the 100% shapes were increased, shape features near the junction of the three segments remained consistent (see 100% and 20% extreme phenotypes in Fig. 4). Those shape changes, in combination with potential blood flow abnormalities, are thought to potentially affect the arterial wall and vascular remodelling [7, 9, 11, 16].

Figure 5 shows the relationship between each CI and the overfitting of the two corresponding LDA axes. The values for each of the 10 runs with each subset of the population are shown. For all three pairs of LDA models, the larger the number of cases used to build the LDA axes, the smaller the overfitting and the

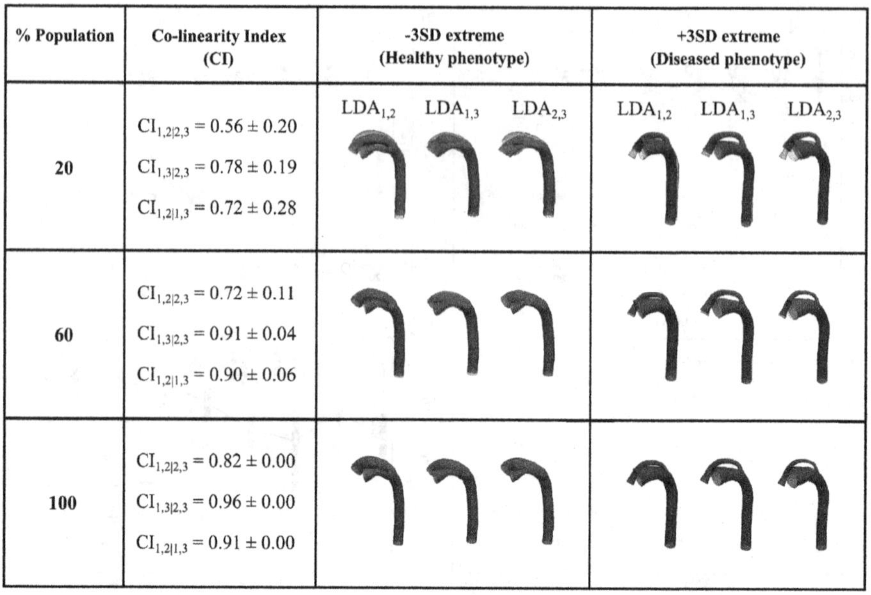

% Population	Co-linearity Index (CI)	-3SD extreme (Healthy phenotype)			+3SD extreme (Diseased phenotype)		
20	$CI_{1,2|2,3} = 0.56 \pm 0.20$ $CI_{1,3|2,3} = 0.78 \pm 0.19$ $CI_{1,2|1,3} = 0.72 \pm 0.28$	$LDA_{1,2}$	$LDA_{1,3}$	$LDA_{2,3}$	$LDA_{1,2}$	$LDA_{1,3}$	$LDA_{2,3}$
60	$CI_{1,2|2,3} = 0.72 \pm 0.11$ $CI_{1,3|2,3} = 0.91 \pm 0.04$ $CI_{1,2|1,3} = 0.90 \pm 0.06$						
100	$CI_{1,2|2,3} = 0.82 \pm 0.00$ $CI_{1,3|2,3} = 0.96 \pm 0.00$ $CI_{1,2|1,3} = 0.91 \pm 0.00$						

Fig. 4. Changes in the extreme phenotypes and co-linearity indexes (CI) with a limited sample size. CI values show the median $\pm$ standard deviation of each of the 10 runs with different population subsets. Extreme phenotypes show the median of those 10 runs. The 100% extreme phenotypes are overlaid with a wireframe in all models. (1) Controls; (2) False positives; (3) Coarctation of the aorta.

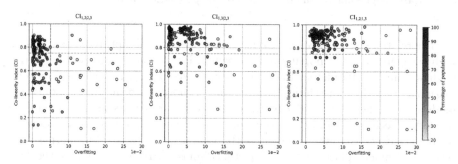

Fig. 5. Relationship between the co-linearity index (CI) between each pair of LDA axes and the overfitting of each of those two axes. Overfitting is computed as the difference between the AUC in RS and LOOCV. The empirical thresholds for acceptable overfitting (< 0.05) and ($CI > 0.75$) are shown with red dashed lines. (Color figure online)

larger the CI as can be seen by how scatter values tend to the top-left corner in all plots. Smaller percentages of the population resulted in a large variability being observed, suggesting that either classification or CI are compromised. With our data, a $CI > 0.75$ and overfitting < 0.05 seems to provide enough confidence for simplification within a single anatomical axis.

4   Discussion and Conclusions

This work explored the confidence in the simplification of mechanisms and classification within a single anatomical axis capturing the different levels of developmental impairment along the disease spectrum of CoA. Our findings suggest that 1) the assumption of a single axis was sensible for our problem, and 2) the combination of classification overfitting, co-linearity metrics, and extreme phenotypes, was useful to test this assumption.

Pairwise classification of the different groups in our cohort resulted in a set of LDA axes all capturing very similar anatomical remodelling patterns (see Fig. 4) with the complete size of the cohort. Such similarity was proven by the high AUC classifying all three groups along all LDA axes (AUC> 0.80), low overfitting (a gap between AUC in RS and LOOCV < 0.04), a good co-linearity index ($CI > 0.80$), and its consistent mechanistic interpretation that was verified by the similarity of the 3D extreme phenotypes.

Highest CI was obtained between $LDA_{1,3}$ and $LDA_{2,3}$ axes. Such result is not surprising as both axes compared cases with and without CoA. Moreover, the finding that good classification performance was obtained with $LDA_{1,2}$ (AUC < 0.84) and a high CI was obtained between $LDA_{1,2}$ and $LDA_{1,3}$ axes confirms the conceptual model of a developmental severity continuum, with two sub-groups within the healthy population: one completely healthy (control group) and one with early signs of developmental impairment as suspected during echocardiography screening (FP group).

The scenario of a limited sample size was then explored. All axes classified cases with good accuracy (median $AUC_{1,2} > 0.75$, $AUC_{1,2} > 0.85$, $AUC_{1,2} > 0.90$ in LOOCV) even with 20% of the population. However, chances of overfitting were highly increased and co-linearity decreased, highlighting the need for larger sample sizes to reach robust findings independent of the population sample. Interestingly, even with a reduced sample size, control and CoA groups were perfectly classified (AUC = 1). Such finding suggests that there is a very clean and large anatomical signature difference between the two extremes of the disease spectrum, and such differences are already captured with a very small sample size.

An unexpected finding was that low co-linearity did not always match with differences in mechanistic interpretations: differences between median shapes with a very limited sample size (20% of the population) and full cohort were observed - however, such differences led to similar topological changes in the junction of the 3 branches and thus similar interpretations of the mechanisms associated with CoA (see Fig. 4) despite having $CI \approx 0.5$. This finding suggests that there are sources of anatomical variation not related to the clinical question of interest that can indeed bring low CI without compromising the interpretation. It would be desirable to set thresholds of overfitting and co-linearity metrics to guide future studies. In our study, a combination of overfitting < 0.05 and $CI > 0.75$ provided reasonable confidence for the simplification with a unique anatomical axis (see Fig. 5). Nevertheless, the threshold of CI will heavily depend on the amount of variability encoded vs. actually related to the question of interest. Further research with more empirical evidence to formalise these questions is needed.

In conclusion, the combination of metrics of overfitting and co-linearity is suggested for testing the assumption of the simplification of disease staging models within a single anatomical axis.

References

1. Antiga, L., Piccinelli, M., Botti, L., Ene-Iordache, B., Remuzzi, A., Steinman, D.A.: An image-based modeling framework for patient-specific computational hemodynamics. Med. Biol. Eng. Comput. **46**(11), 1097–1112 (2008). https://doi.org/10.1007/s11517-008-0420-1

2. Bruse, J.L., et al.: A statistical shape modelling framework to extract 3D shape biomarkers from medical imaging data: assessing arch morphology of repaired coarctation of the aorta. BMC Med. Imaging **16**(1), 1–19 (2016). https://doi.org/10.1186/s12880-016-0142-z

3. Corral-Acero, J., et al.: The 'Digital Twin' to enable the vision of precision cardiology. Eur. Heart J., 1–11 (2020). https://doi.org/10.1093/eurheartj/ehaa159

4. Familiari, A., et al.: Risk factors for coarctation of the aorta on prenatal ultrasound: a systematic review and meta-analysis. Circulation **135**(8), 772–785 (2017). https://doi.org/10.1161/CIRCULATIONAHA.116.024068

5. Gewers, F.L., et al.: Principal component analysis: a natural approach to data exploration. ACM Comput. Surv. **54**(4), 1–33 (2021). https://doi.org/10.1145/3447755

6. Gilbert, K., et al.: Independent left ventricular morphometric atlases show consistent relationships with cardiovascular risk factors: a UK biobank study. Sci. Rep. **9**(1) (2019). https://doi.org/10.1038/s41598-018-37916-6

7. Hermida, U., et al.: Learning the hidden signature of fetal arch anatomy?: a three-dimensional shape analysis in suspected coarctation. J. Cardiovasc. Transl. Res. (2022). https://doi.org/10.1007/s12265-022-10335-9

8. Hermida, U., et al.: Left ventricular anatomy in obstructive hypertrophic cardiomyopathy: beyond basal septal hypertrophy. Eur. Heart J. Cardiovasc. Imaging (In Press) (2022)

9. Hutchins, G.M.: Coarctation of the aorta explained as a branch-point of the ductus arteriosus. Am. J. Pathol. **63**(2), 203–214 (1971)

10. Lewandowski, A.J., et al.: Preterm heart in adult life: cardiovascular magnetic resonance reveals distinct differences in left ventricular mass, geometry, and function. Circulation **127**(2), 197–206 (2013). https://doi.org/10.1161/CIRCULATIONAHA.112.126920

11. Lloyd, D.F., et al.: Analysis of 3-dimensional arch anatomy, vascular flow, and postnatal outcome in cases of suspected coarctation of the aorta using fetal cardiac magnetic resonance imaging. Circ.: Cardiovasc. Imag. (July), 583–593 (2021). https://doi.org/10.1161/CIRCIMAGING.121.012411

12. Lloyd, D.F., et al.: Three-dimensional visualisation of the fetal heart using prenatal MRI with motion-corrected slice-volume registration: a prospective, single-centre cohort study. The Lancet **393**(10181), 1619–1627 (2019). https://doi.org/10.1016/S0140-6736(18)32490-5

13. Matsui, H., Mellander, M., Roughton, M., Jicinska, H., Gardiner, H.M.: Morphological and physiological predictors of fetal aortic coarctation. Circulation **118**(18), 1793–1801 (2008). https://doi.org/10.1161/CIRCULATIONAHA.108.787598

14. Medrano-Gracia, P., et al.: Left ventricular shape variation in asymptomatic populations: the multi-ethnic study of atherosclerosis. J. Cardiovasc. Magn. Reson. **16**(1), 56 (2014). https://doi.org/10.1186/s12968-014-0056-2

15. Rosenthal, E.: Coarctation of the aorta from fetus to adult: curable condition or life long disease process? Heart **91**(11), 1495–1502 (2005). https://doi.org/10.1136/hrt.2004.057182

16. Rudolph, A.M., Heymann, M.A., Spitznas, U.: Hemodynamic considerations in the development of narrowing of the aorta. Am. J. Cardiol. **30**(5), 514–525 (1972). https://doi.org/10.1016/0002-9149(72)90042-2

17. Schroeder, W.J., Martin, K.M.: The visualization toolkit. No. July, Kitware, 4th edn. (2006). https://doi.org/10.1016/B978-012387582-2/50032-0

18. Varela, M., et al.: Novel computational analysis of left atrial anatomy improves prediction of atrial fibrillation recurrence after ablation. Front. Physiol. **8**(FEB), 68 (2017). https://doi.org/10.3389/fphys.2017.00068

19. Vigneswaran, T.V., Zidere, V., Chivers, S., Charakida, M., Akolekar, R., Simpson, J.M.: Impact of prospective measurement of outflow tracts in prediction of coarctation of the aorta. Ultrasound Obstet. Gynecol. **56**(6), 850–856 (2020). https://doi.org/10.1002/uog.21957

Point2Mesh-Net: Combining Point Cloud and Mesh-Based Deep Learning for Cardiac Shape Reconstruction

Marcel Beetz[1]([✉]), Abhirup Banerjee[1,2][iD], and Vicente Grau[1][iD]

[1] Institute of Biomedical Engineering, Department of Engineering Science,
University of Oxford, Oxford OX3 7DQ, UK
`marcel.beetz@eng.ox.ac.uk`
[2] Division of Cardiovascular Medicine, Radcliffe Department of Medicine,
University of Oxford, Oxford OX3 9DU, UK

Abstract. Cine magnetic resonance imaging (MRI) is the gold standard modality for the assessment of cardiac anatomy and function. However, a standard cine acquisition typically consists of only a set of intersecting 2D image slices to represent the true 3D geometry of the human heart, thus limiting its utility in various clinical and research settings. In this work, we present a novel geometric deep learning method, *Point2Mesh-Net*, to directly and efficiently transform a set of 2D MRI slices into 3D cardiac surface meshes. Its architecture consists of an encoder and a decoder, which are based on recent advances in point cloud and mesh-based deep learning, respectively. This allows the network to not only directly process point cloud data, which represents the sparse MRI contours obtained from image segmentation, but also to output 3D triangular surface meshes, which are highly suitable for a variety of follow-up tasks. Furthermore, the Point2Mesh-Net's hierarchical setup with multiple downsampling and upsampling steps enables multi-scale feature learning and helps the network to successfully overcome the two main challenges of cardiac surface reconstruction: data sparsity and slice misalignment. We evaluate the model on a synthetic dataset derived from a 3D MRI-based statistical shape model and find surface distances between reconstructed and gold standard meshes below the underlying image resolution for multiple anatomical substructures of the heart. In addition, we apply the pre-trained Point2Mesh-Net as part of a multi-step pipeline to cine MRI acquisitions of the UK Biobank dataset and observe realistic mesh reconstructions with various clinical metrics in line with corresponding findings of large-scale population studies.

Keywords: Cardiac surface reconstruction · Geometric deep learning · Cine MRI · Graph convolutions · Mesh sampling · Point cloud neural networks · Slice misalignment correction

1 Introduction

Cardiac magnetic resonance imaging (MRI) is the gold standard modality for the assessment of cardiac anatomy and function [30]. In current clinical practice,

O. Camara et al. (Eds.): STACOM 2022, LNCS 13593, pp. 280–290, 2022.
https://doi.org/10.1007/978-3-031-23443-9_26

most cine MRI protocols acquire a set of 2D image slices that intersect the underlying anatomy at multiple spatial locations and orientations. However, this only provides an approximate representation of the true 3D shape of the human heart, which limits the accuracy of cardiac disease diagnosis [12,14,21,31]. Accordingly, many previous works have investigated methods to reconstruct complete 3D cardiac surfaces from the acquired 2D slices [3,5,19,20,32]. This is a challenging task primarily due to the high levels of sparsity in the 2D image acquisitions and the presence of slice misalignment induced by patient, respiratory, and cardiac motions [4]. Previous approaches have tackled these challenges by first segmenting the acquired 2D images in the cardiac substructures of interest, and then fitting a template mesh [17,19,20,29] or optimized surface mesh [3,5,32] to the resulting contours in a per-case regularized optimization procedure.

More recently, deep learning techniques have increasingly been employed in various parts of the reconstruction process with the key benefits of faster execution speeds and easier scalability after termination of network training. However, these methods either rely on highly inefficient voxelgrid representations of anatomical surface data [11,34], lack validation on real datasets [34], require additional preprocessing and postprocessing steps that are complex and error-prone [6,11], can only process single image inputs [33,36], or have only been evaluated on a small number of real cases [6]. In this work, we present *Point2Mesh-Net*, a novel geometric deep learning approach to convert sparse and misaligned MRI contours to dense cardiac surface meshes in a fully automated and efficient manner. It combines a point cloud-specific encoder and a mesh-specific decoder in a hierarchical design, for effective multi-scale feature learning on highly efficient representations of the anatomical surface data in a multi-domain setting. The tailored encoder architecture allows to directly process 3D point cloud representations of MRI contours and hence, can be smoothly integrated into a multi-step surface reconstruction pipeline. Furthermore, the decoder outputs high-resolution cardiac meshes with shared vertex connectivity across the dataset, which are suitable for a variety of follow-up 3D cardiac modeling tasks [7–10,12,14,18,20,28].

2 Methods

2.1 Overview

We present an overview of the proposed Point2Mesh-Net combined with its training data preparation and its application as part of a multi-step cardiac reconstruction pipeline in Fig. 1.

We first create a synthetic dataset based on high-resolution 3D MRI acquisitions to obtain both ground truth meshes and sparse and misaligned point cloud contours (Fig. 1-A) (Sect. 2.2). We then use this dataset to train the Point2Mesh-Net and conduct an initial validation (Fig. 1-B) (Sect. 2.3). Finally, we apply the pre-trained Point2Mesh-Net as the key component of a three-step cardiac shape reconstruction pipeline from raw cine MRI acquisitions in cross-domain transfer setting (Fig. 1-C) (Sect. 3.2).

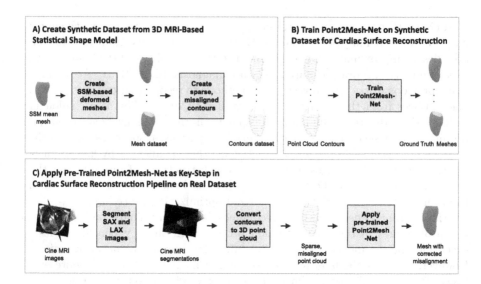

Fig. 1. Overview of the purpose-built training dataset and the proposed Point2Mesh-Net embedded in a multi-step cardiac surface reconstruction pipeline. First, we create a synthetic dataset of sparse and misalinged point cloud MRI contours and the corresponding high-resolution cardiac surface meshes from a 3D MRI-based statistical shape model (SSM) (A) (Sect. 2.2). Next, we use this dataset to train the proposed Point2Mesh-Net (B) (Sect. 2.3). Finally, we apply the pre-trained Point2Mesh-Net as the key component of a three-step pipeline for efficient 3D cardiac shape reconstruction from cine MR images (C) (Sect. 3.2).

2.2 Datasets and Preprocessing

We use both a synthetic dataset for network training and an initial evaluation and a real dataset in this work. The synthetic dataset is derived from a statistical shape model (SSM) based on 3D MRI acquisition of over 1000 subjects [1]. We first randomly generate 250 meshes from the SSM [1]. For each mesh, we randomly sample from a standard normal distribution along the first 99 modes of variation and apply the resulting deformations to the mean template mesh. Since the original 3D magnetic resonance (MR) images have a relatively high voxel resolution of $1.25 \times 1.25 \times 2$ mm compared to the UK Biobank's cine MRI protocol ($1.8 \times 1.8 \times 8.0$ mm), we consider the resulting SSM-based meshes as our high-resolution ground truth for network training in this work. To obtain the sparse and misaligned input contours, we first determine the typical location and orientation of the short-axis, 2-chamber long-axis (LAX), and 4-chamber LAX planes of a standard cine MRI protocol for each of the deformed meshes. We then introduce random translation and rotation to each plane in a way that mimics motion-induced misalignment of real acquisitions. We repeat this procedure 10 times for each of the 250 deformed meshes, resulting in a total of 2500 input point clouds in our SSM dataset. We use a random dataset split of 75%/5%/20% as train, validation, and test datasets, respectively, on the 250 meshes and assign the

10 misaligned point clouds to the dataset of their corresponding mesh. Finally, we extract three separate datasets for the anatomical substructures left ventricular (LV) endocardium, LV epicardium, and right ventricular (RV) endocardium.

As our real dataset, we choose 1000 cine MRI acquisitions of the UK Biobank study [24] with equal representation of female and male cases. We then pass the raw images through the first two steps of our reconstruction pipeline (Fig. 1-C) to obtain point cloud contours in a suitable format for the Point2Mesh-Net. To this end, we first apply the pre-trained fully convolutional neural networks from [2] to each of the SAX and 4-chamber LAX images, and a conditional generative adversarial network with a U-Net generator trained on an in-house annotated UK Biobank dataset to each of the 2-chamber LAX images, to delineate the contours of the LV endocardium, LV epicardium, and RV endocardium. In the second pipeline step, we combine the resulting contours from all views and slices and place them in 3D space as separate point clouds for each anatomical substructure [3], in a similar way as the input data used by the network on the SSM dataset.

2.3   Network Architecture

The architecture of the proposed Point2Mesh-Net combines recent advances in point cloud and mesh-based geometric deep learning in a hierarchical encoder-decoder structure (Fig. 2).

The network inputs are sparse and misaligned MRI contours represented as point clouds in 3D space with 900 points. They are passed into an encoder consisting of two stacked PointNet layers, which are inspired by the PointNet [25], PointNet++ [26], and Point Completion Network [35] architectures, and a multilayer perceptron (MLP) to allow for step-wise multi-scale feature extraction directly on point cloud data. More specifically, we first apply two 1D convolutions connected by a batch normalization layer and a rectified linear unit activation function to the input point clouds as part of a Point Conv Block. Then, a per-point max pooling and an expansion operation are applied, and the output is concatenated with the result of the first Point Conv Block. Next, a second Point Conv Block followed by a point-wise maxpool layer is used before the final output of the point cloud encoder is generated by the shared MLP. The resulting latent space vector of size 1×128 acts as a low-dimensional representation of the respective cardiac input shape and is passed to the network decoder. Its structure combines an initial MLP with four levels of spectral graph convolutions [13] in a hierarchical setup. Each graph convolution layer is followed by a rectified linear unit activation function, and four mesh upsampling layers [27] are used to increase the mesh resolution with each successively higher level in the decoder. This setup enables effective automatic feature learning on both a local and global scale and results in a gradual conversion of the low-dimensional latent space vector to a high-resolution 3D surface mesh as a dense representation of cardiac anatomy with corrected misalignment. Each output mesh consists of 1780 vertices and a vertex-to-vertex connectivity consistent across the entire dataset. We use the Chebyshev polynomial approximation [13] with order 5 for all spectral graph convolutions and quadric error minimization [27] to determine

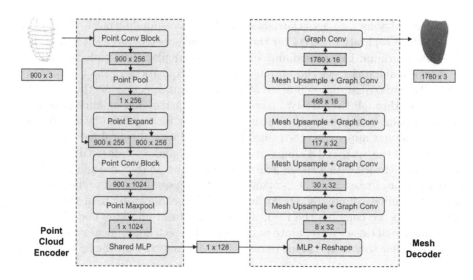

Fig. 2. Architecture of the proposed Point2Mesh-Net. A sparse and misaligned point cloud with 900 points and a dense 3D mesh with corrected misalignment constitute the network inputs and outputs, respectively. The architecture consists of an encoder and a decoder connected by low-dimensional latent space vector. Both the encoder and decoder are specifically designed for direct and efficient geometric deep learning-based processing of point cloud and mesh data, respectively.

all mesh upsampling operations. In total, the network has approximately 4×10^5 trainable parameters.

2.4   Training and Implementation

We train the Point2Mesh-Net with a vertex-wise mean squared error loss function using the Adam optimizer [16] and a batch size of 8, until no improvement on the validation dataset is observed for 10 epochs. The deep learning code is implemented using the PyTorch [22] and PyTorch Geometric frameworks [15]. All training and evaluation steps are run on a CPU with 8 GB memory. We record an average training time of the network of approximately 5 h and an average per-case inference time of 0.2 s.

3   Experiments and Results

3.1   Surface Reconstruction on Synthetic Dataset

As our first experiment, we aim to assess the ability of the proposed Point2Mesh-Net to accurately reconstruct different cardiac shapes with different types of misalignment on the synthetic SSM dataset. We therefore train three separate networks for each of the three cardiac substructures on the respective training datasets and then apply each of them to the corresponding unseen test dataset. Figure 3 depicts the results obtained by the network for three sample cases of the LV endocardial, LV epicardial, and the RV endocardial data, respectively.

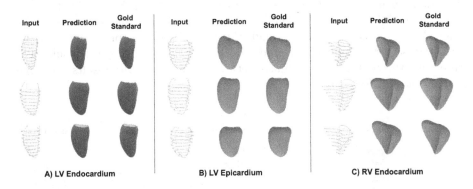

A) LV Endocardium | B) LV Epicardium | C) RV Endocardium

Fig. 3. Qualitative reconstruction results of 3 sample cases from the SSM dataset.

We observe that the predicted meshes closely resemble the corresponding ground truth ones on both a local and global level. The results are consistent across different cardiac shapes, types of misalignment and sparsity, and the three different cardiac substructures. The largest reconstruction errors are typically found in the basal area of the ventricular anatomy.

In addition to a qualitative comparison, we also want to quantify the reconstruction quality achieved by our network. To this end, we select the median surface distance, the mean surface distance, and the Hausdorff distance between the predicted and ground truth meshes as our evaluation metrics and report the results on the unseen test dataset for each of the three cardiac substructures in Table 1. All metrics are calculated directly on the meshes output by the network without applying any post-processing steps. For each subject, we use the average scores of all 10 variations of input contours in our calculations.

Table 1. Reconstruction results of Point2Mesh-Net on the synthetic dataset.

Cardiac structure	Median surface distance (mm)	Mean surface distance (mm)	Hausdorff (mm)
LV endocardium	0.85 (±0.18)	1.04 (±0.25)	4.59 (±1.29)
LV epicardium	0.80 (±0.21)	1.04 (±0.28)	4.73 (±1.76)
RV endocardium	0.98 (±0.20)	1.17 (±0.27)	4.50 (±0.97)

Values represent mean/median (± standard deviation/quartile deviation).

We find the median and mean surface distances below the pixel size of the underlying image acquisitions and only small differences in network performances between the different anatomical substructures.

3.2 Surface Reconstruction Pipeline on Real Dataset

After the evaluation of the Point2Mesh-Net on the synthetic SSM dataset, we also want to analyze its applicability to real data as part of a multi-step reconstruction pipeline. Accordingly, we execute steps 1 and 2 of the pipeline for 1000 cases of the UK Biobank dataset and then pass the resulting sparse and misaligned contour point clouds through the Point2Mesh-Net pre-trained on the SSM dataset. We conduct this procedure separately for each cardiac substructure with the pertinent pre-trained networks and depict the results for three sample cases in Fig. 4.

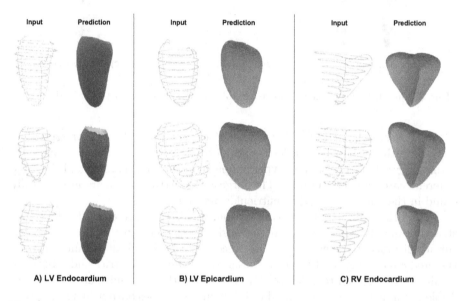

Fig. 4. Qualitative reconstruction results of 3 sample cases from the UK Biobank dataset.

We observe realistic and smooth 3D mesh reconstructions without noticeable misalignment that accurately capture the various different shapes of the corresponding input point clouds, both globally and locally for a variety of misalignment types. Reconstruction quality is similar across the different cardiac substructures.

In order to also conduct a quantitative evaluation of the reconstruction performance of our method on the UK Biobank dataset despite its lack of ground truth 3D shapes, we calculate multiple commonly-used clinical metrics and compare the results with two large-scale population studies. More specifically, we compute the LV volume, LV mass, and the RV volume based on the 3D meshes reconstructed by the Point2Mesh-Net for 500 female and 500 male cases of the UK Biobank dataset (Table 2). As our benchmarks, we report the results

Table 2. Comparison of clinical metrics calculated with different approaches.

Sex	Clinical metric	Petersen et al. [23]	Bai et al. [1]	Point2Mesh-Net
Female	LV volume (ml)	124 (±21)	138 (±24)	127 (±20)
	LV mass (g)	70 (±13)	96 (±16)	83 (±15)
	RV volume (ml)	130 (±24)	-	144 (±39)
Male	LV volume (ml)	166 (±32)	178 (±36)	161 (±30)
	LV mass (g)	103 (±21)	128 (±24)	107 (±17)
	RV volume (ml)	182 (±36)	-	186 (±44)

Values represent mean (± standard deviation) in all cases.

obtained by a 2D slice-based calculation [23] and a 3D MRI-based computation approach [1] in Table 2.

We find a high degree of similarity between the results obtained by our method and the two clinical benchmarks for both mean and standard deviation values of all three metrics. Volume and mass differences between female and male cases are accurately reflected by the Point2Mesh-Net reconstructions with scores larger than the 2D slice-based approach and smaller than the 3D MRI-based calculation in all but one metric.

4 Discussion and Conclusion

In this work, we have proposed and evaluated the Point2Mesh-Net as a novel geometric deep learning approach for cardiac surface reconstruction from cine MRI contours. It achieves small average reconstruction errors below the underlying pixel resolution on a large synthetic dataset for multiple cardiac substructures, with high degrees of similarity between reconstruction and ground truth surfaces. This not only demonstrates the high suitability of its architectural design for the task at hand, but also its ability to cope with a variety of cardiac shapes, misalignment types, and sparsity levels on both a local and global level. When applying the networks pre-trained on the SSM dataset to the real UK Biobank dataset as part of a multi-step reconstruction pipeline, we also find highly realistic reconstructions both individually and on a population level, with widely-used clinical metrics in line with previous large-scale studies. This not only shows that the synthesized dataset adequately reflects the acquisition conditions found in real datasets, but also that the features learned by the Point2Mesh-Nets on the SSM dataset successfully transfer to the real domain and can harmoniously interact with other preprocessing steps in a multi-step pipeline. This is crucial for a more wide-spread applicability of the technique in various clinical and research settings and enables a more detailed analysis of 3D cardiac shape variability of the population that goes beyond purely volume-based metrics.

The results are achieved despite the challenging combination of point cloud inputs and mesh outputs, indicating that the modality-specific design of the encoder and decoder branches can handle both the reconstruction and modality-transfer tasks with high degrees of accuracy. In addition, the output meshes exhibit vertex correspondence, which is an important requirement for many follow-up tasks, such as shape analysis with principal component analysis or graph neural networks. These characteristics are in contrast to previous deep learning approaches [6,34] that require transformations to inefficient voxelgrids or meshes as separate processing steps with considerable negative effects on complexity, execution times, and memory requirements. While we used separate networks to reconstruct the different anatomical substructures in this work, we hypothesize that the presented architecture with its compact latent space representation can also be expanded to decode multiple cardiac structures at the same time.

Compared to per-subject optimization approaches to cardiac shape reconstruction [3,17,19,20,29,32], the Point2Mesh-Net drastically reduces the execution time, since the feature optimization is already conducted during the training phase based on population-wide shape information. Furthermore, our approach can successfully incorporate long-axis information, which is especially beneficial for an accurate reconstruction near the apical and basal regions of the ventricles and further sets it apart from many previous techniques such as [17].

Acknowledgments. This research has been conducted using the UK Biobank Resource under Application Number '40161'. The authors express no conflict of interest. The work of M. Beetz is supported by the Stiftung der Deutschen Wirtschaft (Foundation of German Business). A. Banerjee is a Royal Society University Research Fellow and is supported by the Royal Society (Grant No. URF\R1\221314). The work of A. Banerjee and V. Grau is supported by the British Heart Foundation (BHF) Project under Grant PG/20/21/35082. The work of V. Grau is supported by the CompBioMed 2 Centre of Excellence in Computational Biomedicine (European Commission Horizon 2020 research and innovation programme, grant agreement No. 823712).

References

1. Bai, W., et al.: A bi-ventricular cardiac atlas built from 1000+ high resolution MR images of healthy subjects and an analysis of shape and motion. Med. Image Anal. **26**(1), 133–145 (2015)
2. Bai, W., et al.: Automated cardiovascular magnetic resonance image analysis with fully convolutional networks. J. Cardiovasc. Magn. Reson. **20**(65), 1–12 (2018)
3. Banerjee, A., et al.: A completely automated pipeline for 3D reconstruction of human heart from 2D cine magnetic resonance slices. Philos. Trans. R. Soc. A Math. Phys. Eng. Sci. **379**(2212), 20200257 (2021)
4. Banerjee, A., Zacur, E., Choudhury, R.P., Grau, V.: Optimised misalignment correction from cine MR slices using statistical shape model. In: Papież, B.W., Yaqub, M., Jiao, J., Namburete, A.I.L., Noble, J.A. (eds.) MIUA 2021. LNCS, vol. 12722, pp. 201–209. Springer, Cham (2021). https://doi.org/10.1007/978-3-030-80432-9_16
5. Banerjee, A., Zacur, E., Choudhury, R.P., Grau, V.: Automated 3D whole-heart mesh reconstruction from 2D cine MR slices using statistical shape model. In: 44th

Annual International Conference of the IEEE Engineering in Medicine and Biology Society (EMBC), pp. 1702–1706 (2022)

6. Beetz, M., Banerjee, A., Grau, V.: Biventricular surface reconstruction from cine MRI contours using point completion networks. In: 2021 IEEE 18th International Symposium on Biomedical Imaging (ISBI), pp. 105–109 (2021)

7. Beetz, M., Banerjee, A., Grau, V.: Generating subpopulation-specific biventricular anatomy models using conditional point cloud variational autoencoders. In: Puyol Antón, E., et al. (eds.) STACOM 2021. LNCS, vol. 13131, pp. 75–83. Springer, Cham (2022). https://doi.org/10.1007/978-3-030-93722-5_9

8. Beetz, M., Banerjee, A., Grau, V.: Multi-domain variational autoencoders for combined modeling of MRI-based biventricular anatomy and ECG-based cardiac electrophysiology. Front. Physiol. **13**, 991 (2022)

9. Beetz, M., Banerjee, A., Sang, Y., Grau, V.: Combined generation of electrocardiogram and cardiac anatomy models using multi-modal variational autoencoders. In: 2022 IEEE 19th International Symposium on Biomedical Imaging (ISBI), pp. 1–4 (2022)

10. Beetz, M., Ossenberg-Engels, J., Banerjee, A., Grau, V.: Predicting 3D cardiac deformations with point cloud autoencoders. In: Puyol Antón, E., et al. (eds.) STACOM 2021. LNCS, vol. 13131, pp. 219–228. Springer, Cham (2022). https://doi.org/10.1007/978-3-030-93722-5_24

11. Chen, X., et al.: Shape registration with learned deformations for 3D shape reconstruction from sparse and incomplete point clouds. Med. Image Anal. **74**, 102228 (2021)

12. Corral Acero, J., et al.: Understanding and improving risk assessment after myocardial infarction using automated left ventricular shape analysis. JACC Cardiovasc. Imaging **15**, 1563–1574 (2022)

13. Defferrard, M., Bresson, X., Vandergheynst, P.: Convolutional neural networks on graphs with fast localized spectral filtering. In: Proceedings of the 30th International Conference on Neural Information Processing Systems, pp. 3844–3852 (2016)

14. Di Folco, M., Moceri, P., Clarysse, P., Duchateau, N.: Characterizing interactions between cardiac shape and deformation by non-linear manifold learning. Med. Image Anal. **75**, 102278 (2022)

15. Fey, M., Lenssen, J.E.: Fast graph representation learning with PyTorch geometric. In: ICLR Workshop on Representation Learning on Graphs and Manifolds (2019)

16. Kingma, D.P., Ba, J.: Adam: a method for stochastic optimization. arXiv preprint arXiv:1412.6980 (2014)

17. Lamata, P., et al.: An automatic service for the personalization of ventricular cardiac meshes. J. R. Soc. Interface **11**(91), 20131023 (2014)

18. Li, L., Camps, J., Banerjee, A., Beetz, M., Rodriguez, B., Grau, V.: Deep computational model for the inference of ventricular activation properties. arXiv preprint arXiv:2208.04028 (2022)

19. Mauger, C., et al.: An iterative diffeomorphic algorithm for registration of subdivision surfaces: application to congenital heart disease. In: 2018 40th Annual International Conference of the IEEE Engineering in Medicine and Biology Society (EMBC), pp. 596–599. IEEE (2018)

20. Mauger, C., et al.: Right ventricular shape and function: cardiovascular magnetic resonance reference morphology and biventricular risk factor morphometrics in UK Biobank. J. Cardiovasc. Magn. Reson. **21**(1), 1–13 (2019)

21. O'Dell, W.G.: Accuracy of left ventricular cavity volume and ejection fraction for conventional estimation methods and 3D surface fitting. J. Am. Heart Assoc. **8**(6), e009124 (2019)

22. Paszke, A., et al.: PyTorch: an imperative style, high-performance deep learning library. In: Proceedings of the 33rd International Conference on Neural Information Processing Systems, pp. 8026–8037 (2019)

23. Petersen, S.E., et al.: Reference ranges for cardiac structure and function using cardiovascular magnetic resonance (CMR) in Caucasians from the UK Biobank population cohort. J. Cardiovasc. Magn. Reson. **19**(18), 1–19 (2017)

24. Petersen, S.E., et al.: Imaging in population science: cardiovascular magnetic resonance in 100,000 participants of UK Biobank - rationale, challenges and approaches. J. Cardiovasc. Magn. Reson. **15**(46), 1–10 (2013)

25. Qi, C.R., Su, H., Mo, K., Guibas, L.J.: PointNet: deep learning on point sets for 3D classification and segmentation. In: Proceedings of the IEEE Conference on Computer Vision and Pattern Recognition, pp. 652–660 (2017)

26. Qi, C.R., Yi, L., Su, H., Guibas, L.J.: PointNet++: deep hierarchical feature learning on point sets in a metric space. In: Advances in Neural Information Processing Systems, pp. 5099–5108 (2017)

27. Ranjan, A., Bolkart, T., Sanyal, S., Black, M.J.: Generating 3D faces using convolutional mesh autoencoders. In: Ferrari, V., Hebert, M., Sminchisescu, C., Weiss, Y. (eds.) ECCV 2018. LNCS, vol. 11207, pp. 725–741. Springer, Cham (2018). https://doi.org/10.1007/978-3-030-01219-9_43

28. Rodero, C., et al.: Linking statistical shape models and simulated function in the healthy adult human heart. PLoS Comput. Biol. **17**(4), e1008851 (2021)

29. Sinclair, M., Bai, W., Puyol-Antón, E., Oktay, O., Rueckert, D., King, A.P.: Fully automated segmentation-based respiratory motion correction of multiplanar cardiac magnetic resonance images for large-scale datasets. In: Descoteaux, M., Maier-Hein, L., Franz, A., Jannin, P., Collins, D.L., Duchesne, S. (eds.) MICCAI 2017. LNCS, vol. 10434, pp. 332–340. Springer, Cham (2017). https://doi.org/10.1007/978-3-319-66185-8_38

30. Stokes, M.B., Roberts-Thomson, R.: The role of cardiac imaging in clinical practice. Aust. Prescr. **40**(4), 151 (2017)

31. Suinesiaputra, A., et al.: Statistical shape modeling of the left ventricle: myocardial infarct classification challenge. IEEE J. Biomed. Health Inform. **22**(2), 503–515 (2017)

32. Villard, B., Grau, V., Zacur, E.: Surface mesh reconstruction from cardiac MRI contours. J. Imaging **4**(1), 16 (2018)

33. Wang, Z.-Y., Zhou, X.-Y., Li, P., Theodoreli-Riga, C., Yang, G.-Z.: Instantiation-Net: 3D mesh reconstruction from single 2D image for right ventricle. In: Martel, A.L., et al. (eds.) MICCAI 2020. LNCS, vol. 12264, pp. 680–691. Springer, Cham (2020). https://doi.org/10.1007/978-3-030-59719-1_66

34. Xu, H., Zacur, E., Schneider, J.E., Grau, V.: Ventricle surface reconstruction from cardiac MR slices using deep learning. In: Coudière, Y., Ozenne, V., Vigmond, E., Zemzemi, N. (eds.) FIMH 2019. LNCS, vol. 11504, pp. 342–351. Springer, Cham (2019). https://doi.org/10.1007/978-3-030-21949-9_37

35. Yuan, W., Khot, T., Held, D., Mertz, C., Hebert, M.: PCN: point completion network. In: 2018 International Conference on 3D Vision (3DV), pp. 728–737 (2018)

36. Zhou, X.-Y., Wang, Z.-Y., Li, P., Zheng, J.-Q., Yang, G.-Z.: One-stage shape instantiation from a single 2D image to 3D point cloud. In: Shen, D., et al. (eds.) MICCAI 2019. LNCS, vol. 11767, pp. 30–38. Springer, Cham (2019). https://doi.org/10.1007/978-3-030-32251-9_4

Post-Infarction Risk Prediction
with Mesh Classification Networks

Marcel Beetz[1]([✉]), Jorge Corral Acero[1], Abhirup Banerjee[1,2]iD, Ingo Eitel[3,4,5],
Ernesto Zacur[1], Torben Lange[6,7], Thomas Stiermaier[3,4,5], Ruben Evertz[6,7],
Sören J. Backhaus[6,7], Holger Thiele[8,9], Alfonso Bueno-Orovio[10],
Pablo Lamata[11], Andreas Schuster[6,7]iD, and Vicente Grau[1]iD

[1] Institute of Biomedical Engineering, Department of Engineering Science,
University of Oxford, Oxford, UK
marcel.beetz@eng.ox.ac.uk
[2] Division of Cardiovascular Medicine, Radcliffe Department of Medicine,
University of Oxford, Oxford, UK
[3] University Heart Center Lübeck, Medical Clinic II, Cardiology, Angiology,
and Intensive Care Medicine, Lübeck, Germany
[4] University Hospital Schleswig-Holstein, Lübeck, Germany
[5] German Centre for Cardiovascular Research, Partner Site Lübeck,
Lübeck, Germany
[6] University Medical Center Göttingen, Department of Cardiology and Pneumology,
Georg-August University, Göttingen, Germany
[7] German Centre for Cardiovascular Research, Partner Site Göttingen,
Göttingen, Germany
[8] Department of Internal Medicine/Cardiology, Heart Center Leipzig
at University of Leipzig, Leipzig, Germany
[9] Leipzig Heart Institute, Leipzig, Germany
[10] Department of Computer Science, University of Oxford, Oxford, UK
[11] Department of Biomedical Engineering, King's College London, London, UK

Abstract. Post-myocardial infarction (MI) patients are at risk of major
adverse cardiac events (MACE), with risk stratification primarily based
on global image-based biomarkers, such as ejection fraction, in current
clinical practice. However, these metrics neglect more subtle and localized shape differences in 3D cardiac anatomy and function, which limit
predictive accuracy. In this work, we propose a novel geometric deep
learning approach to directly predict MACE outcomes within 1 year after
the infarction event from high-resolution 3D cardiac anatomy meshes. Its
architecture is specifically designed for direct and efficient processing of
surface mesh data with a hierarchical, multi-scale structure to enable
both local and global feature learning. We evaluate the binary MACE
prediction capabilities of the proposed mesh classification network on a
multi-center dataset of post-MI patients. Our results show that the proposed method outperforms corresponding clinical benchmarks by ∼16%
and ∼6% in terms of area under the receiver operating characteristic
(AUROC) curve for 3D shape and 3D contraction inputs, respectively.
Furthermore, we visually analyze both 3D cardiac shapes and 3D contraction patterns with regards to their MACE predictability and demonstrate how task-specific information learned by the network on a balanced

O. Camara et al. (Eds.): STACOM 2022, LNCS 13593, pp. 291–301, 2022.
https://doi.org/10.1007/978-3-031-23443-9_27

dataset successfully generalizes to increasing levels of class imbalance. Finally, we compare our approach to both clinical and machine learning benchmarks on our original highly-imbalanced dataset of post-MI patients and find average improvements in AUROC scores of ∼9% and ∼3%, respectively.

Keywords: Major adverse cardiac event prediction · 3D cardiac shape · 3D cardiac contraction · Graph neural networks · Mesh pooling · Post-myocardial infarction · Cardiac MRI · Geometric deep learning

1   Introduction

Ischemic heart disease is the most common cause of death worldwide with cases continuing to increase [18]. While progress has recently been made in predicting myocardial infarction (MI) outcomes, post-MI patients are still at risk of recurrent major adverse cardiac events (MACE) [15,18,23,25]. In order to correctly stratify post-MI patients into risk groups and determine suitable preventative measures, cardiac magnetic resonance imaging (MRI) is considered as the gold standard imaging modality as it enables the accurate calculation of widely used image-based biomarkers, such as ejection fraction, and the characterization of left ventricular (LV) anatomy and function [8,15,23,28]. However, such biomarkers only provide a global assessment in the form of a single value, thus neglecting more localized patterns in the complex 3D cardiac physiology, which have been shown to play an important role in the management of post-MI patients [9]. Accordingly, previous works have focused on discovering novel predictive biomarkers by analyzing more subtle aspects of the 3D cardiac shapes [9,14,29].

In the recent past, deep learning has revolutionized the field of medical image analysis and outperformed previous approaches in a variety of different tasks, including clinical outcome prediction [6]. Specifically for anatomical surface data, geometric deep learning methods have been shown to be highly suitable for creating more detailed models of cardiac anatomy and function, using both graph [12,21,22,24] and point cloud-based [1–5,7] processing. Building on these developments, we present in this work a novel mesh classification network to predict MACE outcomes based on 3D shape information of cardiac anatomy and function. Its architecture utilizes recent advances in geometric deep learning to enable direct and targeted processing of high-resolution mesh data and efficient feature learning at both local and global scales to facilitate the usage of more advanced image-based biomarkers. To the best of our knowledge, this is the first deep learning approach for direct 3D shape-based MACE prediction in post-MI patients.

2   Methods

2.1   Overview

We present an overview of our proposed approach for the prediction of MACE outcomes with mesh classification networks in Fig. 1.

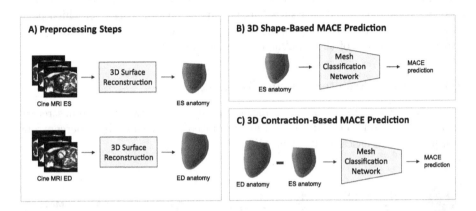

Fig. 1. Overview of the proposed mesh-based deep learning approach for prediction of major adverse cardiac events (MACE). First, 3D surface meshes of the cardiac anatomy at ED and ES are reconstructed from cine MRI acquisitions in a preprocessing step (A). Next, 3D cardiac anatomies at ES are input into a mesh classification network for 3D shape-based MACE prediction (B). Finally, the vertex-wise differences between anatomy meshes at ED and ES are used as representations of 3D contractions to predict MACE outcomes with a separate mesh classification network (C).

We first select cardiac magnetic resonance (MR) images at both the end-diastolic (ED) and end-systolic (ES) phases of the cardiac cycle as our input data. In order to obtain high-resolution 3D representations of cardiac shapes, we apply a fully automatic multi-step surface reconstruction pipeline as a preprocessing step (Fig. 1-A) (Sect. 2.2). We then use the reconstructed meshes at ES and the vertex-wise differences between ED and ES meshes to train and evaluate separate mesh classification networks for 3D shape-based (Fig. 1-B) and 3D contraction-based (Fig. 1-C) binary MACE prediction, respectively (Sect. 2.3).

2.2   Dataset and Preprocessing

The dataset used in this work consists of cine MR images of 1021 post-MI patients acquired as part of two multi-center trials, TATORT-NSTEMI (Thrombus Aspiration in Thrombus Containing Culprit Lesions in Non-ST-Elevation Myocardial Infarction; NCT01612312) [30] and AIDA-STEMI (Abciximab Intra-coronary Versus Intravenously Drug Application in ST-Elevation Myocardial Infarction; NCT00712101) [16]. Major adverse cardiac events (either reinfarction, new congestive heart failure, all-cause death, or any combination of these) 12 months post-MI was chosen as the study endpoint, and 74 patients were recorded with this outcome. In order to obtain the 3D LV anatomy meshes required for our proposed 3D shape-based MACE classification networks, we first pass the MR images of the dataset through a multi-step surface reconstruction pipeline [9]. The pipeline consists of both a segmentation step, which uses deep convolutional neural networks with customized data augmentation to extract the LV contours from the raw short-axis MRI slices [10,11], and a mesh fitting step which positions and deforms a 3D LV template mesh to fit the extracted contours [20]. A detailed description of the pipeline can be found in [9].

2.3 Network Architecture and Training

The architecture of our proposed mesh classification network combines recent advances in graph convolutions and mesh sampling with multilayer perceptrons in a hierarchical cascade structure (Fig. 2).

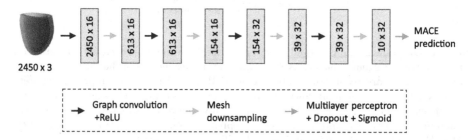

Fig. 2. Architecture of the proposed mesh classification network. High-resolution input meshes are processed by a cascade of graph convolution and mesh downsampling layers followed by a multi-layer perceptron with dropout and a sigmoid activation function to enable efficient multi-scale feature learning for binary MACE prediction.

The network inputs are 3D triangular meshes with 2450 vertices and shared vertex connectivity across the dataset. The meshes either represent the LV surfaces at ES or the per-vertex deformation between ED and ES. The inputs are passed through four levels of spectral graph convolution and mesh downsampling layers to enable efficient multi-scale feature learning directly on mesh data. A multi-layer perceptron followed by a sigmoid function is then applied to obtain the predicted output probabilities for the binary MACE classification task. We compute all graph convolutions with the Chebyshev polynomial approximation [13] of order 5 and subsequently apply rectified linear units (ReLU) as activation functions. All mesh pooling layers are implemented using quadric error minimization [27]. Furthermore, we include drop-out layers after each fully connected layer as a regularizer to improve generalizability. We train on a CPU using the Adam optimizer [19] with a batch size of 32 and a binary cross entropy loss function, until loss convergence is achieved on the validation dataset. The mesh classification networks are implemented using the PyTorch [26] and Pytorch Geometric [17] frameworks.

3   Experiments and Results

3.1   End-Systolic Shape-Based Prediction

We first aim to assess the ability of the mesh classification network to predict binary MACE outcomes based solely on 3D cardiac surface meshes at ES. As a clinical benchmark, we choose ES volume to encapsulate cardiac shape at ES and input the values as the independent variable into a logistic regression model with binary MACE outcomes as the dependent variable. To facilitate effective feature learning for both approaches, we select a dataset with a balanced representation of MACE and Non-MACE cases by randomly sampling a subset of Non-MACE cases (n = 74) from the original dataset and apply 3-fold cross validation in our experiment. We report in Table 1 the averaged results across the respective unseen test folds in terms of widely used binary classification metrics.

Table 1. ES Shape-based MACE prediction results on balanced dataset.

Input	AUROC	Accuracy	Precision	Recall	F1-score
ES volume	0.624	0.605	0.604	0.633	0.614
ES 3D shape	0.721	0.721	0.718	0.710	0.717

We find that the mesh classification network achieves considerably higher scores than the clinical benchmark across all metrics, including a ~15.5% performance improvement in cross-validated area under the receiver operating characteristic (AUROC) curve.

Next, we want to investigate which types of 3D meshes cause the network to give accurate predictions. To this end, we calculate the absolute differences between the network's predicted output probabilities and the respective ground truth encodings of either MACE or Non-MACE. Hereby, a small difference value indicates a very good prediction by the network for both outcome possibilities and vice versa for large difference values. Based on these prediction error values, we select and visualize in Fig. 3 three sample cases corresponding to the network's good prediction of MACE, good prediction of Non-MACE, bad prediction of MACE, and bad prediction of Non-MACE.

We observe that the network typically predicts MACE outcomes for hearts with larger volumes and smaller myocardial thickness. Bad network predictions are more likely to occur if this association is less clear or reversed for the particular case, e.g. a smaller LV volume with a MACE outcome.

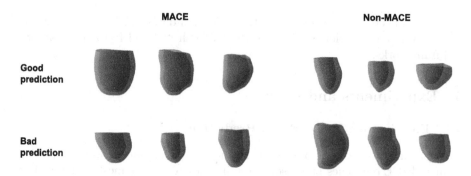

Fig. 3. Input ES meshes that resulted in good (row 1) and bad (row 2) predictions for ground truth MACE (column 1) and Non-MACE (column 2) cases respectively by the shape-based mesh classification network.

3.2   Contraction-Based Prediction

In addition to ES 3D shape, we also want to study the utility of 3D contractions for MACE prediction. We therefore repeat the same experiment of the last section but pass meshes that represent vertex-wise differences between ED and ES shapes as inputs to a separate mesh classification network. Since this gives the network access to both ED and ES shapes, we also choose a different clinical metric, ejection fraction, as the independent variable in our logistic regression. The averaged results of the respective 3-fold cross validation experiments are depicted for both methods in Table 2.

Table 2. Contraction-based MACE prediction results on balanced dataset.

Input	AUROC	Accuracy	Precision	Recall	F1-score
Ejection fraction	0.696	0.666	0.705	0.581	0.633
3D contraction	0.738	0.726	0.732	0.712	0.719

We find that, although the contraction-based clinical benchmark, i.e. ejection fraction, improves the classification performance significantly from the shape-based clinical benchmark, i.e. ES volume, the mesh classification network outperforms the clinical benchmark across all metrics and achieves a ~6% increase in cross-validated AUROC scores.

Similar to our analysis of ES shapes, we also want to visualize 3D contraction patterns of cases that result in good and bad predictions by the network. To this end, we follow the same procedure as for the ES shape inputs to determine the network's prediction error for each case, and present in Fig. 4 the ED and ES 3D anatomies of patients that resulted in both good and bad predictions for both MACE and Non-MACE outcomes.

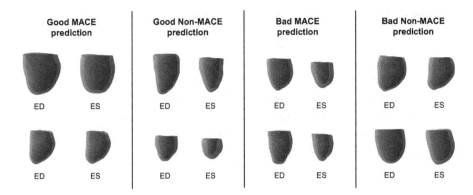

Fig. 4. Examples of input ED and ES meshes (in rows) that result in good (panels 1 and 2) and bad (panels 3 and 4) predictions of true MACE (panels 1 and 3) and Non-MACE (panels 2 and 4) cases by the contraction-based mesh classification network.

We observe that greater reductions in volumes and larger thickening of the myocardium between ED and ES phases lead to good predictions of Non-MACE outcomes. Similar to the ES shape results, bad predictions often occur when these associations between contraction and outcome are less present or reversed.

3.3   Effect of Class Imbalance

Since MACE outcomes are considerably less common than Non-MACE outcomes in our original dataset, we analyze next whether the patterns learned on the balanced dataset generalize well to more imbalanced datasets. To this end, we first create multiple datasets of both ES shapes and 3D contractions, each with larger levels of class imbalance between Non-MACE and MACE cases (2:1, 5:1, and 10:1). We achieve this by randomly sampling successively larger subsets from the remaining original dataset and then merging these subsets with the unseen test set of the balanced dataset used in the previous sections. We then apply the mesh classification networks and clinical benchmarks trained on the balanced dataset to each of the imbalanced datasets.

We find that while this results in a small decrease in AUROC scores for the ES shape-based mesh classification network compared to the balanced dataset, the scores for the 3D contraction-based mesh classification network and both clinical benchmarks show little change for higher levels of class imbalance. Finally, we repeat the same procedure for the full dataset (class imbalance of $\sim$13:1) and compare the results with both the clinical benchmarks and a previous machine learning-based approach [9] in Table 3.

We find that the mesh classification network achieves the highest AUROC scores for both ES shape and 3D contraction inputs with average improvements of $\sim$9% and $\sim$3% compared to the clinical and machine learning-based benchmarks, respectively.

Table 3. MACE prediction results on the original highly imbalanced dataset.

Input	Method	AUROC
ES volume	Regression	0.624
ES 3D shape	PCA + regression [9]	0.681
ES 3D shape	Mesh classification network	0.705
Ejection fraction	Regression	0.696
3D contraction	PCA + regression [9]	0.716
3D contraction	Mesh classification network	0.734

4   Discussion and Conclusion

In this work, we have presented a novel geometric deep learning approach to predict MACE outcomes from 3D cardiac anatomy and function data and found it capable of outperforming corresponding clinical benchmarks by considerable margins. On the one hand, this indicates that the full 3D shape representations of the heart, which enable the combined assessment of both global and local 3D shape features, contain more MACE-related information than the respective clinical metrics which are limited to a global cardiac function assessment with a single value. On the other hand, it demonstrates that the mesh classification network is adequately designed to successfully extract such multi-scale features for the purpose of MACE prediction. The outperformance of the respective clinical benchmarks by the pertinent networks was larger for ES shape inputs compared to 3D contraction inputs. We hypothesize that the considerably higher parameterization of the network compared to the regression model allows it to discover more intricate patterns in ES shapes, which implicitly provide relevant information about cardiac function and, in turn, help it achieve considerably improved prediction performance closer to the ones observed when modeling contraction explicitly.

The mesh-specific design of the network allows it to achieve these results by directly processing 3D surface mesh data in an efficient manner, and enables all network training to be conducted on a standard CPU. This is in contrast to voxelgrid-based deep learning approaches, which typically require high-powered GPU infrastructure for similar use cases.

In the visual analysis of the input heart shapes, we observe that small endo-cardial volume changes between ED and ES meshes typically lead to a correct MACE classification and vice versa for Non-MACE predictions. This is plausible from a clinical perspective as it is indicative of impaired cardiac contraction which in turn increases the risk of a future MACE outcome. Similarly, a large ES endocardial volume and small myocardial thickness at ES often cause the shape-based network to correctly classify cases as MACE. We hypothesize that such hearts can also implicitly be associated with poor contractive ability since both volume and thickness would typically be expected to have opposite characteristics at ES for healthy subjects. Both the shape- and contraction-based networks make bad predictions when these associations that are present in the majority of cases are reversed. While this shows that the network has indeed learned useful patterns and applies them in a consistent and robust way, it also indicates that for some cases additional biomarkers apart from shape descriptors might be beneficial to further boost prediction performance.

We also find that the mesh networks trained on balanced datasets generalize well when applied to higher levels of class imbalance. This indicates that most of the 3D shape and contraction variability required for MACE classification are already present and extractable in the smaller balanced subset. The considerable outperformance over the clinical benchmark is also maintained by the mesh classification network in the original dataset with a large imbalance of about 13:1. This is particularly important for the applicability of the method in clinical practice where class imbalances between different outcomes are commonly encountered.

The mesh classification network also achieves higher AUROC scores for both ES shapes and contraction inputs of the fully imbalanced dataset than a machine learning-based approach, which combines principal component analysis with a binomial sigmoid regression model [9]. We attribute this finding to the higher number of trainable parameters and the presence of non-linearities in the deep learning network, giving it the ability to learn more complex feature relationships in the data. However, we note that this should only be interpreted as an approximate comparison, as the objective of the machine learning benchmark was not only to achieve the highest possible classification accuracy but also to retain high interpretability and clinical applicability. While our proposed deep learning approach comes with some potential comparative disadvantages, such as a more complex optimization process, longer training times, and possibly lower robustness, we find in our experiments that a careful design of the network architecture and training procedure tailored to the dataset at hand is sufficient to address these challenges and achieve the observed outperformance of both clinical and machine learning benchmarks. Hereby, the usage of a large mini-batch size and the introduction of dropout with an empirically determined probability of 0.3 are particularly important to regularize the network and improve training stability and generalizability to unseen test data. Furthermore, we believe that a larger number of MACE cases or the usage of data augmentation would likely further help the network to improve predictive performance and reach its full potential.

Acknowledgments. The authors express no conflict of interest. The work of MB is supported by the Stiftung der Deutschen Wirtschaft (Foundation of German Business). AB is a Royal Society University Research Fellow and is supported by the Royal Society (Grant No. URF\R1\221314). The work of AB and VG is supported by the British Heart Foundation (BHF) Project under Grant PG/20/21/35082. The work of VG is supported by the CompBioMed 2 Centre of Excellence in Computational Biomedicine (European Commission Horizon 2020 research and innovation programme, grant agreement No. 823712). The work of JCA is supported by the EU's Horizon 2020 research and innovation program under the Marie Sklodowska-Curie (g.a. 764738) and the EPSRC Impact Acceleration Account (D4D00010 DF48.01), funded by

UK Research and Innovation. ABO holds a BHF Intermediate Basic Science Research Fellowship (FS/17/22/32644). The work is also supported by the German Center for Cardiovascular Research, the British Heart Foundation (PG/16/75/32383), and the Wellcome Trust (209450/Z/17).

References

1. Beetz, M., Banerjee, A., Grau, V.: Biventricular surface reconstruction from cine MRI contours using point completion networks. In: 2021 IEEE 18th International Symposium on Biomedical Imaging (ISBI), pp. 105–109. IEEE (2021)
2. Beetz, M., Banerjee, A., Grau, V.: Generating subpopulation-specific biventricular anatomy models using conditional point cloud variational autoencoders. In: Puyol Antón, E., et al. (eds.) STACOM 2021. LNCS, vol. 13131, pp. 75–83. Springer, Cham (2022). https://doi.org/10.1007/978-3-030-93722-5_9
3. Beetz, M., Banerjee, A., Grau, V.: Multi-domain variational autoencoders for combined modeling of MRI-based biventricular anatomy and ECG-based cardiac electrophysiology. Front. Physiol., 991 (2022)
4. Beetz, M., Banerjee, A., Sang, Y., Grau, V.: Combined generation of electrocardiogram and cardiac anatomy models using multi-modal variational autoencoders. In: 2022 IEEE 19th International Symposium on Biomedical Imaging (ISBI), pp. 1–4 (2022)
5. Beetz, M., Ossenberg-Engels, J., Banerjee, A., Grau, V.: Predicting 3D cardiac deformations with point cloud autoencoders. In: Puyol Antón, E., et al. (eds.) STACOM 2021. LNCS, vol. 13131, pp. 219–228. Springer, Cham (2022). https://doi.org/10.1007/978-3-030-93722-5_24
6. Bello, G.A., et al.: Deep-learning cardiac motion analysis for human survival prediction. Nat. Mach. Intell. **1**(2), 95–104 (2019)
7. Chang, Y., Jung, C.: Automatic cardiac MRI segmentation and permutation-invariant pathology classification using deep neural networks and point clouds. Neurocomputing **418**, 270–279 (2020)
8. Corral Acero, J., et al.: The 'digital twin' to enable the vision of precision cardiology. Eur. Heart J. **41**(48), 4556–4564 (2020)
9. Corral Acero, J., et al.: Understanding and improving risk assessment after myocardial infarction using automated left ventricular shape analysis. JACC: Cardiovasc. Imaging (2022)
10. Corral Acero, J., et al.: Left ventricle quantification with cardiac MRI: deep learning meets statistical models of deformation. In: Pop, M., et al. (eds.) STACOM 2019. LNCS, vol. 12009, pp. 384–394. Springer, Cham (2020). https://doi.org/10.1007/978-3-030-39074-7_40
11. Corral Acero, J., et al.: SMOD - data augmentation based on statistical models of deformation to enhance segmentation in 2D cine cardiac MRI. In: Coudière, Y., Ozenne, V., Vigmond, E., Zemzemi, N. (eds.) FIMH 2019. LNCS, vol. 11504, pp. 361–369. Springer, Cham (2019). https://doi.org/10.1007/978-3-030-21949-9_39
12. Dalton, D., Lazarus, A., Rabbani, A., Gao, H., Husmeier, D.: Graph neural network emulation of cardiac mechanics. In: Proceedings of the 3rd International Conference on Statistics: Theory and Applications (ICSTA 2021), pp. 127-1-8 (2021)
13. Defferrard, M., Bresson, X., Vandergheynst, P.: Convolutional neural networks on graphs with fast localized spectral filtering. In: Proceedings of the 30th International Conference on Neural Information Processing Systems, pp. 3844–3852 (2016)

14. Di Folco, M., Moceri, P., Clarysse, P., Duchateau, N.: Characterizing interactions between cardiac shape and deformation by non-linear manifold learning. Med. Image Anal. **75**, 102278 (2022)
15. Eitel, I., et al.: Left ventricular global function index assessed by cardiovascular magnetic resonance for the prediction of cardiovascular events in ST-elevation myocardial infarction. J. Cardiovasc. Magn. Reson. **17**(1), 1–9 (2015)
16. Eitel, I., et al.: Intracoronary compared with intravenous bolus abciximab application during primary percutaneous coronary intervention in ST-segment elevation myocardial infarction: cardiac magnetic resonance substudy of the AIDA STEMI trial. J. Am. Coll. Cardiol. **61**(13), 1447–1454 (2013)
17. Fey, M., Lenssen, J.E.: Fast graph representation learning with PyTorch Geometric. In: ICLR Workshop on Representation Learning on Graphs and Manifolds (2019)
18. Ibanez, B., et al.: 2017 ESC guidelines for the management of acute myocardial infarction in patients presenting with ST-segment elevation: The task force for the management of acute myocardial infarction in patients presenting with ST-segment elevation of the european society of cardiology (ESC). Eur. Heart J. **39**(2), 119–177 (2018)
19. Kingma, D.P., Ba, J.: Adam: a method for stochastic optimization. arXiv preprint arXiv:1412.6980 (2014)
20. Lamata, P., et al.: An automatic service for the personalization of ventricular cardiac meshes. J. Roy. Soc. Interface **11**(91), 20131023 (2014)
21. Lu, P., Bai, W., Rueckert, D., Noble, J.A.: Modelling cardiac motion via spatio-temporal graph convolutional networks to boost the diagnosis of heart conditions. In: Puyol Anton, E., et al. (eds.) STACOM 2020. LNCS, vol. 12592, pp. 56–65. Springer, Cham (2021). https://doi.org/10.1007/978-3-030-68107-4_6
22. Lu, P., Bai, W., Rueckert, D., Noble, J.A.: Multiscale graph convolutional networks for cardiac motion analysis. In: Ennis, D.B., Perotti, L.E., Wang, V.Y. (eds.) FIMH 2021. LNCS, vol. 12738, pp. 264–272. Springer, Cham (2021). https://doi.org/10.1007/978-3-030-78710-3_26
23. Marcos-Garcés, V., et al.: Risk score for early risk prediction by cardiac magnetic resonance after acute myocardial infarction. Int. J. Cardiol. **349**, 150–154 (2022)
24. Meister, F., et al.: Graph convolutional regression of cardiac depolarization from sparse endocardial maps. In: Puyol Anton, E., et al. (eds.) STACOM 2020. LNCS, vol. 12592, pp. 23–34. Springer, Cham (2021). https://doi.org/10.1007/978-3-030-68107-4_3
25. Nestelberger, T., et al.: Predicting major adverse events in patients with acute myocardial infarction. J. Am. Coll. Cardiol. **74**(7), 842–854 (2019)
26. Paszke, A., et al.: PyTorch: an imperative style, high-performance deep learning library. In: Proceedings of the 33rd International Conference on Neural Information Processing Systems, pp. 8026–8037 (2019)
27. Ranjan, A., Bolkart, T., Sanyal, S., Black, M.J.: Generating 3D faces using convolutional mesh autoencoders. In: Proceedings of the European Conference on Computer Vision (ECCV), pp. 704–720 (2018)
28. Reindl, M., Eitel, I., Reinstadler, S.J.: Role of cardiac magnetic resonance to improve risk prediction following acute ST-elevation myocardial infarction. J. Clin. Med. **9**(4), 1041 (2020)
29. Suinesiaputra, A., et al.: Statistical shape modeling of the left ventricle: myocardial infarct classification challenge. IEEE J. Biomed. Health Inf. **22**(2), 503–515 (2017)
30. Thiele, H., et al.: Effect of aspiration thrombectomy on microvascular obstruction in NSTEMI patients: the TATORT-NSTEMI trial. J. Am. Coll. Cardiol. **64**(11), 1117–1124 (2014)

Statistical Shape Modeling of Biventricular Anatomy with Shared Boundaries

Krithika Iyer[1,2(✉)], Alan Morris[2], Brian Zenger[2,4], Karthik Karanth[1,2], Benjamin A. Orkild[2,3], Oleksandre Korshak[1,2], and Shireen Elhabian[1,2]

[1] School of Computing, University of Utah, Salt Lake City, UT, USA
krithika.iyer@utah.edu
[2] Scientific Computing and Imaging Institute, University of Utah,
Salt Lake City, UT, USA
{amorris,karthik,oleks,shireen}@sci.utah.edu
[3] Department of Biomedical Engineering, University of Utah,
Salt Lake City, UT, USA
ben.orkild@utah.edu
[4] School of Medicine, University of Utah, Salt Lake City, UT, USA
brian.zenger@hsc.utah.edu

Abstract. Statistical shape modeling (SSM) is a valuable and powerful tool to generate a detailed representation of complex anatomy that enables quantitative analysis and the comparison of shapes and their variations. SSM applies mathematics, statistics, and computing to parse the shape into a quantitative representation (such as correspondence points or landmarks) that will help answer various questions about the anatomical variations across the population. Complex anatomical structures have many diverse parts with varying interactions or intricate architecture. For example, the heart is a four-chambered anatomy with several shared boundaries between chambers. Coordinated and efficient contraction of the chambers of the heart is necessary to adequately perfuse end organs throughout the body. Subtle shape changes within these shared boundaries of the heart can indicate potential pathological changes that lead to uncoordinated contraction and poor end-organ perfusion. Early detection and robust quantification could provide insight into ideal treatment techniques and intervention timing. However, existing SSM approaches fall short of explicitly modeling the statistics of shared boundaries. In this paper, we present a general and flexible data-driven approach for building statistical shape models of multi-organ anatomies with shared boundaries that captures morphological and alignment changes of individual anatomies and their shared boundary surfaces throughout the population. We demonstrate the effectiveness of the proposed methods using a biventricular heart dataset by developing shape models that consistently parameterize the cardiac biventricular structure and the interventricular septum (shared boundary surface) across the population data.

Keywords: Statistical shape modeling · Biventricular · Cardiac MRI · Particle-based shape modeling · Interventricular septum

© The Author(s), under exclusive license to Springer Nature Switzerland AG 2022
O. Camara et al. (Eds.): STACOM 2022, LNCS 13593, pp. 302–316, 2022.
https://doi.org/10.1007/978-3-031-23443-9_28

1   Introduction

Statistical shape modeling (SSM) is an important computational tool employed to discover significant shape parameters directly from medical data (such as MRI and CT scans) that can fully describe complex anatomy in the context of a population. SSM is used in biomedical research to visualize organs [18], bones [16], and tumors [15], aid surgical planning and guidance [1], monitor disease progression [9,25], and implant design [11].

Traditionally, SSM approaches have focused on generating organ or disease-specific models of single-organ anatomy. However, the human body consists of complex organs and systems that are functionally, spatially, and physically connected [2,17,20]. Recent research in computational anatomy has shifted focus towards modeling multi-organ anatomies [6]. The motivation for modeling multiple organs stems from the need to consider joint statistical shape analysis to quantify meaningful shape variations and contextual information when studying the group differences and identifying the shape differences occurring due to a particular pathology affecting multiple interacting organs. Such comprehensive analysis of multiple organs and their interactions can be incredibly beneficial in diagnosing and providing timely therapeutic assistance [12,14,21]. Specifically, in the case of cardiology, the interventricular septum (IVS) has been shown to change shape during various cardiomyopathies. Others have described the flattening and reversal of curvature in patients with significant right ventricular pathologies [24]. Therefore, it is crucial to model the left and right ventricles together and the changes at the interventricular septum or the shared boundary.

Shapes can be represented using an implicit (deformation fields [8], level set methods [19]) or explicit (landmarks/points) representation. Explicit parameterization, such as landmarks, is one of the most popular techniques used to represent shapes because of its simplicity and ability to represent multiple objects easily [6]. Hence, in this work, we focus on point distribution models (PDM) for representation, as PDMs are suitable for the statistical analysis of models with shared boundaries. To enable comparison and to obtain shape statistics from an ensemble of shapes, points of the same anatomical position must be established consistently across shape populations. These points are called *correspondences*. Multiple methods for correspondence generation have been proposed, which include non-optimized landmark estimation, parametric, and non-parametric correspondence optimization. Non-optimized methods entail manually annotating the reference shape and warping these landmarks on the population data, and they employ hard surface constraints to distribute points on a shape. Parametric methods use fixed geometrical basis (e.g., spheres) [22] to parameterize objects and generate correspondences. The correspondence model obtained using manual or parametric techniques is not optimal and can prove to be incapable of handling complex anatomies. On the other hand, non-parametric methods provide a robust and general framework as they use a PDM without relying on a specific geometric basis. Methods that follow group-wise approach find the correspondence by considering the variability of entire data in the optimization process (e.g., particle-based optimization [3], Minimum Description Length -

MDL [7]). The group-wise SSM approaches have been extended to model multi-organ anatomies. These approaches either parameterize each object separately, sacrificing anatomical integrity [6], or minimize the combined cost function to generate correspondences assuming a global statistical model [4,8].

To the best of our knowledge, none of the existing SSM methods have explicitly incorporated nuanced interactions, such as shared surfaces between multiple anatomies that can reveal key features that might not be observable on the individual anatomies when modeled independently. To address this issue, we propose a mesh grooming pipeline for extracting shared boundary surfaces and a correspondence-based optimization scheme to consistently parameterize multi-organ anatomies and their shared surfaces. We demonstrate the entire pipeline with a cardiac biventricular dataset, where we model the right ventricle (RV), left ventricle wall (LVW), and interventricular septum (IVS). We use the group-wise non-parametric particle-based optimization method proposed by Cates et al., [3,5] to generate PDM and modify the framework to support multi-organ anatomies with shared boundaries.

2   Methods

Constructing a statistical shape model for multi-organ anatomies with shared boundaries requires consistent point distribution on the shared boundary across the multi-organ anatomies and explicitly modeling the statistics of both the contour and the interior of the shared boundary. To fulfill these requirements, we first need tools to detect and extract shared boundaries and their edges (i.e., contour information) from two adjoining anatomies. The steps and methods for the proposed general pipeline for shared boundary extraction are explained in Sect. 2.2. Second, we need to fit a PDM that includes joint statistics of the multi-organ anatomies, shared boundary interior and contour. Herein we leverage the particle-based shape modeling (PSM) approach [3,5] for automatically constructing PDMs by optimizing point (or particle) distributions over a cohort of shapes using an entropy-based optimization method. The PSM method uses a system of interacting particles with mutually repelling forces that learn the most compact statistical descriptors of the anatomy [3]. For consistent parameterization on the shared boundary, the surface sampling objective of the PSM method has to be modified to accommodate the interaction between the anatomies and the shared surface. A brief overview of the PSM entropy optimization method for single anatomy is provided in Sect. 2.1 and the proposed surface cost function modifications for multi-organ anatomies with shared boundary surfaces is provided in Sect. 2.3.

2.1   Background: Particle-Based Shape Modeling

Consider a cohort of shapes $\mathcal{S} = \{\mathbf{z}_1, \mathbf{z}_2, ..., \mathbf{z}_N\}$ of N surfaces, each with its set of M corresponding particles $\mathbf{z}_n = [\mathbf{x}_1, \mathbf{x}_2, ..., \mathbf{x}_M] \in \mathbb{R}^{dM}$ where each particle

$\mathbf{x}_m \in \mathbb{R}^d$ lives in d–dimensional Cartesian (i.e., configuration) space. The ordering of the particles implies correspondence among shapes. Each correspondence particle is constrained to lie on the shape's surface. Collectively, the set of M particles is known as the *configuration*, and the space of all possible configurations is known as the *configuration space*. The particle positions are samples (i.e., realizations) of a random variable $\mathbf{X} \in \mathbb{R}^d$ in the configuration space with an associated probability distribution function (PDF) $p(\mathbf{X} = \mathbf{x})$. Each configuration of M particles can be mapped to a single dM–dimensional *shape space* by concatenating the correspondence coordinate positions into a single vector $\mathbf{z}_n$ which is modeled as an instance of random variable $\mathbf{Z}$ in the shape space with PDF $p(\mathbf{Z} = \mathbf{z})$ assuming shapes are Gaussian distributed in the shape space, i.e., $\mathbf{Z} \sim \mathcal{N}(\boldsymbol{\mu}, \boldsymbol{\Sigma})$. The optimization to establish correspondence minimizes the energy function

$$Q = H(\mathbf{Z}) - \sum_{k=1}^{N} H(\mathbf{X}_k) \tag{1}$$

where H is an estimation of differential entropy. The differential entropy of $p(\mathbf{X})$ is given as

$$H(\mathbf{X}) = -\int_S p(\mathbf{X}) \log p(\mathbf{X}) dx = -E\{\log p(\mathbf{X})\} \approx -\frac{1}{M} \sum_{i=1}^{M} \log p(\mathbf{x}_i) \tag{2}$$

Minimization of the first term in Q from Eq. (1) produces a compact distribution of samples in shape space and encourages particles to be in correspondence across shapes. The second term seeks uniformly-distributed correspondence positions on the shape surfaces for accurate shape representation [3,5]. Further details regarding the optimization and gradient updates can be found in [3,5].

2.2 Shared Boundary Extraction

To demonstrate the shared boundary extraction pipeline, consider two adjoining organs, A and B, with a shared boundary. The steps for shared boundary extraction entail the following:

1. **Isotropic Explicit Re-meshing:** This generates a new mesh triangulation that conforms to the original data but contains more uniformly sized triangles. This also has the benefit of ensuring equivalent average edge lengths across the two shapes, which is useful in ensuing steps [23].
2. **Extracting Shared Boundary:** In this step, we perform extraction on the two original shapes and output three new shapes, two of which correspond to the original shapes and one for the shared boundary. Let us designate the original meshes as A_o and B_o (Fig. 1.a and 1.b) then:
   (a) Find all the triangles in A_o that are close to B_o and construct a mesh with these triangles called A_s. A triangle with vertices v_0, v_1 and v_2 is considered close to a mesh if the shortest Euclidean distance to the mesh

for all three vertices is below a small threshold. The threshold must be experimentally tuned to ensure the extracted shared surfaces are clinically relevant. We similarly find all the triangles in B_o that are close to A_o and designate this mesh as B_s

(b) Find the remainder of the mesh in A_o after removing the triangles in A_s and designate this as A_r. Similarly, we designate the remainder of the mesh in B_o after removing the triangles in B_s as B_r.

(c) Arbitrary designate B_s as the shared surface M

(d) Move all the points on the boundary loop of A_r to the boundary loop of M and return three new shapes A_r, M, and B_r (Fig. 1.c).

3. **Laplacian Smoothing:** At this point, the resulting triangulation typically contains jagged edges. We apply Laplacian smoothing to correct for this [10].

4. **Extract Contour:** The boundary loop of the shared surface M is computed using LibIGL *boundary_loop* tool [13] and designate this contour as C (Fig. 1.d).

The input consisting of two adjoining organs with a shared surface has been converted into input with four separate parts, the organs A and B, the shared surface, and the contour using the pipeline (Fig. 1.d).

Fig. 1. Extracting shared boundary between two meshes. The regions in green have Euclidean distances that fall within the threshold and are extracted as a shared boundary as per step 2. The green arrows show the distances within the threshold, and the red arrows show distances greater than the threshold. The contour is extracted from the green region as per step 4. Note: the meshes are farther apart, and the threshold is larger for visualization purposes. (Color figure online)

2.3 Particle-Based Shape Modeling with Shared Boundaries

In Sect. 2.2, two adjoining organs with shared surfaces were separated into four parts. Now a shape model has to be built that can faithfully capture the joint statistics of all the organs while representing the individual parts consistently. The first requirement for a shared boundary shape model is that particle-based

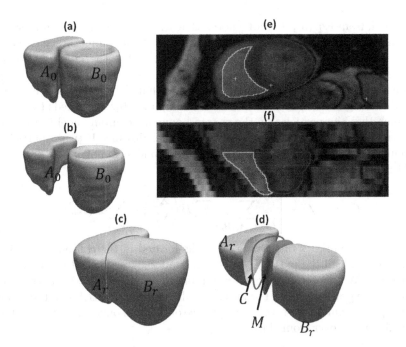

Fig. 2. An example of output obtained after shared boundary extraction. Meshes representing (a) RV and LVW show that they have a shared boundary surface, and (b) RV and LVW meshes are pried apart. The meshes and contour obtained after shared boundary extraction (c) RV, LVW, shared surface and contour (d) all outputs pried apart for visualization. The red color indicates the contour. The image shows the endocardial segmentation for the RV (blue) and epicardial segmentation for the LV (violet) at the end diastole in the (e) axial view and (f) coronal view (Color figure online)

optimization should handle multi-organ anatomies. The optimization set up in Eq. (1) was extended for multiple organs by treating all the organs as a single structure [4]. From the original formulation, it is important to note that $p(\mathbf{x}_i)$ in Eq. 2 was estimated from the particle position using non-parametric kernel density estimation method [3,5]. This results in a set of points on the surface that repel each other with Gaussian-weighted forces. Therefore, for multiple organ anatomy, if one organ has distinct identities, the spatial interactions between particles on different organs are decoupled, and particles are constrained to lie on a single organ (surface). The covariance $\boldsymbol{\Sigma}$ includes all particle positions across the multiple organs so that optimization takes place in the joint, multi-organ shape space and the shape statistics remain coupled [4]. For D organs in anatomy, the cost function is

$$Q = H(\mathbf{Z}) - \sum_{j=1}^{D} \left[\sum_{k=1}^{N} H(\mathbf{X}_k^j) \right] \tag{3}$$

where $\mathbf{X}_k^j$ represents the kth particle on j^{th} organ.

Second, from Eq. (3), it can be noted that the second term, which represents the sampling objective, is summed over all the shape samples, such that the sampling is restricted to the particles contained within the individual organs. As a result, when two organs have a shared boundary and sampling is done independently, there is no mechanism to ensure that the particles do not clutter around the edges of the organs. Hence, the sampling objective needs to be modified such that the particles on the shared boundary contour repel the particles of other mesh objects. This will result in a buffer distance between particles of the multiple organs leading to a uniform correspondence model.

The proposed objective function is:

$$Q = H(\mathbf{Z}) - \left[\sum_{j \in (A_r, M, B_r)} \sum_{k=1}^{N} H\left(\frac{\mathbf{X}_k^j}{\mathbf{X}_k^C}\right) + \sum_{k=1}^{N} H(\mathbf{X}_k^C) \right] \tag{4}$$

where $\mathbf{X}^c$ is the matrix of particle positions located in the contour. Effectively, this means that all the particles on A_r, M, and B_r are repelled by the contour C particles. We do not change the sampling objective for the contour. This is because large gradients from the meshes could cause the particles on the contour to swap places. Since there is only one degree of freedom on a contour, it is almost impossible to recover from this situation.

3   Experiments and Results

Dataset: We evaluate our method on a cardiac biventricular dataset based on how the resulting correspondence model captures variability in shape for cardiovascular clinic patients and healthy volunteer groups. The dataset consists of MRIs of 6 healthy volunteers and 23 patients treated at a cardiovascular clinic. In the patient group, tricuspid regurgitation was secondary to pulmonary hypertension in one patient, congestive heart failure (CHF) in 10 patients, and other causes (atrial fibrillation, pacemaker lead injury, pacemaker implantation, congenital heart disease) in 12 patients. The healthy volunteers had no diagnosis of cardiac disease and no cardiovascular risk factors.

Initially, the RV and LVW segmentation images were generated by converting end-diastole CINE MRI to volume stack. From each CINE short-axis time stack, an image of the heart at the end diastole was extracted to create a volume image stack. Image extraction was performed using a custom MATLAB image processing code. The volume stacks were then segmented using the open-source Seg3D software (SCI Institute, University of Utah, SLC UT). The segmentations were then isotropically resampled and converted to meshes using the open software ShapeWorks. In order to align the shapes, the meshes were centered and rigidly aligned to a representative reference sample selected from the population. The rigid alignment was done by calculating the transformations only using the RV meshes of the population due to their complex shapes. These transformations were then applied to the RV and the LVW meshes. The average edge length of the right ventricle meshes was 0.8224 ± 0.3987, left ventricle wall meshes was 0.9438 ± 0.3399, the IVS meshes 0.5196 ± 0.4047, and the contours 21.469 ± 26.205.

Shape Model Construction: We used ShapeWorks, an open-source software that implements the particle-based entropy optimization [3,5] described in Sect. 2.1. We modified the optimization with the proposed cost function (Eq. 4) to support multi-organ anatomies with shared boundaries. First, the shared boundary surface and contour were extracted for building a shape model using the tool described in Sect. 2.2. Figure 2 shows an example output for one sample. Then, a shape model was built using 512 particles for the RV and LVW, and 64 particles were used for the IVS surface and contour. From this PDM, mean shapes and differences were computed.

Discussion: The shape model was used to identify group-level shape differences of the RV, LVW, and IVS. Figure 3 shows the mean shape of each group and the color-coded group differences. There is a marked difference in the curvature of IVS of the healthy group as compared to the patient group. The curvature of the IVS was also captured in the modes of variation of the shape model. In Fig. 5, we visualize the modes of variation obtained using principal component analysis of the IVS shared surface and contour obtained after building the shape model with all the four parts - RV, LVW, IVS shared surface, and contour. The RV and LVW are excluded only for visualization. Since the curvature is not a linear feature, a single PCA mode is not enough to capture it. We show the top four PCA modes that capture the curvature in various directions. In order to study statistically significant geometric differences, we performed linear discrimination of variation. The particle-wise mean shapes of both groups were compared, and a difference vector was generated. The group means for the cardiac patients is set as -1, and controls are set as 1. Each shape is mapped to a single scalar value (or a "shape-based-score") that places subject-specific anatomy on a group-based shape difference statistically derived from the shape population. Figure 4 shows the mapping for all shapes of the two groups. Selected shapes correspond to individual points on the graph. The shapes at the extreme ends of the mapping also confirm that the shape model appropriately identified the curvature of IVS as a significant geometrical difference between the two groups. The shape in Fig. 4 also shows free wall bulging and narrowing of the base of RV for cardiac patients. For modes of variation of the shape model, see Appendix A.1.

Since the number of samples in the patient group and control group are not the same, we performed hypothesis testing to identify if the shape-based score assigned to each sample is statistically significant and agnostic to the data imbalance. We generated the shape-based scores for each sample using the statistics of 6 randomly selected samples from the patient group and all six control group samples and repeated the experiment 1000 times. The shape-based scores from the experiment were then compared to those generated using the complete dataset. We use t-test to test for the null hypothesis that the expected value (mean) of a sample of independent observations from the 1000 trials is equal to the given population mean, i.e., the scores generated using the complete dataset. Figure 6 shows the box-and-whisker plot of the distribution of scores of each sample obtained from the experiment, and the color indicates the p-values. We select

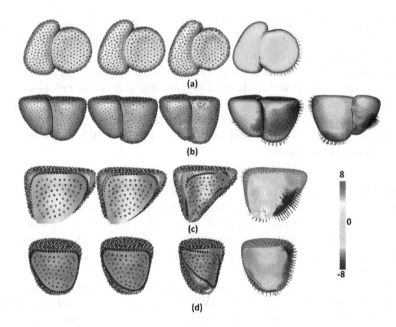

Fig. 3. Columns 1, 2, and 3 show the mean shape of the patient group, overall mean, and mean shape of the control group, respectively. Columns 4 and 5 show the difference between the group-mean shapes (two views). The arrows indicate the direction of group differences, and the color represents the magnitude of the group difference. The PDM for bivenctricle data (a) top-view and (b) front view. The same shapes after (c) excluding the left ventricular wall and (d) excluding the right ventricle for visualizing the IVS.

the alpha value to be 0.01. Hence, if the p-values are smaller than 0.01, the null hypothesis holds (shown in green), and if the p-value is greater than 0.01, we can reject the null hypothesis and assume that the scores are affected by the imbalance (shown in red). It can be seen from Fig. 6 that the imbalance does not affect the shape-based scores for the majority of samples.

These results confirm what has been observed in the cardiology literature: a decrease in interventricular septal curvature during prolonged right ventricular dysfunction. A healthy heart has a significant pressure gradient between the right and left ventricles. However, in many cardiac diseases, the pressure gradient dissipates because right ventricular pressure increases. As the pressure increases, a distortion occurs at the interventricular septum, and the original septal curvature matching the left ventricular becomes flattened. This signifies the structural remodeling that occurs with severe right-sided cardiac pathologies.

Despite these observations being made previously, the clinical utility of septal curvature has been minimal because of inadequate tools for precise and accurate measurements. Structural remodeling initially occurs to compensate for acute changes in cardiac physiology. As acute changes become chronic, the

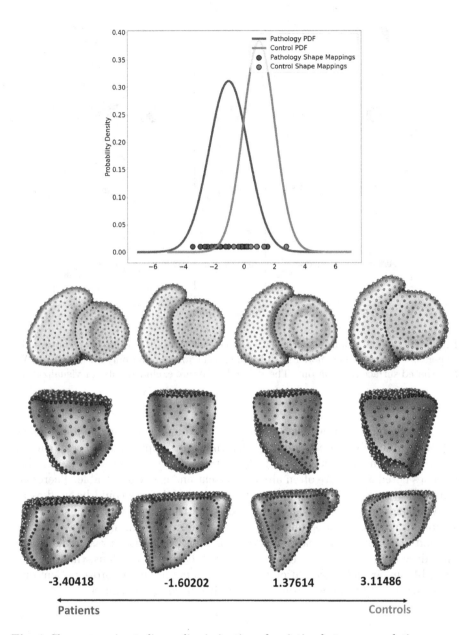

Fig. 4. Shape mapping to linear discrimination of variation between population means for the groups of patients and controls. The first row represents the biventricle shared boundary shapes. The second row represents the same biventricle shared boundary shapes after excluding the left ventricular wall, and the third row after excluding the right ventricle for visualizing the IVS. The number below each shape denotes the "shape-based score" of each anatomy derived from the shape population.

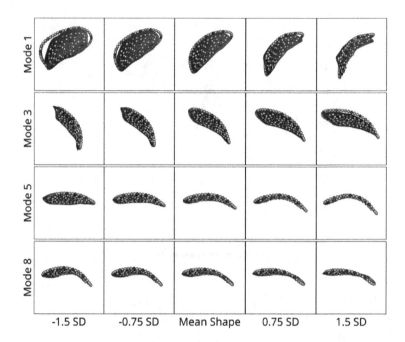

Fig. 5. Modes of variation of the IVS shared surface and contour that capture the change in curvature. The model was obtained using all four anatomy parts - RV, LVW, IVS shared surface and contour. The RV and LVW are excluded only for visualization.

cardiac structural adaptations become permanent and cause long-term detrimental effects. Initially, patients do not feel significant symptoms because of cardiac tissue's excellent adaptability and plasticity. However, structural changes, like the ones noted above, are often already present and easily detectable. Therefore, shape analysis of interventricular septal curvature changes could be used as an early prognosticator of cardiac dysfunction prior to patients reporting significant symptoms. Notably, shape analysis can be performed using non-invasive imaging and does not require cardiac catheterization, a routine, invasive diagnostic procedure typically used for detecting cardiac dysfunction. Thus, the proposed shared-boundary SSM generation technique can potentially improve patient outcomes with early diagnosis using non-invasive imaging procedures.

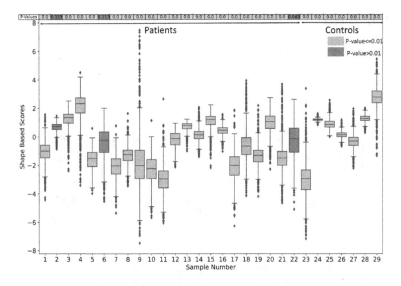

Fig. 6. Statistical test for testing the effect of dataset imbalance on the shape-based scores. Box and whisker plot showing the distribution of shape-based scores for each sample obtained using a subset of the patient group and all the control samples repeated 1000 times. Each box is color-coded based on the p-values: green - samples with p-values ≤ 0.01 and red -samples with p-value> 0.01. (Color figure online)

4   Conclusion

Our method provides a novel way of extracting and generating shape models of multi-organ anatomies with shared boundary surfaces. We showed our method provides a consistent and robust representation of the shared boundary without compromising the integrity of the multiple-organ PDM. We applied our method to a cardiac biventricular dataset and showed unique shape changes of the IVS that is not captured when modeling the ventricles alone. This pipeline could pave the way for using shape analysis from non-invasive imaging for early diagnosis and prognostication of pathologies affecting multiple organs and further our understanding of interactions between any anatomical system with shared boundaries.

Acknowledgement. This work was supported by the National Institutes of Health under grant numbers NIBIB-U24EB029011, NIAMS R01AR076120, NHLBI-R01HL135568, NIBIB-R01EB016701, NIGMS-P41GM103545, NIGMS-R24GM136986 (MacLeod), and NHLBI-F30HL149327 (Zenger). We thank the University of Utah Division of Cardiovascular Medicine and the ShapeWorks team.

A   Appendix

A.1   Modes of Variation

(See Fig. 7).

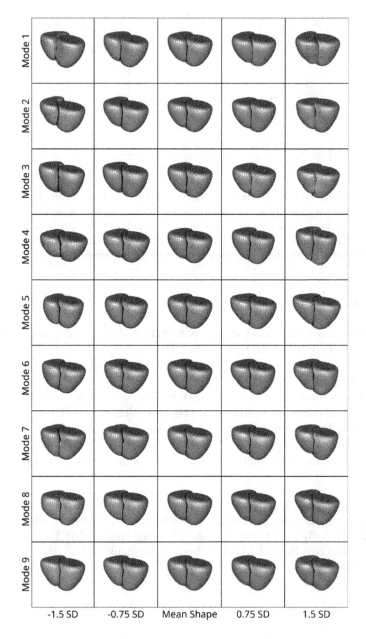

Fig. 7. Modes of variation discovered by the shape model of the biventricle dataset with the proposed optimization for multi-organ anatomies with shared boundaries.

References

1. Borghi, A., et al.: A population-specific material model for sagittal craniosynostosis to predict surgical shape outcomes. Biomech. Model. Mechanobiol. **19**(4), 1319–1329 (2020)
2. Cates, J., et al.: Computational shape models characterize shape change of the left atrium in atrial fibrillation. Clin. Med. Insights: Cardiol. **8**, CMC-S15710 (2014)
3. Cates, J., Elhabian, S., Whitaker, R.: Shapeworks: particle-based shape correspondence and visualization software. In: Statistical Shape and Deformation Analysis, pp. 257–298. Elsevier (2017)
4. Cates, J., Fletcher, P.T., Styner, M., Hazlett, H.C., Whitaker, R.: Particle-based shape analysis of multi-object complexes. In: Metaxas, D., Axel, L., Fichtinger, G., Székely, G. (eds.) MICCAI 2008. LNCS, vol. 5241, pp. 477–485. Springer, Heidelberg (2008). https://doi.org/10.1007/978-3-540-85988-8_57
5. Cates, J., Fletcher, P.T., Styner, M., Shenton, M., Whitaker, R.: Shape modeling and analysis with entropy-based particle systems. In: Karssemeijer, N., Lelieveldt, B. (eds.) IPMI 2007. LNCS, vol. 4584, pp. 333–345. Springer, Heidelberg (2007). https://doi.org/10.1007/978-3-540-73273-0_28
6. Cerrolaza, J.J., et al.: Computational anatomy for multi-organ analysis in medical imaging: a review. Med. Image Anal. **56**, 44–67 (2019)
7. Davies, R.H.: Learning shape: optimal models for analysing natural variability. The University of Manchester (United Kingdom) (2002)
8. Durrleman, S., et al.: Morphometry of anatomical shape complexes with dense deformations and sparse parameters. NeuroImage **101**, 35–49 (2014)
9. Faber, B.G., et al.: Subregional statistical shape modelling identifies lesser trochanter size as a possible risk factor for radiographic hip osteoarthritis, a cross-sectional analysis from the osteoporotic fractures in men study. Osteoarthritis Cartilage **28**(8), 1071–1078 (2020)
10. Field, D.A.: Laplacian smoothing and delaunay triangulations. Commun. Appl. Numer. Methods **4**(6), 709–712 (1988)
11. Goparaju, A., et al.: Benchmarking off-the-shelf statistical shape modeling tools in clinical applications. Med. Image Anal. **76**, 102271 (2022)
12. Hensel, J.M., et al..: Development of multiorgan finite element-based prostate deformation model enabling registration of endorectal coil magnetic resonance imaging for radiotherapy planning. Int. J. Radiat. Oncol.* Biol.* Phys. **68**(5), 1522–1528 (2007)
13. Jacobson, A., Panozzo, D., et al.: libigl: A simple C++ geometry processing library (2018). https://libigl.github.io/
14. Kobatake, H.: Future cad in multi-dimensional medical images:-project on multi-organ, multi-disease cad system-. Comput. Med. Imaging Graph. **31**(4–5), 258–266 (2007)
15. Krol, Z., Skadlubowicz, P., Hefti, F., Krieg, A.H.: Virtual reconstruction of pelvic tumor defects based on a gender-specific statistical shape model. Comput. Aided Surg. **18**(5–6), 142–153 (2013)
16. Lenz, A.L., et al.: Statistical shape modeling of the talocrural joint using a hybrid multi-articulation joint approach. Sci. Rep. **11**(1), 1–14 (2021)
17. Marrouche, N.F., et al.: Association of atrial tissue fibrosis identified by delayed enhancement mri and atrial fibrillation catheter ablation: the decaaf study. Jama **311**(5), 498–506 (2014)

18. Orkild, B.A., et al.: All roads lead to rome: diverse etiologies of tricuspid regurgitation create a predictable constellation of right ventricular shape changes. Front. Physiol., 1092 (2022)
19. Samson, C., Blanc-Féraud, L., Aubert, G., Zerubia, J.: A level set model for image classification. Int. J. Comput. Vision **40**(3), 187–197 (2000)
20. Sanfilippo, A.J., et al.: Atrial enlargement as a consequence of atrial fibrillation: a prospective echocardiographic study. Circulation **82**(3), 792–797 (1990)
21. Si, W., Heng, P.A., et al.: Point-based visuo-haptic simulation of multi-organ for virtual surgery. Dig. Med. **3**(1), 18 (2017)
22. Styner, M., et al.: Framework for the statistical shape analysis of brain structures using spharm-pdm. Insight J. (1071), 242 (2006)
23. Surazhsky, V., Alliez, P., Gotsman, C.: Isotropic remeshing of surfaces: a local parameterization approach. Ph.D. thesis, INRIA (2003)
24. Tanaka, H., et al.: Diastolic bulging of the interventricular septum toward the left ventricle: an echocardiographic manifestation of negative interventricular pressure gradient between left and right ventricles during diastole. Circulation **62**(3), 558–563 (1980)
25. Uetani, M.: Statistical shape model of the liver and its application to computer-aided diagnosis of liver cirrhosis. Electr. Eng. Jpn. **190**(4), 37–45 (2015)

Computerized Analysis of the Human Heart to Guide Targeted Treatment of Atrial Fibrillation

Roshan Sharma[1], Andy Lo[1], Zhaohan Xiong[1], Xiaoxiao Zhuang[1], James Kennelly[1], Anuradha Kulathilaka[1], Marta Nuñez-Garcia[2,3], Vadim V. Fedorov[4], Martin K. Stiles[5], Mark L. Trew[1], Christopher P. Bradley[1], and Jichao Zhao[1 (✉)]

[1] Auckland Bioengineering Institute, Auckland, New Zealand
j.zhao@auckland.ac.nz
[2] Université de Bordeaux, Bordeaux, France
[3] IHU Liryc, Electrophysiology and Heart Modeling Institute, Fondation Bordeaux Université, Pessac, France
[4] Department of Physiology and Cell Biology, The Ohio State University Wexner Medical Center, Columbus, USA
[5] Waikato Clinical School, Faculty of Medical and Health Sciences, University of Auckland, Auckland, New Zealand

Abstract. Current treatment for atrial fibrillation (AF) remains suboptimal due to a lack of understanding of the atrial substrate which sustains AF in human atria. Comparing atrial structural characteristics (fibrosis, myofibers and wall thickness) across different hearts is challenging due to the complexity and thinness of the atrial wall. There is also a need for quantitative tools to guide treatment strategies in clinical settings. Using 111 late gadolinium-enhanced MRI (LGE-MRI) scans taken from patients with AF from two clinical centers, we developed novel convolutional neural networks to perform automatic bi-atrial segmentation. Then a standardized 2D representation of atrial anatomical structures was used for analyzing and comparing patient data. We also developed an algorithm to define rule-based fibers in volumetric human atrial models using existing knowledge about atrial fibers. Finally, realistic fibrosis and the generated fibers were integrated into voxelized atrial geometries sourced from the LGE-MRIs. Computer simulations were performed on these geometries, using techniques that made efficient use of computational memory. The techniques also allowed for rapid and reliable reentry formation, reducing computational time. Our analysis of three flattened atria concisely showed that the lateral right atrium and the inferior pulmonary veins accrued substantial fibrosis. The fiber generation method gave smooth and finely tunable results. The simulations showed that fibrosis and fibers substantially affect AF dynamics. The proposed pipeline can potentially identify biomarkers and conduct computer simulations of the human heart to guide targeted ablation treatment.

Keywords: Atrial segmentation · Atrial anatomy · Computer modeling · Atrial fibrillation · Fiber orientation · MRI · Atrial flattening

O. Camara et al. (Eds.): STACOM 2022, LNCS 13593, pp. 317–329, 2022.
https://doi.org/10.1007/978-3-031-23443-9_29

1   Introduction

Atrial fibrillation (AF) is the most common heart rhythm disturbance [1]. The current treatment for AF remains suboptimal due to a lack of basic understanding of the underlying atrial substrate which sustains AF directly in human atria. In addition, there is a need for quantitative tools to investigate optimal treatment strategies in clinical/experimental settings [2, 3].

AF is driven by complex substrates which are distributed widely throughout both atria [4, 5]. Atrial myofiber orientation plays an important role in healthy and pathophysiological conditions; an electrical signal travels considerably faster in the direction of fibers, as opposed to across them [6]. The development and progression of atrial fibrosis, on the other hand, is widely accepted as the most important substrate for AF perpetuation by inducing marked local conduction abnormalities. Clinical studies using late gadolinium-enhanced MRI (LGE-MRI) support that atrial fibrosis is an important contributor to AF, knowledge of which is key to improved treatments [2].

LGE-MRIs are widely used to provide views of atrial structure including atrial anatomy and key biomarkers including dilatation, wall thickness and fibrosis [2]. With the rise in popularity of convolutional neural networks (CNN), robust CNN-based techniques are being developed for automatic segmentation of LGE-MRIs. These methods have drastically improved on the traditional atlas- or shape-based segmentation approaches [3]. In addition, there is a lack of standardized tools to compare structural components among different hearts. Finally, diffusion or structure tensor imaging is a quantitative approach for finding fiber orientation [9, 10]. Both approaches determine fibers from co—lor intensity gradients and have been applied to high-resolution ex-vivo MRIs. However, they are unsuitable for in-vivo MRIs due to low imaging resolution and thin atrial walls [11]. Hence to find fiber orientation for in-vivo MRIs, rule-based techniques with the help of atrial fiber atlases [12, 13] are developed. Computer models incorporated with accurate atrial anatomy, realistic fibrosis distribution and 3D fiber orientation are a powerful tool for quantitative dissection of the fundamental mechanisms and investigating their individual roles in AF [9]. However, current computing power of the simulation solvers limits their usage in clinical settings.

In this paper, we propose and evaluate a computational pipeline to process 3D LGE-MRIs from patients with AF (Fig. 1). The aim of this pipeline is to provide biomarkers and conduct computer simulations of human atria to guide targeted ablation treatment. It includes a double CNN approach for robust segmentation and reconstruction of the left atrium (LA) and right atrium (RA) without human intervention, and a semi-automatic approach for mapping 3D heart atria into standardized 2D representations for comparison of atrial structures across different hearts. Finally, we developed a rule-based fiber generation approach with help from human atrial fiber atlases and used it in the computer modeling analysis.

2   Methods

2.1   LGE-MRI Data and Automatic Segmentation

A total of 100 3D LGE-MRIs from patients with AF and their LA segmentations (provided by the University of Utah) were used for developing and training our AI approach,

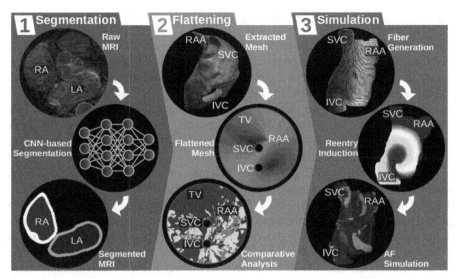

Fig. 1. Full pipeline summarized with the three main components shown sequentially from left to right: segmentation, flattening and simulation. Their sub-components are sequentially shown from top to bottom. RA/LA – left/right atrium CNN—convolutional neural network, RAA – RA appendage, SVC/IVC – superior/inferior vena cava, TV*tricuspid valve, AF – atrial fibrillation.

which is briefly described here (for more details see [2]). The RA segmentations were manually performed by our team based on the protocols used for LA segmentation to achieve consistency across both atrial chambers [8]. Our machine learning pipeline consisted of two 2D CNNs used in a sequential manner. The first CNN performed coarse segmentation on a down-sampled version LGE-MRI to construct an approximate segmentation of the atria, and the second CNN then performed slice-by-slice regional detailed segmentation on the output of the first CNN.

In this study, we used our trained AI to directly segment atrial chambers from 11 3D LGE-MRI scans sourced from patients with AF at Waikato Hospital (Waikato, New Zealand) between June 2017 and February 2020. The Waikato LGE-MRIs were split into 36% training and 64% testing. To robustly estimate the 3D atrial wall thickness in each atrial chamber, coupled partial differential equations, formed by coupling the Laplace solution with two surface trajectory functions were solved [7]. To estimate fibrosis in the atria, we used the fibrosis threshold approach developed by the University of Utah [8]. The method worked layer-by-layer for each slice in the z-axis of the MRI scan. The mean and standard deviation of the pixel intensity was calculated for each slide. Healthy tissue was defined as pixels with an intensity between the 2nd and 40th percentile. Fibrotic tissue was defined as pixels with an intensity greater than some standard deviations above the healthy tissue mean. In this study, 3.85 and 4.25 standard deviations were used for the LA and RA, respectively.

2.2   Standardized Atrial Geometry Fattening

Here the flattening process is briefly described (see [14, 15] for more details and the associated code). First, for each segmented geometry, LA and RA surface meshes are generated using the marching cubes algorithm [16]. Before flattening, the meshes undergo
simplifications. For RA meshes, the superior vena cava (SVC), inferior vena cava (IVC)
and tricuspid valve (TV) are manually removed using a single flat cut (we used Paraview, Kitware Inc., for the removal). For LA meshes, the pulmonary veins (PVs), LA
appendage (LAA) and mitral valve (MV) are removed semi-automatically within the
LA flattening pipeline described below.

For the RA, first, its SVC, IVC and TV holes are closed. Next, three seed points are
placed on the geometry (see the three red dots in Fig. 2B). Flattening morphs the atrium
into a 2D disk as shown in Fig. 2C–E. The hole boundaries in the 3D atrial chambers
get mapped to predetermined locations on the 2D disk, acting as anchors, while the rest
of the mesh vertices conform to these anchoring points. The TV boundary is mapped
to the circumference of this disk. The SVC gets mapped to a small circular hole in the
center and the IVC is positioned below the SVC, midway to the TV. The RA appendage
(RAA) tip is anchored to the top right of the SVC.

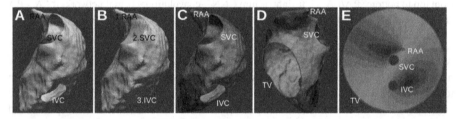

Fig. 2. The flattening procedure for the RA. (**A**) A clipped 3D RA mesh. (**B**) The RA mesh
with its SVC, IVC and TV holes covered. In addition, it shows three seed points (red dots) with
numbered labels indicating placement order. (**C**) Lateral and (**D**) anterior views of the 3D mesh.
(**E**) Final flattened geometry. The colors show the distance to a specific point on the RA to aid
the comparison between the original 3D and the flattened RA. RA – right atrium, RAA – RA
appendage, SVC/IVC – superior/inferior vena cava, TV – tricuspid valve. (Color figure online)

The LA process begins with the removal of the PVs, LAA and MV. The algorithm performs these removals based on five seed points placed on the tips of the PVs and LAA,
as shown in Fig. 3A. Next, all of the non-MV holes are covered, and nine more seeds are
manually placed onto the surface. The first five are placed centrally on the hole covers of
the PVs and LAA, the remaining four are placed on the MV contour (Fig. 3B and C). Like
the RA, the hole boundaries get mapped to predetermined locations on the 2D LA disk as
shown in Fig. 2D–F. Compared to the RA, the LA's more complex structure necessitates
extra anchoring structures in the 2D LA disk. These structures are in the form of geodesics
(shortest distance paths along the surface of the atrial mesh, found using Dijkstra's Algorithm [17]) that connect specific pairs of seed points (yellow lines in Fig. 2D–F). As illustrated by Fig. 3D–F the MV is mapped to the outer circumference of the 2D disk. The PVs

and the four geodesics connecting them get mapped to the center in a rectangular formation. Above the projected left superior PV hole is the projected LAA (Fig. 3F). All of the
geodesics on the 3D mesh get transformed into straight lines on the 2D disk.

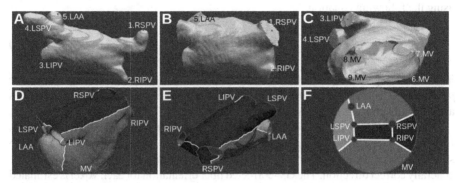

Fig. 3. The flattening procedure for the LA. (**A**) Five clipping seed points placed on the LA. (**B**) and
(**C**) show nine region-separating seed points. The seed placement order is indicated by the numbered
seed labels. (**D**) The posterior and (**E**) superior view of the 3D LA. Yellow lines are the geodesics that
connect specific seed points. (**F**) The final flattened LA geometry. Different atrial regions are color-
coded. LA—left atrium, LAA—LA appendage, LSPV/RSPV—left/right superior pulmonary vein,
LIPV/RIPV—left/right inferior pulmonary vein, MV—mitral valve.

2.3 Semi-automatic Rule-Based Fiber Generation

A simple and intuitive method was developed to generate 3D fibers for 3D human atria
derived from LGE-MRIs. This is a semi-automatic approach in which vectors are placed
onto the epi- or endocardium, referred to as guide vectors (GV) from here onward. Each
GV indicates the local fiber orientation in its vicinity. Given the low resolution of in-vivo
MRI scans and the thinness of atrial walls [11], non-transmurality was assumed for the
fibers. After the desired number of GVs are placed, the fiber orientation for each voxel
is estimated by

$$\mathbf{F} = \sum_{i=1}^{n} \left(\overline{\mathbf{G}}_i / d_i^p \right) \tag{1}$$

Here $\mathbf{F}$ is the vector representing the fiber, n is the number of GVs, $\overline{\mathbf{G}}_i$ is the normalized
ith GV, d_i is the distance from the voxel containing $\mathbf{F}$ to the midpoint of $\mathbf{G}_i$, p is a key
parameter discussed further below. After $\mathbf{F}$ is found, it is then normalized.

Variable p has a major impact on the generated fibers. At $p = 0.1$, each voxel gets a
practically unweighted mean of the GVs; hence the same fiber orientation gets computed
for all the voxels in the atrial geometry. With $p = 10$ this method approaches the nearest-
neighbor algorithm [18]. At $p = 1$, the resultant fibers blend more gradually, but if there
are many distant GVs, their combined effect noticeably alters the orientation intended by
more local GVs. With $p = 3$, the distant GVs become negligible, but near GVs excessively
influence the local fibers, giving unnaturally sharp changes. One way to choose p is to

compromise with a medium value, for instance $p = 1.5$, as shown in Fig. 4. When this value is applied to the RA geometry, there is still considerable influence from the numerous distant GVs. Lower values for p are more suitable for near GVs, while higher values work better on distant GVs. For the atria, a d_i dependent p was implemented using the following equation:

$$p = \frac{8d_i}{\sqrt{l^2 + w^2 + h^2}} \tag{2}$$

Here l, w and h are the geometry's bounding box dimensions. With this, local GVs use $p < 1.5$, and distant GVs use $p > 3$, resulting in fibers with a smooth blend of local GVs and negligible impact from distant ones.

This method can also be adapted to complement other fiber generation techniques. If there exists a noisy fiber region in a preexisting fiber layout (e.g. calculated using structure tensor imaging), then a GV can be used to guide the noisy fibers towards the generally correct direction. The GV-corrected fiber orientation for each voxel is estimated for each voxel using the following.

$$\mathbf{F}_{new} = \mathbf{F}_{old} + (\overline{\mathbf{G}} - \mathbf{F}_{old})/\left(1 + \exp\left(\frac{10d}{L} - 5\right)\right) \tag{3}$$

Here, $\mathbf{F}_{new}$ is the updated fiber orientation of the voxel, $\mathbf{F}_{old}$ is the initial fiber orientation, $\overline{\mathbf{G}}$ is the normalized GV, d is the distance from the voxel containing $\mathbf{F}$ to the midpoint of $\mathbf{G}$, and L is the length of the GV.

While the fiber orientation in simulations is independent of sign, for the GVs used to generate the fibers, the sign plays an important role in controlling the flow and convergence of the resultant fiber orientation. Figure 4A demonstrates this interaction between GVs (black arrows) and fibers (white arrows and color) through a 2D example. Figure 4B demonstrates the approach on a hollow cylinder using six GVs. Figure 4C shows how a GV can patch a defective region in some preexisting fiber map.

2.4 Efficient Electrophysiology Computer Modeling

To showcase the full pipeline, some cursory example simulations were run using a voxelized 3D atrial geometry (Fig. 5A). The human atrial cellular model used was an adapted Courtemanche-Ramirez-Nattel (CRN) model [19, 20] with AF modifications [21]. Fibrosis and fibers were estimated using the techniques from Sects. 2.1 and 2.3, and incorporated into the 3D atrial geometry, the fibrotic tissue was considered unexcitable [22]. A monodomain finite-difference-based approach was used with an isotropic space-step of 0.625 mm. Both the cellular model and the diffusion equation were solved with the same time-step of 0.05 ms. The space-step was limited by the resolution of the MRIs, and the time-step was chosen to produce quick results without excessively affecting accuracy. While this level of discretization was adequate for these simple example simulations, for properly detailed simulations, the CRN model should ideally be used with discretizations of approximately 0.01 ms for time and 0.4 mm for space. To account for the MRI-resolution limited space-step, the atrial geometry can be isotropically interpolated to

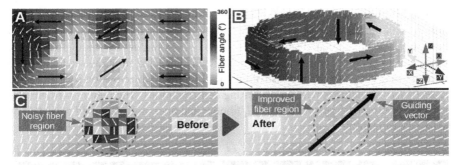

Fig. 4. The fiber generation approach was illustrated using simple geometries. (**A**) A 2D example of the approach. (**B**) A 3D example of the approach. (**C**) Before and after images demonstrating a noisy fiber region, and its subsequent improvement with the help of a GV. The black arrows are the manually placed GVs, while the white arrows and colors represent the resultant fibers. For the coloring, 3D fiber angles [0°, 360°] along the x, y and z axis are linearly mapped in the red, green and blue color channels. GV – guide vector. (Color figure online)

give a voxel side length of 0.4 mm (at most). The fibers were implemented with an anisotropy ratio of 9:1 [23].

To improve the computational memory efficiency, an adjacency list approach was used rather than the standard 3D arrays. Depending on tissue volume, this alone reduced computer RAM usage by up to 90%. Along with lowering the memory usage occupied on shared high-performance computers it also allows simulations to be performed on a standard PC. This advantage becomes more noticeable on high-resolution geometries.

To investigate the dynamics of AF, reentry needs to be induced. One method is burst pacing [24]; this mimics the method used in ex-vivo experiments but effective pacing sites for reentry initiation are reliant on the fibrosis distribution, and it does not guarantee reentry formation [25]. Its reentry formation process also takes considerable computational time. Another approach is cross-field stimulation [26], it is typically used in 2D simulations, but has seen some usage on 3D atrial geometries [27]. It forms reentries more easily, but the initiation site is hard to control. In addition, it can also generate unintentional secondary reentries on the opposite side of the atrium. Our proposed approach reliably forms reentries, and does so faster than both the previously mentioned methods, while also giving better control of the reentry initiation site than either of them. In addition, it limits the formation of unwanted secondary reentries (Fig. 5).

In our adapted 3D cross-field stimulation, first, three seed points (P_1, P_2 and P_3) are placed on the surface. P_1 and P_2 indicate the direction of voltage propagation (from P_1 towards P_2), and P_3 determines the turning direction which is placed next to P_2, approximately perpendicular to the line formed by P_1 and P_2. The points form three planes (L_1, L_2 and L_3). L_1 has the normal vector of P_2-P_1 and contains point P_2. L_2 and L_3 both have the same normal vector P_2-P_3 and contain P_2 and P_3, respectively. Voxels on the P_1 side of L_1, between L_2 and L_3, are stimulated. Voxels on the non-P_2 side of L_3 are made refractory [28], and the remaining voxels are set as resting cells.

A demonstration of reentry formation is shown in Fig. 5, it shows the three seed points, and the stimulated region they form (in yellow). These points also form a smaller secondary stimulated region on the other side of the TV (Fig. 5B), its propagation can interfere with the behavior of the primary reentry, so the final step in this method is to eliminate all extra stimulated sites. This is done by first finding all stimulated regions in the geometry, and then the stimulated regions unconnected from the seed points are returned to a resting state. In the case of Fig. 5B, there are two activation sites, of which the secondary stimulus is returned to resting cell conditions. The reentry formed with the primary stimulus is shown in Fig. 5C, fibrosis and fibers were disabled to best show the reentry formation.

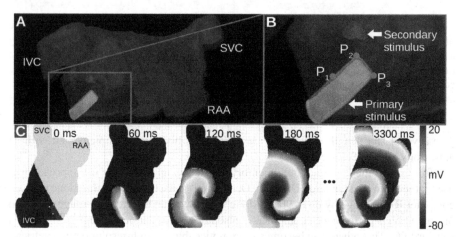

Fig. 5. Our proposed 3D pacing protocol. **A**) An example of the initial state of a simulation, yellow indicates the stimulus. **B**) A zoomed-in view of the stimulus from (**A**), showing the placement of three seed points and the formed primary and secondary stimuli. **C**) Resulting reentry from the primary stimulus in (**B**), shown at time-stamps of 0 ms to 180 ms, and eventually 3300 ms. RA – right atrium, RAA – RA appendage, SVC/IVC – superior/inferior vena cava

3   Results

3.1   CNN-Based LGE-MRI Segmentation

Three-dimensional visualization of the segmentation results shows that the proposed double CNN approach successfully captured the overall geometry of the LA and RA endocardium with a particularly impressive Dice score of 93%, which is in line with other literature using LGE-MRIs. More detailed results can be found in [8, 29].

3.2   Atrial Flattening

Fibrosis distributions in the RA and LA were found for three (#01, #10 and #11) of the eleven Waikato hearts, shown in Fig. 6, featuring both a view of the 3D atrial mesh and its final flattened form. For the three hearts, fibrosis percentages were 23.5%, 15.7% and 21.5% in the RA, and 29.5%, 24.9% and 21.3% in the LA, respectively.

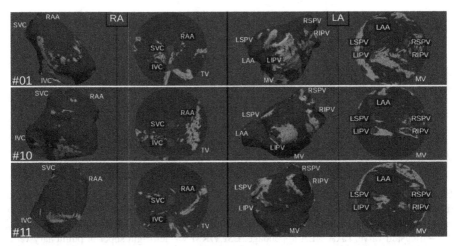

Fig. 6. Fibrosis distributions in the LA and RA of the three hearts (#01, #10 and #11). The first and third columns show fibrosis on the 3D RA and LA, respectively. The second and fourth columns show the flattened view of the atria. Blue indicates healthy tissue, and the shades of red represent the fibrosis projected onto the mesh. RA/LA – left/right atria, RAA/LAA – RA/LA appendage, SVC/IVC – superior/inferior vena cava, TV – tricuspid valve, LSPV/RSPV – left/right superior pulmonary vein, LIPV/RIPV – left/right inferior pulmonary vein, MV – mitral valve. (Color figure online)

A clear advantage of standardized 2D projection can be seen in Fig. 7, namely being able to make rapid quantitative comparisons by analyzing fibrosis distribution between three RA (Fig. 7A) and LA (Fig. 7B). In addition, the standardization implemented here for the 2D data further simplifies processing. These comparisons can scale up to include more hearts and other cardiac characteristics.

The 2D view clearly and concisely shows that fibrosis overlap in RA occurs most prominently on the lateral wall, followed by the septal wall. For the LA, there is substantially more fibrosis overlap surrounding the inferior PVs, relative to the superior PVs. There is also considerable overlap next to the MV on the left lateral and anterior atrial walls.

3.3   Three-Dimensional Fiber Orientation Generation

Figure 8 shows fibers generated in the RA of heart #01 from Fig. 6. In total, 148 GVs were used, their orientations were based on the atlas in [13]. The figure compares the results of varying p in Eq. 1. Of note, color gradients demonstrate the effect of p. Figure 8A (like Fig. 4) uses $p = 1.5$ and shows the influence of the numerous distant GVs through a bluer IVC (more positive z-component). Figure 8B uses $p = 3$, giving more local influence, but the fibers change more abruptly. The final generated fibers can be seen in Fig. 8C and D, which use a distance (d_i) dependent p (Eq. 2), giving both the localized GV influence seen in Fig. 8B and the smooth blending of fibers between GVs (Fig. 8A).

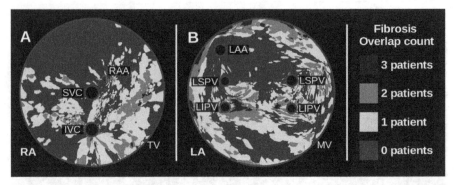

Fig. 7. Superimposed overview of the fibrosis distributions of the RA (**A**) and LA (**B**) of three different hearts, with four different colors indicating the differing inter-heart fibrosis overlap count. RA/LA – left/right atrium, RAA – RA appendage, SVC/IVC – superior/inferior vena cava, TV – tricuspid valve, LAA – LA appendage, LSPV/RSPV – left/right superior pulmonary vein, LIPV/RIPV – left/right inferior pulmonary vein, MV – mitral valve.

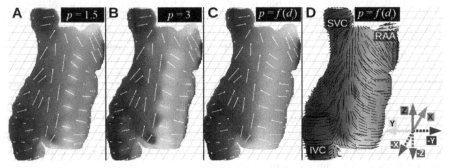

Fig. 8. Generated 3D fibers for an RA geometry with varying p. **A**) and **B**) use a constant $p = 1.5$ and $p = 3$, respectively (Eq. 11. **C**) and **D**) use p as a function of distance (d_i) (Eq. 1 and Eq. 2). **D**) displays the same results as (**C**) but with removed GVs and added black lines that help visualize fibers. White lines (with circular heads) represent the GVs. The colors and black lines show the generated fibers. For the coloring, 3D fiber angles $[0°, 360°]$ along the x, y and z axis are linearly mapped in the red, green and blue color channels. RA—right atrium, SVC/IVC—superior/inferior vena cava, RAA—RA appendage, GV—guide vector.

3.4 Modeling

Fibrosis and fibers were implemented into the geometry of heart #01's RA from Fig. 6, and some typical cursory simulations were run as examples to show the full pipeline. Figure 9 shows stills from four computer simulations of the RA, demonstrating the influence of fibrosis and fibers by varying their presence. The simulations were identically paced, and the stills were taken at the same in-simulation time of 3 s. The first case (Fig. 9A) is without fibrosis and fibers, in which a single reentry was formed. The second case excluded fibrosis but integrated fibers (Fig. 9B). As a result, we get unimpeded traversal of the reentry wave. In the third case, fibrosis was integrated but fibers were disabled, in which fibrosis led the reentry to dissipate. Finally, the fourth simulation with integrated fibrosis and fibers led to the reentry becoming patchy and chaotic (Fig. 9C).

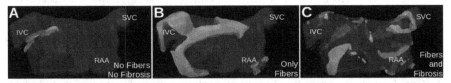

Fig. 9. Different computer simulations of the RA in the presence or absence of fibrosis and fibers at 3 s since the introduction of a stimulus. The atrial tissue is shown as red, while yellow represents depolarized wavefront regions. (**A**) no fibrosis and fibers, (**B**) only fibers, (**C**) both fibrosis and fibers. RA—right atrium, SVC/IVC—superior/inferior vena cava, RAA—RA appendage.

4   Conclusion

The proposed computational pipeline can potentially provide biomarkers and conduct computer simulations of the human heart to guide targeted ablation treatment. The CNN robustly segmented atria from LGE-MRIs. The standardized flattening allowed for efficient qualitative and quantitative inter-heart comparison, helping with finding possible patient-specific AF substrates. In addition, potential substrate and ablation targets can be investigated efficiently using computer modeling techniques. These are integrated with rule-based fibers and personalized atrial anatomy and fibrosis extracted from in-vivo scans.

Acknowledgements. We would like to acknowledge the Health Research Council of New Zealand, the Heart Foundation New Zealand, and US National Institutes of Health (HL115580, HL135109) for funding this research.

References

1. Nattel, S.: New ideas about atrial fibrillation 50 years on. Nature **415**(6868), 219–226 (2002). https://doi.org/10.1038/415219a
2. Xiong, Z., Fedorov, V.V., Fu, X., Cheng, E., Macleod, R., Zhao, J.: fully automatic left atrium segmentation from late gadolinium enhanced magnetic resonance imaging using a dual fully convolutional neural network. IEEE Trans. Med. Imaging **38**(2), 515–524 (2019). https://doi.org/10.1109/tmi.2018.2866845
3. Tobon-Gomez, C., et al.: Benchmark for algorithms segmenting the left atrium from 3D CT and MRI datasets. IEEE Trans. Med. Imaging **34**(7), 1460–1473 (2015). https://doi.org/10.1109/tmi.2015.2398818
4. Hirsh, B.J., Copeland-Halperin, R.S., Halperin, J.L.: Fibrotic atrial cardiomyopathy, atrial fibrillation, and thromboembolism. J. Am. Coll. Cardiol. **65**(20), 2239–2251 (2015). https://doi.org/10.1016/j.jacc.2015.03.557
5. Anter, E., Josephson, M.E.: Bipolar voltage amplitude: What does it really mean? Heart Rhythm **13**(1), 326–327 (2016). https://doi.org/10.1016/j.hrthm.2015.09.033

6. Zhao, J., et al.: Image-based model of atrial anatomy and electrical activation: A computational platform for investigating atrial arrhythmia. IEEE Trans. on Med. Imaging. **32**(1), 18–27 (2013). https://doi.org/10.1109/TMI.2012.2227776
7. Wang, Y., et al.: A robust computational framework for estimating 3D Bi-Atrial chamber wall thickness. Comput. Biol. Med. **114**, 103444 (2019). https://doi.org/10.1016/j.compbiomed.2019.103444
8. Xiong, Z., et al.: A global benchmark of algorithms for segmenting late gadolinium-enhanced cardiac MRI. Med. Image Anal. **67**(101832) (2021). https://doi.org/10.1016/j.media.2020.101832
9. Zhao, J., et al.: An image-based model of atrial muscular architecture. Circ. Arrhythmia Electrophysiol. **5**(2), 361–370 (2012). https://doi.org/10.1161/circep.111.967950
10. Mekkaoui, C., Reese, T.G., Jackowski, M.P., Bhat, H., Sosnovik, D.E.: Diffusion MRI in the heart. NMR Biomed. **30**(3), e3426 (2015). https://doi.org/10.1002/nbm.3426
11. Fastl, T.E., et al.: Personalized computational modeling of left atrial geometry and transmural myofiber architecture. Med. Image Anal. **47**, 180–190 (2018). https://doi.org/10.1016/j.media.2018.04.001
12. Roney, C.H., et al.: Constructing a Human Atrial Fibre Atlas. Ann. Biomed. Eng. **49**(1), 233–250 (2020). https://doi.org/10.1007/s10439-020-02525-w
13. Hoermann, J.M., Pfaller, M.R., Avena, L., Bertoglio, C., Wall, W.A.: Automatic mapping of atrial fiber orientations for patient-specific modeling of cardiac electromechanics using image registration. Int. J. Num. Methods Biomed. Eng. e3190 (2019). https://doi.org/10.1002/cnm.3190
14. Nuñez-Garcia, M., Bernardino, G., Doste, R., Zhao, J., Camara, O., Butakoff, C.: Standard quasi-conformal flattening of the right and left atria. In: Coudière, Y., Ozenne, V., Vigmond, E., Zemzemi, N. (eds.) FIMH 2019. LNCS, vol. 11504, pp. 85–93. Springer, Cham (2019). https://doi.org/10.1007/978-3-030-21949-9_10
15. Nunez-Garcia, M., et al.: Fast quasi-conformal regional flattening of the left atrium. IEEE Trans. Visual Comput. Graphics **26**(8), 2591–2602 (2020). https://doi.org/10.1109/tvcg.2020.2966702
16. Lorensen, W., Cline, H.: Marching cubes: A high resolution 3D surface construction algorithm. In: ACM SIGGRAPH Computer Graphics (2022). https://doi.org/10.1145/37402.37422
17. Dijkstra, E.W.: A note on two problems in connexion with graphs. Numer. Math. **1**(1), 269–271 (1959). https://doi.org/10.1007/bf01386390
18. Fix, E., Hodges, J.L.: Discriminatory analysis. nonparametric discrimination: Consistency properties on JSTOR (1951). https://www.jstor.org/stable/1403797
19. Ni, H., Whittaker, D.G., Wang, W., Giles, W.R., Narayan, S.M., Zhang, H.: Synergistic anti-arrhythmic effects in human atria with combined use of sodium blockers and acacetin. Front. Physiol. **8** (2017). https://doi.org/10.3389/fphys.2017.00946
20. Courtemanche, M., Ramirez, R., Nattel, S.: Ionic mechanisms underlying human atrial action potential properties: insights from a mathematical model, Am. J. Physiol.-Heart Circ. Physiol. (2020). https://doi.org/10.1152/ajpheart.1998.275.1.h301
21. Workman, A.: The contribution of ionic currents to changes in refractoriness of human atrial myocytes associated with chronic atrial fibrillation. Cardiovasc. Res. **52**(2), 226–235 (2001). https://doi.org/10.1016/s0008-6363(01)00380-7
22. Hansen, B.J. et al.: Unmasking arrhythmogenic hubs of reentry driving persistent atrial fibrillation for patient-specific treatment. J. Am. Heart Assoc. **9**(19) (2020). https://doi.org/10.1161/jaha.120.017789
23. Zhao, J., et al.: Three-dimensional integrated functional, structural, and computational mapping to define the structural 'fingerprints' of heart-specific atrial fibrillation drivers in human heart ex vivo. J. Am. Heart Assoc. **6**(8) (2017). https://doi.org/10.1161/jaha.117.005922

24. Virag, N., et al.: Study of atrial arrhythmias in a computer model based on magnetic resonance images of human atria. Chaos: Interdiscip. J. Nonlin. Sci. **12**(3), 754–763 (2002). https://doi.org/10.1063/1.1483935

25. McDowell, K.S., Zahid, S., Vadakkumpadan, F., Blauer, J., MacLeod, R.S., Trayanova, N.A.: Virtual electrophysiological study of atrial fibrillation in fibrotic remodeling. PLoS ONE **10**(2), e0117110 (2015). https://doi.org/10.1371/journal.pone.0117110

26. Pravdin, S.F., Epanchintsev, T.I., Panfilov, A.V.: Overdrive pacing of spiral waves in a model of human ventricular tissue. Sci. Rep. **10**(1) (2020). https://doi.org/10.1038/s41598-020-77314-5

27. Majumder, R., Mohamed Nazer, A.N., Panfilov, A.V., Bodenschatz, E., Wang, Y.: Electrophysiological characterization of human atria: The understated role of temperature. Front. Physiol. **12** (2021). https://doi.org/10.3389/fphys.2021.639149

28. Denes, P., Wu, D., Dhingra, R., Pietras, R.J., Rosen, K.M.: The effects of cycle length on cardiac refractory periods in man. Circulation **49**(1), 32–41 (Jan.1974). https://doi.org/10.1161/01.cir.49.1.32

29. Xiong, Z., et al.: Fully automatic 3D bi-atria segmentation from late gadolinium-enhanced MRIs using double convolutional neural networks. In: Lecture Notes in Computer Science, vol. 11395 LNCS, pp. 63–71 (2020)

3D Mitral Valve Surface Reconstruction from 3D TEE via Graph Neural Networks

Matthias Ivantsits[1(✉)], Boris Pfahringer[1], Markus Huellebrand[1,2],
Lars Walczak[1,2], Lennart Tautz[2], Olena Nemchyna[3], Serdar Akansel[3],
Jörg Kempfert[3], Simon Sündermann[1], and Anja Hennemuth[1,2,4]

[1] Charité – Universitätsmedizin Berlin, Augustenburger Pl. 1, 13353 Berlin, Germany
matthias.ivantsits@charite.de
[2] Fraunhofer MEVIS, Am Fallturm 1, 28359 Bremen, Germany
[3] German Heart Center Berlin, Augustenburger Pl. 1, 13353 Berlin, Germany
[4] DZHK (German Centre for Cardiovascular Research), Berlin, Germany

Abstract. Mitral valve insufficiency is a condition in which the valve
does not close properly, and blood leaks back into the atrium from the
ventricle. Valve assessment for surgery planning is typically performed
with 3D transesophageal echocardiography (TEE). The simulation of
the resulting valve dynamics can support selecting the most promising
surgery strategy. These simulations require an accurate reconstruction
of the open valve as a 3D surface model. 3D mitral valve reconstruction
from image data is challenging due to the fast-moving and thin valve
leaflets, which might appear blurred and covered by very few voxels
depending and spatio-temporal resolution. State-of-the-art voxel-based
CNN segmentation methods need an additional processing step to reconstruct a 3D surface from this voxel-based representation which can introduce unwanted artifacts. We propose an end-to-end deep-learning-based
method to reconstruct a 3D surface model of the mitral valve directly
from 3D TEE images. The suggested method consists of a CNN-based
voxel encoder and decoder inter-weaved with a graph neural network-
based (GNN) multi-resolution mesh decoder. This GNN samples feature
vectors from the CNN-decoder at different resolutions to deform a prototype mesh. The model was trained on 80 sparsely annotated 3D TEE
images ($1\,mm^3$ voxel resolution) of the valve during end-diastole. Each
time frame was annotated by two cardiovascular experts on nine planes
rotated around the axis through the apex of the left ventricle and the
center of the mitral valve. Our method's average bidirectional point-to-
point distance is 1.1 mm, outperforming the inter-observer point-to-point
distance of 1.8 mm.

Keywords: Mitral valve reconstruction · Deep learning · Graph
neural network

1 Introduction

The mitral valve regulates blood flow between the left atrium and left ventricle. It
consists of the anterior and posterior leaflet, separated by two indentations—the

O. Camara et al. (Eds.): STACOM 2022, LNCS 13593, pp. 330–339, 2022.
https://doi.org/10.1007/978-3-031-23443-9_30

anterior and posterior commissure (Fig. 1A). These two leaflets are attached to the annulus and are pulled down during the diastole via thin chords attached to the papillary muscles, which are connected to the left myocardium. During systole, the myocardium contracts, and the mitral valve closes (Fig. 1A). Mitral valve regurgitation is the most common mitral valve disease—with approximately 2% affected people worldwide [1]—causing blood flow from the ventricle into the atrium during systole. This regurgitation can be classified into three main types—type I, II, III—defined by Carpentier [2,3]. These mitral valve regurgitation types need different methods for repair, or replacement [4]. A common method to assess the pathology of the mitral valve is via 3D transesophageal echocardiography (TEE) [5,6]. Commercially available 3D TEE mitral valve tools typically analyze the mitral valve pathology or morphology in a closed state [7–9]. This allows the calculation of the annulus perimeter, annulus area, anterior and posterior leaflet area and length, and tenting height [10] (Fig. 1B). Assessing the mitral valve leaflets in a closed state is challenging due to their connection in the coaptation area. To enable the assessment of the valve leaflets and the characteristics of the closed valve, recent approaches reconstruct the valve model in an open state and propagate the valve model to the closed state [11,12]. The model can be changed according to different treatment strategies to select the most promising surgery procedure. This approach requires an initial reconstruction of the mitral valve from image data as a 3D surface. This reconstruction is challenging because the fast-moving and thin valve leaflets might be covered by only 1–2 voxels and blurred depending on the spatiotemporal sampling. State-of-the-art voxel-based CNN approaches [13–16] or interactive methods to delineate the valve [17,18] need an additional processing step to reconstruct a 3D surface from this voxel-based representation, which can introduce unwanted artifacts. We propose an end-to-end deep-learning-based method to reconstruct a 3D surface model of the open mitral valve directly from 3D TEE images.

2   Materials and Method

2.1   Dataset

The dataset was acquired utilizing the GE Vingmed Ultrasound Vivid E9 system in the German Heart Centre Berlin (DHZB). The dataset consists of 3D TEE images of 38 patients examined before and after surgery. 81 end-diastolic image volumes were used for model training and validation. The temporal resolution is between 12 and 65 ms in-plane resolution between 0.47 and 1.43 mm, and through-plane resolution between 0.33 and 0.98 mm. The image volumes were sparsely annotated by two cardiologists with four and 17 years of experience as a cardiologist, utilizing the method proposed by Tautz et al. [19]. The leaflets were annotated on nine planes rotated around the axis through the apex of the left ventricle and the center of the mitral valve (Fig. 1B). Additionally, the annulus and the orifice of the mitral valve were annotated. From these leaflet poly-lines,

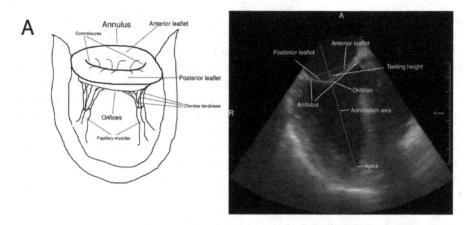

Fig. 1. An illustration of the relevant anatomical structures of the mitral valve and supporting structures (A). B illustrates relevant anatomical structures plus clinically important parameters.

a full 3D surface mesh was reconstructed for the final valve geometry, with each mesh consisting of 280 to 1200 nodes.

2.2   Method

Our proposed method is derived from the Voxel2Mesh [20] architecture, which is an extension to Pixel2Mesh [21] and Pixel2Mesh++ [22]. This method takes a 3D image in the voxel domain as input and produces the valve as a 3D surface model without any post-processing steps. The model consists of three main blocks—a voxel encoder and decoder and a mesh decoder—which are illustrated in Fig. 2. The voxel encoder and decoder are basic blocks, as proposed in the UNet paper [23], which compress the data into a latent space and reverse the process. The mesh decoder takes a prototype mesh as additional input, which is successively deformed. During each step of the mesh decoder, features around a fixed sphere of the mesh node positions are sampled and aggregated by 1D convolution. These latent node features plus the actual mesh node positions are concatenated and passed to a graph neural network [24], which produces a 3D offset vector for each mesh node. This delta is added to the previous mesh to produce an intermediate, low-resolution mesh. Next, this mesh is up-sampled by adding mesh nodes in each edge's center, creating four new faces. This process is repeated until the full resolution of the voxel representation is reached.

We extended the Voxel2Mesh architecture by adapting the mesh decoder to deform a topological annulus (we are referring to a topological/geometrical object here) instead of a sphere, which is, aside from a torus, the only feasible topology for a mitral valve surface reconstruction, due to the preservation of the prototype topology. The proposed method uses a uniform up-sampling strategy instead of the adaptive unpooling method suggested in the Voxel2Mesh

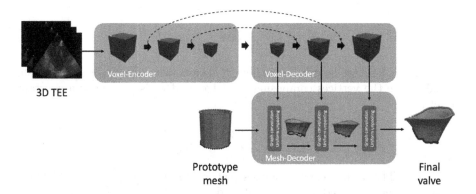

Fig. 2. An overview of the proposed architecture, including the three main blocks—a voxel-encoder, a voxel-decoder, and a mesh-decoder. The model takes a 3D TEE image as input and produces a 3D surface model of the mitral valve. The mesh-decoder samples features from the latent voxel-space and successively refines the prototype mesh at increasing resolutions. During the deformation, the orientation of the edge vertices is preserved, thus reconstructing the mitral valve with known annulus and orifice vertices. (Color figure online)

architecture [20], which does not preserve the topology of an annulus. Furthermore, we implement a dedicated loss function for mitral valve reconstruction. This loss function ensures an unfolded 3D surface and guarantees a valve model with labeled nodes, where the annulus and the orifices of the final mesh are known. This assignment of anatomical structures is illustrated in Fig. 2, where the annulus vertices (yellow) and orifice vertices (blue) are known in the prototype mesh, as well as in the final reconstructed mitral valve surface, resulting in a known orientation.

The final loss function L is a linear combination of multiple loss terms γ weighted with their respective hyperparameters λ at different resolutions R plus the binary cross-entropy loss (illustrated in Eq. 1.6). The set of loss terms Γ includes the chamfer loss γ_{MV}, defined as the average minimum distance between two point clouds. Furthermore, the same loss function is applied to the annulus and orifice vertices to ensure the correct orientation of the final 3D surface. Additionally, the loss contains constraints for the edge length γ_E, normal consistency of neighboring faces γ_{NC}, Laplacian smoothing objective γ_L, and a face normal loss γ_{MVN}, which ensures similar normal vectors of the predicted and ground-truth faces. These additional constraints ensure a smooth and non-self-folding solution to the 3D surface reconstruction.

$$\gamma_{MV} = \sum_{v_p \epsilon S_p} \min_{v_g \epsilon S_g} \|v_p - v_g\|_2^2 + \sum_{v_p \epsilon S_g} \min_{v_g \epsilon S_p} \|v_p - v_g\|_2^2 \qquad (1)$$

Equation 1: The chamfer loss function used for the mitral valve, the annulus, and the orifices. S_p and S_g denote the predicted and ground-truth mitral valve surface models, where v_p and v_g designate vertices sampled from the surface.

$$\gamma_E = \frac{1}{|E|} \sum_{(v_0,v_1)\epsilon E} (\|v_0 - v_1\|_2 - L_t)^2 \tag{2}$$

Equation 2: The edge length loss, where E denotes the set of edges of a mesh. v_0 and v_1 are the vertices composing the edge, and L_t is defined as a constant target length.

$$\gamma_{NC} = \frac{1}{|N_F|} \sum_{(n_0,n_1)\epsilon N_F} 1 - \frac{n_0 \cdot n_1}{\|n_0\| \cdot \|n_1\|} \tag{3}$$

Equation 3: The normal consistency loss, where N_F denotes the set of neighboring faces of the predicted mesh. n_0 and n_1 designate the normal vectors of these neighboring faces.

$$\gamma_L = \frac{1}{|V|} \sum_{v\epsilon V} \| \sum_{v'\epsilon N(v)} v - v'\| \tag{4}$$

Equation 4: The Laplacian smoothing constraint, where V denotes the set of vertices of the predicted mitral valve surface. $N(v)$ depicts the set of neighboring vertices of vertex v.

$$\gamma_{MVN} = \frac{1}{|F_p|} \sum_{f_p\epsilon F_p} 1 - \frac{n_p \cdot n_g}{\|n_p\| \cdot \|n_g\|} \tag{5}$$

Equation 5: The face normal loss, where F_p depicts the set of faces of the predicted mitral valve. n_p denotes the normal vector of face f_p and n_g the normal vector of the closest face f_g to the predicted face f_p.

$$L = \lambda_{CE}\gamma_{CE} + \sum_{\gamma\epsilon\Gamma} \sum_{r\epsilon R} \lambda_{\gamma r}\gamma_r \tag{6}$$

Equation 6: The loss function of the proposed method, including terms for the cross-entropy loss γ_{CE}, the chamfer loss of the mitral valve γ_{MV}, the edge length loss γ_E, the normal consistency loss γ_{NC}, the Laplacian smoothing objective γ_L, and a face normal loss γ_{MVN}. Additionally, each loss term is weighted by a hyperparameter λ.

3   Results

We performed all experiments on an Intel(R) Core(TM) i7-8700K CPU @ 3.70 GHz with 16 GBs RAM and an Nvidia RTX 2080 Ti GPU with 11 GB memory. 63 3D TEE volumes were used for model training, and the remaining 18 for model testing. We performed a 5-fold cross-validation (CV) to find optimal hyperparameters for the number of down- and up-sampling steps R and the loss weights λ. Before model training and testing, all images were re-sampled to an isotropic voxel-spacing of $1\,\mathrm{mm}^3$. We performed random affine, blur, and Gaussian noise augmentation during model training. Furthermore, a grid search on the loss hyperparameters was performed, combined with visual examinations, to avoid folded surface reconstructions. The final loss hyperparameters were uniformly weighted across all resolutions and weighted with: $\lambda_{MV} = 1$, $\lambda_A = 1$, $\lambda_O = 1$, $\lambda_{MVN} = 0.2$, $\lambda_L = 0.2$, $\lambda_{NC} = 0.2$, $\lambda_E = 0.2$, and $\lambda_{CE} = 1$.

We performed an interobserver variability analysis for the 81 3D volumes annotated by both experts. Figure 3 illustrates the average and 95th-percentile bidirectional point-to-point distance of the annotations to the reconstructed mitral valve mesh. We observe an average distance of 1.1 mm and 95th-percentile of 2.13 mm outperforming the inter-observer metrics of 1.86 mm and 3.82 mm. This figure highlights the average and 95th-percentile distance of the annulus and orifices, with an average distance of 1.29 mm and an average distance of the orifices of 1.67 mm, compared to the inter-observer metrics of 2.45 mm and 3.29 mm. The 95th-percentile results in 2.23 mm for the annulus and 3.19 mm for the orifices compared to the experts' variability of 4.06 mm and 6.71 mm.

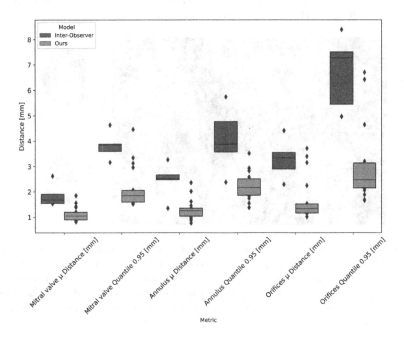

Fig. 3. An illustration of the bidirectional inter-observer and surface reconstruction metrics. This comparison includes the average and 95th-percentile point-to-point distance of the reconstructed valve to the ground truth and the distances for the annulus and orifices.

Additionally, we performed a linear regression on the target and predicted mitral valve area plus the maximum annulus diameter to identify any bias in the proposed model. We observe an R^2 value of 0.91 for the mitral valve area and 0.88 for the annulus diameter. The average difference in the mitral valve area

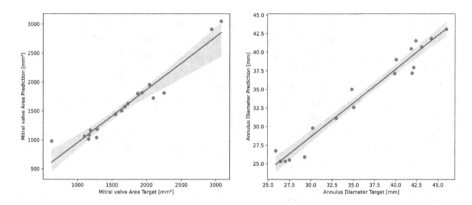

Fig. 4. An illustration of the linear regression of the mitral valve surface area and the maximum annulus diameter. The linear regression of the surface area results in an R^2 value of 0.91 and a bias of $-103\,\text{mm}^2$. The linear regression of the maximum annulus diameter results in an R^2 value of 0.88 and a bias of $-1.9\,\text{mm}$.

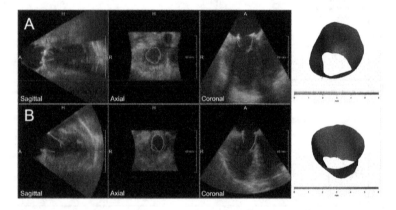

Fig. 5. An illustration of two 3D surface mitral valve reconstructions produced by our model. The orange contours represent the reconstructed mitral valves produced by the proposed method, and the red contours the reconstruction created by the cardiologists. Row A depicts a result with an average point-to-point distance of 0.37 mm. Row B illustrates a reconstruction with an average point-to-point distance of 1.23 mm. (Color figure online)

is $-103\,\text{mm}^2$. The average difference of the maximum annulus diameter results $-1.9\,\text{mm}$.

Figure 5 illustrates two 3D surface reconstruction results of the proposed method. Row A illustrates a mitral valve reconstruction that aligns very well with the reconstruction by the cardiologist. The average bidirectional point-to-point distance of this result is 0.37 mm. The sagittal, coronal, and axial planes show almost perfect alignment. Row B depicts an example with an average point-to-point distance of 1.23 mm, where the sagittal and coronal planes reveal large deviations towards the orifices. The axial plane illustrates a large discrepancy in the annulus.

4 Discussion and Conclusion

We have presented an end-to-end deep-learning method to reconstruct a 3D surface model of the mitral valve from 3D TEE images without post-processing. By directly reconstructing the 3D surface, we avoid artifacts introduced in conventional approaches. The inter-observer study performed and illustrated in Fig. 3 exhibits an average bidirectional point-to-point distance of 1.1 mm, surpassing the average distance by two cardiologists of 1.86 mm. The analysis shown in Fig. 4 reveals a bias with an under-estimation of the reconstructed surface area of $-103\,\text{mm}^2$ and $-1.9\,\text{mm}$ of the maximum annulus diameter. Furthermore, Fig. 5 B illustrates an reconstruction with an average point-to-point distance of 1.23 mm. The sagittal and coronal views reveal large discrepancies towards the orifices of the annulus.

To improve the proposed method, it is very likely that an increase in sample size can help better reconstruct the 3D surface model of the mitral valve. The inter-observer variability analysis highlights that an accurate annotation of the mitral valve from 3D TEE images is complex, especially the distinction between the chordae tendineae and the leaflets. Since acquiring ground truths for the valve is difficult, an extension of the training set with synthetic data might be advantageous. Future extensions to this method might also include a vertex classification of the reconstructed 3D surface. This extension will lead to an extended semi-registered model, where the individual segments of the leaflets are learned end-to-end.

References

1. Douedi, S., Douedi, H.: Mitral regurgitation. In: StatPearls. StatPearls Publishing, Treasure Island (FL) (2022). https://www.ncbi.nlm.nih.gov/books/NBK553135/. Accessed 8 Nov 2021
2. Apostolidou, E., Maslow, A.D., Poppas, A.: Primary mitral valve regurgitation: update and review. Glob. Cardiol. Sci. Pract. **2017**(1), e201703 (2017). https://doi.org/10.21542/gcsp.2017.3
3. Lancellotti, P., Moura, L., Pierard, L.A., et al.: European association of echocardiography recommendations for the assessment of valvular regurgitation. Part 2: mitral and tricuspid regurgitation (native valve disease). Eur. J. Echocardiogr. **11**(4), 307–332 (2010). https://doi.org/10.1093/ejechocard/jeq031
4. Allen, N., O'Sullivan, K., Jones, J.M.: The most influential papers in mitral valve surgery; a bibliometric analysis. J. Cardiothorac. Surg. **15**, 175 (2020). https://doi.org/10.1186/s13019-020-01214-y
5. Nishimura, R.A., Otto, C.M., Bonow, R.O., et al.: 2017 AHA/ACC focused update of the 2014 AHA/ACC guideline for the management of patients with valvular heart disease: a report of the American college of cardiology/American heart association task force on clinical practice guidelines. Circulation **135**(25), e1159–e1195 (2017). https://doi.org/10.1161/CIR.0000000000000503

6. de Groot-de Laat, L.E., McGhie, J., Ren, B., Frowijn, R., Oei, F.B., Geleijnse, M.L.: A modified echocardiographic classification of mitral valve regurgitation mechanism: the role of three-dimensional echocardiography. J. Cardiovasc. Imaging **27**(3), 187–199 (2019). https://doi.org/10.4250/jcvi.2019.27.e29
7. Sandra, L., et al.: A new diagnostic tool for effective regurgitant orifice quantification in mitral regurgitation. Echocardiography **35**(11), 1812–1817 (2018). https://doi.org/10.1111/echo.14114
8. Veronesi, F., Lie, G., Rabben, S.: 4D auto MVQ. GE Healthcare (2017)
9. Tomtec. 4D MV-Assessment (2018)
10. Ryan, L., et al.: Quantification and localization of mitral valve tenting in ischemic mitral regurgitation using real-time three-dimensional echocardiography. Eur. J. Cardiothorac. Surg. **31**(5), 839–844 (2007). https://doi.org/10.1016/j.ejcts.2007.01.050
11. Walczak, L., et al.: Using position-based dynamics for simulating mitral valve closure and repair procedures. Comput. Graph. Forum **41**, 270–287 (2022). https://doi.org/10.1111/cgf.14434
12. Walczak, L., et al.: Interactive editing of virtual chordae tendineae for the simulation of the mitral valve in a decision support system. Int. J. Comput. Assist. Radiol. Surg. **16**(1), 125–132 (2020). https://doi.org/10.1007/s11548-020-02230-y
13. Carnahan, P., Moore, J., Bainbridge, D., Eskandari, M., Chen, E.C.S., Peters, T.M.: DeepMitral: fully automatic 3D echocardiography segmentation for patient specific mitral valve modelling. In: de Bruijne, M., et al. (eds.) MICCAI 2021. LNCS, vol. 12905, pp. 459–468. Springer, Cham (2021). https://doi.org/10.1007/978-3-030-87240-3_44
14. Costa, E., et al.: Mitral valve leaflets segmentation in echocardiography using convolutional neural networks. In: 2019 IEEE 6th Portuguese Meeting on Bioengineering (ENBENG), pp. 1–4 (2019). https://doi.org/10.1109/ENBENG.2019.8692573
15. Kerfoot, E., Clough, J., Oksuz, I., Lee, J., King, A.P., Schnabel, J.A.: Left-ventricle quantification using residual U-net. In: Pop, M., et al. (eds.) STACOM 2018. LNCS, vol. 11395, pp. 371–380. Springer, Cham (2019). https://doi.org/10.1007/978-3-030-12029-0_40
16. Liu, S., et al.: 3D anisotropic hybrid network: transferring convolutional features from 2D images to 3D anisotropic volumes. In: Frangi, A.F., Schnabel, J.A., Davatzikos, C., Alberola-López, C., Fichtinger, G. (eds.) MICCAI 2018. LNCS, vol. 11071, pp. 851–858. Springer, Cham (2018). https://doi.org/10.1007/978-3-030-00934-2_94
17. Carnahan, P., et al.: Interactive-automatic segmentation and modelling of the mitral valve. In: Coudière, Y., Ozenne, V., Vigmond, E., Zemzemi, N. (eds.) FIMH 2019. LNCS, vol. 11504, pp. 397–404. Springer, Cham (2019). https://doi.org/10.1007/978-3-030-21949-9_43
18. Jassar, A.S., Brinster, C.J., Vergnat, M., et al.: Quantitative mitral valve modeling using real-time three-dimensional echocardiography: technique and repeatability. Ann. Thorac. Surg. **91**(1), 165–171 (2011). https://doi.org/10.1016/j.athoracsur.2010.10.034
19. Tautz, L., et al.: Extraction of open-state mitral valve geometry from CT volumes. Int. J. Comput. Assist. Radiol. Surg. **13**(11), 1741–1754 (2018). https://doi.org/10.1007/s11548-018-1831-6
20. Wickramasinghe, U., Remelli, E., Knott, G., Fua, P.: Voxel2Mesh: 3D mesh model generation from volumetric data. In: Martel, A.L., et al. (eds.) MICCAI 2020. LNCS, vol. 12264, pp. 299–308. Springer, Cham (2020). https://doi.org/10.1007/978-3-030-59719-1_30

21. Wang N., Zhang Y., Li Z., Fu Y., Liu W., Jiang Y.: Pixel2Mesh: generating 3D mesh models from single RGB images. In: European Conference on Computer Vision (2018)
22. Wen C., Zhang Y., Li Z., Fu Y.: Pixel2Mesh++: multi-view 3D mesh generation via deformation. In: International Conference on Computer Vision (2019)
23. Ronneberger, O., Fischer, P., Brox, T.: U-net: convolutional networks for biomedical image segmentation. In: Navab, N., Hornegger, J., Wells, W.M., Frangi, A.F. (eds.) MICCAI 2015. LNCS, vol. 9351, pp. 234–241. Springer, Cham (2015). https://doi.org/10.1007/978-3-319-24574-4_28
24. Kipf, T., Welling, M.: Semi-supervised classification with graph convolutional networks. In: 5th International Conference on Learning Representations, ICLR 2017, Toulon, France, 24–26 April 2017, Conference Track Proceedings (2017)

Efficient MRI Reconstruction with Reinforcement Learning for Automatic Acquisition Stopping

Ruru Xu[(⊠)] and Ilkay Oksuz

Computer Engineering Department, Istanbul Technical University, Istanbul, Turkey
xu21@itu.edu.tr

Abstract. Magnetic resonance imaging (MRI) is accelerated through subsampling of the associated Fourier domain in current clinical practice. The decisions on subsampling strategies and acceleration factors are provided heuristically before the acquisition. In this paper, we propose a reinforcement learning strategy for automatically deciding a subsampling strategy and acceleration factor for cardiac image acquisition. We build an environment that has a set of actions, including which k-space line to select next and when to stop the acquisition. We propose to use a reward term that penalizes extra line acquisitions and favours improved image quality. Experiments on cardiac MRI with different weightings of the reward function have shown that our method can achieve better image quality results without increasing the acquisition time and can automatically stop the k-space sampling process.

Keywords: Cardiac MRI · Reinforcement learning · Reconstruction

1 Introduction

MRI is safe, painless, and radiation-free, which is especially important for scanning and imaging internal organs and other soft tissues. However, MRI acquisition is inherently slow, and subsampling is used to accelerate the MR acquisition. This in turn reduces the image quality, and MRI reconstruction is needed to remove artifacts and obtain high-quality images. Deep learning-based methods have been the standard way to solve the inverse problem of image reconstruction [7,9,14]. Even though image reconstruction methods are capable of generating high-quality images with reduced scan times, they require fixed acceleration rates and pre-defined subsampling patterns. In the case of cardiac MRI, the heart's movement makes sampling more difficult, which requires longer scanning time, which is unbearable for patients. In order to reduce the scanning time, currently undersampled k-space is used [13], and a variety of learning-based subsampling strategies have been proposed. In this paper, we propose a reinforcement learning strategy that can decide to stop the acquisition.

We propose a sampling policy based on reinforcement learning for k-space that can apply different sampling schemes for different slices and automatically

O. Camara et al. (Eds.): STACOM 2022, LNCS 13593, pp. 340–348, 2022.
https://doi.org/10.1007/978-3-031-23443-9_31

stop when a satisfactory image quality is reached. We build on the work of Baker et al. [2] and include a mechanism to stop the MRI acquisition automatically when a satisfactory image quality is reached. This way, a fixed acceleration rate definition before the acquisition is not necessary, and satisfactory image quality is the only guiding mechanism for increasing the acquisition time.

In the Cartesian sampling scenario, a new k-space line is sampled in each step and the reinforcement learning process will determine which line needs to be sampled in the next step [2]. Without a fixed acceleration rate, k-space can be continuously sampled without additional image quality gain from the reconstruction method. We aim to use the reconstructed image in the pipeline for guidance on when the acquisition of new k-space lines is redundant. To be able to stop the acquisition automatically, we designed a novel reward which penalizes the acquisition of new k-space lines and encourages improved image quality. We experiment with a variety of weights between the terms to see the influence on automatic stopping and image quality. We validated our findings on the OCMR [4] dataset, which is a public dataset for cardiac image reconstruction.

Our contributions are two-fold. First, we propose a reinforcement learning strategy for MRI subsampling that can automatically stop sampling k-space at some point. Second, we validate the subsampling strategy using k-space under-sampling on cardiac MRI for the first time in the literature to the best of our knowledge.

2   Related Works

Cardiac MRI Reconstruction: Cardiac MRI reconstruction with deep learning and designing subsampling patterns has been an active area of research. To improve the quality of reconstructed images, the most common method is to train a CNN reconstruction network from a low-quality zero-filled image to a high-quality reconstructed image. However, image domain to image domain reconstruction, like in UNet-based networks [5], does not make full use of the information in the k-space frequency domain. Although the cascade-based networks [14] incorporate k-space information, they only use a fixed initialization undersampled k-space and subsequent operations do not utilize the original k-space information.

Optimized for Kspace Undersampling: Learning-based k-space sampling is an active area of research [7,8,15,16,18,20]. The usual method is to fix some frequency information in the middle of the k-space and then randomize sampling in the remaining region according to a fixed acceleration ratio set manually. Unfortunately, all slices can only use fixed k-space acceleration ratios [9,10,14, 16], which cannot be flexibly applied to different patients. In addition, random sampling of k-space cannot obtain the most valuable information in k-space. Gozcu et al. proposed optimization of k-space measurement trajectory based on compressed sensing [6], but it uses fixed trajectories in its inference. Dynamic acquisition trajectory optimization of k-space based on deep learning has been proposed [11,19]. Bahadir et al. proposed a scheme to jointly reconstruct MRI and learn subsampling mask [1,3]. Reinforcement learning has emerged as a viable option to design optimal subsampling patterns [2,12,17], but all methods require manual intervention or a fixed acceleration rate to stop sampling.

3   Methods

Our method is divided into two steps: the first step is to train a reconstruction network; the second step is to freeze the parameters of the reconstruction network and train a policy network to guide the sampling of k-space.

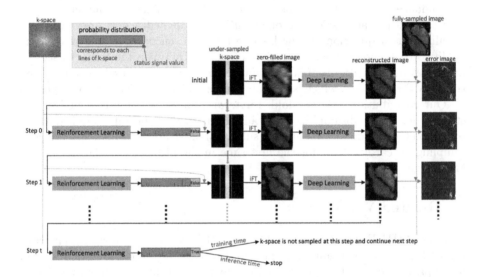

Fig. 1. A diagram shows the entire pipeline of the reconstruction process. With a zero-filled image as the input, the reconstruction network can output a high-quality reconstructed image. The output of the policy network is a vector of 257 values, where the first 256 values represent the probability that the next k-space line will be sampled and the last value represents the state signal. If the status signal value is false, the sampling of k-space and the next step of reconstruction work will continue. If the state signal value is true, the next step of sampling is continued during the training time while the reconstruction of this slice is stopped during the inference time.

3.1   Reconstruction Network

The input of this reconstruction network is zero-filled images, which is obtained by the inverse Fourier transform of under-sampled k-space, and the output of the network is the reconstructed image. As shown in Fig. 1 in the reconstruction stage, only the process labeled "initial" in the first row is carried out, and reinforcement learning is not involved. We use the standard 64-channel U-Net as our reconstruction network architecture and L2 loss for training.

3.2   Policy Network

The model obtained in the reconstruction network stage is used as our reconstruction model, but in the policy network stage, the reconstruction network only predicts and does not participate in updating parameters. Policy network is used to guide k-space dynamic sampling. Our goal is to train a policy network that can automatically sample and automatically stop sampling k-space.

The input of the policy network is the single-channel reconstructed image obtained in the previous stage, and the output is a vector containing $l + 1$ values, of which $l = 256$ corresponds to lines of frequency information in k-space, and the additional value is the status signal value. We call the vector of the policy network output the probability distribution, and then, referring to the probability value, we're going to randomly sample a position as the next line to sample k-space. In the output section of the policy network shown in Fig. 1, in each step, if the state signal value is not selected, we mark it as false. That is, if one of the other 256 values is selected, we will sample a line at the corresponding position in k-space. During the training time, if the state signal value is selected, we continue the reconstruction process. During the inference time, if the state signal value is selected, we stop the entire reconstruction work.

When training policy network, our reward is represented by Eq. 1, and it consists of two parts. We will calculate the SSIM error of the reconstructed image and the fully-sampled image, and at the same time, we will calculate the error of the mask before and after reconstruction. λ is a hyperparameter to balance the penalty of new acquisitions with the reward of increased image quality. λ can be tuned to choose between image quality and acquisition time automatically. In the continuous sampling process, the current step samples one more line of k-space data than the previous step, so the value of $\Delta mask$ is equal to 1. When we stop sampling new k-space lines, the current step is the same as the under-sampled k-space of the previous step, so the value of $\Delta mask$ is equal to 0 and rewards will become $rewards = \Delta SSIM$.

$$rewards = \Delta SSIM - \lambda \Delta mask \qquad (1)$$

where $\Delta SSIM = new\_SSIM - old\_SSIM$. We denoted the SSIM in the previous step as old_SSIM, and the SSIM in the current step is new_SSIM. Where $\Delta mask = new\_mask - old\_mask$. The k-space sampling mask in the current step is denoted as new_mask, Similarly, the mask in the previous step is denoted as old_mask.

4   Experimental Results

Datasets: In our work, we used the OCMR [4] cardiac data set, which contains multi-coil k-space data. It includes 74 fully sampled and 212 prospectively under-sampled cardiac datasets. In this work, we only used 74 fully-sampled 4D cardiac MRI files in .h5 format. We first convert the .h5 file of 4D information ([kx, ky, phase, slice]) into the .mat file of single coil 3D information ([kx, ky, phase]), and then we obtain 2D information ([kx, ky]) data from the .mat file containing 3D information. We randomly divide it into training sets (1243 2D slices), validation sets (409 2D slices), and testing sets (783 2D slices), respectively. In the experiment of this paper, the data is processed in the form of slices.

Reconstruction Network: We use 64-channel U-net as the baseline for reconstruction, similar to [2]. First we crop the size of the data set to be loaded to 256×256 pixels, corresponding to 256 lines in k-space, and then we will sample k-space line by line. The input to the network is zero-filled images, which is the inverse Fourier transform of the under-sampled k-space. The output of the network is a single channel of high-quality reconstructed images. We set batch_size to be 64, the learning_rate to be 0.001, the loss function to be L2, and the epoch to be 50.

Policy Network: At training time, we first fixed 32 k-space lines in the middle position as the under-sampled k-space, then obtained the zero-filled image through the inverse Fourier transform, and then reconstructed images were obtained after the reconstruction process. We take the reconstructed image and the corresponding undersampled k-space as the initial state of the policy network. When the state signal value is selected, k-space will not be sampled in this step, and continue to the next step. At inference time, stop the prediction when the state signal value is first selected. We conducted the experiment by setting the values of λ to 0, 0.0005, 0.0002 We also used the pg_mri model [2] to conduct an experiment on our OCMR cardiac data set with 256 output values of the policy network. We trained 50 epochs on the V100 GPU using greedy learning, which means updating network parameters after every sampling. The batch_size was set to 128.

Evaluation: At inference time, the steps stopped for each slice are when the signal value is selected, so the number of slices under each step is different. We utilize SSIM and PSNR (the average value of each reconstruction step) to show the trade-off between acquisition time and image quality.

4.1   Results Analysis

Automatically Stop Dynamic Sampling Analysis: The predicted results on 783 testing slices for SSIM and PSNR are shown in Fig. 2. 'Stop' means stopping the prediction process when the state signal value is first selected. 'nostop' means that the prediction of subsequent steps will continue after the state signal value is first selected. Our initial state is masked_kspace with 32 fixed lines in the middle, and then we get the initial SSIM/PSNR. Then a new k-space line is sampled at each step and added to masked_kspace. Meanwhile, calculate the average SSIM/PSNR of all slices at each step. We compare three pairs of curves that have the same λ value but contain the labels "stop" and "nostop" respectively. The 'stop' curve and the 'nostop' curve overlap only at the very beginning, because they have the same initialization state. The three curves in the label with the keyword "nostop" tend to be smooth, indicating

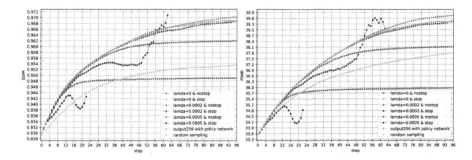

Fig. 2. The mean SSIM and PSNR of all slices tested at each step when the policy network has 257 output values with different λ. The horizontal axis is steps. We also did a comparative experiment: the same reconstruction network U-Net and different sampling methods The left picture is the average SSIM of each step. The right picture is the average PSNR of each step.

that after the first selection of the state signal, if the reconstruction of the slice is not stopped, most of the subsequent steps will continue to select the state signal, so their mask is the same, and the slight increase in PSNR is due to the effect of the reconstruction network. Focusing on the three lines in the label with the "stop" keyword. We found that the smaller the value of λ, the larger the average SSIM, which also requires more steps. There is no significant smoothing of "lamda = 0 & stop" shown in the Fig. 2 because, for most slices, the signal vectors are first selected after step 96. In addition, we used the policy network to guided sampling and random sampling of the next k-space respectively for comparison. We found that both SSIM and PSNR obtained by using the policy network were larger than those obtained by random sampling. Our conclusion is that our method can automatically stop sampling without losing too much SSIM and PSNR, and a larger λ value can cause the slice's reconstruction process to stop early.

Table 1. Comparison of test time for the six curves in Fig. 2. We calculated the average time saved per slice.

λ	0		0.0002		0.0005	
Stop?	No	Yes	No	Yes	No	Yes
Test time	688.61 s	655.34 s	686.09 s	256.84 s	683.54 s	101.21 s
Total time saved	33.27		429.25		582.33	
Average time saved per slice	0.0425 s		0.5482 s		0.7437 s	

Time Analysis: On 783 testing sets, we calculate the time spent in the cases of automatic dynamic stop sampling and 2x (96 steps) acceleration sampling, respectively. Table 1 shows that, compared with the fixed 2x acceleration sampling, our automatic dynamic stopping sampling method can save a lot of time, and the difference in acceleration time depends on the size of λ value.

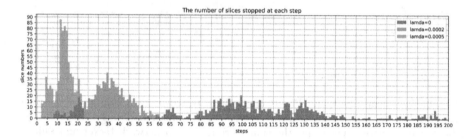

Fig. 3. The horizontal axis represents the steps when the prediction process is stopped. The vertical axis represents the number of slices that were stopped under this step. 0, 0.0002, 0.0005, they are the λ of $rewards = \Delta SSIM - \lambda \Delta mask$.

Validity of Status Signal Value: The 783 slices were tested and the steps were recorded when the state signal value was selected for the first time, resulting in Fig. 3. For each slice, the horizontal axis represents the step when sampling stops. The vertical axis represents the number of slices when sampling is stopped at this step. We can see that the smaller the value of λ, the more crucial it is for the acquisition steps. Higher values of λ will generate faster acquisition automatically. We also compared our policy network with 257 actions for $\lambda = 0$, which tend to stop at very late stages due to the lack of penalty for additional acquisitions. This further proves that the status signal values work, and they can stop the k-space dynamic sampling process automatically.

Reconstructed Image Quality Analysis: The comparative experimental results of our method and pg_mri are shown in Fig. 4. When $\lambda = 0.0002$, the acceleration of automatic stop sampling is 3.2. So we did an experiment on pg_mri when the acceleration was 3.2. The SSIM and PSNR for both models were on par. Similarly, when $\lambda = 0.0005$, the acceleration when the reconstruction process is stopped automatically is 5.95. We conducted experiments on pg_mri when the acceleration of k-space under-sampling was 5.95. By comparing their SSIM and PSNR, we also found that they were very similar. In summary, compared with pg_mri, our method has no loss of accuracy, and it can automatically stop dynamic sampling.

5   Discussion and Conclusions

We designed a policy network with a state signal value that can stop sampling k-space at a certain point and then consequently stop automatically. We added a penalty term to the reward and conducted experiments by setting different

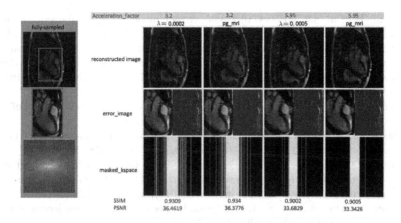

Fig. 4. The first row shows the reconstructed image. In the second row, the heart region is enlarged, and we calculate the error image between the reconstructed heart region and the fully sampled heart region. The third row shows the under-sampled k-space. The middle 32 lines are fixed, and the other lines are selected based on the policy network.

penalty item coefficients. The experimental results showed the capability of our method to balance the trade-off between acquisition time and image quality automatically.

We also ran the experiment without stopping sampling after selecting the state signal value for the first time. We found that the following steps would always select the state signal value, which proved that the current under-sampled k-space was already optimal. In future, we aim to design specific reward functions for downstream tasks (e.g. segmentation) to optimize the subsampling and stopping criteria.

Acknowledgments. This paper has been produced benefiting from the 2232 International Fellowship for Outstanding Researchers Program of TUBITAK (Project No: 118C353). However, the entire responsibility of the publication/paper belongs to the owner of the paper. The financial support received from TUBITAK does not mean that the content of the publication is approved in a scientific sense by TUBITAK.

References

1. Bahadir, C.D., Dalca, A.V., Sabuncu, M.R.: Learning-based optimization of the under-sampling pattern in MRI. In: Chung, A.C.S., Gee, J.C., Yushkevich, P.A., Bao, S. (eds.) IPMI 2019. LNCS, vol. 11492, pp. 780–792. Springer, Cham (2019). https://doi.org/10.1007/978-3-030-20351-1_61
2. Bakker, T., Hoof, H., Welling, M.: Experimental design for MRI by greedy policy search. Adv. Neural. Inf. Process. Syst. **33**, 18954–18966 (2020)
3. Bahadir, C., Wang, A., Dalca, A., Sabuncu, M.: Deep-learning-based optimization of the under-sampling pattern in MRI. IEEE Trans. Comput. Imaging **6**, 1139–1152 (2020)

4. Chen, C., et al.: OCMR (v1. 0)-Open-access multi-coil k-space dataset for cardiovascular magnetic resonance imaging. ArXiv Preprint ArXiv:2008.03410 (2020)
5. Ghodrati, V., et al.: MR image reconstruction using deep learning: evaluation of network structure and loss functions. Quant. Imaging Med. Surg. **9**, 1516 (2019)
6. Gözcü, B., et al.: Learning-based compressive MRI. IEEE Trans. Med. Imaging **37**, 1394–1406 (2018)
7. Hammernik, K., et al.: Learning a variational network for reconstruction of accelerated MRI data. Magn. Reson. Med. **79**, 3055–3071 (2018)
8. Haldar, J., Kim, D.: OEDIPUS: an experiment design framework for sparsity-constrained MRI. IEEE Trans. Med. Imaging **38**, 1545–1558 (2019)
9. Huang, Q., Yang, D., Wu, P., Qu, H., Yi, J., Metaxas, D.: MRI reconstruction via cascaded channel-wise attention network. In: 2019 IEEE 16th International Symposium on Biomedical Imaging (ISBI 2019), pp. 1622–1626 (2019)
10. Hyun, C., Kim, H., Lee, S., Lee, S., Seo, J.: Deep learning for undersampled MRI reconstruction. Phys. Med. Biol. **63**, 135007 (2018)
11. Jin, K., Unser, M., Yi, K.: Self-supervised deep active accelerated MRI. ArXiv Preprint ArXiv:1901.04547 (2019)
12. Pineda, L., Basu, S., Romero, A., Calandra, R., Drozdzal, M.: Active MR k-space sampling with reinforcement learning. In: Martel, A.L., et al. (eds.) MICCAI 2020. LNCS, vol. 12262, pp. 23–33. Springer, Cham (2020). https://doi.org/10.1007/978-3-030-59713-9_3
13. Qin, C., Schlemper, J., Caballero, J., Price, A., Hajnal, J., Rueckert, D.: Convolutional recurrent neural networks for dynamic MR image reconstruction. IEEE Trans. Med. Imaging **38**, 280–290 (2018)
14. Schlemper, J., Caballero, J., Hajnal, J.V., Price, A., Rueckert, D.: A deep cascade of convolutional neural networks for MR image reconstruction. In: Niethammer, M., et al. (eds.) IPMI 2017. LNCS, vol. 10265, pp. 647–658. Springer, Cham (2017). https://doi.org/10.1007/978-3-319-59050-9_51
15. Seeger, M., Nickisch, H., Pohmann, R., Schölkopf, B.: Optimization of k-space trajectories for compressed sensing by Bayesian experimental design. Magn. Reson. Med.: Off. J. Int. Soc. Magn. Reson. Med. **63**, 116–126 (2010)
16. Wang, S., et al.: Accelerating magnetic resonance imaging via deep learning. In: 2016 IEEE 13th International Symposium on Biomedical Imaging (ISBI), pp. 514–517 (2016)
17. Yin, T., Wu, Z., Sun, H., Dalca, A., Yue, Y., Bouman, K.: End-to-end sequential sampling and reconstruction for MR imaging. ArXiv Preprint ArXiv:2105.06460 (2021)
18. Zhang, P., Wang, F., Xu, W., Li, Yu.: Multi-channel generative adversarial network for parallel magnetic resonance image reconstruction in k-space. In: Frangi, A.F., Schnabel, J.A., Davatzikos, C., Alberola-López, C., Fichtinger, G. (eds.) MICCAI 2018. LNCS, vol. 11070, pp. 180–188. Springer, Cham (2018). https://doi.org/10.1007/978-3-030-00928-1_21
19. Zhang, Z., Romero, A., Muckley, M., Vincent, P., Yang, L., Drozdzal, M.: Reducing uncertainty in undersampled MRI reconstruction with active acquisition. In: Proceedings of the IEEE/CVF Conference on Computer Vision and Pattern Recognition, pp. 2049–2058 (2019)
20. Zijlstra, F., Viergever, M., Seevinck, P.: Evaluation of variable density and data-driven k-space undersampling for compressed sensing magnetic resonance imaging. Invest. Radiol. **51**, 410–419 (2016)

Unsupervised Cardiac Segmentation Utilizing Synthesized Images from Anatomical Labels

Sihan Wang[1], Fuping Wu[1], Lei Li[2], Zheyao Gao[1], Byung-Woo Hong[3], and Xiahai Zhuang[1($\boxtimes$)]

[1] School of Data Science, Fudan University, Shanghai, China
zxh@fudan.edu.cn
[2] Institute of Biomedical Engineering, University of Oxford, Oxford, UK
[3] Computer Science Department, Chung-Ang University, Seoul, Korea
https://www.sdspeople.fudan.edu.cn/zhuangxiahai/

Abstract. Cardiac segmentation is in great demand for clinical practice. Due to the enormous labor of manual delineation, unsupervised segmentation is desired. The ill-posed optimization problem of this task is inherently challenging, requiring well-designed constraints. In this work, we propose an unsupervised framework for multi-class segmentation with both intensity and shape constraints. Firstly, we extend a conventional non-convex energy function as an intensity constraint and implement it with U-Net. For shape constraint, synthetic images are generated from anatomical labels via image-to-image translation, as shape supervision for the segmentation network. Moreover, augmentation invariance is applied to facilitate the segmentation network to learn the latent features in terms of shape. We evaluated the proposed framework using the public datasets from MICCAI2019 MSCMR Challenge, and achieved promising results on cardiac MRIs with Dice scores of 0.5737, 0.7796, and 0.6287 in Myo, LV, and RV, respectively.

Keywords: Unsupervised segmentation · Cardiac anatomical segmentation · Image-to-image translation

1 Introduction

Image segmentation is the core problem of medical image analysis, providing pixel-wise classification and precise location for further image analysis and clinic decision [19,20]. Recently, great progress on semantic segmentation has been driven by the improvement of Convolution Neural Network (CNN) with the supervision of numerous annotated images [15,18]. However, manual delineation is rather time-consuming and laborious. Hence, unsupervised segmentation algorithms are desired. There are plenty of efforts on unsupervised binary segmentation [9,11]. However, extending these methods directly to multi-class situations could be difficult in both implementation and achieving satisfactory performance. In this work, we study the multi-class unsupervised segmentation (MCUS) problem with both intensity and shape constraints.

O. Camara et al. (Eds.): STACOM 2022, LNCS 13593, pp. 349–358, 2022.
https://doi.org/10.1007/978-3-031-23443-9_32

For MCUS, without supervision from annotated images, the design of loss function would be the main determinant of segmentation performance. In the literature, many well-designed conventional energy functions have been proposed for pixel clustering [5,13,14], such as the famous level-set based Mumford-Shah functional [12], which can be implemented via neural network. Nevertheless, there are two inherent challenges among such efforts. Firstly, the number of segments, is an unknown parameter optimized by themselves [17] and their non-convexity typically traps themselves in a local minimum, crippling the applicability to tackle multi-class segmentation [2,10]. For example, Cai et al. [1] utilized a clustering technique as post-processing on binary segmentation from the Mumford-Shah functional to generate multi-class results. To tackle this challenge, Vese et al. [17] extended the energy function by simply representing multi-class results with multiply binary segments. In this work, we introduce a reasonable representing method utilizing the inherent relation among target objects to penalize the number of segments and implement it via U-Net.

Secondly, as mere intensity and length of regions are taken into account by these loss functions [2], intensity-similar regions, such as left and right ventricles, would be difficult to distinguish without any shape constraint. For example, Kim et al. [8] provided the shape constraint with extra annotated images in a semi-supervised training strategy. Inspired by image-to-image translation in unsupervised domain adaptation [3,16], we propose the idea of providing synthesized annotated images for the segmentation network with a semi-supervised training strategy. Specifically, we adopted the well-known Multimodal Unsupervised Image-to-image Translation (MUNIT) model [7] to predict synthetic images from given anatomical labels with adversarial loss compared with real MRIs.

However, the indelible gap between synthesized images and real images cripples the generalization ability of the segmentation network [16], leading to undesired predictions on real images. Hence, the network is desired to learn the latent content features under different appearance. Inspired by SimCLR [4], we introduce an augmentation module to penalize the similarity of predictions from images in different perspective views as shape constraint.

In this paper, we propose an unsupervised segmentation framework tackling the above-mentioned two challenges. The main contributions of our approach can be summarized as follows. 1) We extend Mumford-Shah functional for multi-class segmentation with relation constraint and implement it via neural network. 2) We propose a novel unsupervised segmentation training strategy, providing synthetic supervision from synthetic annotated images via an explicit image-to-image translation strategy. 3) We introduce intensity constraint and spatial constraint based on vanilla MUNIT for more realistic and precise predictions. 4) We introduce an augmentation invariant strategy to facilitate the model to generate the complex and multi distributions inherent in the shape of the data. and 5) We validate its performance with Cardiac MR images, and achieved promising results.

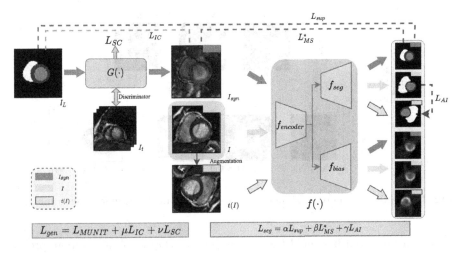

Fig. 1. Overall structure of the proposed unsupervised segmentation framework. (Color figure online)

2 Method

In this work, we propose an unsupervised segmentation framework for multi-class delineation with both intensity and shape constraints. As illustrated in Fig. 1, an intensity-oriented unsupervised loss function L^*_{MS} is proposed based on Mumford-Shah functional [12], while synthetic annotated images (I_{syn}, I_L) from generator $G(\cdot)$ provide shape supervision for segmentation network $f(\cdot)$. Moreover, self-learning module is also applied to emphasize structure.

2.1 Unsupervised Segmentation Network with Intensity Constraint

Let $I(x) \in R^{H \times W \times C}$ be an input image. We introduce an embedding function $\phi_n : \Omega \mapsto [0, 1]$ for each class, and denote $\Phi = [\phi_1, \phi_2, ..., \phi_N]$ as the encoding of prediction, where N is the number of target classes. The unsupervised loss function L_{un} for segmentation network contains three components,

$$L_{un} = L^*_{MS} + \eta L_{IR} + \varepsilon L_{smooth}, \tag{1}$$

where L^*_{MS} refers to variant Mumford-Shah functional, L_{IR} is an inclusion regularization and L_{smooth} is total variation loss as smooth regularization. Specifically, L^*_{MS} is to penalize the intensity variance within a object,

$$L^*_{MS} = \sum_{n=1}^{N} \int |I(x) - b(x)c_n|^2 \phi_n(x) dx, \tag{2}$$

where c_n is intensity representation of class n. A reduced form of it is the average intensity within the object. With regard of the heterogeneity of intensity within

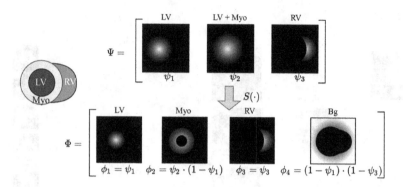

Fig. 2. The illustration of mask mapping $S(\cdot)$ utilized in the Mumford-Shah function.

a class, a predicted bias field $b(x)$ is imposed on $I(x)$ to obtain a more precise intensity constraint and then c_n is calculated as follows.

$$c_n = \frac{\int I(x)b(x)\phi_n(x)dx}{\int b(x)\phi_n(x)dx}. \tag{3}$$

The estimation of the bias field can be easily implemented in the segmentation network with extra convolution layers before the last segmentation layer (refer to f_{bias} in Fig. 1), which is activated by a sigmoid function. Moreover, L_{smooth} is to penalize the smoothness of generated bias field and the length of partitioning boundary [2], and can be formulated as $L_{smooth} = \sum_{n=1}^{N} \int |\nabla\phi_n(x)|dx + \int |\nabla b(x)|dx$.

As L_{MS}^* and L_{smooth} require the pre-definition of foreground subjects, for multi-class segmentation, we need to design an appropriate set of foregrounds, which should be compatible with each other, and easy to derive the final prediction via simple addition or multiplication. To this end, we denote the output of the network as Ψ with channel size $N-1$, i.e., $\Psi = [\psi_1, \psi_2, ..., \psi_{N-1}]$. Specifically, for cardiac ventricle segmentation, as illustrated in Fig. 2, ψ_1, ψ_2 and ψ_3 are defined as LV, LV+Myo and Rv, respectively. Then, with $\phi_1 = \psi_1$ and $\phi_3 = \psi_3$, the prediction for Myo and background can be respectively represented by $\phi_2 = \psi_2 \cdot (1 - \psi_1)$ and $\phi_4 = (1 - \psi_2) \cdot (1 - \psi_3)$.

Furthermore, when solely Mumford-Shah function is utilized, it is possible for $f(\cdot)$ to generate opposite ψ_1 and ψ_2. Hence, we introduce an inclusion regularization (IR) to penalize the inclusion relation between each foreground in Ψ,

$$L_{IR} = ||\psi_1 \cdot \psi_2 - \psi_1||_2^2, \tag{4}$$

which constrains $\psi_1 \cap \psi_2$ to be ψ_1, and penalizes heavily when $\psi_1 \cap \psi_2 = \emptyset$.

2.2 Strong Shape Constraints

We introduce the label-to-image translation technique and augmentation invariance for shape constrains.

Label-to-Image Translation. Let $i_l \in I_L$ be cardiac anatomic label map, and $i_t \in I_t$ be target CMR image. Note that the I_L and I_t are unpaired. Our goal is to estimate $p(i_t|i_l)$ with image-to-image translation models $G(\cdot)$, which generates target image sample I_{syn}. Our idea for shape constraint is that given I_L, the generated pair (I_{syn}, I_L) could help provide supervision for the segmentation model $f_{seg}(\cdot)$. For more precise prediction, we introduce intensity constraint (IC) and spatial constraint (SC) based on MUNIT [7]. Firstly, assume that pixels within an object should share similar intensity, then we can introduce an intensity constraint (IC) to emphasize such correlation as follows,

$$L_{IC} = \int |I_{syn} - C|^2 \cdot I_L dx \tag{5}$$

where C is calculated similar with Eq. (3).

Moreover, because of the significant variance in appearances and structures across slices of different positions in an image volume, we introduce a spatial constraint (SC) for the generator to force the outputs to be more discriminative. We categorize the slices into 5 classes according to their positions, denoted as y. An additional fully-connection (FC) layers connected to vanilla content encoder E^c [7] is then adopted to classify the position of the generated image, that is,

$$L_{SC} = -\sum_i^C y_i \log(\hat{y}_i) \tag{6}$$

where $\hat{y}$ represents the position prediction.

The total loss for the generator is

$$L_{gen} = L_{MUNIT} + \mu L_{IC} + \nu L_{SC} \tag{7}$$

where L_{MUNIT} refers to the loss function of vanilla MUNIT, and μ, ν are hyperparameters. By incorporating IC and SC, the generator can be penalized to be aware of the multi distribution and generate more realistic images in different positions.

Augmentation Invariance. The distance between the appearance of real image and I_{syn} cripples the performance of the segmentation network. Hence, the augmentation invariant module is introduced to facilitate the network to sense the shape behind different appearances.

A set of stochastic data augmentation modules $\mathcal{T}$ is applied to randomly transform the input image, resulting in images in another perspective, $i.e.$, $\tilde{I} = \mathcal{T}(I)$. $\mathcal{T}$ contains simple transformations, such as random rotation, random horizontal/vertical flip and random color distortions. Then the segmentation model $f(\cdot)$ is applied on the above pair $(I, \tilde{I})$. We assume that predictions of these homologue images remain the same, despite different orientations and contrasts. Augmentation invariant loss, L_{AI} is utilized to encourage these predictions to be the same,

$$L_{AI} = \frac{1}{H \times W}(f(I) - t^{-1}(f(\tilde{I})))^2. \tag{8}$$

Table 1. The quantitative results of the ablation study and compared algorithm on CMR dataset. #1* refers to #1 with shape constraint via GAN. Syn refers to the network trained with synthetic cardiac images I_{syn} and AI refers to the augmentation invariant module. Note that the Dice score of network with vanilla Mumford-Shah functional is not presented since only binary segmentation is achieved.

Method	Module			Dice			Avg dice
	L_{MS}^*	Syn	AI	Myo	LV	RV	
#1	✓	×	×	0.2028 ± 0.0991	0.2914 ± 0.1118	0.1414 ± 0.0188	0.2119
#1*	✓	×	×	0.2110 ± 0.0204	0.3152 ± 0.0859	0.1746 ± 0.0313	0.2236
#2	✓	✓	×	0.5204 ± 0.0606	0.6799 ± 0.1170	0.3760 ± 0.1330	0.5254
MS_CNN	–	–	–	0.2430 ± 0.0619	0.122 ± 0.1164	0.2703 ± 0.1165	0.2118
#3	×	✓	✓	0.5270 ± 0.2218	0.7049 ± 0.22438	0.1691 ± 0.1758	0.4670
Proposed	✓	✓	✓	**0.5737 ± 0.0870**	**0.7796 ± 0.1075**	**0.6287 ± 0.1320**	**0.6610**

2.3 Overall Architecture

As shown in Fig. 1, there are two kinds of input images for segmentation network (blue part), i.e., target image $I(x)$ and synthetic annotated images I_{syn}, which is sampled from the generator (green part). The variant Mumford-Shah functional is applied as unsupervised supervision for segmentation. To utilize both synthetic annotated image (I_{syn}, I_L) and unlabeled data $I(x)$, the total loss function for the segmentation network is as follows:

$$L_{seg} = \alpha L_{sup} + \beta L_{un} + \gamma L_{AI}, \tag{9}$$

where $L_{sup} = L_{DSC}(f_{seg}(I_{syn}), I_L)$ is the Dice loss for the prediction, α and β are hyper-parameters with $\alpha = 0$ for unlabeled data.

3 Experiment

3.1 Materials and Experimental Setups

We evaluated the proposed algorithm on two datasets, i.e., synthetic images and public cardiac MR images. For synthetic datasets, we randomly generated images with square and circle in various size and position to evaluate robustness. For real cardiac image, we collected 45 bSSFP MRI subjects from the MICCAI2019 MSCMR challenge [20], each of subject contains 8–12 slices. We randomly divided these cases into 4 groups, i.e., 15 unlabeled training images $I(x)$, 10 anatomical labels I_L, 10 real target images I_t, and 10 testing images. The 2D slices extracted from original volumes were cropped into 128×128 and normalized via Z-score as the network inputs. Data augmentation including random resized crop, random rotation and flipping were applied.

The framework was implemented in Pytorch. 1) For the image translation network, we implemented the proposed module based on the official implementation of MUNIT [7]. The hyper-parameters μ, ν in Eq. (7) were both set to

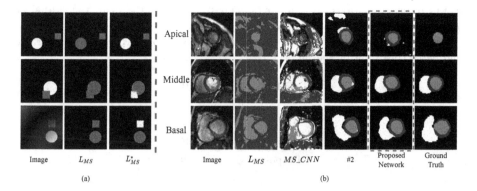

Fig. 3. (a) Visualization of prediction on synthetic datasets. The third row represents the performance on image multiplied by bias field. (Since ground truths of synthetic images are obvious, we omit them.) (b) Examples from different spatial position are selected.

be 0 for the first 5k iterations, and 10 for the rest 15k iterations, while other hyper-parameters were kept the same as in [7]. The batch size for training was 1. 2) For the segmentation network, the network was pre-trained with synthetic images for 70 iterations with the batch size of 8, and then trained with both synthetic images and unlabeled images. The hyper-parameters β, γ in Eq. (1) were 0.0001 and 0.001, respectively, while η, ε in Eq. (9) were both 1. The training was optimized with adam optimizer. The learning rate was set to 10^{-4}. All experiments were implemented on one 24G NVIDIA TITAN RTX GPU.

3.2   Performance of Segmentation Network

We first provide a visual illustration of segmentation results on synthetic images to evaluate the performance of proposed L^*_{MS}. Space limited, the ground-truths of synthetic images are omitted due to their obviousness. The prediction of vanilla L_{MS} and L^*_{MS} are presented in the second and third rows in Fig. 3 respectively. As one can see, merely binary predictions were generated with L_{MS} while promising multi-class results could be achieved simply with L^*_{MS}. Moreover, it presents robustness of the segmentation network with images multiplied with bias field shown in the third row of Fig. 3(a).

Table 1 and Fig. 3(b) present performance of the whole framework on CMR images. Besides the ablation study, we compared our framework with MS_CNN [8], which solved the non-convexity of Mumford-Shah functional theoretically and implemented it via U-Net. Our proposed algorithm achieved Dice scores of 0.5737, 0.7796, 0.6287 in Myo, LV and RV, respectively, which improved the performance by 40% compared to MS_CNN. Since the intensity of RV is similar to that of LV, it is difficult for the network to distinguish them without strong shape constraint, which explains the unsatisfactory dice score in model #1, #2. With augmentation invariance (AI), there were improvements for the proposed

Table 2. The quantitative results of the ablation study and compared algorithm of generator. We evaluete the performance of the generator by Fréchet Inception Distance (FID). #1 refers to basic MUNIT. Module IS and SC represent intensity constraint and patial constraint present in Sect. 2.2.

Method	Module		FID
	IC	SC	
#1	–	–	355
#2	√	–	345
Proposed	√	√	**320**

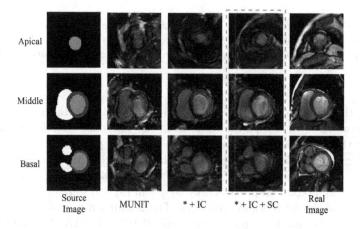

Fig. 4. Visualization of the predictions from slices in different position. * refers to MUNIT while IC and SC refers to intensity constraint and spatial constraint module, respectively.

network, especially in RV by almost 30%. Moreover, as shown by #3, without L_{MS}^*, the segmentor presents poor generalization and weak on distinguishing LV and RV.

As shown in Fig. 3(b), network with vanilla Mumford-Shah functional only predicted binary segments while multi-class segmentation could be predicted with the proposed L_{MS}^*. However, it was incapable of distinguishing outliers in similar intensity and had difficulty in distinguishing LV and RV. With the supervision of synthetic images, reasonable anatomic segmentation results could be achieved. Moreover, L_{AI} dramatically improved the results, especially on RV. It is worth noting that the proposed framework could predict existing regions omitted in manual ground truth caused by expert experience or manual regulation, as shown in the first row of Fig. 3(b).

3.3   Performance of Generator

We utilize the proposed Fréchet Inception Distance (FID) [6] to quantitatively evaluate the performance of the generator. FID is to measure the similarity between two datasets of images and the lower its value, the more similarities there are between real and generated images. As shown in Table 2, the proposed generator with IC and SC constraint presents best performance. The visual performance of shape-to-image translation module is shown in Fig. 4. There were obvious drawbacks lying in the outputs of vanilla MUNIT, that is, the huge heterogeneity of intensity in the same region and the blurry boundary between myocardium and ventricle. Moreover, MUNIT presented a disability in generating apical and basal slices with clear structure. The application of IC module highly improved the intensity heterogeneity problem especially in the region of the myocardium. Additionally, spatial constraint dramatically improved the prediction in the apex.

4   Conclusion

In this work, we introduce a multi-class unsupervised segmentation framework with both intensity and shape constraints for cardiac anatomical segmentation. We firstly propose an intensity-based loss function capable for multi-class based on Mumford-Shah functional. In terms of shape constraint, two modules are introduced. Augmentation invariance penalizes the variance between predictions of the same image in different perspective views, and facilitates the segmentation network to learn the latent features preserving the information of shapes. A special and explicit label-to-image translation is proposed, generating synthetic images directly from given labels. This framework can provide strong shape constraint with synthetic supervision for the segmentation network. We evaluated the proposed framework on a synthetic dataset and real CMR images from a public dataset, and obtained promising Dice scores of 0.5737, 0.7796, and 0.6287 in Myo, LV, and RV, respectively.

References

1. Cai, X., Chan, R., Zeng, T.: Image segmentation by convex approximation of the Mumford-Shah model. UCLA CAM Report, pp. 12–20 (2012)
2. Caselles, V., Kimmel, R., Sapiro, G.: Geodesic active contours. Int. J. Comput. Vision **22**(1), 61–79 (1997)
3. Chen, C., et al.: Unsupervised multi-modal style transfer for cardiac MR segmentation. In: Pop, M., et al. (eds.) STACOM 2019. LNCS, vol. 12009, pp. 209–219. Springer, Cham (2020). https://doi.org/10.1007/978-3-030-39074-7_22
4. Chen, T., Kornblith, S., Norouzi, M., Hinton, G.: A simple framework for contrastive learning of visual representations. In: International Conference on Machine Learning, pp. 1597–1607. PMLR (2020)
5. Freedman, D., Zhang, T.: Interactive graph cut based segmentation with shape priors. In: 2005 IEEE Computer Society Conference on Computer Vision and Pattern Recognition (CVPR 2005), vol. 1, pp. 755–762. IEEE (2005)

6. Heusel, M., Ramsauer, H., Unterthiner, T., Nessler, B., Hochreiter, S.: GANs trained by a two time-scale update rule converge to a local nash equilibrium. Adv. Neural Inf. Process. Syst. **30** (2017)
7. Huang, X., Liu, M.Y., Belongie, S., Kautz, J.: Multimodal unsupervised image-to-image translation. In: Proceedings of the European Conference on Computer Vision (ECCV), pp. 172–189 (2018)
8. Kim, B., Ye, J.C.: Mumford-shah loss functional for image segmentation with deep learning. IEEE Trans. Image Process. **29**, 1856–1866 (2019)
9. Kim, D., Hong, B.W.: Unsupervised segmentation incorporating shape prior via generative adversarial networks. In: Proceedings of the IEEE/CVF International Conference on Computer Vision, pp. 7324–7334 (2021)
10. Massari, U., Tamanini, I.: On the finiteness of optimal partitions. Annali dell'Università'di Ferrara **39**(1), 167–185 (1993)
11. Melas-Kyriazi, L., Rupprecht, C., Laina, I., Vedaldi, A.: Finding an unsupervised image segmenter in each of your deep generative models. arXiv preprint arXiv:2105.08127 (2021)
12. Mumford, D.B., Shah, J.: Optimal approximations by piecewise smooth functions and associated variational problems. Commun. Pure Appl. Math. (1989)
13. Osher, S., Sethian, J.A.: Fronts propagating with curvature-dependent speed: algorithms based on Hamilton-Jacobi formulations. J. Comput. Phys. **79**(1), 12–49 (1988)
14. Panjwani, D.K., Healey, G.: Markov random field models for unsupervised segmentation of textured color images. IEEE Trans. Pattern Anal. Mach. Intell. **17**(10), 939–954 (1995)
15. Ronneberger, O., Fischer, P., Brox, T.: U-net: convolutional networks for biomedical image segmentation. In: Navab, N., Hornegger, J., Wells, W.M., Frangi, A.F. (eds.) MICCAI 2015. LNCS, vol. 9351, pp. 234–241. Springer, Cham (2015). https://doi.org/10.1007/978-3-319-24574-4_28
16. Tajbakhsh, N., Jeyaseelan, L., Li, Q., Chiang, J.N., Wu, Z., Ding, X.: Embracing imperfect datasets: a review of deep learning solutions for medical image segmentation. Med. Image Anal. **63**, 101693 (2020)
17. Vese, L.A., Chan, T.F.: A multiphase level set framework for image segmentation using the Mumford and Shah model. Int. J. Comput. Vision **50**(3), 271–293 (2002)
18. Yue, Q., Luo, X., Ye, Q., Xu, L., Zhuang, X.: Cardiac segmentation from LGE MRI using deep neural network incorporating shape and spatial priors. In: Shen, D., et al. (eds.) MICCAI 2019. LNCS, vol. 11765, pp. 559–567. Springer, Cham (2019). https://doi.org/10.1007/978-3-030-32245-8_62
19. Zhuang, X., et al.: A framework combining multi-sequence MRI for fully automated quantitative analysis of cardiac global and regional functions. In: Metaxas, D.N., Axel, L. (eds.) FIMH 2011. LNCS, vol. 6666, pp. 367–374. Springer, Heidelberg (2011). https://doi.org/10.1007/978-3-642-21028-0_47
20. Zhuang, X., et al.: Cardiac segmentation on late gadolinium enhancement MRI: a benchmark study from multi-sequence cardiac MR segmentation challenge. arXiv preprint arXiv:2006.12434 (2020)

PAT-CNN: Automatic Segmentation and Quantification of Pericardial Adipose Tissue from T2-Weighted Cardiac Magnetic Resonance Images

Zhuoyu Li[1], Camille Petri[2,3], James Howard[3], Graham Cole[3], and Marta Varela[3(✉)]

[1] Department of Metabolism, Digestion and Reproduction, Imperial College London, London, UK

[2] Department of Computing, Imperial College London, London, UK

[3] National Heart and Lung Institute, Imperial College London, London, UK

marta.mvarela@imperial.ac.uk

Abstract. Background: Increased pericardial adipose tissue (PAT) is associated with many types of cardiovascular disease (CVD). Although cardiac magnetic resonance images (CMRI) are often acquired in patients with CVD, there are currently no tools to automatically identify and quantify PAT from CMRI. The aim of this study was to create a neural network to segment PAT from T2-weighted CMRI and explore the correlations between PAT volumes (PATV) and CVD outcomes and mortality.

Methods: We trained and tested a deep learning model, PAT-CNN, to segment PAT on T2-weighted cardiac MR images. Using the segmentations from PAT-CNN, we automatically calculated PATV on images from 391 patients. We analysed correlations between PATV and CVD diagnosis and 1-year mortality post-imaging.

Results: PAT-CNN was able to accurately segment PAT with Dice score/ Hausdorff distances of $0.74 \pm 0.03/27.1 \pm 10.9$ mm, similar to the values obtained when comparing the segmentations of two independent human observers ($0.76 \pm 0.06/21.2 \pm 10.3$ mm). Regression models showed that, independently of sex and body-mass index, PATV is significantly positively correlated with a diagnosis of CVD and with 1-year all cause mortality (p-value < 0.01).

Conclusions: PAT-CNN can segment PAT from T2-weighted CMR images automatically and accurately. Increased PATV as measured automatically from CMRI is significantly associated with the presence of CVD and can independently predict 1-year mortality.

Keywords: Cardiac MRI · Segmentation · Pericardial fat volume · Cardiovascular disease risk factors · Convolutional neural network

1 Introduction

Large volumes of pericardial adipose tissue, PAT, and endocardial adipose tissue, EAT, predispose to a number of cardiovascular conditions [24] such as atrial

O. Camara et al. (Eds.): STACOM 2022, LNCS 13593, pp. 359–368, 2022.
https://doi.org/10.1007/978-3-031-23443-9_33

fibrillation [8]. This is due to the high metabolic activity of adipose tissue which can stimulate remodelling in cardiac muscle, leading to fibrosis, chamber enlargement and inflammation, among others. PAT can also regulate cardiac tissue innervation [16] likely pro-arrhythmically. Obese patients usually have large amounts of PAT, but correlations between cardiac fat volume and body mass index (BMI) are relatively weak [8,10].

So far, PAT has been predominantly segmented and quantified using computed tomography (CT) [1]. In these images, fat's low Hounsfield number allows for comparatively straightforward identification and segmentation using simple intensity threshold-based methods such as region growing and active contours [22] or more complicated approaches like neural networks [5]. Fat is in general more difficult to segment in Cardiac Magnetic Resonance Images (CMRI), where its intensity is much more variable and similar to other tissues'. Quantification of PAT volume (PATV) from MRI has so far only been reported using time-consuming manual segmentation of PAT [6] or dedicated CMRI sequences [7] such as Dixon imaging [10,12]. This prevents widespread quantification of PATV from more common types of CMRI, despite CMRI's increasing prominence in the diagnosis and assessment of cardiovascular disease (CVD). A tool to rapidly and accurately quantify PAT from CMRI would allow rapid quantification of PATV in the clinic, to aid CVD risk assessment.

MRI Segmentation. Convolutional neural networks (CNNs) have shown great ability to segment MR images, enabling easy quantification of several imaging biomarkers whose analysis would otherwise be too time consuming. They consistently outperform non-learned techniques such as thresholding [20], region growing [19] and probability graph models [9]. Once trained, CNNs can perform image analysis in real time in the clinic. To date, and to the best of our knowledge, there are no CNN-based tools for the segmentation of PAT from T2-weighted CMR images.

Aims. In this project, we aim to estimate PATV from clinical CMRI. We focus on axial T2-weighted rapid gradient-echo scans which are routinely acquired in the clinic and in which CNNs have been used to accurately segment several cardiac structures [11]. We initially create a non-learned semi-automatic image processing protocol to segment PAT and generate training data for the proposed CNN. We then train and test a dedicated CNN to segment PAT in these images. Finally, using data from 391 patients, we investigate correlations between: a) PATV and the presence of CVD and b) PATV and 1-year all-cause mortality.

2   Methods

Data. We uses fully anonymized T2-prepared spoiled GRE axial stacks acquired at 1.5T (TE: 1.56 ms, TR: 3.1 ms, recon voxel size: $1.4 \times 1.4 \times 6.0$ mm^3) in 391 patients with known or suspected CVD, under ethical approval. The number of slices in each volumetric scan ranged from 23 to 55.

Preprocessing. Ground truth PAT segmentation was semi-automatically performed in 70 training images (Fig 1). As the first step, we identified the cardiac chambers and left ventricular (LV) myocardium using HRNet, an existing CNN trained in T2-weighted images with a similar contrast [11]. Using these segmentations, we created a bounding box around the heart. We then performed a 2-class Otsu segmentation [21] on the cropped images. In most cases, one of the classes roughly coincided with subcutaneous fat, allowing the rapid segmentation of PAT after some further manual corrections. These semi-automatic PAT segmentations were reviewed by a second independent observer.

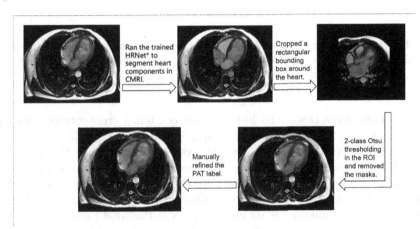

Fig. 1. Semi-automatic pipeline for ground truth PAT segmentation. ROI: Region of Interest

Model: Architecture. To allow fully automatic PAT labelling, we trained PAT-CNN, a 3D Res-UNet CNN [13]. The network (shown in Fig 2) included a contracting path with four convolutional layers, an expanding path with four upsampling layers and concatenate connections. A residual unit was added to each layer. In residual units, batch normalization and parametric rectified linear unit (PReLU) were implemented to improve the feature selection results. The output layer used a sigmoid function as the activation function and was trained to classify each pixel as: PAT (label 1) or background/other tissues (label 0).

Model: Dataset Setup. 70 scans with manual PAT labels were randomly divided into a training+validation dataset (60 images) and a testing dataset (10 images). A second independent observer independently segmented PAT in the test dataset to assess inter-observer segmentation metrics.

Model: Data Augmentation. To enhance PAT-CNN's invariance and generalization, as well as expand the size of the dataset, we applied data augmentation during training using the image analysis library SimpleITK [25]. We randomly applied rotation, crop, flip, blur and Rayleigh noise addition to each input image of the network in each epoch. The probability of each transformation was 0.1.

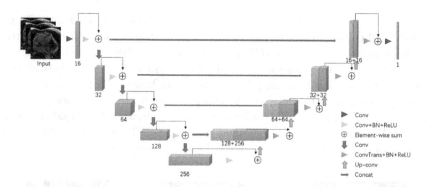

Fig. 2. Structure diagram of PAT-CNN. The kernel sizes of convolutions and transpose convolutions are $3 \times 3 \times 3$. The numbers represent filters numbers in each layer. The triangle arrows represents the residual units.

Model: Loss Function. The loss function combined cross-entropy loss and Dice loss terms [26]. The Dice loss term help improves PAT-CNN's accuracy in the presence of the large imbalance between the volumes of PAT and voxels labelled as background/other organs. The cross-entropy loss term minimises the likely instability of the gradients of a pure Dice-score-based loss function. The total loss function can be represented as below, where x_n is the ground truth label of the nth sample, y_n is its prediction, N is the total number of samples (and the small term ϵ in the Dice coefficient is used to avoid the denominator becoming 0):

$$Loss(X,Y) = 1 - \frac{2\sum_{n=1}^{N} x_n y_n + \epsilon}{\sum_{n=1}^{N} x_n + \sum_{n=1}^{N} y_n + \epsilon} - \frac{1}{N}\sum_{n=1}^{N} x_n log(y_n) \qquad (1)$$

Model: Hyperparameters and Performance Assessment. Through our experimental results, the optimal batch size was 8. We used Adam [15] and set the learning rate to 10^{-4}, training for 60 epochs. This combination performed best on the validation dataset. We used the Dice Score and Hausdorff Distance to evaluate PAT-CNNs performance on the test dataset. Using these metrics, we additionally compared PAT-CNN to the vanilla U-Net [23] commonly using for segmenting medical images.

Statistical Analysis: Data. We used PAT-CNN to estimate PATV (in cm³) in all 391 volumetric T2-weighted axial CMR images. From the anonymized imaging metadata and radiology reports, we extracted the following demographic information (*mean* $\pm$ *std*): age (55 ± 18 years), sex (42% female), BMI ($27.7 \pm 5.9\,\mathrm{kg/m^2}$) and 1-year post-scan mortality ('Deceased', 8.2%). We also compiled diagnostic information, which was synthesised into 73 binary clinical diagnostic labels (presence/absence of: dilated cardiomyopathy, left/right ventricular failure, left/right ventricular hypertrophy, myocardial infarction, ischaemia, etc.).

The sum of these diagnostic labels was compiled into a single variable 'CVD Diagnosis' (2.9 ± 1.9).

Statistical Analysis: Methods. We investigated the relationship between the regressor variables: PATV, sex, age and BMI and the dependent variables: 'CVD diagnosis', 'Deceased' and PATV, using univariate Pearson/phi correlation analysis. Those regressors that were significantly correlated ($\alpha = 0.01$) with the independent variables were then used as inputs to a multivariate analysis using ordinary least squares (for the 'CVD Diagnosis' and 'PATV' dependent variables) and a logistic regression model (for the dichotomous variable 'Deceased'). We note that the clinical dataset used did not include time-to-mortality information and survival analyses were therefore not performed.

3   Results

PAT-CNN Performance. On the test dataset, PAT was segmented with an average Dice score/Hausdorff distance of $0.74 \pm 0.03/27.1 \pm 10.9$ mm. The model coped well with variations in PAT shape, size and location (Fig 3) and the presence of pathological changes. PAT-CNN performed better than U-Net and its results compared well with the agreement between two human operators on the same dataset (Table 1).

Table 1. Performance of PAT-CNN, U-Net and a second human observer on the test images. (The segmentations from human Observer 1 were treated as the ground truth for this analysis.)

	PAT-CNN	U-Net	Observer 2
Dice Score	0.74 ± 0.03	0.71 ± 0.03	0.76 ±± 0.06
Hausdorff Distance (mm)	27.1 ± 10.9	32.8 ± 11.2	21.2 ± 10.3

Statistical Analysis. Across the 391 analysed patients, PATV had a mean of $139.60\,cm^3$ and a standard deviation of $80.24\,cm^3$, in excellent agreement with PAT quantification from CT [5]. PATV is significantly correlated with patients' age ($\rho_{Pearson} = 0.38$), sex and BMI ($\rho_{Pearson} = 0.34$), but in a multivariate analysis, only sex and BMI remain significant (Table 2). PATV and age were the only regressor variables found to be significantly correlated to both CVD diagnostic labels and 1-year mortality post-imaging in both univariate and multivariate analyses (Table 2). This is in contrast to BMI which was not an independent predictor of either mortality or CVD diagnosis in univariate analysis (Table 2).

4   Discussion

We present PAT-CNN, a CNN able to identify PAT accurately in T2-weighted CMRI across subjects of various sizes and in the presence of different types of

Table 2. Regression coefficients from 3 multivariate regression analyses (n = 391), for the dependent variables listed in each column. Each row shows each regressor considered in the model. (For the dichotomous variable 'Sex', females are represented by 1 and males by 0.) Only regressors found to be statistically significant in univariate analyses were considered. Entries are the regression coefficients in the specified units, with the asterix (*) denoting significant correlations at $\alpha = 0.01$. PATV: pericardial adipose tissue volume; BMI: body-mass index; CVD: cardiovascular disease.

	PATV (cm^3)	Deceased	CVD Diagnosis
Sex	-50.2*		-0.30
Age $(years)$	1.70	0.03*	0.02*
BMI (kg/m^2)	3.92*		
PATV (cm^3)		0.01*	0.01*

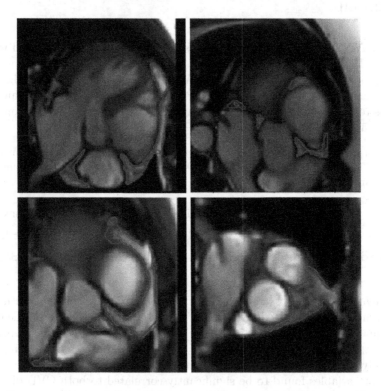

Fig. 3. Representative examples of T2-weighted slices from four different patients overlaid with the contours for the ground truth (red) and PAT-CNN (green) PAT labels. (Color figure online)

pathology. PATV estimated using PAT-CNN's segmentation of CMRI is significantly correlated to the presence of cardiac pathology and is an independent predictor of 1-year mortality, as seen in studies from the literature in which PAT was segmented manually [8].

PAT Segmentation. PAT is distributed around in the heart in several non-contiguous patches with various shapes, making PAT segmentation more complex than the segmentation of contiguous and less variable structures such as the cardiac chambers themselves. Moreover, when present, pericardial fluid can easily be mistakenly labelled as PAT, as fluid and fat have a similar intensity and appearance on T2-weighted CMR images.

Despite the difficulty of the task, PAT-CNN was able to segment PAT with an average Dice score of 0.74, outperforming U-Net and performing only marginally worse than the average Dice score of 0.76 obtained when comparing the segmentations from different human observers (see Table 1).

PAT-CNN uses a 3D Res-UNet previously used to segment heart chambers [14,18], with a modified combined Dice-Cross-Entropy loss that outperformed Dice-only and Cross-Entropy-only losses for this task. We therefore show that the 3D Res-UNet has a good performance in CMRI even when segmenting irregular, non-connected regions such as PAT.

Statistical Analysis of PATV

PATV estimated automatically using PAT-CNN shows similar correlations to clinical variables as PATV estimated from CT and echocardiography [2,17]. Females, elderly people and those with higher BMI are more likely to have more PATV. Statistically significant correlations were found between PATV and a diagnosis of cardiovascular disease at the time of scan, as has been reported in other studies [2]. PATV was also an independent predictor of one-year all-cause mortality in the analysed 319 subjects, as also found on CT [4]. Interestingly, BMI was not a predictor for mortality or the presence of CVD, although it was significantly but weakly correlated with PAT ($\rho_{Pearson} = 0.34$), also as seen in other studies [2,10].

Our analysis used 1-year all-cause mortality, as we could not obtain information about CVD-related deaths or temporal information about the time to death. Body surface area (BSA) is more commonly used than BMI for the indexing of cardiac function variables. BSA would have been an interesting variable to include in our regression models, but it was not available in our dataset.

Future Work. PAT segmentations, such as those provided by PAT-CNN, provide information beyond that of PATV alone. In the current study, we did not analyse the clinical importance of the location or spatial distribution of PAT, which is likely to have clinical value beyond PAT volume only. We also did not attempt to index PATV by body area or heart size, which may yield biomarkers with further clinical value [3].

Alongside PAT, epicardial adipose tissue (EAT) is also likely to play an important role in cardiovascular pathophysiology and it should be identifiable in CMRI. EAT, however, is expected to deposit in regions of smaller volume and its identification is limited by partial volume effects at MRI's typical spatial resolution. EAT's reliable identification would require MRI at much higher spatial resolution or the use of specialised CMRI sequences such as Dixon techniques [10,12]. Dixon technique CMR images are not usually acquired as part of clinical examinations, stressing the importance of our approach to automatically segment PAT from commonly acquired T2-weighted clinical scans.

Pending training and testing across images from different sites and scanners, PAT-CNN can be deployed as a scanner-side analysis tool to automatically estimate PATV. It can also be used to explore the mechanisms of different CVDs by highlighting potential relationships between PAT volume and location and CVD manifestations.

5   Conclusions

We propose PAT-CNN, a neural network to automatically segment PAT from CMR images. We show that PAT segmentations are reliable, making PAT-CNN ready to be deployed on clinical images to yield biomarkers of potential clinical interest. Using data from 391 patients, we find that PAT volume, estimated using PAT-CNN, is significantly correlated with both a diagnosis of CVD and 1-year all-cause mortality, independently of sex and BMI.

Acknowledgements. This work was supported by the British Heart Foundation Centre of Research Excellence at Imperial College London (RE/18/4/34215).

References

1. Benčević, M., Galić, I., Habijan, M., Pižurica, A.: Recent progress in epicardial and pericardial adipose tissue segmentation and quantification based on deep learning: a systematic review. Appl. Sci. **12**(10), 5217 (2022). https://doi.org/10.3390/app12105217
2. Britton, K.A., Massaro, J.M., Murabito, J.M., Kreger, B.E., Hoffmann, U., Fox, C.S.: Body fat distribution, incident cardiovascular disease, cancer, and all-cause mortality. J. Am. Coll. Cardiol. **62**(10), 921–925 (2013). https://doi.org/10.1016/J.JACC.2013.06.027
3. Cai, S., et al.: Cardiac MRI measurements of pericardial adipose tissue volumes in patients on in-centre nocturnal hemodialysis. J. Nephrol. **33**(2), 355–363 (2019). https://doi.org/10.1007/s40620-019-00665-4
4. Cheng, V.Y., et al.: Pericardial fat burden on ECG-gated noncontrast CT in asymptomatic patients who subsequently experience adverse cardiovascular events. JACC Cardiovasc. Imaging **3**(4), 352–360 (2010). https://doi.org/10.1016/J.JCMG.2009.12.013
5. Commandeur, F., et al.: Deep learning for quantification of epicardial and thoracic adipose tissue from non-contrast CT. IEEE Trans. Med. Imaging **37**(8), 1835–1846 (2018). https://doi.org/10.1109/TMI.2018.2804799
6. Davidovich, D., Gastaldelli, A., Sicari, R.: Imaging cardiac fat. Eur. Heart J. - Cardiovasc. Imaging **14**(7), 625–630 (2013). https://doi.org/10.1093/EHJCI/JET045
7. Ding, X., et al.: Automated pericardial fat quantification from coronary magnetic resonance angiography: feasibility study. J. Med. Imaging **3**(1), 014002 (2016). https://doi.org/10.1117/1.jmi.3.1.014002
8. Fitzgibbons, T.P., Czech, M.P.: Epicardial and perivascular adipose tissues and their influence on cardiovascular disease: basic mechanisms and clinical associations. J. Am. Heart Assoc. **3**(2) (2014). https://doi.org/10.1161/JAHA.113.000582

9. Grosgeorge, D., Petitjean, C., Dacher, J.N., Ruan, S.: Graph cut segmentation with a statistical shape model in cardiac MRI. Comput. Vis. Image Underst. **117**(9), 1027–1035 (2013). https://doi.org/10.1016/j.cviu.2013.01.014

10. Henningsson, M., Brundin, M., Scheffel, T., Edin, C., Viola, F., Carlhäll, C.J.: Quantification of epicardial fat using 3D cine Dixon MRI. BMC Med. Imaging **20**(1), 1–9 (2020). https://doi.org/10.1186/s12880-020-00478-z

11. Howard, J.P., et al.: Automated analysis and detection of abnormalities in transaxial anatomical cardiovascular magnetic resonance images: a proof of concept study with potential to optimize image acquisition. Int. J. Cardiovasc. Imaging **37**(3), 1033–1042 (2020). https://doi.org/10.1007/s10554-020-02050-w

12. Kellman, P., et al.: Multiecho dixon fat and water separation method for detecting fibrofatty infiltration in the myocardium. Magn. Reson. Med. **61**(1), 215–221 (2009). https://doi.org/10.1002/mrm.21657

13. Kerfoot, E., Clough, J., Oksuz, I., Lee, J., King, A.P., Schnabel, J.A.: Left-ventricle quantification using residual U-net. In: Pop, M., et al. (eds.) STACOM 2018. LNCS, vol. 11395, pp. 371–380. Springer, Cham (2019). https://doi.org/10.1007/978-3-030-12029-0_40

14. Kerfoot, E., Puyol Anton, E., Ruijsink, B., Clough, J., King, A.P., Schnabel, J.A.: Automated CNN-based reconstruction of short-axis cardiac MR sequence from real-time image data. In: Stoyanov, D., et al. (eds.) RAMBO/BIA/TIA -2018. LNCS, vol. 11040, pp. 32–41. Springer, Cham (2018). https://doi.org/10.1007/978-3-030-00946-5_4

15. Kingma, D., Ba, J.: Adam: a method for stochastic optimization. Comput. Sci. (2014)

16. Lavie, C.J., Pandey, A., Lau, D.H., Alpert, M.A., Sanders, P.: The present and future obesity and atrial fibrillation prevalence, pathogenesis, and prognosis effects of weight loss and exercise. Technical report (2017). https://doi.org/10.1016/J.JACC.2017.09.002

17. Liu, J., et al.: Pericardial adipose tissue, atherosclerosis, and cardiovascular disease risk factors the Jackson heart study. Diabetes Care **33**(7), 1635–1639 (2010). https://doi.org/10.2337/DC10-0245

18. Lourenço, A., et al.: Left atrial ejection fraction estimation using SEGANet for fully automated segmentation of CINE MRI. In: Puyol Anton, E., et al. (eds.) STACOM 2020. LNCS, vol. 12592, pp. 137–145. Springer, Cham (2021). https://doi.org/10.1007/978-3-030-68107-4_14

19. Mancas, M., Gosselin, B., Macq, B.: Segmentation using a region-growing thresholding. In: Dougherty, E.R., Astola, J.T., Egiazarian, K.O. (eds.) Image Processing: Algorithms and Systems IV, vol. 5672, pp. 388–398. International Society for Optics and Photonics, SPIE (2005). https://doi.org/10.1117/12.587995

20. Norouzi, A., et al.: Medical image segmentation methods, algorithms, and applications. IETE Tech. Rev. **31**(3), 199–213 (2014). https://doi.org/10.1080/02564602.2014.906861

21. Otsu, N.: A threshold selection method from gray-level histograms. IEEE Trans. Syst. Man Cybern. **9**(1), 62–66 (1979). https://doi.org/10.1109/TSMC.1979.4310076

22. Rodrigues, O., Morais, F.F., Morais, N.A., Conci, L.S., Neto, L.V., Conci, A.: A novel approach for the automated segmentation and volume quantification of cardiac fats on computed tomography. Comput. Methods Program. Biomed. **123**, 109–128 (2016). https://doi.org/10.1016/J.CMPB.2015.09.017

23. Ronneberger, O., Fischer, P., Brox, T.: U-net: convolutional networks for biomedical image segmentation. In: Navab, N., Hornegger, J., Wells, W.M., Frangi, A.F. (eds.) MICCAI 2015. LNCS, vol. 9351, pp. 234–241. Springer, Cham (2015). https://doi.org/10.1007/978-3-319-24574-4_28
24. Weschenfelder, C., Quadros, A., Santos, J., Garofallo, S.B., Marcadenti, A.: Adipokines and adipose tissue-related metabolites, nuts and cardiovascular disease. Metabolites **10**(1), 32 (2020)
25. Yaniv, Z., Lowekamp, B.C., Johnson, H.J., Beare, R.: SimpleITK image-analysis notebooks: a collaborative environment for education and reproducible research. J. Digit. Imaging **31**(3), 290–303 (2017). https://doi.org/10.1007/s10278-017-0037-8
26. Yeung, M., Sala, E., Schönlieb, C.B., Rundo, L.: Unified focal loss: generalising dice and cross entropy-based losses to handle class imbalanced medical image segmentation. Comput. Med. Imaging Graphics **95**, 102026 (2022). https://doi.org/10.1016/J.COMPMEDIMAG.2021.102026

Deep Computational Model for the Inference of Ventricular Activation Properties

Lei Li[1](✉), Julia Camps[2], Abhirup Banerjee[1,3], Marcel Beetz[1],
Blanca Rodriguez[2], and Vicente Grau[1]

[1] Department of Engineering Science, University of Oxford, Oxford, UK
lei.li@eng.ox.ac.uk
[2] Department of Computer Science, University of Oxford, Oxford, UK
[3] Division of Cardiovascular Medicine, Radcliffe Department of Medicine,
University of Oxford, Oxford, UK

Abstract. Patient-specific cardiac computational models are essential for the efficient realization of precision medicine and in-silico clinical trials using digital twins. Cardiac digital twins can provide non-invasive characterizations of cardiac functions for individual patients, and therefore are promising for the patient-specific diagnosis and therapy stratification. However, current workflows for both the anatomical and functional twinning phases, referring to the inference of model anatomy and parameter from clinical data, are not sufficiently efficient, robust, and accurate. In this work, we propose a deep learning-based patient-specific computational model, which can fuse both anatomical and electrophysiological information for the inference of ventricular activation properties, i.e., conduction velocities and root nodes. The activation properties can provide a quantitative assessment of cardiac electrophysiological function for the guidance of interventional procedures. We employ the Eikonal model to generate simulated electrocardiograms (ECGs) with ground truth properties to train the inference model, where patient-specific information has also been considered. For evaluation, we test the model on the simulated data and obtain generally promising results with fast computational time.

Keywords: Deep computational models · Ventricular activation properties · ECG simulation · Digital twin

1 Introduction

Cardiovascular diseases are one of the most common pathologies globally, resulting in changes in cardiac anatomy, structure, and function [24]. Ventricular activation properties offer a valuable quantitative description of electrical activation and propagation, which is essential for identifying arrhythmias, localizing diseased tissue, and stratifying patients at risk [11,22]. For example, the location

O. Camara et al. (Eds.): STACOM 2022, LNCS 13593, pp. 369–380, 2022.
https://doi.org/10.1007/978-3-031-23443-9_34

of the Purkinje endocardial root nodes (RNs), i.e., earliest activation sites, can provide important information for the selection of optimal implantation sites of pacing leads [30]. Conduction velocities (CVs) can describe the speed and direction of electrical propagation through the heart, and its alterations play an important role in the generation and maintenance of cardiac arrhythmias [23].

Electrocardiograms (ECGs) can provide a substantial amount of information about the heart rhythm and reveal abnormalities related to the conduction system. For instance, the QRS morphology in 12-lead ECG can indicate the origin of ventricular activation, which can be used to guide the clinicians to the potential ablation targets in real time [29]. However, there exist large inter-subject anatomical variations that may modify the ECG patterns for specific activation properties. Besides, ECG cannot be used to locate and characterize diseases, such as arrhythmias. The cardiac structural and functional information from imaging data, such as ultrasound, computed tomography or cardiac magnetic resonance (CMR), could be complementary to the information provided by ECG. Computational models combining ECG and imaging data can be used to estimate ventricular activation properties for therapy guidance and variability interpretation among different patients [25,28]. However, it is complicated to accurately localize RNs, as there is limited knowledge about the actual topology of Purkinje activation networks [19]. Moreover, the localization is often computationally expensive due to the complexity of structural and spatial variations of such networks. The estimation of CVs is also fundamentally challenging owing to the underlying mechanisms of complex, nonlinear, and heterogeneous myocardial activation [20]. The simultaneous inference of CVs and RNs could be more challenging considering the existence of continuous and discrete mixed-type parameter space.

Nevertheless, there exist several CV and RN estimation techniques in the literature [12]. For CV estimation, Bayly et al. [5] utilized inverse-gradient techniques to predict CVs from epicardial mapping data for the understanding and description of reentrant arrhythmias. Chinchapatnam et al. [14] employed an adaptive algorithm for the estimation of local CVs from a noncontact mapping of the endocardial surface potential. Instead of predicting endocardial CV, Good et al. [20] considered both epicardial and volumetric CVs and examined triangulation-based, inverse-gradient-based, and streamline-based techniques for comparison. Compared to CV estimation, the localization of RNs received limited attention so far with only a few works, some of which solely used ECG signals for the localization [21]. The simultaneous optimization of CVs and RNs has been explored, but it is generally achieved via conventional iterative algorithms [11,18,22,31]. Recently, deep learning based methods have achieved promising performance for cardiac activation modeling [1,26]. For example, Bacoyannis et al. [1] proposed a β-conditional variational autoencoder to predict activation maps for various cardiac geometries with corresponding simulated body surface potentials. Meister et al. [26] utilized a graph convolutional regression network to predict the activation time maps from ECG and CMR images.

In this work, we have proposed a patient-specific deep computational model (PS-DCM) for an efficient and simultaneous estimation of ventricular activation properties, i.e., RNs and CVs. As the activation properties are unavailable in healthy subjects, we create "virtual cohorts" by simulation via Eikonal models for known ground truth values. We consider additional physiological information as conditions to model the variations among specific sub-populations. Also, we analyze the relationship between sampling latent space and the predictions, i.e., anatomy, signal, and activation parameters. To the best of our knowledge, this is the first deep learning based computational model for ventricular activation property estimation.

2   Methodology

Figure 1 provides an overview of the proposed PS-DCM, consisting of a conditional variational autoencoder (c-VAE) and an inference model. Here, the c-VAE includes one encoder and two separated decoders for the structural and ECG signal reconstructions, respectively, while the conditional inference network (c-InfNet) aims to predict the ventricular activation properties based on the low-dimensional features from c-VAE. In Sect. 2.1, the mesh generation from 2D end-diastolic CMR images is explained. The generation of simulated data as the ground truth of the CVs and RNs is described in Sect. 2.2. Finally, Sect. 2.3 presents the details of the computational model for the prediction of two activation properties.

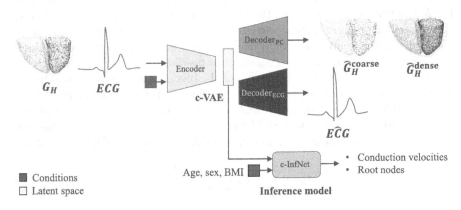

Fig. 1. The proposed patient-specific deep computational model (PS-DCM) for the inference of ventricular activation properties. The conditional variational autoencoder (c-VAE) aims to learn anatomical and electrical information that is used to assist the conditional inference network (c-InfNet) in predicting the activation parameters.

2.1   Geometrical Triangular Mesh Generation

To obtain patient-specific 3D anatomical information, we generate smooth 3D biventricular tetrahedral meshes from 2D multi-view CMR images via a completely end-to-end automatic pipeline [3]. The pipeline starts by performing a deep learning based ventricle segmentation [2] on the automatically selected slices from long- and short-axis CMR images. Considering the potential misalignments due to respiration and cardiac motions, we employ the intensity and contour information as well as a statistical shape model for in-plane and out-plane misalignment corrections, respectively [4]. Finally, we perform the surface mesh reconstruction from cardiac contours, mitigating the remaining discrepancies between sparse 3D contours. For the following simulation of electrical activity in the heart, we further generate the 3D volumetric tetrahedral mesh from the 3D biventricular surface mesh using mesh generators [16,32].

2.2   Simulated Data Generation via Eikonal Models

As the ventricular activation properties can not be measured for clinical data, we built a synthetic dataset from real clinical data via Eikonal model-based simulation [11]. Specifically, the Eikonal model is simulated over the generated 3D tetrahedral meshes in Sect. 2.1 and can be defined as,

$$\sqrt{(\nabla d^T \cdot \nabla d)} = 1. \tag{1}$$

Here, $\nabla d = v \nabla t$, where ∇t is the traveling time passing through a node and v denotes the CVs of the fiber, sheet (transmural), and sheet-normal directions. For fast optimization, we regard the tetrahedral mesh as a graph, where electric current can enter directly from two connected nodes through the edges connecting them. Consequently, one could use Dijkstra's algorithm to predict Eikonal's activation time. Note that we consider the position of the RN as the starting point with $t = 0$, while the other nodes of the biventricular network are the destinations for which we need to compute t for final activation time maps (ATMs). Subsequently, one could calculate ECG signals from the ATMs simulated from Eikonal models via the pseudo-ECG equation with the electrode locations from torso geometry and orientation. The pseudo-ECG equation is defined as follows,

$$\Phi_e = \sum_{j=1}^{N} -(\nabla V_m)_j \cdot \left[\nabla \frac{S_j}{r_j} \right], \tag{2}$$

where j is the index of tetrahedral element, N is the number of tetrahedral elements, ∇V_m is an estimated gradient, S is the normalized volume, and r is distance calculated using the centroid of each element. Here, we estimate the gradients by assigning $V_m^i = 1$ if the node i is activated, otherwise $V_m^i = 0$, to generate a scaled amplitude ECG signal. The pseudo-ECG method can efficiently generate normalized ECG signals without significant loss of morphological information compared to the bidomain simulation [27]. As a result, with the Eikonal models one could generate virtual subjects by setting different CVs and RNs.

2.3 Patient Specific Deep Computational Model

Figure 2 presents the network architecture details of the proposed PS-DCM. The c-VAE aims to reconstruct both anatomical and electrophysiological information. The outputs of Decoder$_{PC}$ are coarse and dense point clouds (PCs) to simultaneously learn global shape and local structures of ventricles [6]. The PC reconstruction loss function is defined as follows,

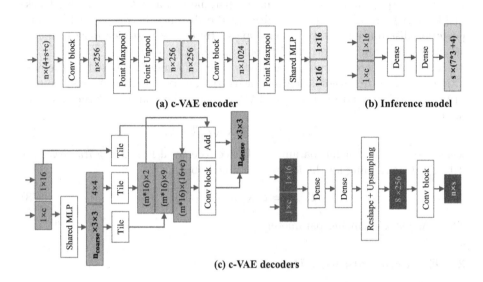

(a) c-VAE encoder

(b) Inference model

(c) c-VAE decoders

Fig. 2. The network architecture of the proposed PS-DCM, where the output of each module is labeled using bold text. Here, n is the number of nodes of input point cloud (PC) which is a 4D vector (three point coordinates and a class label); n_{coarse} and n_{dense} are the numbers of nodes in the coarse and dense output PCs, respectively; s is the number of simulated ECGs for each mesh (n is set as the same value as the size of ECG signal for the convenience of input PC and ECG concatenation); c is the number of conditions that has been reshaped to match the size of input PC for concatenation.

$$\mathcal{L}_{PC}^{rec} = \sum_{i=1}^{K} \left(\mathcal{L}_{i,coarse}^{EMD} + \alpha \mathcal{L}_{i,dense}^{EMD} \right), \tag{3}$$

where K is the number of classes, α is the weight term between the two PCs, and $\mathcal{L}^{EMD}$ is the earth mover's distance (EMD) [34] between the ground truth G_H and predicted heart geometries $\hat{G}_H$,

$$\mathcal{L}^{EMD}(G_H, \hat{G}_H) = \arg\min_{\xi:G_H \to \hat{G}_H} \sum_{p \in G_H} \|p - \xi(p)\|_2, \tag{4}$$

where p is a node of point cloud and $\xi : G_H \to \hat{G}_H$ is a one-to-one correspondence mapping. Decoder$_{ECG}$ predicts the reconstructed ECG signals by minimizing the mean absolute error (MAE) between the ground truth and predicted ECG (denoted as EĈG),

$$\mathcal{L}_{\text{ECG}}^{\text{rec}} = \mathcal{L}^{\text{MAE}}(\text{ECG}, \text{EĈG}). \tag{5}$$

Therefore, the loss function for training the c-VAE is calculated as,

$$\mathcal{L}_{\text{c-VAE}} = \lambda_{\text{PC}}\mathcal{L}_{\text{PC}}^{\text{rec}} + \lambda_{\text{ECG}}\mathcal{L}_{\text{ECG}}^{\text{rec}} + \lambda_{\text{KL}}\mathcal{L}^{\text{KL}}, \tag{6}$$

where λ_{PC}, λ_{ECG} and λ_{KL} are balancing parameters, and $\mathcal{L}^{\text{KL}}$ is the Kullback-Leibler (KL) divergence loss to mitigate the distance between the prior and posterior distributions of the latent space. Here, the posterior distribution is assumed as a standard normal distribution, i.e., $\mathcal{N}(0,1)$.

As for the inference model, we predict the CVs and RNs based on the conditions and low-dimensional features learned from the c-VAE. We also employ MAE and velocity-normalized (vnMAE) loss for the training of the inference model,

$$\mathcal{L}_{\text{inf}} = \mathcal{L}_{\text{RN}}^{\text{MAE}} + \lambda_{\text{CV}}\mathcal{L}_{\text{CV}}^{\text{vnMAE}}, \tag{7}$$

where λ_{CV} is a balancing parameter. Hence, the total loss of the framework is defined by combining all the losses mentioned above,

$$\mathcal{L}_{\text{total}} = \mathcal{L}_{\text{c-VAE}} + \lambda_{\text{inf}}\mathcal{L}_{\text{inf}}, \tag{8}$$

where λ_{inf} is a balancing parameter.

3   Experiments and Results

3.1   Materials

Data Acquisition and Pre-processing. We collect 100 healthy subjects with cine CMR images and ECGs from the UK Biobank study [9]. Besides, we employ biological information, i.e., age, sex, and body mass index (BMI), as the conditions of cVAE, as they may affect the ECG morphology [17,33]. For anatomical reconstruction, we resample all PCs into the PC with the same number of nodes, i.e., $n = 2048$, and the numbers of nodes in the coarse and dense output PCs are set as 1024 and 4096. For simulation, CVs are randomly selected within physiological ranges, i.e., fiber-directed, sheet-directed, sheet normal-directed, and endocardial-directed CVs are within the ranges [50, 88], [32, 49], [29, 45], and [120, 179] cm/s, respectively. Moreover, we constrain the fiber-directed speed to be larger than the sheet-directed one, which in turn is larger than the sheet normal-directed speed [10]. In contrast, RNs are set to seven fixed homologous locations (3 in the right ventricle and 4 in the left ventricle) for realistic application [13]. Note that we did not consider ventricle-specific protocols for RN placement to ensure more flexible extension in the future. For each mesh, we generate 10 virtual subjects with different ECGs but the same ventricle anatomy for the training of the PS-DCM. We randomly split the data into three sets, i.e., 60 subjects for training, 10 subjects for validation and the remaining 30 subjects for the test.

Gold Standard and Evaluation. We employ specific ventricular activation properties to simulate data, and then real resampled PCs, simulated ECGs and corresponding properties are regarded as ground truths of this model. We calculate the EMD from coarse and fine PCs with corresponding ground truth PCs as the accuracy of PC reconstruction. As there exist length variations among different leads of the simulated ECG signals, we perform zero-padding to unify the length of ECG. Therefore, in the evaluation of ECG reconstruction in this study, we only consider the non-zero signals when employing L1 loss as the discrepancy metric for the ECG data. As for the inference model evaluation, we employ the average distance of root nodes and a velocity-normalized error metric.

Implementation. The framework was implemented in PyTorch, running on a computer with 3.50 GHz Intel(R) Xeon(R) E-2146G CPU and an NVIDIA GeForce RTX 3060. We use the Adam optimizer to update the network parameters (weight decay = 1e−3). The initial learning rate is set to 1e−4 and multiplied by 0.7 every 30000 iterations. The balancing parameters in Sect. 2.3 are set as follows: $\alpha = 0.1$, $\lambda_{PC} = 0.1$, $\lambda_{ECG} = 0.1$, $\lambda_{KL} = 0.02$, $\lambda_{CV} = 0.2$, and $\lambda_{inf} = 0.01$. The training of the model took about 2 h (1000 epochs in total), while the inference of the networks required about 9 s to process one test image.

3.2 Results

Inference Accuracy of Ventricular Activation Properties. Table 1 presents the quantitative results of different methods for RN and CV inference. One can see that the proposed PS-DCM method obtains the best inference results compared to the other two schemes without conditions or PC reconstruction constraints, respectively. It reveals the importance of anatomy and physical information for the patient-specific activation property prediction. Even though we consider this information in our proposed framework, it is too challenging to predict the positions of RNs, as shown in Fig. 3. One can see that the predicted RNs are generally located inside the middle (or even outside) of the ventricles instead of locally matching with the ground truth RNs. It indicates that the anatomical constraints are desired for the deep learning based RN inference in the future. In contrast, the prediction of CVs is more promising with comparable results as the conventional method [11], and there exist some accuracy variations for different directions of CVs. Specifically, the sheet and endocardial-directed

Table 1. Summary of the quantitative results of ventricular activation properties. LV: left ventricle; RV: right ventricle; C: conditions.

Method	Root node error (cm)		Conduction velocity error (%)			
	LV	RV	Fiber	Sheet	Sheet-normal	Endocardial
PS-DCM w/o C	3.83 ± 1.10	3.90 ± 1.07	18.3 ± 4.47	22.2 ± 6.19	22.3 ± 7.27	12.1 ± 1.91
PS-DCM w/o PC	4.13 ± 1.07	3.75 ± 1.06	24.3 ± 6.26	29.7 ± 14.7	38.8 ± 20.6	14.8 ± 4.79
PS-DCM	2.63 ± 0.91	2.56 ± 0.88	13.9 ± 2.70	9.53 ± 3.13	15.1 ± 2.86	11.7 ± 2.23

CVs are better identified than the CVs in fiber and sheet-normal directions by the proposed model. For healthy subjects, the sheet and endocardial-directed CVs have been regarded as the dominant factors in the activation sequence patterns, while the impact from fiber and sheet-normal CVs is negligible [15]. Consequently, there exist performance differences in CV inference from ECG data under healthy sinus rhythm conditions. The fiber-directed CV may play a more important role in pathological conditions, which, however, is out of the scope of this study.

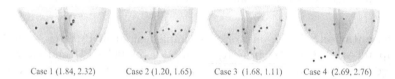

Case 1 (1.84, 2.32)   Case 2 (1.20, 1.65)   Case 3 (1.68, 1.11)   Case 4 (2.69, 2.76)

Fig. 3. Visualization of ground truth (yellow spheres) and predicted root nodes (cyan-blue spheres) by the proposed method from four randomly selected cases. Here, we put the average RN prediction errors on the LV and RV in brackets. (Color figure online)

Reconstruction Quality of Point Cloud and ECG. The average PC reconstruction errors of the proposed method are 4.34 ± 2.23 cm and 4.07 ± 2.08 cm for coarse and fine PCs, respectively. Figure 4 presents the ECG visualization results from eight leads. Note that in this study we only consider the eight independent leads instead of complete 12-lead ECG, as the remaining leads are linear combinations of the other leads [11]. One can see that, compared to the ground truth, the reconstructed ECGs are generally matched with input ECGs in several leads even though not quite smooth. There exists reconstruction accuracy variance among different leads, namely the prediction of leads V1, V2, and V3 had larger misalignment compared to other leads. We argue that simultaneous feature encoding of PCs and ECGs may be too challenging, resulting poor reconstruction results. In the future, we may consider employing separated encoders to extract features from PCs and ECGs [7,8].

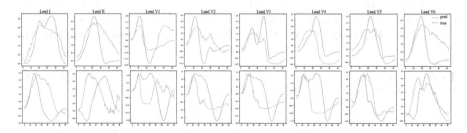

Fig. 4. Illustration of reconstructed ECGs (labeled in pink) with corresponding simulated ground truth ECGs (labeled in cyan-blue). (Color figure online)

Correlation Study. To evaluate the robustness of the proposed inference scheme to the reconstruction error, we analyze the relationship between the reconstruction and inference errors by the proposed method. We plot these two values for each test data as two-dimensional scatter points along with the fitted linear regression, as presented in Fig. 5. The R^2 values are estimated as 0.731, 0.000, 0.008, and 0.003 for PC-RN, PC-CV, ECG-RN, and ECG-CV correlations, respectively, indicating low linear correlations between inference and reconstruction accuracy except for PC-RN. It implies that the CV inference by the proposed method does not rely on accurate PC/ECG reconstruction results. In contrast, the RN inference may require accurate anatomy information from PC reconstruction but without the high demand for electrophysiological information. This is reasonable, as RNs belong to ventricle positions that are locally distributed on the endocardium while CVs are more general properties.

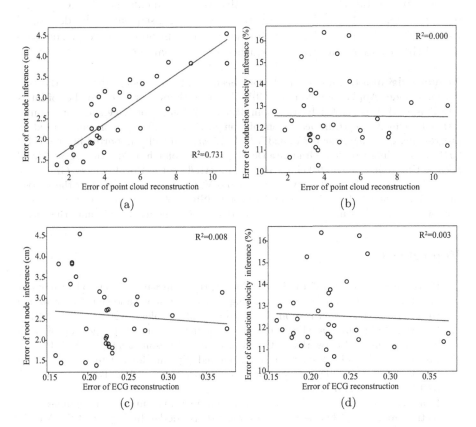

Fig. 5. Scatter plots along with the fitted linear regression lines, depicting the correlations between inference and reconstruction errors.

4 Conclusion

In this work, we have presented an end-to-end deep learning-based computational model for simultaneous inference of RNs and CVs combining the ECG and CMR images. The proposed algorithm has been applied to 100 simulated ECGs with the corresponding anatomies obtained from the UK biobank dataset. The results have demonstrated the potential of an efficient activation property inference based on the physical information and low-dimensional features from the c-VAE. Note that this is only a quite preliminary study with several limitations, such as assuming a known set of RNs and anisotropic CVs on specific directions (fiber, sheet, sheet-normal, and endocardial directions). Moreover, currently we only consider cardiac anatomical information and ignore the torso geometry, which provides important information during the propagation of electrical conduction. In the future, we will extend this work by including a more realistic representation of the cardiac conduction system. Consequently, the developed models and techniques will enable further research in non-invasive personalization of the ventricular activation sequences for real patients.

Acknowledgement. This research has been conducted using the UK Biobank Resource under Application Number '40161'. The authors express no conflict of interest. This work was funded by the CompBioMed 2 Centre of Excellence in Computational Biomedicine (European Commission Horizon 2020 research and innovation programme, grant agreement No. 823712). L. Li was partially supported by the SJTU 2021 Outstanding Doctoral Graduate Development Scholarship. A. Banerjee is a Royal Society University Research Fellow and is supported by the Royal Society Grant No. URF\R1\221314. The work of A. Banerjee and V. Grau was supported by the British Heart Foundation (BHF) Project under Grant HSR01230. The work of M. Beetz was supported by the Stiftung der Deutschen Wirtschaft (Foundation of German Business).

References

1. Bacoyannis, T., Ly, B., Cedilnik, N., Cochet, H., Sermesant, M.: Deep learning formulation of electrocardiographic imaging integrating image and signal information with data-driven regularization. EP Europace **23**(Supplement_1), i55–i62 (2021)
2. Bai, W., et al.: Automated cardiovascular magnetic resonance image analysis with fully convolutional networks. J. Cardiovasc. Magn. Reson. **20**(1), 1–12 (2018)
3. Banerjee, A., et al.: A completely automated pipeline for 3D reconstruction of human heart from 2D cine magnetic resonance slices. Phil. Trans. R. Soc. A **379**(2212), 20200257 (2021)
4. Banerjee, A., Zacur, E., Choudhury, R.P., Grau, V.: Optimised misalignment correction from cine MR slices using statistical shape model. In: Papież, B.W., Yaqub, M., Jiao, J., Namburete, A.I.L., Noble, J.A. (eds.) MIUA 2021. LNCS, vol. 12722, pp. 201–209. Springer, Cham (2021). https://doi.org/10.1007/978-3-030-80432-9_16
5. Bayly, P.V., KenKnight, B.H., Rogers, J.M., Hillsley, R.E., Ideker, R.E., Smith, W.M.: Estimation of conduction velocity vector fields from epicardial mapping data. IEEE Trans. Biomed. Eng. **45**(5), 563–571 (1998)

6. Beetz, M., Banerjee, A., Grau, V.: Generating subpopulation-specific biventricular anatomy models using conditional point cloud variational autoencoders. In: Puyol Antón, E., et al. (eds.) STACOM 2021. LNCS, vol. 13131, pp. 75–83. Springer, Cham (2022). https://doi.org/10.1007/978-3-030-93722-5_9

7. Beetz, M., Banerjee, A., Grau, V.: Multi-domain variational autoencoders for combined modeling of MRI-based biventricular anatomy and ECG-based cardiac electrophysiology. Front. Physiol. 991 (2022)

8. Beetz, M., Banerjee, A., Sang, Y., Grau, V.: Combined generation of electrocardiogram and cardiac anatomy models using multi-modal variational autoencoders. In: 2022 IEEE 19th International Symposium on Biomedical Imaging (ISBI), pp. 1–4. IEEE (2022)

9. Bycroft, C., et al.: The UK Biobank resource with deep phenotyping and genomic data. Nature 562(7726), 203–209 (2018)

10. Caldwell, B.J., Trew, M.L., Sands, G.B., Hooks, D.A., LeGrice, I.J., Smaill, B.H.: Three distinct directions of intramural activation reveal nonuniform side-to-side electrical coupling of ventricular myocytes. Circul. Arrhythmia Electrophysiol. 2(4), 433–440 (2009)

11. Camps, J., et al.: Inference of ventricular activation properties from non-invasive electrocardiography. Med. Image Anal. 73, 102143 (2021)

12. Cantwell, C.D., Roney, C.H., Ng, F.S., Siggers, J.H., Sherwin, S.J., Peters, N.S.: Techniques for automated local activation time annotation and conduction velocity estimation in cardiac mapping. Comput. Biol. Med. 65, 229–242 (2015)

13. Cardone-Noott, L., Bueno-Orovio, A., Mincholé, A., Zemzemi, N., Rodriguez, B.: Human ventricular activation sequence and the simulation of the electrocardiographic QRS complex and its variability in healthy and intraventricular block conditions. EP Europace 18(suppl_4), iv4–iv15 (2016)

14. Chinchapatnam, P., et al.: Model-based imaging of cardiac apparent conductivity and local conduction velocity for diagnosis and planning of therapy. IEEE Trans. Med. Imaging 27(11), 1631–1642 (2008)

15. Durrer, D., Van Dam, R.T., Freud, G., Janse, M., Meijler, F., Arzbaecher, R.: Total excitation of the isolated human heart. Circulation 41(6), 899–912 (1970)

16. Engwirda, D.: Conforming restricted Delaunay mesh generation for piecewise smooth complexes. Proc. Eng. 163, 84–96 (2016)

17. Fraley, M., Birchem, J., Senkottaiyan, N., Alpert, M.: Obesity and the electrocardiogram. Obes. Rev. 6(4), 275–281 (2005)

18. Giffard-Roisin, S., et al.: Transfer learning from simulations on a reference anatomy for ECGI in personalized cardiac resynchronization therapy. IEEE Trans. Biomed. Eng. 66(2), 343–353 (2018)

19. Gillette, K., et al.: Automated framework for the inclusion of a His-Purkinje system in cardiac digital twins of ventricular electrophysiology. Ann. Biomed. Eng. 49(12), 3143–3153 (2021)

20. Good, W.W., et al.: Estimation and validation of cardiac conduction velocity and wavefront reconstruction using epicardial and volumetric data. IEEE Trans. Biomed. Eng. 68(11), 3290–3300 (2021)

21. Grandits, T., Effland, A., Pock, T., Krause, R., Plank, G., Pezzuto, S.: GEASI: Geodesic-based earliest activation sites identification in cardiac models. Int. J. Numer. Methods Biomed. Eng. 37(8), e3505 (2021)

22. Grandits, T., et al.: An inverse Eikonal method for identifying ventricular activation sequences from epicardial activation maps. J. Comput. Phys. 419, 109700 (2020)

23. Han, B., Trew, M.L., Zgierski-Johnston, C.M.: Cardiac conduction velocity, remodeling and arrhythmogenesis. Cells **10**(11), 2923 (2021)
24. Kaptoge, S., et al.: World Health Organization cardiovascular disease risk charts: revised models to estimate risk in 21 global regions. Lancet Glob. Health **7**(10), e1332–e1345 (2019)
25. Martinez-Navarro, H., Zhou, X., Bueno-Orovio, A., Rodriguez, B.: Electrophysiological and anatomical factors determine arrhythmic risk in acute myocardial ischaemia and its modulation by sodium current availability. Interface Focus **11**(1), 20190124 (2021)
26. Meister, F., et al.: Graph convolutional regression of cardiac depolarization from sparse endocardial maps. In: Puyol Anton, E., et al. (eds.) STACOM 2020. LNCS, vol. 12592, pp. 23–34. Springer, Cham (2021). https://doi.org/10.1007/978-3-030-68107-4_3
27. Mincholé, A., Zacur, E., Ariga, R., Grau, V., Rodriguez, B.: MRI-based computational torso/biventricular multiscale models to investigate the impact of anatomical variability on the ECG QRS complex. Front. Physiol. 1103 (2019)
28. Niederer, S.A., Lumens, J., Trayanova, N.A.: Computational models in cardiology. Nat. Rev. Cardiol. **16**(2), 100–111 (2019)
29. Park, K.M., Kim, Y.H., Marchlinski, F.E.: Using the surface electrocardiogram to localize the origin of idiopathic ventricular tachycardia. Pacing Clin. Electrophysiol. **35**(12), 1516–1527 (2012)
30. Pashaei, A., Romero, D., Sebastian, R., Camara, O., Frangi, A.F.: Fast multiscale modeling of cardiac electrophysiology including Purkinje system. IEEE Trans. Biomed. Eng. **58**(10), 2956–2960 (2011)
31. Pezzuto, S., et al.: Reconstruction of three-dimensional biventricular activation based on the 12-lead electrocardiogram via patient-specific modelling. EP Europace **23**(4), 640–647 (2021)
32. Si, H.: TetGen, a Delaunay-based quality tetrahedral mesh generator. ACM Trans. Math. Softw. (TOMS) **41**(2), 1–36 (2015)
33. Taneja, T., Windhagen Mahnert, B., Passman, R., Goldberger, J., Kadish, A.: Effects of sex and age on electrocardiographic and cardiac electrophysiological properties in adults. Pacing Clin. Electrophysiol. **24**(1), 16–21 (2001)
34. Yuan, W., Khot, T., Held, D., Mertz, C., Hebert, M.: PCN: point completion network. In: 2018 International Conference on 3D Vision, pp. 728–737. IEEE (2018)

CMRxMotion Challenge Papers

Semi-supervised Domain Generalization for Cardiac Magnetic Resonance Image Segmentation with High Quality Pseudo Labels

Wanqin Ma[1(✉)], Huifeng Yao[1], Yiqun Lin[1], Jiarong Guo[1], and Xiaomeng Li[1,2]

[1] Department of Electronic and Computer Engineering,
The Hong Kong University of Science and Technology, Hong Kong, China
{wmaag,hyaoad,yiqun.lin,jguoaz}@connect.ust.hk, eexmli@ust.hk
[2] Shenzhen Research Institute, The Hong Kong University of Science and Technology, Shenzhen, China

Abstract. Developing a deep learning method for medical segmentation tasks heavily relies on a large amount of labeled data. However, the annotations require professional knowledge and are limited in number. Recently, semi-supervised learning has demonstrated great potential in medical segmentation tasks. Most existing methods related to cardiac magnetic resonance images only focus on regular images with similar domains and high image quality. A semi-supervised domain generalization method was developed in [2], which enhances the quality of pseudo labels on varied datasets. In this paper, we follow the strategy in [2] and present a domain generalization method for semi-supervised medical segmentation. Our main goal is to improve the quality of pseudo labels under extreme MRI Analysis with various domains. We perform Fourier transformation on input images to learn low-level statistics and cross-domain information. Then we feed the augmented images as input to the double cross pseudo supervision networks to calculate the variance among pseudo labels. We evaluate our method on the CMRxMotion dataset [1]. With only partially labeled data and without domain labels, our approach consistently generates accurate segmentation results of cardiac magnetic resonance images with different respiratory motions. Code will be available after the conference.

Keywords: Semi-supervised learning · Domain generalization · Medical segmentation

1 Introduction

In practice medical cases, accurate segmentation result is highly demanded as they can give important structural instructions on disease treatment and diagnosis. With the development of many convolutional neural networks [5–8,10–12] for medical image segmentation, deep learning methods are widely applied in varied

O. Camara et al. (Eds.): STACOM 2022, LNCS 13593, pp. 383–391, 2022.
https://doi.org/10.1007/978-3-031-23443-9_35

medical tasks. Medical segmentation on cardiac magnetic resonance images is one of the most important issues. Most existing methods focus on pure cardiac segmentation in regular MRI with a similar domain. In practical cases, cardiac magnetic resonance (CMR) images are obtained from different patients with various equipments, provided under unstable imaging environments, affected by population shifts and unexpected human behaviors. This paper focuses on solving the issue raised in the CMRxMotion challenge [1]. The CMRxMotion dataset is collected from 45 patients with four different respiratory motion stages, which means the data is varied and leads to the problem of domain generalization. The example images of the CMRxMotion dataset are shown in Fig. 1. However, the above methods are designed to solve medical segmentation tasks in regular CMR images and generate inferior results in varied CMR image domains. Hence, these methods are not suitable for solving problems raised in the CMRxMotion challenge [1].

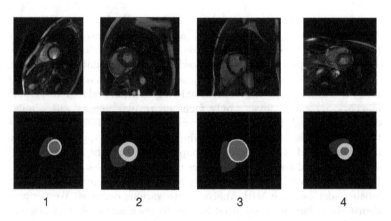

Fig. 1. These are representative 2D slices of images and masks from CMRxMotion Dataset [1]. The pictures are randomly selected from different patients with different stages. The original pictures are in the first row. The masks are in the second row. Labels 1, 2, 3, 4 refer to different breath-hold stages: full breath-hold stage, half breath-hold stage, free breath stage, and intensive breath stage, respectively. There are three different colors in masks and refer to different structures: red refers to left ventricle blood pool, blue refers to the right ventricle blood pool, green refers to the left ventricular myocardium. (Color figure online)

To deal with the problems caused by domain generalization, many methods have been proposed. Li et al. [9] developed representative methods: skin lesion classification, and spinal cord gray matter segmentation. The method in [9] is aimed to learn consistent semantic information among multiple domains by capturing feature space. The above methods demand a fully-labeled dataset in each domain. However, in CMRxMotion dataset [1], some images are unlabeled because of bad quality. Hence, the above method does not apply to CMRxMotion dataset [1]. Generating pseudo labels is an effective method to use the dataset, including unlabeled data fully. Chen et al. [4] introduced a cross pseudo supervision method by generating two segmentation networks. The two networks supervised each other by the corresponding pseudo labels. The unlabeled data in [4] is from the same distribution as the labeled data; thus, pseudo labels from the same distribution can be used to improve another segmentation network directly. However, in CMRxMotion dataset [1], the unlabeled images are from unknown distribution because of the variance of patients and different respiratory motion, leading to a biased pseudo label.

Yao et al. [2] first attempt at combining Fourier transformation and cross pseudo supervision [4] to improve the quality of pseudo labels caused by domain generalization. For [2], Fourier transformation is used to proceed with data augmentation and obtain important low-level statistics from the data, and a double network of cross pseudo supervision can train the model effectively by using pseudo labels. Specifically, it first randomly selects a sample to augment by Fourier transformation. Then, they use the original and augmented samples as input and pass them to the two cross pseudo supervision networks with the same structure. For the first network, they get prediction of augmented sample and the original sample. After that, they count the variance of the two predictions and generate the first one-hot vector. For the second network, the authors repeat the same steps as the first network and obtain the second one-hot vector. With two one-hot vectors, the two networks can supervise each other and improve the performance of the model.

Following this philosophy, in this work, we ensemble the semi-supervised framework consisting of Fourier transformation and cross pseudo supervision, which is proposed in [2]. The ensemble network is aimed to enhance the quality of pseudo labels, then can facilitate the whole training process. We conduct experiments on the Extreme Cardiac MRI Analysis Challenge under the Respiratory Motion dataset [1]. We find that, under the condition of extreme cardiac MRI Analysis, our method can always generate accurate segmentation results of cardiac magnetic resonance images with different respiratory motions.

2 Methodology

2.1 Data Augmentation Based on Fourier Transformation

The overall methods are demonstrated in Fig. 2. During the training process, we input source images to the model without the score of the quality, the serial number of patient and respiratory motion stage. Then we obtain an amplitude spectrum S and a phase image P by performing Fourier transformation F on source images. Specifically, we can collect low-level statistics from amplitude spectrum and high-level semantics from phase images. After that, we repeat the steps above on another randomly selected sample image I' and obtain a new amplitude spectrum S'. With the information collected from the second image I', we can augment the first image I by:

$$I_{new} = (1 - \lambda)S * (1 - M) + \lambda S' * M \qquad (1)$$

where I_{new} is the augmented phase image; λ is a parameter to control the proportion of the phase information of I and I'; and M is a binary mask to exchange the spatial range of amplitude spectrum. To obtain low-frequency information, we set M as the central region of the amplitude spectrum. Then we perform F^{-1} inverse Fourier transformation on new sample to transform it from frequency domain to image domain. After that, we obtain the image sample X which includes low-level information from another sample by Fourier transformation:

$$X = F^{-1}(I_{new}, P) \tag{2}$$

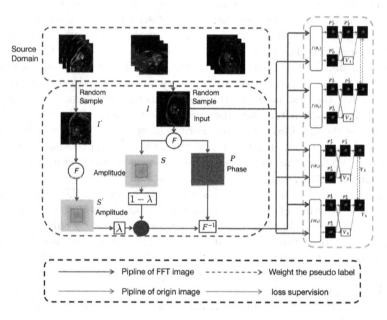

Fig. 2. The overall architecture of the proposed double confidence-aware cross-pseudo supervision network. The yellow area refers to the source domain. The orange area describes the steps of Fourier transformation. The green area is the network of the first CACPS model with DeepLabv3+ backbone. The blue area is the network of the second CACPS model with Resnet101 backbone. (Color figure online)

2.2   Ensemble of Two Confidence-Aware Cross Pseudo Supervision

In this section of the training process, we follow the strategy in [2], which introduced method Confidence-Aware Cross Pseudo Supervision (CACPS) to build the segmentation networks. We use double Confidence-Aware Cross Pseudo Supervision with the same structure. We train two CACPS models separately. In each CACPS, we build two parallel segmentation networks $f1$ and $f2$ (or $f3$ and $f4$) with the same structure but different initialized weights. Then, we take original image I and the augmented image X into the two networks above to obtain the predicted pseudo label of both images by:

$$P_F^{1(3)} = f(\theta_{1(3)})(X) \qquad (3)$$

$$P_O^{1(3)} = f(\theta_{1(3)})(I) \qquad (4)$$

$$P_F^{2(4)} = f(\theta_{2(4)})(X) \qquad (5)$$

$$P_O^{2(4)} = f(\theta_{2(4)})(I) \qquad (6)$$

where $P_F^{1(3)}$ and $P_F^{2(4)}$ are pseudo labels of the augmented image; $P_O^{1(3)}$ and $P_O^{2(4)}$ are pseudo labels of the origin image; X is the augmented image, I is the origin image.

After that, we use the pseudo labels from one network to supervise the pseudo labels from another network because we do not use labels in supervised part to generate supervision signals for the unlabeled data. This method is from cross pseudo supervision [4]. The model cross pseudo supervision [4] can generate pseudo label with high quality because of its consistent training data. However, the Extreme Cardiac MRI Analysis Challenge [1] data is from different domains with different levels of qualities. This varied dataset may cause low-quality prediction. Method Confidence-Aware Cross Pseudo Supervision [2] improves quality of pseudo label by counting variance between different input data. As showed in Fig. 2, we count the predictions from the original image and transformed image, $P_O^{1(3)}$ and $P_F^{1(3)}$. Then we calculate the average value $P_E^{1(3)}$ of $P_O^{1(3)}$ and $P_F^{1(3)}$ by:

$$P_E^{1(3)} = (P_O^{1(3)} + P_F^{1(3)})/2 \qquad (7)$$

Similarly, we obtain the $P_E^{2(4)}$ by:

$$P_E^{2(4)} = (P_O^{2(4)} + P_F^{2(4)})/2 \qquad (8)$$

To test the quality of the pseudo labels, we then calculate the variance $V_{1(3)}$ of predictions $P_O^{1(3)}$ and $P_F^{1(3)}$, the variance $V_{2(4)}$ of predictions $P_O^{2(4)}$ and $P_F^{2(4)}$:

$$V_{1(3)} = E[P_F^{1(3)} \log(\frac{P_F^{1(3)}}{P_O^{1(3)}})] \qquad (9)$$

$$V_{2(4)} = E[P_F^{2(4)} \log(\frac{P_F^{2(4)}}{P_O^{2(4)}})] \qquad (10)$$

where E is the expectation value. If the variance value is large, this indicates the difference between two predictions is large and the quality of pseudo label is low. Then, we obtain one-hot vectors $Y_{1(3)}$ and $Y_{2(4)}$ from the probability maps $P_E^{1(3)}$ and $P_E^{2(4)}$. We define the confidence-aware loss function $L_{cacps} = L_a + L_b$ to improve the models by counting cross supervision loss:

$$L_a = E[e^{-V_{1(3)}}L_{ce}(P_E^{2(4)}, Y_{1(3)}) + V_{1(3)}] \qquad (11)$$

$$L_b = E[e^{-V_{2(4)}} L_{ce}(P_E^{1(3)}, Y_{2(4)}) + V_{2(4)}] \tag{12}$$

where L_{ce} is the cross-entropy loss. We use dice loss as loss function in the supervised part. We formulate the supervision loss L_s.

$$L_s = E[L_{Dice}(P_O^{1(3)}, G_{1(3)}) + L_{Dice}(P_O^{2(3)}, G_{2(4)}] \tag{13}$$

where L_{Dice} denotes the dice loss function and $G_{1(3)}(G_{2(4)})$ is the ground truth. Hence, the whole loss function is:

$$L = L_s + \beta * L_{cacps} \tag{14}$$

where β is the CACPS weight. This parameter is help to keep balance between the two losses; L_{cacps} is the confidence-aware loss. We use the combination of two separated model's predictions as final results for the whole Condidence-Aware Cross Pseudo Supervision section.

As we actually use two CACPS models, in test procedure, we use the average of two Confidence-Aware Cross Pseudo Supervision Models' prediction as final result.

3   Experiments and Results

3.1   Dataset and Preprocessing

We evaluate our methods on a public medical image segmentation dataset from MICCAI 2022, the Extreme Cardiac MRI Analysis Challenge [1]. The dataset contains 360 3D clinical CMR scans from 45 healthy volunteers. A set contains 4 subsets obtained from a single volunteer under four levels of respiratory motion, including breath-hold state, halve the breath-hold state, breath freely states, and breathe intensively state. There are 2 CMR scans in a single subset. Among 360 CMR images, a set of 200 CMR scans are for training and validation, and the other 160 images are for the test. The annotations include three structures of the heart: left ventricle, left ventricle myocardium, and right ventricle.

Regarding the preprocessing, we generate 2D slices from the original 3D CMR images. Then we resize the slices of all 3D images to keep the correspondence in shape 512×512.

3.2   Experiment Setting

We ran the experiments using four Nvidia Geforce RTX 3090 GPUs with 96 GB RAM on the Ubuntu20.04 system. Similar to [2], we implemented the both CACPS models on PyTorch1.8. For the first CACPS model, we use DeepLabv3+ [3] as the backbone. For the second one, we use the ResNet101 [13] as backbone.

We can only access training and validation data during the challenge period. To train and test our model normally, we proceed following steps:

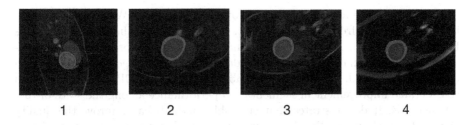

Fig. 3. These are representative segmentation results of Validation data. The pictures are randomly selected. Labels 1, 2, 3, 4 refer to different breath-hold stages: full breath-hold stage, half breath-hold stage, free breath stage, and intensive breath stage, respectively.

There are 160 CMR scans generated by 20 volunteers in total. We divide the training dataset into three parts according to its quality scores one, two or three. We name the group of images with quality score of one as domain 1. Similarly, we name domain 2 and domain 3. The ratio of the number of images in domain 1, domain 2, and domain 3 is 70:69:21. We notice that some images in the CMRxMotion Dataset is without masks. Hence, we use all training data to train the model through semi-supervised learning, and we select label data for validation. For Validation data, we use it to test our model.

We choose AdamW as the optimizer of the network. The initial learning rate is 0.00001 and the number of epochs is 100. We use CosineAnnealingLR in pytorch to change learning rate effectively with cosine period in each epoch. And the batch size is 16. The weight of CACPS is 1.5.

3.3 Experiment Results

We proceed with the experiment on three different methods. DeepLabv3+, a trained Confidence Aware Cross Pseudo Supervision model fine-tuning (Single CACPS (DeepLabv3+)) and the ensemble of double CACPS models (Double CACPS (DeepLabv3+Resnet101)). We use the dice scores given by CMRxMotion Challenge Website. The visualization of the segmentation results using our double CACPS models are showed in Fig. 3.

Table 1. The table reveals the dices scores of all methods on validation dataset. LV-dice scores, MYO-dice scores, RV-dice scores refer to the accuracy of segmentation on left ventricle blood pool, the right ventricle blood pool, and the left ventricular myocardium, respectively.

Methods	LV-dice scores	MYO-dice scores	RV-dice scores	Avg-dice scores
DeepLabv3+	0.799	0.060	0.781	0.547
CACPS (DeepLabv3+)	0.804	0.552	0.802	0.720
CACPS (DeepLabv3+Resnet101)	0.809	0.594	0.810	0.738

Table 1 lists the average dice scores of all the methods. We find some meaningful information from this table: First, the use of the Confidence Aware Cross Pseudo Supervision network can enhance the quality of pseudo labels significantly. It shows that using cross pseudo supervision can get predictions with high quality. Second, we notice that combining the two CACPS networks gains a considerable improvement in segmentation performance among the state-of-the-art methods. It demonstrates that ensemble methods can improve the quality of pseudo labels more. Besides, it is noticeable that the performance of only DeepLabv3+ is not good. According to our analysis, the main reasons are as follows: First, some images in the CMRxMotion challenge are of bad quality under extreme respiratory motion. Although DeepLabv3+ is strong, it still can't learn data features effectively without augmentation. Second, we observe that the decline of the loss of model is prolonged after around 50 epochs when only using DeepLab3+. Without double CACPS networks, the learning efficiency of the model is low.

After the challenge, we receive the results of the average dice scores of double Confidence Aware Cross Pseudo Supervision network on the test dataset. The details of the result are list in Table 2.

Table 2. The table reveals the dices scores of only CACPS (DeepLabv3+Resnet101) on test dataset. LV-dice scores, MYO-dice scores, RV-dice scores refer to the accuracy of segmentation on left ventricle blood pool, the right ventricle blood pool, and the left ventricular myocardium, respectively.

Methods	LV-dice scores	MYO-dice scores	RV-dice scores	Avg-dice scores
CACPS (DeepLabv3+Resnet101)	0.866	0.802	0.815	0.828

4 Conclusion

It's always very challenged to develop deep learning method for medical segmentation with limited annotation and various domain. Aiming at tackling this challenge, in this work, we develop a semi-supervised method to learn key information under domain generalization. Further, we use two strategies, Fourier transformation and cross pseudo supervision to improve the quality of prediction. Experiments on the Extreme Cardiac MRI Analysis Challenge under Respiratory Motion dataset [1] with partially labeled data reveals that when combining two confidence aware cross pseudo supervision. The resulting semi-supervised learning achieves accurate segmentation results on cardiac magnetic resonance images with different respiratory motions.

References

1. Wang, S., et al.: The extreme cardiac mri analysis challenge under respiratory motion (CMRxMotion). arXiv preprint arXIv: 2210.06385 (2022)

2. Yao, H., Hu, X., Li, X.: Enhancing pseudo label quality for semi-supervised domain-generalized medical image segmentation. arXiv preprint arXiv:2201.08657 (2022)
3. Chen, L.-C., Papandreou, G., Schroff, F., Adam, H.: Rethinking atrous convolution for semantic image segmentation. arXiv preprint arXiv:1706.05587 (2017)
4. Chen, X., Yuan, Y., Zeng, G., Wang, J.: Semi-supervised semantic segmentation with cross pseudo supervision. In: IEEE/CVF Conference on Computer Vision and Pattern Recognition, pp. 2613–2622 (2021)
5. Linardos, A., Kushibar, K., Walsh, S.: Federated learning for multi-center imaging diagnostics: a simulation study in cardiovascular disease. Sci. Rep. **12**, 3551 (2022)
6. Ren, H., Raj, A., El-Khamy, M., Lee, J.: SUW-learn: joint supervised, unsupervised, weakly supervised deep learning for monocular depth estimation. In: IEEE/CVF Conference on Computer Vision and Pattern Recognition, pp. 750–751 (2020)
7. Bernard, O., et al.: Deep learning techniques for automatic MRI cardiac multi-structures segmentation and diagnosis: is the problem solved? IEEE Trans. Med. Imaging **37**(11), 2514–2525 (2018)
8. Isensee, F., Jaeger, P.F., Full, P.M., Wolf, I., Engelhardt, S., Maier-Hein, K.H.: Automatic cardiac disease assessment on cine-MRI via time-series segmentation and domain specific features. In: Pop, M., et al. (eds.) STACOM 2017. LNCS, vol. 10663, pp. 120–129. Springer, Cham (2018). https://doi.org/10.1007/978-3-319-75541-0_13
9. Li, H., Wang, Y., Wan, R., Wang, S., Li, T.-Q., Kot, A.C.: Domain generalization for medical imaging classification with linear-dependency regularization. arXiv preprint arXiv:2009.12829 (2022)
10. Li, X., Yu, L., Chen, H., Fu, C.W., Xing, L., Heng, P.A.: Transformation consistent self-ensembling model for semi-supervised medical image segmentation. arXiv preprint arXiv:1903.0034 (2019)
11. Lin, Y., Yao, H., Li, Z., Zheng, G., Li, X.: Calibrating label distribution for class-imbalanced barely-supervised knee segmentation. arXiv preprint arXiv:2205.03644 (2022)
12. Li, X., Yu, L., Chen, H., Fu, C.W., Heng, P.A.: Semi-supervised skin lesion segmentation via transformation consistent self-ensembling model. arXiv preprint arXiv:1808.03887 (2018)
13. Iakubovskii, P.: Segmentation models Pytorch. GitHub repository (2019). https://github.com/qubvel/segmentation_models.pytorch

Cardiac Segmentation Using Transfer Learning Under Respiratory Motion Artifacts

Carles Garcia-Cabrera[1]([✉]) [iD], Eric Arazo[1] [iD], Kathleen M. Curran[2] [iD],
Noel E. O'Connor[1] [iD], and Kevin McGuinness[1] [iD]

[1] Dublin City University, Dublin, Ireland
carles.garciacabrera6@mail.dcu.ie
[2] University College Dublin, Dublin, Ireland

Abstract. Methods that are resilient to artifacts in the cardiac magnetic resonance imaging (MRI) while performing ventricle segmentation, are crucial for ensuring quality in structural and functional analysis of those tissues. While there has been significant efforts on improving the quality of the algorithms, few works have tackled the harm that the artifacts generate in the predictions. In this work, we study fine tuning of pretrained networks to improve the resilience of previous methods to these artifacts. In our proposed method, we adopted the extensive usage of data augmentations that mimic those artifacts. The results significantly improved the baseline segmentations (up to 0.06 Dice score, and 4 mm Hausdorff distance improvement).

Keywords: Deep learning · Medical imaging · Segmentation · Transfer learning

1 Introduction

Cardiac-related diseases (CVDs) are the most prominent cause of death globally [9]. To accelerate, cheapen and enhance the quality of the diagnostics and treatment of the patient with cardiovascular diseases, the related techniques and technologies should be accurate and efficient. Non-invasive imaging modalities such as cardiac magnetic resonance imaging (CMRI) are particularly useful in the clinical assessment due to being capable of providing detailed data from patients. Extracting morphological and functional information from these data is tedious and intensive and can lead to observer bias [12].

The automation of these tasks has attracted the attention of scientists due to its high impact on daily clinical workflows. In the last few years, the innovation of deep learning techniques, and in particular convolutional neural networks, have brought more interest in the topic and have demonstrated great potential [3,5,7]. While recent efforts address approaches to resolve the quality of the segmentation of all the chambers within the heart (right ventricle, left ventricle, and myocardium) and multiple views of the tissue (long-axis and short-axis) [2]; sub-optimal segmentations, e.g. those affected by respiratory artifacts, are still under-explored.

© The Author(s), under exclusive license to Springer Nature Switzerland AG 2022
O. Camara et al. (Eds.): STACOM 2022, LNCS 13593, pp. 392–398, 2022.
https://doi.org/10.1007/978-3-031-23443-9_36

To address this, in this work, we propose a method to benefit from pretrained models that are publicly available. Using the pretrained weights not only hasted the training process, but also enhanced the predictions of the network. In particular, the results were substantially better for the right ventricle (RV) and the myocardium (MYO), leading to an increase in DICE and a reduction in Hausdorff. Results on the left ventricle (LV), however, remained unchanged. This work was done in the context of the Extreme Cardiac MRI Analysis Challenge under Respiratory Motion (CMRxMotion), where we focused on Task 2 (the segmentation task).

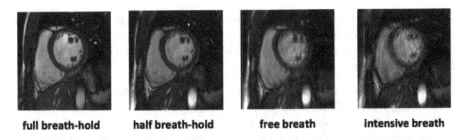

full breath-hold   **half breath-hold**   **free breath**   **intensive breath**

Fig. 1. The four different breathing intensities resulting in motion artifacts present in the CMRxMotion Challenge data (official challenge images).

2   Method

Our proposed method consists of: (1) resampling, preprocessing, and normalising the data, (2) loading a pretrained model and setting the appropriate training parameters, and (3) training the loaded network with data that have been augmented with a number of different deformation and intensity changes.

We refer to our 3D data as volumes and our short-axis slices as images. The data resampling, preprocessing and normalisation were done in 3D, while the data augmentation was performed over the slices.

Our method focuses on fine tuning an encoder during the segmentation task (with a segmentation head and decoder).

2.1   Data Resampling, Preprocessing and Normalisation

First, we reoriented the volumes to the canonical orientation, which were then resampled. A crop was then applied to the region of interest. The data were subsequently split into training and validation subsets, where validation represented 20% of the available subjects.

Second, the intensity was normalized using a histogram that we obtain out of the training samples. We perform the histogram standardization [8] of all sets with the mentioned histogram.

2.2   Architecture Study

In this work, we tested two different versions of the widely used U-Net architecture [13] to experiment if pretrained weights from a different problem and data domain could lead to improved results. The two approaches are:

- a U-Net trained from scratch;
- a ResNet-based [6] U-Net architecture with weights pretrained for ImageNet [4] classification.

2.3   Data Augmentation

An important part of our study consisted of applying four different types of augmentation techniques, which have previously been shown to enhance the quality of the predictions on scans without the above-mentioned artifacts [1]. These augmentations were:

- Random Motion: simulates movement in the tissue during the scan acquisition, and follows Shaw et al. [14].
- Random Ghosting: since respiratory motion is a cause for "ghost" artifacts this augmentation simulates the effect along the phase-encoded direction, altering the intensity of the imaged structures.
- Random Bias Field: this field is modelled as a linear combination of polynomial functions, as in Van Leemput et al. [15]. These fields create inhomogeneity in the intensity of low frequencies throughout the image.
- Random Gamma: this intensity transform consists of a random change on the contrast of an image by raising its values.

2.4   Cardiac MRI Dataset

The dataset for our study was provided by the Extreme Cardiac MRI Analysis Challenge under Respiratory Motion [16]. In particular, we employed the data from the segmentation challenge (task 2).

Training data represented a cohort of 20 subjects, which were scanned four different times, each under a different grade of respiratory intensity as in Fig. 1, described as follows: (1) full breath hold, (2) half breath hold, (3) free breathing, and (4) intensive breathing. The evaluation data represented a cohort of five subjects with the same four different intensities of breathing. Lastly, the test data represented a cohort of 15 subjects, unavailable for participants, just used in the offline testing phase. For all subjects and breathing intensities, an expert radiologist recorded and labelled the end-diastolic (ED) and end-systolic (ES) phases.

3   Experiment Settings

The experiment settings that were set during the training and inference of our method are described in the following section.

First, the architectures in Sect. 2.2 had the following details:

- U-Net trained from scratch: 32 filters in the first out of five pairs of convolutional layers and a max-pooling layer after each of the four first pairs of convolutional layers. The convolutional layers used the ReLU activation and batch normalization.
- U-Net pretrained: used a ResNet101 [6] backbone as an encoder, pretrained with ImageNet [4].

The general learning rate was set at 10^{-3} except for fine tuning, where the learning rate for the trained encoder was set at 10^{-4}. The learning rate was scheduled to change on plateau with patience of 100 epochs, reducing the learning rate to half its previous value. The optimizer chosen was Adam.

The steps detailed in Sects. 2.1 and 2.3 were done using the TorchIO library [11]. The pretrained network was downloaded from PyTorch Segmentation Models [10].

The augmentation policy that was applied consisted of applying always one of the techniques described in Sect. 2.3. In particular, we found that making the random motion augmentation three times as much as the others corresponded to a model more resilient to respiratory motion artifacts, leads to better segmentation performance.

4 Results

We present our results in three different sections: (1) validation results, (2) evaluation results which correspond to the results provided by the challenge platform, and (3) the test results released during the final stage of the challenge.

4.1 Validation Results

Table 1 shows the results on the validation split of the training set. Four different models are listed, indicating whether the training included augmentations, and if the weights were trained from scratch or pretrained on ImageNet.

Table 1. Validation results (DICE). Augs indicates the additional data augmentation. Best results in bold.

	DICE			
	LV	MYO	RV	ALL
U-Net (scratch)	0.97	0.945	0.963	0.959
U-Net (scratch) Augs	0.974	0.949	0.965	0.963
U-Net (ImageNet)	0.97	0.948	0.966	0.962
U-Net (ImageNet) Augs	**0.976**	**0.952**	**0.97**	**0.966**

4.2   Evaluation Results

Table 2 shows the results of the inference of the evaluation data on the challenge platform. The four models are the same used in the previous section.

Table 2. Evaluation results (DICE and Hausdorff 95). Augs indicates the additional data augmentation. Best results in bold.

	DICE			Hausdorff (mm)		
	LV	MYO	RV	LV	MYO	RV
U-Net (scratch)	0.88	0.768	0.789	11.78	7.64	11.37
U-Net (scratch) Augs	0.88	0.771	0.782	**10.57**	7.74	11.87
U-Net (ImageNet)	0.879	0.796	0.826	11.4	6.2	8.71
U-Net (ImageNet) Augs	**0.883**	**0.797**	**0.851**	11.04	**5.64**	**7.77**

4.3   Test Results

Table 3 shows the results of the inference of the test data, where the data were processed on the challenge platform through the submission of our algorithm. The submitted model corresponded to the pretrained version of the U-net including the augmentation strategy.

Table 3. Test results (DICE and Hausdorff 95).

	DICE			Hausdorff (mm)		
	LV	MYO	RV	LV	MYO	RV
U-Net (ImageNet) Augs	0.897	0.837	0.842	6.17	5.03	7.69

5   Conclusions

In this paper, we proposed starting the training from weights that were obtained in a classification problem in another data domain. In addition, we proposed an augmentation policy consisting of four different augmentations with random motion being applied three times more than the rest.

From the quantitative analysis that was done in the validation split of the training data, and the evaluation data from the official platform, we can say that both additions corresponded to an increase of the quality of segmentation. Moreover, the adoption of the pretrained weights also accelerated training times.

Acknowledgment. This publication has emanated from research conducted with the financial support of Science Foundation Ireland under Grant number 18/CRT/6183.

References

1. Arega, T.W., Legrand, F., Bricq, S., Meriaudeau, F.: Using MRI-specific data augmentation to enhance the segmentation of right ventricle in multi-disease, multi-center and multi-view cardiac MRI. In: Puyol Antón, E. (ed.) STACOM 2021. LNCS, vol. 13131, pp. 250–258. Springer, Cham (2022). https://doi.org/10.1007/978-3-030-93722-5_27

2. Campello, V.M., Palomares, J.F.R., Guala, A., Marakas, M., Friedrich, M., Lekadir, K.: Multi-centre, multi-vendor & multi-disease cardiac image segmentation challenge (2020). https://doi.org/10.5281/zenodo.3715890

3. Chen, C., et al.: Deep learning for cardiac image segmentation: a review. Front. Cardiovasc. Med. **7** (2020). https://doi.org/10.3389/fcvm.2020.00025, https://dx.doi.org/10.3389/fcvm.2020.00025

4. Deng, J., Dong, W., Socher, R., Li, L.J., Li, K., Fei-Fei, L.: ImageNet: a large-scale hierarchical image database. In: 2009 IEEE Conference on Computer Vision and Pattern Recognition, pp. 248–255 (2009). https://doi.org/10.1109/CVPR.2009.5206848

5. Garcia-Cabrera, C., Curran, K.M., E. O'Connor, N., McGuinness, K.: Semi-supervised learning of cardiac MRI using image registration. In: Irish Pattern Recognition & Classification Society Conference Proceedings (2021). https://doras.dcu.ie/26161/

6. He, K., Zhang, X., Ren, S., Sun, J.: Deep residual learning for image recognition (2015). https://doi.org/10.48550/ARXIV.1512.03385, https://arxiv.org/abs/1512.03385

7. Isensee, F., Jaeger, P.F., Kohl, S.A., Petersen, J., Maier-Hein, K.H.: nnU-Net: a self-configuring method for deep learning-based biomedical image segmentation. Nat. Methods **18**, 203–211 (2021). https://doi.org/10.1038/s41592-020-01008-z

8. Nyul, L., Udupa, J., Zhang, X.: New variants of a method of MRI scale standardization. IEEE Trans. Med. Imaging **19**(2), 143–150 (2000). https://doi.org/10.1109/42.836373

9. World Health Organization: World health organization. World Health Statistics (2018)

10. Paszke, A., et al.: PyTorch: an imperative style, high-performance deep learning library. In: Advances in Neural Information Processing Systems, vol. 32, pp. 8024–8035. Curran Associates, Inc. (2019). http://papers.neurips.cc/paper/9015-pytorch-an-imperative-style-high-performance-deep-learning-library.pdf

11. Pérez-García, F., Sparks, R., Ourselin, S.: TorchIO: a python library for efficient loading, preprocessing, augmentation and patch-based sampling of medical images in deep learning. Comput. Methods Program. Biomed. **208**, 106236 (2021). https://doi.org/10.1016/j.cmpb.2021.106236

12. Petitjean, C., Dacher, J.N.: A review of segmentation methods in short axis cardiac MR images. Med. Image Anal. **15**, 169–184 (2011). https://doi.org/10.1016/j.media.2010.12.004

13. Ronneberger, O., Fischer, P., Brox, T.: U-net: convolutional networks for biomedical image segmentation. In: Navab, N., Hornegger, J., Wells, W.M., Frangi, A.F. (eds.) MICCAI 2015. LNCS, vol. 9351, pp. 234–241. Springer, Cham (2015). https://doi.org/10.1007/978-3-319-24574-4_28

14. Shaw, R., Sudre, C., Ourselin, S., Cardoso, M.J.: MRI k-space motion artefact augmentation: model robustness and task-specific uncertainty. In: Cardoso, M.J., et al. (eds.) Proceedings of the 2nd International Conference on Medical Imaging with Deep Learning. Proceedings of Machine Learning Research, vol. 102, pp. 427–436. PMLR (2019). https://proceedings.mlr.press/v102/shaw19a.html

15. Van Leemput, K., Maes, F., Vandermeulen, D., Suetens, P.: Automated model-based tissue classification of MR images of the brain. IEEE Trans. Med. Imaging **18**(10), 897–908 (1999)

16. Wang, S., et al.: The extreme cardiac MRI analysis challenge under respiratory motion (CMRxMotion) (2022). https://doi.org/10.48550/ARXIV.2210.06385, https://arxiv.org/abs/2210.06385

Deep Learning Based Classification and Segmentation for Cardiac Magnetic Resonance Imaging with Respiratory Motion Artifacts

Alejandro Mora-Rubio$^{(\boxtimes)}$ ⓘ, Michelle Noga ⓘ,
and Kumaradevan Punithakumar ⓘ

Department of Radiology and Diagnostic Imaging, University of Alberta, Edmonton,
AB T6G 2B7, Canada
{morarubi,mnoga,punithak}@ualberta.ca

Abstract. Cardiac Magnetic Resonance (CMR) is key in the evaluation of heart anatomy and function, and the diagnosis of multiple diseases. However, it requires extensive manual analysis by medical specialists, which slows the diagnostic process, and creates an additional burden for professionals. Different computational techniques have been proposed to automate and accelerate the segmentation of different heart structures within the images, and with the rise of Deep Learning (DL) techniques in the last decade, the performance has improved significantly. Nevertheless, there are still some limitations to be addressed in the automatic processing of CMR, being the respiratory motion artifacts the focus of this paper. This paper presents a DL-based approach for the task of image quality classification and segmentation of heart structures in the context of the CMRxMotion public challenge.

Keywords: Automatic medical image segmentation · Convolutional neural networks · Image quality · Magnetic resonance imaging

1 Introduction

Magnetic Resonance Imaging (MRI) is a non-invasive medical imaging technique based on detecting energy and alignment changes of protons in the presence of a strong magnetic field [8]. MRI produces highly detailed three-dimensional images that allow physicians to evaluate soft-tissue structure and organ function without using harmful ionizing radiation on patients. In particular, Cardiac Magnetic Resonance (CMR) is considered the gold-standard technique to visualize and assess cardiovascular anatomy, volumes, and function [11]. Nevertheless, the analysis of CMR relies on computing different indices, such as chamber and stroke volumes, ejection fraction, and cardiac output [9], which requires the images to be manually segmented by experienced experts, resulting in intensive

Supported by Mitacs (Globalink Research Internship).

and time-consuming labor that increases the time for an image to be processed and a diagnosis is given to the patient.

In this context, Deep Learning (DL) algorithms, namely Convolutional Neural Networks (CNN), have been implemented for the task of automatic segmentation of CMR images. For this task, the UNet architecture [10], originally introduced in 2015 for biomedical image segmentation, has been the most used model [6,16]. For example, Wong et al. [15] tackled the task of the heart's chamber segmentation, by implementing a preprocessing stage where different degrees of Gaussian blur are applied to a single image and are then combined as an input to the UNet model for segmentation. In a different approach, Sun et al. [12] modified the base UNet architecture to incorporate two inputs, one being the MRI slice to segment, and the other being a stack of the previous and following slices, where the stack could be created spatially (consecutive slices at a timestamp) or over time (same slice at different timestamps). Recently, with the rapid development, and positive results, of attention mechanisms and transformers models in the field of Natural Language processing, some authors have reported good performance of these techniques in tasks such as CMR image segmentation [3,5].

Despite the reported good results of DL algorithms to perform automatic segmentation of CMR images, some limitations are to be addressed to generate a robust and general CMR segmentation system; the main ones being labeled data availability, different annotation protocols on available data, data source mismatch in terms of equipment and acquisition conditions, and artifacts that affect image quality. In the past few years, there have been public challenges aimed to tackle some of these limitations, such as the Automated Cardiac Diagnosis Challenge (ACDC) [1], and last year's Multi-Centre, Multi-Vendor, and Multi-Disease Cardiac Segmentation (M&Ms) [2] with a comprehensive dataset of images from different healthcare institutions. There is a new challenge called Extreme Cardiac MRI Analysis Challenge under Respiratory Motion (CMRx-Motion), presented as part of Statistical Atlases and Computational Modeling of the Heart (STACOM, https://stacom.github.io/stacom2022/) workshop for 2022. The CMRxMotion challenge is centered around assessing the effect of respiratory motion artifacts in CMR image quality for diagnosis, as well as in the performance of segmentation algorithms.

This paper presents a DL-based methodology for the classification and segmentation of CMR scans in the context of the CMRxMotion challenge. The rest of the paper is organized as follows: Sect. 2 presents the dataset, image processing, and DL algorithms involved; Sect. 3 describes the performed experiments and the corresponding results; Sect. 4 summarizes the results and describes the contribution, as well as limitations found; finally, Sect. 5 presents the conclusion of the work.

2   Materials and Methods

2.1   CMRxMotion Data

For this challenge, the dataset contains CMR clinical scans from 45 healthy volunteers, captured with the Siemens 3T MRI scanner (MAGNETOM Vida). As mentioned before, the aim of this challenge is to evaluate model robustness under respiratory motion, to this end, each participant undertakes a 4-stage scan in a single visit: 1) adhere to the breath-hold instructions; 2) halve the breath-hold period; 3) breathe freely; 4) breathe intensively. With this in mind, and the fact that just the End of Diastole (ED) and End of Systole (ES) frames are released, there are a total of 200 CMR scans for training and validation purposes (25 participants, 4 breathing patterns, 2 frames), and 160 CMR scans for testing (20 participants, 4 breathing patterns, 2 frames). For more information refer to [14].

Regarding annotations, the challenge presents two different tasks:

- **Task 1:** Image quality assessment of respiratory motion artefacts For this classification task, each image was evaluated by expert radiologists who gave a score representing the diagnostic quality. These scores were then assigned to one of three labels: Label 1 for mild motion; Label 2 for intermediate motion; and Label 3 for severe motion.
- **Task 2:** CMR image segmentation with respiratory motion artifacts Those images defined with diagnostic quality in the first task, are then segmented by an experienced radiologist in 3D Slicer (www.slicer.org), including contours for the left (LV) and right ventricle (RV) blood pools, as well as for the left ventricular myocardium (MYO). Labels are: 1 (LV), 2 (MYO) and 3 (RV).

2.2   Deep Learning Algorithms

Classification. In an image classification task, the network receives an image as the input and outputs a single number that represents the class or label for that image. Usually, these networks are formed by a set of convolutional layers for feature extraction, and some fully connected layers to make the final prediction. In this work, we implemented EfficientNet B7, the biggest of a series of CNNs that present a novel way of model scaling that achieves better performance with fewer parameters than similar architectures. The idea is to scale, at the same time, the three dimensions of neural networks: depth (number of layers), width (number of channels or filters per layer) and resolution (size of the input images and feature maps) [13].

Segmentation. in an image segmentation task, the network receives an image and outputs a segmentation mask of the same size. This mask has a value of one for pixels inside the region of interest (ROI) and zero for the rest. It is also possible to define segmentation tasks with multiple independent ROIs, such as this task that requires to segment LV, RV and MYO for each image. In the context of biomedical image segmentation, the UNet [10] model has been tested in a wide set of applications, achieving good results. This model consists of a encoder-decoder architecture based on convolutional layers. The encoder, or downsampling path, is composed by convolutional and max pooling layers that extract relevant features of the image, which are then passed to the decoder, or upsampling path, that takes this new representation of the image and using the transposed convolution operation generates the segmentation mask of the same size of the original image. In this work, the nnUnet [4] framework was used, since it has been shown to work well in previous CMR segmentation challenges.

3   Results

3.1   Experimental Setup

All models, operations, and training procedures were built on top of nnUnet [4], and MONAI [7] frameworks. The first one was used for Task 2, and the only requirement is data organization into certain folders, after that, the data processing pipeline and training are handled by the framework. On the other hand, MONAI [7] framework is similar to a library, built on top of PyTorch which has several models and methods available for medical image processing.

After the initial exploratory analysis of the training data, which includes 160 3D images with a median size of $432 \times 512 \times 11$, the images were divided into the training and validation sets, with an 80/20 split performed by patient, to avoid biased results. This way, the training set was composed of 128 images (16 patients), and the validation set of 32 images (4 patients). Given the anisotropic voxel spacing in the images, it was decided to decompose the images into 2D slices for training and evaluating the models. Moreover, given the small size of the remaining training set, data augmentation operations were introduced into the training pipeline to further improve results by varying spatial and intensity features of the input images; in particular, random 90° rotations, flips, zooms were applied, as well as random Gaussian sharpen and smoothing operations to simulate the effect of respiratory motion artifacts.

3.2   Results

Classification. Besides the mentioned data augmentation operations applied to the input images, these were resized to 224×224 pixels and normalized in intensity to the range $[0,1]$. Table 1 presents the evaluation metrics for different experiments performed using the EfficientNet B7 architecture, computed over the development validation set composed of 32 images from the training set. In all experiments, the epochs were set to 200, the loss function was CrossEntropy, and the batch size was set to four 3D images which translate into 40 to 50 2D slices per batch. For the weighted loss experiment, the mistakes the network makes for each class are weighted according to the number of samples of that class with respect to the total number of samples, which translates to a higher loss value when the network makes incorrect predictions of the class with fewer samples, this method helps with the class imbalance problem. After analyzing the training curves, such as those presented in Fig. 1, a learning rate scheduler was implemented to reduce the learning rate at certain points during training, aiming to get a smoother learning procedure. The Base, Weighted loss, and Learning rate schedule experiments used the Adam optimizer with a learning rate of 1×10^{-4}.

Finally, the best performing model was achieved using the NAdam optimizer and a learning rate of 2×10^{-3}. This model achieved an accuracy of 67.5% and a Cohen Kappa of 0.4117 over the blind validation set, as evaluated by the challenge submission platform. Figure 1 presents the evolution of the loss, validation accuracy, and validation Cohen Kappa during training.

Table 1. Classification results for different experimental setups

Experiment	Validation accuracy	Validation cohen kappa
Base	65.43%	0.4533
Weighted loss	67.28%	0.4744
Learning rate schedule	66.66%	0.4404
NAdam optimizer	**67.90%**	**0.5019**

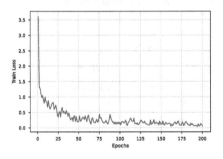

(a) Loss during training   (b) Metrics during training

Fig. 1. Training curves for best performing EfficientNet

In the end, our methodology achieved a test set accuracy of 60.83% and a Cohen's Kappa coefficient of 0.4316 over 120 image volumes.

Segmentation. For the segmentation task, the nnUNet framework handles the preprocessing operations (resampling, normalization, patch sizes) and the training hyperparameters, such as the batch size, according to the input data characteristics and computing resources available. In this case, the inferred loss function is a combination of the CrossEntropy and Dice losses. Having in mind that each component of the inferred loss function optimizes different behaviours of the segmentation model, the gradual transition experiment is proposed, where only the CrossEntropy loss is used in the initial stage of the training, then both losses are combined in a second stage, and only Dice loss is used in the final stage. This way, the training of the network focuses first on predicting the correct distribution of classes (CrossEntropy), and then on aligning the predicted segmentation regions (Dice). The no smooth experiment consists of setting the smoothing parameter, which helps avoid division by zero in particular cases but affects the result, of Dice loss to 0.0.

On the other hand, the Batch Normalization experiment consists of changing the Instance Normalization layers on the network for Batch Normalization. In our testing, this methodology paired with the default loss function performed the best. This approach achieved a Dice Coefficient of 92.04%, 83.45% and 90.34% for the LV, MYO and RV, respectively.

Table 2 presents the Dice Coefficient for each class, computed over the development validation set created by nnUNet with 20% of the training data.

Table 2. Segmentation results for different experimental setups

Experiment	Dice coefficient		
	LV	MYO	RV
nnUNet inferred configuration	94.06%	89.58%	94.26%
Loss function gradual transition	93.82%	89.42%	**94.28%**
Loss function no smooth	94.14%	89.68%	94.14%
Batch normalization	**94.30%**	**89.76%**	94.21%

Figure 2 presents a sample prediction for a CMR scan captured performing breath hold.

Figure 3 presents a sample prediction for a CMR scan captured during free breathing.

In the end, our methodology achieved the results presented in Table 3.

4   Discussion

In this paper, a DL-based methodology for the classification and segmentation of CMR scans is presented. In the context of the CMRxMotion challenge, the classification task consists of classifying the images into adequate or non-adequate for diagnosis according to the impact of respiratory motion artifacts; and the

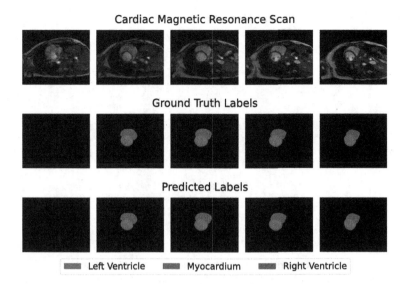

Fig. 2. Segmentation results for a breath-hold scan

Table 3. Segmentation results over the test set

Metric	Results (mean ± std)
LV Dice	93.90306% ± 3.719059%
MYO Dice	87.36886% ± 3.360553%
RV Dice	92.21828% ± 6.067618%
LV HD95	3.049473 ± 3.540949
MYO HD95	2.196029 ± 2.099378
RV HD95	3.531699 ± 3.552028

segmentation task consists of the segmentation of the myocardium, left and right ventricles in scans captured with different breathing patterns. Regarding the dataset for this challenge, it is worth mentioning that only the ED and ES frames were released, as opposed to the full scan with complete temporal information; this way, it was not possible to take advantage of temporal consistency for a more robust analysis, and work possibilities were limited to image processing techniques, in which CNNs architectures have been shown to achieve good results in biomedical applications, even in small data scenarios, which led us to create our approaches around them.

For the classification task, the proposed methodology involves the Efficient-Net B7 model architecture, trained with 2D slices and data augmentation operations, achieving an overall validation accuracy of 67% and a Cohen Kappa coefficient of 0.44. For the segmentation task, the nnUNet framework was used, achieving a validation Dice Coefficient of 94% for the left ventricle, 89% for

406   A. Mora-Rubio et al.

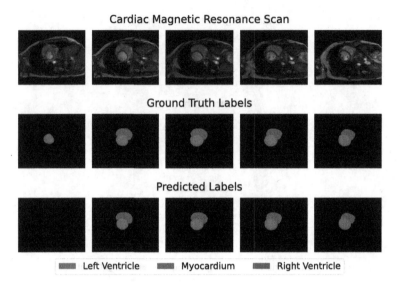

Fig. 3. Segmentation results for an intensive breathing scan

the myocardium, and 94% for the right ventricle. By studying the segmentation
results per type of scan (breathing pattern) presented in Table 4, the changes in
model performance are according to expected behavior. For scan types 3 (free
breathing) and 4 (intensive breathing), the Dice coefficient is lower in comparison with scan types 1 and 2, which implies that respiratory motion artifacts
do have an impact on the performance of automatic segmentation algorithms,
although it doesn't seem to be a critical factor affecting the potential real-life
performance of the systems.

Table 4. Segmentation results for different scan types

Scan type	Average dice coefficient [%]		
	LV	MYO	RV
1: breath-hold	94.43	90.94	94.62
2: halve the breath-hold period	94.91	90.33	94.74
3: free breathing	93.86	88.94	93.59
4: intensive breathing	93.87	88.37	93.71

This type of challenges aim to tackle important challenges for the design of
effective and trustworthy applications of artificial intelligence, in particular, by
improving the performance of current algorithms for CMR classification, and
designing more robust segmentation systems we could: accelerate and alleviate
the medical expert workload, allowing more time for the analysis of complicated

cases, and reducing the amount of scans that need to be repeated because of the artifact introduced due to respiratory motion; opt for objective segmentations, reducing intraobserver and interobserver variability in the procedure. And more generally, increase the knowledge base for medical imaging applications of artificial intelligence, where techniques and procedures developed in this particular application can be the source of further development in related fields.

5 Conclusion

This paper presents promising results for the automatic classification and segmentation of Cardiac Magnetic Resonance scans in presence of respiratory motion artifacts. Future work would be aimed at compiling bigger, and more representative, real-life datasets to evaluate and improve Deep Learning model's ability to generalize. On the other hand, developing new techniques and procedures based on human-expert knowledge poses a great challenge, as well as great opportunity, to further improve model performance.

Acknowledgment. This research was supported by Mitacs as part of the Globalink Research Internship program in 2022, and was enabled in part by support provided by WestGrid (www.westgrid.ca) and the Digital Research Alliance of Canada (alliancecan.ca)

References

1. Bernard, O., et al.: Deep learning techniques for automatic MRI cardiac multi-structures segmentation and diagnosis: is the problem solved? IEEE Trans. Med. Imaging **37**, 2514–2525 (2018). https://doi.org/10.1109/TMI.2018.2837502
2. Campello, V.M., et al.: Multi-centre, multi-vendor and multi-disease cardiac segmentation: the MMS challenge. IEEE Trans. Med. Imaging **40**, 3543–3554 (2021). https://doi.org/10.1109/TMI.2021.3090082
3. Hatamizadeh, A., et al.: UNETR: transformers for 3D medical image segmentation. arXiv (2021). https://doi.org/10.48550/arxiv.2103.10504
4. Isensee, F., Jaeger, P.F., Kohl, S.A., Petersen, J., Maier-Hein, K.H.: nnU-net: a self-configuring method for deep learning-based biomedical image segmentation. Nat. Methods **18**(2), 203–211 (2020). https://doi.org/10.1038/s41592-020-01008-z
5. Liu, H., et al.: MEA-Net: multilayer edge attention network for medical image segmentation. Sci. Rep. **12**(1), 1–15 (2022). https://doi.org/10.1038/s41598-022-11852-y
6. Lu, Y., Zhao, Y., Chen, X., Guo, X.: A novel U-net based deep learning method for 3D cardiovascular MRI segmentation. Comput. Intell. Neurosci. 1–11 (2022). https://doi.org/10.1155/2022/4103524
7. MONAI Consortium: MONAI: Medical open network for AI (2022). https://doi.org/10.5281/ZENODO.6639453
8. National Institute of Biomedical Imaging and Bioengineering: Magnetic resonance imaging (MRI). https://www.nibib.nih.gov/science-education/science-topics/magnetic-resonance-imaging-mri

9. Peng, P., Lekadir, K., Gooya, A., Shao, L., Petersen, S.E., Frangi, A.F.: A review of heart chamber segmentation for structural and functional analysis using cardiac magnetic resonance imaging. Magn. Reson. Mater. Phys. Biol. Med. **29**(2), 155–195 (2016). https://doi.org/10.1007/S10334-015-0521-4

10. Ronneberger, O., Fischer, P., Brox, T.: U-net: convolutional networks for biomedical image segmentation. arXiv (2015). https://doi.org/10.48550/arxiv.1505.04597

11. Schulz-Menger, J., et al.: Standardized image interpretation and post-processing in cardiovascular magnetic resonance - 2020 update: society for cardiovascular magnetic resonance (SCMR): board of trustees task force on standardized post-processing. J. Cardiovasc. Magn. Reson. **22**, 1–22 (2020). https://doi.org/10.1186/S12968-020-00610-6/FIGURES/10

12. Sun, X., Garg, P., Plein, S., van der Geest, R.J.: SAUN: stack attention U-Net for left ventricle segmentation from cardiac cine magnetic resonance imaging. Med. Phys. **48**, 1750–1763 (2021). https://doi.org/10.1002/MP.14752

13. Tan, M., Le, Q.V.: Efficientnet: rethinking model scaling for convolutional neural networks. arXiv (2019). https://doi.org/10.48550/arxiv.1905.11946

14. Wang, S., et al.: The extreme cardiac MRI analysis challenge under respiratory motion (CMRxMotion). arXiv (2022). https://doi.org/10.48550/arxiv.2210.06385

15. Wong, K.K., Zhang, A., Yang, K., Wu, S., Ghista, D.N.: GCW-UNet segmentation of cardiac magnetic resonance images for evaluation of left atrial enlargement. Comput. Methods Programs Biomed. **221**, 106915 (2022). https://doi.org/10.1016/J.CMPB.2022.106915

16. Yang, F., et al.: A deep learning segmentation approach in free-breathing real-time cardiac magnetic resonance imaging. BioMed Res. Int. **2019** (2019). https://doi.org/10.1155/2019/5636423

Multi-task Swin Transformer for Motion Artifacts Classification and Cardiac Magnetic Resonance Image Segmentation

Michal K. Grzeszczyk[1]($\boxtimes$), Szymon Płotka[1,2], and Arkadiusz Sitek[3]

[1] Sano Centre for Computational Medicine, Cracow, Poland
m.grzeszczyk@sanoscience.org
[2] Informatics Institute, University of Amsterdam, Amsterdam, The Netherlands
[3] Massachusetts General Hospital, Harvard Medical School, Boston, MA, USA

Abstract. Cardiac Magnetic Resonance Imaging is commonly used for the assessment of the cardiac anatomy and function. The delineations of left and right ventricle blood pools and left ventricular myocardium are important for the diagnosis of cardiac diseases. Unfortunately, the movement of a patient during the CMR acquisition procedure may result in motion artifacts appearing in the final image. Such artifacts decrease the diagnostic quality of CMR images and force redoing of the procedure. In this paper, we present a Multi-task Swin UNEt TRansformer network for simultaneous solving of two tasks in the CMRxMotion challenge: CMR segmentation and motion artifacts classification. We utilize both segmentation and classification as a multi-task learning approach which allows us to determine the diagnostic quality of CMR and generate masks at the same time. CMR images are classified into three diagnostic quality classes, whereas, all samples with non-severe motion artifacts are being segmented. Ensemble of five networks trained using 5-Fold Cross-validation achieves segmentation performance of DICE coefficient of 0.871 and classification accuracy of 0.595.

Keywords: Multi-task learning · Cardiac magnetic resonance imaging · Segmentation · Image quality assessment

1 Introduction

Cardiac Magnetic Resonance Imaging (CMR) is often used to assess cardiac anatomy and/or function [10]. As a noninvasive method, CMR gained popularity due to the generation of high-quality images enabling the diagnosis of multiple diseases of the human heart. Unfortunately, Magnetic Resonance Imaging (MRI) acquisition procedure is long and is susceptible to many artifacts [16]. Motion artifacts are prevalent in CMR due to the high blood flow in the field of view (FOV), breathing, or patient physical motion.

The delineation of the left ventricle (LV) and right ventricle (RV) blood pools together with left ventricular myocardium (MYO) can be useful for the diagnosis of cardiac diseases [19]. However, manual segmentation of cardiac structures

© The Author(s), under exclusive license to Springer Nature Switzerland AG 2022
O. Camara et al. (Eds.): STACOM 2022, LNCS 13593, pp. 409–417, 2022.
https://doi.org/10.1007/978-3-031-23443-9_38

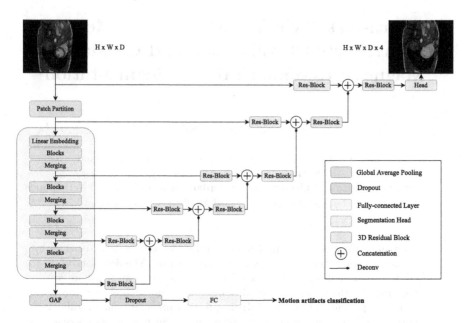

Fig. 1. The overview of the multi-task Swin UNETR (inspired by [7]) for CMR segmentation and diagnostic quality assessment. Swin Transformer is utilized as the backbone, encoder network of the U-Net architecture. The outputs of the encoder are passed to the convolutional decoder for segmentation purposes and to the classification branch for motion artifacts classification.

is time-consuming and requires expert's knowledge. Thus, several methods for automated CMR segmentation have been proposed. In recent years Deep Learning (DL) allowed automated software to achieve human-level segmentation accuracy [2]. Convolutional Neural Networks (CNNs) have been heavily applied for such tasks [1,14]. Unfortunately, the performance of such models shown in many publications often cannot be matched when applied to data populations different from the one used for training. The image properties differ between institutions due to different imaging procedures and equipment used for data acquisition. Motion artifacts are also important factors that prevent achieving accurate segmentations.

In this paper, we present a Multitask Swin UNEt TRansformers (Swin UNETR) network that enables the classification of motion artifacts of the CMR acquired under different breathing conditions and simultaneous segmentation of LV, RV and MYO. We present the performance of the model on the CMRxMotion challenge organized alongside the Medical Image Computing and Computer Assisted Intervention (MICCAI) 2022 conference.

The rest of this paper is structured as follows. In the next section, we present works related to multi-task learning and segmentation with DL. Section 3 describes our model and the dataset. In Sect. 4, results for tasks of motion artifacts classification and CMR segmentation are shown.

2   Related Works

Since the introduction of AlexNet in 2012 [9] deep neural networks (especially CNNs) have become the go-to tool for solving computer vision problems. One of the most popular CNN architecture utilized for image segmentation is U-Net [15]. U-Net was initially developed for biomedical image segmentation and comprises of CNN encoder and decoder with skip connections between different levels of those components. This approach was later improved and changed in various ways [8,12].

Deep learning is also being applied in other domains. For example, Transformer architecture was proposed for the text translation task in the Natural Language Processing area [17]. Transformer-based models achieve state-of-the-art results on the language understanding benchmarks thanks to the utilization of the self-attention mechanism which allows for detecting long-range dependencies between elements of the sequence. This idea was later refined to the image domain when the Vision Transformer (ViT) was introduced [5]. ViT divides images into non-overlapping patches which are then fed into the Transformer Encoder. Unfortunately, ViT requires substantially big datasets to achieve high performance, has the quadratic computation complexity depending on the image size and lacks the inductive bias of CNNs. Therefore, Swin Transformer (Shifted Window Transformer) computing self-attention only within local windows which scales linearly was presented [11]. Additionally, Swin Transformer applies the *shifted window* approach which changes the local window for self-attention computation in each layer.

To overcome the limitations of Transformer architectures while retaining their advantages, hybrid architectures combining CNNs with Transformers are being proposed [6,13]. One of the examples is Swin UNETR model [7]. In this architecture, a 3D Swin Transformer is used as the encoder of the U-Net-like architecture and multiple convolutional Residual Blocks (Res-Blocks) are utilized in the decoder part. In this paper we utilize Swin UNETR architecture as the base for our model.

Multi-task learning (MTL) is the area of DL whose target is to improve the performance of deep neural networks via simultaneous training of the network to solve multiple tasks [20]. It has been shown that MTL can improve the model's performance even if the number of data samples is limited [4]. In this paper, we present the utilization of MTL for motion artifacts classification and CMR segmentation.

3   Method

In this section, we present our approach to CMR segmentation on the CMRx-Motion challenge dataset. We employ Multitask Swin UNETR to solve two tasks in this challenge: 1) motion artifacts classification, and 2) CMR segmentation.

3.1   Multitask Swin UNETR

To solve both problems simultaneously with one network, we employ a Multi-task Swin UNETR (Fig. 1). We follow the architecture proposed by [7]. Swin UNETR is similar to the U-Net architecture [15]. However, instead of utilizing fully convolutional DL model, Swin UNETR uses Swin Transformer as the encoder network. At first, the input to the Swin Transformer is partitioned into patches which are then projected into embedding space. Such tokens are passed to Swin Transformer blocks computing the following equations [11]:

$$
\begin{aligned}
\hat{z}^l &= WMSA(LN(z^{l-1})) + z^{l-1} \\
z^l &= MLP(LN(\hat{z}^l)) + \hat{z}^l \\
\hat{z}^{l+1} &= SWMSA(LN(z^l)) + z^l \\
z^{l+1} &= MLP(LN(\hat{z}^{l+1})) + \hat{z}^{l+1}
\end{aligned}
\tag{1}
$$

Here, MLP is Multilayer Perceptron, LN denotes layer normalization and z^i is the output from previous Swin Transformer blocks or the embedding layer in case of the first block. WMSA and SWMSA are ordinary and shifted-window multi-head self-attentions (MHSA), respectively. Swin Transformer blocks compute MHSA on $M \times M \times M$ windows (extracted from the input patches) instead of the full image, to enable linear computational complexity growth depending on the image size (ordinary self-attention computation on the full image has the quadratic computational cost). The SWMSA is computed with the shifted window approach where the windows for computing MHSA are shifted by ($\lfloor \frac{M}{2} \rfloor$, $\lfloor \frac{M}{2} \rfloor$, $\lfloor \frac{M}{2} \rfloor$) pixels. The self-attention is computed according to the following equation:

$$
Attention(Q, K, V) = Softmax(\frac{QK^T}{\sqrt{d}})V,
\tag{2}
$$

where Q, K, V are queries, keys, values, while d denotes the Q size. After, Swin Transformer blocks patches are merged again.

The Swin UNETR's decoder is created out of convolutional layers. To solve the motion artifacts assessment task, we use high-level feature map representation, and we fed them to the classification branch. The output of the decoder is global average pooled and passed through one Dropout and Fully-connected layers. We set the number of features at each Swin Transformer layer to $k \times 60$, where k is the features factor at each layer.

4   Experiments and Results

4.1   Dataset

CMRxMotion dataset's purpose is to investigate the impact of motion artifacts on the performance of automated CMR segmentation methods [18]. It contains data from 45 volunteers (20 patients in train, 5 in validation and 20 in test datasets). Every volunteer underwent end-diastolic (ED) and end-systolic (ES) acquisition of CMR under *full breath-hold*, *half breath-hold*, *free breathing*, and *intensive breathing* conditions. This results in 160, 40, and 160 images in the training, validation, and test sets. Different acquisition conditions created different levels of motion artifacts from the least in *full breath-hold* to the most in *intensive breathing*. All images were labeled by experts into three classes: *1 -* mild motion, *2 -* intermediate motion, *3 -* severe motion. The *1, 2* classes have sufficient quality for diagnosis. In all samples labelled as such LV, RV and MYO were segmented. The aim of the challenge is to create an algorithm for motion artifacts classification into three classes describing the severity of motion artifacts (task 1) and perform segmentation from samples from diagnostic quality samples *1* and *2* (task 2). Samples with severe motion artifacts were not used in the segmentation task.

4.2   Loss Function

The model is trained to minimize the sum of segmentation and classification losses defined as:

$$\mathcal{L}_{sum} = \lambda_1(\mathcal{L}_{CE_{seg}} + \mathcal{L}_{Dice}) + \lambda_2\mathcal{L}_{CE_{cls}}, \tag{3}$$

where $\lambda_1 = 2.25$ and $\lambda_2 = 1$, $\mathcal{L}_{CE_{seg}}$ is Cross-Entropy loss calculated on the segmentation mask, $\mathcal{L}_{CE_{cls}}$ is loss calculated on the classification labels and $\mathcal{L}_{Dice}$ is DICE coefficient loss computed on non-background classes. The segmentation component of loss function is omitted for samples with severe motion artifacts.

4.3   Implementation Details

All training samples are of shape $H \times W \times D$, where $H \in [400, 512]$, $W \in [594, 731]$, $D \in [12, 18]$. We resize all volumes to the same patch size of $256 \times 256 \times 32$ (the D channel is padded with zeros). The output of the model is of shape $H \times W \times D \times 4$ (background class and three segmentation classes - LV, RV, MYO), where each slice in the final dimension is the segmentation class mask. We train five models using 5-Fold Stratified Cross-validation where the stratification groups are selected based on classification labels to ensure that motion artifacts are evenly distributed across all folds. During training, we apply multiple spatial transformations to avoid overfitting. Among those augmentations there are: random flips for each axis (p = 0.2), random zoom (p = 0.1), random rotate (p = 0.1). Before the augmentations, we normalize the input data. The model is implemented with PyTorch and trained on 2× NVIDIA A100 80 GB GPUs. We train the model for 250 epochs during each fold with AdamW optimizer. We set an initial learning rate to 2e-04 with the Cosine Annealing learning rate scheduler. We use the default parameters of Swin UNETR from the MONAI library [3] unless stated otherwise.

Table 1. Average Accuracy and Cohen's Kappa among folds on CMRxMotion challenge training and validation datasets

Metric	Fold 1	Fold 2	Fold 3	Fold 4	Fold 5	Average
Accuracy - KFold test	0.594	0.6	0.75	0.5	0.531	0.595
Accuracy - validation	-	-	-	-	-	0.525
Accuracy - test	-	-	-	-	-	0.4
Cohen's Kappa - KFold test	0.257	0.286	0.59	0.0	0.0	0.184
Cohen's Kappa - validation	-	-	-	-	-	0.127
Cohen's Kappa - test	-	-	-	-	-	−0.075

Table 2. Average Dice and Hausdorff distance among folds on CMRxMotion challenge training and validation datasets

Metric	Fold 1	Fold 2	Fold 3	Fold 4	Fold 5	Average
Dice - KFold test	0.88	0.858	0.898	0.89	0.831	0.871
Dice - validation	-	-	-	-	-	0.807
Dice - test	-	-	-	-	-	0.863
HD95 - KFold test	2.245	6.347	1.794	3.271	12.187	5.169
HD95 - validation	-	-	-	-	-	9.803
HD95 - test	-	-	-	-	-	5.305

4.4   Evaluation Metrics

To measure the performance of the model we utilize the DICE coefficient and Hausdorff Distance 95 (HD95) as segmentation metrics as well as Accuracy and Cohen's Kappa as classification metrics.

4.5   Motion Artifacts Classification

Motion artifacts classification is a difficult task due to the highly imbalanced dataset - 70 samples with mild motion, 69 with intermediate motion and only 21 with severe motion. The results achieved by Swin UNETR on different folds and validation set are presented in Table 1. The model reaches up to 0.75 accuracy and 0.59 Cohen's Kappa on the third fold's test set. On average, the accuracy on all folds is 0.595 and Cohen's Kappa reaches 0.184.

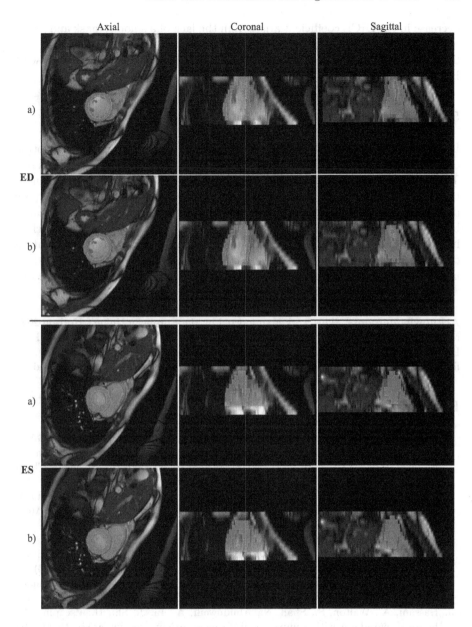

Fig. 2. Results of the segmentation of end-diastolic (ED) and end-systolic (ES) images shown on the three CMR views (Axial, Coronal, Sagittal) a) ground truth, b) Swin UNETR.

4.6 CMR Segmentation

The exemplary results of CMR segmentation are presented in Fig. 2. Swin UNETR manages to delineate RV, LV and MYO with high accuracy. The biggest decrease in the DICE coefficient arises from the lack of a smooth mask and issues with small details in the output. Segmentation metrics are presented in Table 2. Swin UNETR achieves an average DICE coefficient of 0.871 on all folds with HD95 of 5.169.

5   Conclusions

In this paper we presented a Multi-task Swin UNETR network for CMRxMotion challenge tasks solving. The proposed model was evaluated on the CMR dataset. This network is able to perform motion artifacts classification and CMR segmentation in a single forward pass. Such an approach enables the utilization of more data samples in the single network training (which enhances network's capabilities) and the application of only one model for solving two tasks which is important due to lower computational complexity (contrary to the utilization of two networks). In the future, the model's performance can be improved by utilizing more data (for example from other CMR segmentation challenges like ACDC [2]).

Acknowledgements. This work is supported by the European Union's Horizon 2020 research and innovation programme under grant agreement Sano No 857533 and the International Research Agendas programme of the Foundation for Polish Science, co-financed by the European Union under the European Regional Development Fund.

References

1. Al Khalil, Y., Amirrajab, S., Pluim, J., Breeuwer, M.: Late fusion U-Net with GAN-based augmentation for generalizable cardiac MRI segmentation. In: Puyol Antón, E., et al. (eds.) STACOM 2021. LNCS, vol. 13131, pp. 360–373. Springer, Cham (2022). https://doi.org/10.1007/978-3-030-93722-5_39
2. Bernard, O., et al.: Deep learning techniques for automatic MRI cardiac multi-structures segmentation and diagnosis: is the problem solved? IEEE Trans. Med. Imaging **37**(11), 2514–2525 (2018)
3. Consortium, M., et al.: MONAI: medical open network for AI (2020)
4. Dobrescu, A., Giuffrida, M.V., Tsaftaris, S.A.: Doing more with less: a multitask deep learning approach in plant phenotyping. Front. Plant Sci. **11**, 141 (2020)
5. Dosovitskiy, A., et al.: An image is worth 16x16 words: transformers for image recognition at scale. arXiv preprint arXiv:2010.11929 (2020)
6. Hassani, A., Walton, S., Shah, N., Abuduweili, A., Li, J., Shi, H.: Escaping the big data paradigm with compact transformers. arXiv preprint arXiv:2104.05704 (2021)
7. Hatamizadeh, A., Nath, V., Tang, Y., Yang, D., Roth, H., Xu, D.: Swin UNETR: swin transformers for semantic segmentation of brain tumors in MRI images. arXiv preprint arXiv:2201.01266 (2022)
8. Kossaifi, J., Bulat, A., Tzimiropoulos, G., Pantic, M.: T-net: parametrizing fully convolutional nets with a single high-order tensor. In: Proceedings of the IEEE/CVF Conference on Computer Vision and Pattern Recognition, pp. 7822–7831 (2019)

9. Krizhevsky, A., Sutskever, I., Hinton, G.E.: Imagenet classification with deep convolutional neural networks. In: Advances in Neural Information Processing Systems, vol. 25 (2012)

10. Lima, J.A., Desai, M.Y.: Cardiovascular magnetic resonance imaging: current and emerging applications. J. Am. Coll. Cardiol. **44**(6), 1164–1171 (2004)

11. Liu, Z., et al.: Swin transformer: hierarchical vision transformer using shifted windows. In: Proceedings of the IEEE/CVF International Conference on Computer Vision, pp. 10012–10022 (2021)

12. Ottom, M.A., Rahman, H.A., Dinov, I.D.: Znet: deep learning approach for 2D MRI brain tumor segmentation. IEEE J. Transl. Eng. Health Med. **10**, 1–8 (2022). Art no. 1800508. https://doi.org/10.1109/JTEHM.2022.3176737

13. Płotka, S., et al.: BabyNet: residual transformer module for birth weight prediction on fetal ultrasound video. In: Wang, L., Dou, Q., Fletcher, P.T., Speidel, S., Li, S. (eds.) Medical Image Computing and Computer-Assisted Intervention, pp. 350–359. Springer, Cham (2022). https://doi.org/10.1007/978-3-031-16440-8_34

14. Queirós, S.: Right ventricular segmentation in multi-view cardiac MRI using a unified U-net model. In: Puyol Antón, E., et al. (eds.) STACOM 2021. LNCS, vol. 13131, pp. 287–295. Springer, Cham (2022). https://doi.org/10.1007/978-3-030-93722-5_31

15. Ronneberger, O., Fischer, P., Brox, T.: U-Net: convolutional networks for biomedical image segmentation. In: Navab, N., Hornegger, J., Wells, W.M., Frangi, A.F. (eds.) MICCAI 2015. LNCS, vol. 9351, pp. 234–241. Springer, Cham (2015). https://doi.org/10.1007/978-3-319-24574-4_28

16. Smith, T.B.: MRI artifacts and correction strategies. Imaging Med. **2**(4), 445 (2010)

17. Vaswani, A., et al.: Attention is all you need. In: Advances in Neural Information Processing Systems, vol. 30 (2017)

18. Wang, S., et al.: The extreme cardiac MRI analysis challenge under respiratory motion (cmrxmotion). arXiv preprint arXIv: 2210.06385 (2022)

19. White, H.D., Norris, R.M., Brown, M.A., Brandt, P.W., Whitlock, R., Wild, C.J.: Left ventricular end-systolic volume as the major determinant of survival after recovery from myocardial infarction. Circulation **76**(1), 44–51 (1987)

20. Zhang, Y., Yang, Q.: An overview of multi-task learning. Natl. Sci. Rev. **5**(1), 30–43 (2018)

Automatic Quality Assessment of Cardiac MR Images with Motion Artefacts Using Multi-task Learning and K-Space Motion Artefact Augmentation

Tewodros Weldebirhan Arega$^{(\boxtimes)}$, Stéphanie Bricq, and Fabrice Meriaudeau

ImViA Laboratory, Université Bourgogne Franche-Comté, Dijon, France
tewdrosw@gmail.com

Abstract. The movement of patients and respiratory motion during MRI acquisition produce image artefacts that reduce the image quality and its diagnostic value. Quality assessment of the images is essential to minimize segmentation errors and avoid wrong clinical decisions in the downstream tasks. In this paper, we propose automatic multi-task learning (MTL) based classification model to detect cardiac MR images with different levels of motion artefact. We also develop an automatic segmentation model that leverages k-space based motion artefact augmentation (MAA) and a novel compound loss that utilizes Dice loss with a polynomial version of cross-entropy loss (PolyLoss) to robustly segment cardiac structures from cardiac MRIs with respiratory motion artefacts. We evaluate the proposed method on Extreme Cardiac MRI Analysis Challenge under Respiratory Motion (CMRxMotion 2022) challenge dataset. For the detection task, the multi-task learning based model that simultaneously learns both image artefact prediction and breath-hold type prediction achieved significantly better results compared to the single-task model, showing the benefits of MTL. In addition, we utilized test-time augmentation (TTA) to enhance the classification accuracy and study aleatoric uncertainty of the images. Using TTA further improved the classification result as it achieved an accuracy of 0.65 and Cohen's kappa of 0.413. From the estimated aleatoric uncertainty, we observe that images with higher aleatoric uncertainty are more difficult to classify than the ones with lower uncertainty. For the segmentation task, the k-space based MAA enhanced the segmentation accuracy of the baseline model. From the results, we also observe that using a hybrid loss of Dice and PolyLoss can be advantageous to robustly segment cardiac MRIs with motion artefact, leading to a mean Dice of 0.9204, 0.8315, and 0.8906 and mean HD95 of 8.09 mm, 3.60 mm and 6.07 mm for LV, MYO and RV respectively on the official validation set. On the test set, the proposed segmentation method was ranked in second place in the segmentation task of CMRxMotion 2022 challenge.

Keywords: Cardiac MRI · Multi-task learning · Quality control · Aleatoric uncertainty · Segmentation · Deep learning · Motion artefact

O. Camara et al. (Eds.): STACOM 2022, LNCS 13593, pp. 418–428, 2022.
https://doi.org/10.1007/978-3-031-23443-9_39

1   Introduction

Cardiovascular diseases (CVDs) are the number one cause of death globally. More people die annually from CVDs than from any other causes [20]. Advanced medical imaging techniques are employed in clinical practice for the diagnosis and prognosis of cardiac diseases. Cardiac magnetic resonance (CMR) is a set of magnetic resonance imaging used to provide anatomical and functional information about the heart.

Deep learning-based methods have recently achieved promising results in cardiac structures (left ventricular blood pool, myocardium, and right ventricular blood pool) segmentation from CMR images [4,5]. As part of the ACDC challenge [4], Baumgartner et al. [3] used a 2D U-Net with a cross-entropy loss to segment the cardiac structures from CMR images. Isensee et al. [10] implemented an ensemble of 2D and 3D U-Net segmentation models with Dice loss, and Khened et al. [11] proposed 2D Dense U-Net with inception module to segment the cardiac structures. Li et al. [13] utilized a two-stage FCNs method that first localizes the heart region as a region of interest (ROI) and then segments the left ventricular blood pool, myocardium and right ventricular blood pool from the localized region of the CMR image. In addition, [8,16] utilized data augmentation-based solutions, including histogram matching, contrast modification, and image synthesis to tackle domain shift or distribution shift in CMR images segmentation as part of M&Ms challenge [5]. However, in clinical settings, the CMR images are prone to motion artefacts due to respiratory motion or patient movement during CMR acquisition. This degrades the CMR image quality, which can challenge the diagnostic value of the image and can lead to incorrect analysis.

To detect and reduce motion artefact in Cardiac MR images, Lorch et al. (2017) [14] introduced a random forest-based method that used box, line, histogram, and texture features as inputs. However, only artificially created motion artifacts were used to evaluate their approach. Oksuz et al. (2018) [19] proposed a deep learning based method to detect cardiac MR motion artefacts. They utilized 3D spatio-temporal CNNs to identify motion artefacts in 2D+time short axis CMR sequences. To increase their training dataset size, they employed k-space based data augmentation. Lyu et al. (2021) [15] utilized a recurrent generative adversarial network for cardiac MR motion artefact reduction. Oksuz et al. (2020) [18] presented a deep learning framework that jointly detects, corrects, and segments CMR images with motion artefacts.

In this paper, we proposed a fully automatic classification network to detect cardiac MRIs with motion artefact. The cardiac images were acquired using different breath-hold instructions. Noticing that there might be a contextual similarity between predicting the level of motion artefact and breath-hold type (as some breath-hold types may cause more motion artefact than others), we propose a multi-task learning (MTL) based classification network to learn better features for the main task and reduce overfitting through shared representations. To robustly segment cardiac structures from cardiac MRIs with respiratory motion artefact, we propose a fully automatic deep learning segmentation model that benefits from k-space based motion artefact data augmentation to increase training appearance variability and improve the generalization. We evaluated our method on Extreme Cardiac MRI Analysis Challenge under Respiratory Motion (CMRxMotion 2022) challenge dataset. For the classification task, the MTL-based method achieved better results compared to the single-task method, showing the advantage of sharing features between related tasks. For the segmentation task, the motion artefact augmentation and compound loss of Dice and polynomial loss enhanced the segmentation performance of the baseline method.

2   Dataset

The Extreme Cardiac MRI Analysis Challenge under Respiratory Motion (CMRxMotion 2022)[1] consists of 360 cine CMR images which have extreme cases with different levels of respiratory motions. The images were acquired using Siemens 3T MRI scanner (MAGNETOM Vida). To acquire the images, scan parameters with a spatial resolution of 2.0×2.0 mm^2, slice thickness of 8.0 mm, and slice gap 4.0 mm were used. In this challenge, the standard of procedure (SOP) mandates the 45 healthy volunteers to follow specific breath-holds guidelines. Each of the volunteers undergoes a 4-stage scan over the course of a single visit: 1) adhere to the breath-hold instructions; 2) halve the breath-hold period; 3) breathe freely; 4) breathe intensively. For each scan, short-axis CMR acquired at End-Systolic (ES) and End-Diastolic (ED) time frames are provided. From the total 360 (45 volunteers $\times$ 4 scans $\times$ 2 frames) short-axis CMR images, 160 of them were used for training, 40 for validation, and 160 for test [23].

The challenge has two tasks. The first one focuses on the quality assessment of the CMR images under respiratory motion artefacts. The quality of the images is labeled by radiologists using a standard 5-point Likert scale. Based on these scores, three levels of motion artefacts were defined: mild motion artefacts (label 1), intermediate motion artefacts (label 2) and severe motion artefacts (label 3). The second task focuses on the segmentation of cardiac structures from CMR images. All images with diagnostic quality are segmented by an experienced radiologist, including contours for left ventricular blood pool (LV), myocardium (MYO) and right ventricular (RV) blood pool [23].

3   Methods

3.1   Cardiac Image Quality Classification

Multi-task learning aims to solve multiple tasks simultaneously by taking advantage of the commonalities and differences across these tasks. Compared to training the models separately, it can lead to improved regularization, learning efficiency, and prediction accuracy for the task-specific models [6,7]. To leverage the advantages of multi-task learning, we proposed a multi-task deep learning

---

[1] http://cmr.miccai.cloud/.

based classification network to classify the cardiac image quality based on their respiratory motion artefact level. The proposed method is designed to simultaneously conduct two classification tasks: image quality prediction (main task) and patient's breath-hold type prediction (auxiliary task). The main task has three classes: mild motion artefacts (class 1), intermediate motion artefacts (class 2) and severe motion artefacts (class 3). The auxiliary task focuses on predicting the type of breath-hold the patient followed during the CMR acquisition. It has four classes: adhere to the breath-hold instructions (class 1), halve the breath-hold period (class 2), breathe freely (class 3), and breathe intensively (class 4). The loss function of the multi-task model is computed as follows:

$$L_{Total} = L_{MainTask} + \beta L_{AuxTask} \tag{1}$$

where $L_{MainTask}$ is a weighted categorical cross-entropy loss for the main classification task, $L_{AuxTask}$ is a categorical cross-entropy loss for the auxiliary task and β is a hyper-parameter value that controls the contribution of $L_{AuxTask}$ to the total loss (Eq. 1).

For the network architecture, we modified the 3D ResNet-18[2] architecture by adding a second classification branch, as shown in Fig. 1. More specifically, after the global average pooling of ResNet-18, we added a fully connected layer (512) with ReLU activation then, it is followed by two classification branches for the two tasks. During pre-processing, all the volumes were resampled to 0.664mm × 0.664mm × 9.60 mm and the intensity of every volume was normalized to have zero-mean and unit-variance. In order to improve the robustness of the model, we utilized data augmentation such as random cropping (with size of $(300, 300, 8)$), random rotation ($degrees = (-15, 15)$ with probability of 0.5), random flipping (horizontal and vertical flipping with probability of 0.5) and random scaling (range $(0.9, 1.2)$ with probability of 0.5). To enhance the results further and study the data-dependent (aleatoric) uncertainty [2, 22], we employed test-time augmentation (TTA). For TTA, we used M different types of data augmentations during testing. The augmentations utilized for TTA include horizontal flipping, vertical flipping, random rotation ($degrees = (-10, 10)$) and random scaling ($range = (0.75, 1.25)$). Then the mean prediction of the M augmented images is used as the final prediction, and the variance or entropy of these predictions is considered as the uncertainty measure.

3.2 Cardiac Segmentation

For the Cardiac MRI segmentation, we employed a 3D segmentation network which is based on 3D nnU-Net framework [9]. The U-Net's encoder and decoder consist of 12 convolutional layers where each convolution is followed by instance normalization and Leaky ReLU (negative slope of 0.01) activation function.

Dice loss (L_{Dice}) is a region-based loss that directly optimizes the Dice coefficient metric. PolyLoss [12] redesigns loss functions as a linear combination of

---

[2] https://github.com/Project-MONAI.

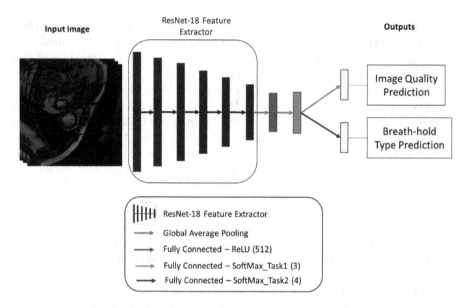

Fig. 1. Overview of the multi-task classification network

polynomial functions. By dropping the higher order polynomials and adding terms that perturb the polynomial coefficients, Leng *et al.* (2022) [12] came up with a simplified version of the polynomial loss, which is called Poly-1. As can be observed in Eq. 2, this loss function modifies the cross-entropy (L_{CE}) by just adding one hyper-parameter (ϵ) [12], where p is the prediction probability of the target class. In this work, we utilized a hybrid loss of Dice loss and a Polynomial version of cross-entropy loss ($L_{DicePolyCE}$) to segment the heart structures from cardiac MRIs, as shown Eq. 3.

$$L_{PolyCE} = L_{CE} + \epsilon(1 - p) \qquad (2)$$

$$L_{DicePolyCE} = L_{Dice} + L_{PolyCE} \qquad (3)$$

To enhance the robustness of the segmentation model in the cardiac MRI dataset with motion artefact, we utilized k-space motion artefact data augmentation in addition to intensity-based data augmentation and spatial data augmentation. We applied brightness, contrast, gamma, Gaussian noise, and blur as an intensity-based augmentation. Elastic deformation, random rotation, random scaling, random flipping, and low resolution simulation are employed as spatial augmentations. During MR image acquisition, signals from the MRI scanner are temporarily stored in the k-space. The inverse Fourier transform of the k-space is then computed to provide the MR image. To generate k-space motion artefact from the cardiac MRIs, the k-space is filled with random rigidly modified versions of the original image. Then the inverse transform of the compound k-space is computed [1, 21]. The k-space motion artefact augmentations were generated using TorchIO library[3]. We applied the k-space motion artefact on randomly selected good-quality cardiac MR training images. The generated images were then added to the training dataset to increase the variety of the training images.

3.3   Training

The segmentation models were trained for 1000 epochs in a five-fold cross-validation scheme. The weights of the network were optimized using Stochastic gradient descent (SGD) with Nesterov momentum ($\mu = 0.99$) with an initial learning rate of 0.01 [9]. The learning rate was decayed using the "poly" learning rate policy. The mini-batch size was 2 for the segmentation model. For the classification task, the dataset was shuffled and randomly split into 65% training, and 35% validation. The models were trained for 300 epochs with an ADAM optimizer and a mini-batch size of 32. We used a value of 1 for epsilon (ϵ) in the PolyLoss (Eq. 2). For the multi-task loss, the weighting factor (β) (in Eq. 1) is empirically chosen to be 0.4. To compute the data-dependent uncertainty, we used a value of 5 for M (number of data augmentations). All the training was done on NVIDIA Tesla V100 GPUs using Pytorch deep learning framework.

4   Results and Discussion

For the evaluation of segmentation results, Dice coefficient and Hausdorff distance 95% percentile (HD95) metrics are employed. To assess the classification results, accuracy and Cohen's kappa metrics are utilized.

As can be seen in Table 1, the proposed multi-task network yielded a significantly better result compared to the single-task network. The performance increase was almost by 10% and 15% in terms of accuracy and Cohen's kappa, respectively. This shows that simultaneously learning both tasks can help the network learn better representations for the main task. Applying TTA to the multi-task model further improved the results, as shown in Table 1. When we ensemble the results of the models which use different loss functions with the results of single-task and multi-task models, it yielded an accuracy of 0.7 and Cohen's kappa of 0.507.

TTA can also be used to estimate heteroscedastic aleatoric uncertainty of images [2]. Comparing the aleatoric uncertainties on our own validation split ($n = 56$) in Fig. 2, one can observe that the input-dependent uncertainty (variance) of the incorrect predictions (0.073) is higher than the correct predictions (0.060). This indicates that images with higher aleatoric uncertainty are harder for the model to classify than images with lower aleatoric uncertainty.

For the segmentation task, the baseline method is the default nnUNet network. It uses light data augmentation and a compound loss that combines Dice loss with cross-entropy Loss (DiceCE). We compared the performance of the proposed DicePolyCE loss with two common segmentation losses: DiceCE

---

[3] https://github.com/fepegar/torchio.

Table 1. Image quality classification results for different methods on the official validation set ($n = 40$) of the challenge. The bold values are the best.

Method	Accuracy	Cohen's Kappa
Single-task (image quality)	0.525	0.209
Multi-task (image quality, breath-hold type)	0.625	0.367
Multi-task +TTA	0.65	0.413
Ensemble	**0.7**	**0.507**

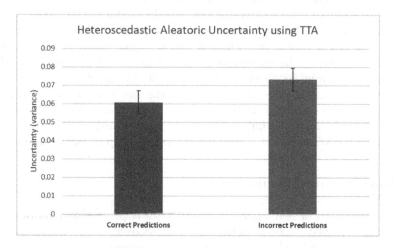

Fig. 2. Comparing the aleatoric uncertainty of correctly and incorrectly predicted images on our local validation split

loss (baseline) and DiceFocal loss [17] (a hybrid loss that combines Dice loss with Focal loss). Regarding data augmentations, we divided the experiments into light data augmentation (baseline), moderate motion artefact augmentation (moderate MAA) and heavy motion artefact augmentation (heavy MAA). The light data augmentation uses elastic deformation ($\alpha : (0., 200.), \sigma : (9., 13.)$, random rotation ($-30, +30$ degrees), scaling ($0.85, 1.25$), mirroring, brightness ($\mu : 0, \sigma : 0.1$), and gamma ($0.7, 1.5$). For moderate MAA, we added ten images with artificial motion artefact (number of transforms $= 2$). For heavy MAA, we increased the number of images to 30 and the number of simulated movements to 6 (number of transforms $= 6$), where larger values generate more distorted images. Some visual examples of motion artefact augmented images are shown in Fig. 3.

Comparing the performance of the data augmentation types, using a moderate MAA achieved the best result in terms of Dice and HD95 of the cardiac structures compared to the baseline and heavy MAA, as can be seen from Table 2. This shows the advantage of adding a moderate number of artificially generated motion artefacts into the training images to improve the model's generalizability on a dataset with respiratory motion artefacts. While using heavy MAA, which generates unrealistic motion artefacts that are more distorted from the original training images, leads to poor segmentation performance. Regarding the performance of the different hybrid losses, DiceFocal loss increased the segmentation accuracy of the baseline (DiceCE). Utilizing the proposed DicePolyCE loss outperformed the other two loss functions by achieving Dice scores of 0.9204, 0.8315, and 0.8906 and HD95 of 8.09 mm, 3.60 mm and 6.07 mm for LV, MYO and RV respectively (Table 2). This shows the robustness of the proposed loss that combines region-based Dice loss with a polynomial version of cross-entropy on cardiac MR segmentation with motion artefacts.

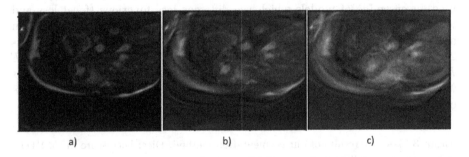

a)   b)   c)

Fig. 3. Visual appearance of a) original CMR image b) moderate motion artefact augmented CMR image c) heavy motion artefact augmented CMR image

Since the challenge allows the usage of an external public dataset, we added ACDC dataset [4] to CMRxMotion dataset to diversify the input images and improve the generalization of the segmentation model. For the ACDC dataset, as a preprocessing step, we resampled the images to the median voxel spacing of CMRxMotion and applied moderate MAA. As shown in Table 2, the addition of ACDC (CMRx+ACDC) improved the result of the baseline. When we changed the loss function from DiceCE to the proposed loss (DicePolyCE), it enhanced the segmentation performance further and achieved the best result with Dice of 0.9236, 0.8341, and 0.8967 and HD95 (in mm) of 8.31, 3.46 and 5.86 for LV, MYO and RV respectively.

Table 2. Comparison of cardiac MR segmentation performances using various data augmentations and compound loss functions on validation set ($n = 40$) of the challenge. Dice: Dice score, HD95: Hausdorff distance 95% percentile in mm, moderate MAA: moderate motion artefact augmentation, heavy MAA: heavy motion artefact augmentation. The bold values are the best. Asterisk * indicates that the dataset used is a mixture of CMRxMotion and ACDC datasets.

Method	Dice LV	Dice MYO	Dice RV	HD95 LV	HD95 MYO	HD95 RV
Baseline	0.9147	0.8298	0.8886	8.64	3.74	6.11
Moderate MAA	0.9168	0.8306	0.8918	8.49	3.70	6.04
Heavy MAA	0.9141	0.8298	0.8927	8.45	3.69	5.83
DiceFocal Loss	0.9181	0.8303	0.8927	8.24	3.67	5.62
DicePolyCE Loss	0.9204	0.8315	0.8906	**8.09**	3.60	6.07
Baseline + ACDC (CMRx+ACDC)*	0.9196	0.8325	0.8961	8.39	3.64	**5.57**
CMRx+ACDC with DicePolyCE Loss*	**0.9236**	**0.8341**	**0.8967**	8.31	**3.46**	5.86

For the final submission of task-1 (cardiac image quality classification), we used an ensemble of models which use different loss functions (focal loss and weighted cross-entropy loss) with single-task and multi-task models, it yielded an accuracy of 0.6083 and Cohen's kappa of 0.4545 on the unseen test set. For task-2 (cardiac image segmentation), we utilized *CMRx+ACDC with DicePolyCE Loss* model and the test result is summarized in Table 3. The proposed methods were ranked in second place for the segmentation task and fifth place in the classification task.

Table 3. Test set results of our segmentation method. Dice: Dice score in %, HD95: Hausdorff distance 95% percentile in mm

Dice LV (%)	Dice MYO (%)	Dice RV (%)	HD95 LV	HD95 MYO	HD95 RV
93.79 ± 3.77	87.08 ± 3.20	91.14 ± 5.93	2.94 ± 3.38	2.10 ± 1.77	4.68 ± 4.02

5   Conclusion

In this paper, we proposed fully automatic deep learning methods to detect motion artefact levels and segment cardiac structures from cardiac MRIs with respiratory motion artefact. For the classification task, the proposed network simultaneously predicts both the motion artefact level (main task) and breath-hold type to leverage the benefits of multi-task learning. From the results, we observed that training a multi-task network encourages the network to extract useful features for both tasks and results in better prediction performance. The proposed method greatly increased the classification result in terms of accuracy and Cohen's kappa compared to the single-task model. Applying test-time augmentation also helped the model to improve the classification result further and to estimate the aleatoric uncertainty of the data. In the segmentation task, we showed that moderate k-space based motion artefact augmentation and hybrid loss of Dice and polynomial loss can enhance the segmentation performance and improve the model's generalization capability on cardiac MRIs with motion artefacts.

Acknowledgements. This work was supported by the French National Research Agency (ANR), with reference ANR-19-CE45-0001-01-ACCECIT. Calculations were performed using HPC resources from DNUM CCUB (Centre de Calcul de l'Université de Bourgogne) and from GENCI-IDRIS (Grant 2022-AD011013506). We also thank the Mesocentre of Franche-Comté for the computing facilities.

References

1. Arega, T.W., Legrand, F., Bricq, S., Meriaudeau, F.: Using MRI-specific data augmentation to enhance the segmentation of right ventricle in multi-disease, multi-center and multi-view cardiac MRI. In: Puyol Antón, E., et al. (eds.) STACOM 2021. LNCS, vol. 13131, pp. 250–258. Springer, Cham (2022). https://doi.org/10.1007/978-3-030-93722-5_27

2. Ayhan, M.S., Berens, P.: Test-time data augmentation for estimation of heteroscedastic aleatoric uncertainty in deep neural networks (2018)

3. Baumgartner, C.F., Koch, L.M., Pollefeys, M., Konukoglu, E.: An exploration of 2D and 3D deep learning techniques for cardiac MR image segmentation. In: Pop, M., et al. (eds.) STACOM 2017. LNCS, vol. 10663, pp. 111–119. Springer, Cham (2018). https://doi.org/10.1007/978-3-319-75541-0_12

4. Bernard, O., et al.: Deep learning techniques for automatic MRI cardiac multi-structures segmentation and diagnosis: is the problem solved? IEEE Trans. Med. Imaging **37**, 2514–2525 (2018)

5. Campello, V.M., et al.: Multi-centre, multi-vendor and multi-disease cardiac segmentation: the M&MS challenge. IEEE Trans. Med. Imaging **40**, 3543–3554 (2021)

6. Caruana, R.: Multitask learning. Mach. Learn. **28**, 41–75 (2004)

7. Chen, C., Bai, W., Rueckert, D.: Multi-task learning for left atrial segmentation on GE-MRI. In: Pop, M., et al. (eds.) STACOM 2018. LNCS, vol. 11395, pp. 292–301. Springer, Cham (2019). https://doi.org/10.1007/978-3-030-12029-0_32

8. Full, P.M., Isensee, F., Jäger, P.F., Maier-Hein, K.: Studying robustness of semantic segmentation under domain shift in cardiac MRI. In: Puyol Anton, E., et al. (eds.) STACOM 2020. LNCS, vol. 12592, pp. 238–249. Springer, Cham (2021). https://doi.org/10.1007/978-3-030-68107-4_24

9. Isensee, F., Jaeger, P., Kohl, S., Petersen, J., Maier-Hein, K.: nnU-Net: a self-configuring method for deep learning-based biomedical image segmentation. Nat. Methods **18**, 1–9 (2021). https://doi.org/10.1038/s41592-020-01008-z

10. Isensee, F., Jaeger, P.F., Full, P.M., Wolf, I., Engelhardt, S., Maier-Hein, K.H.: Automatic cardiac disease assessment on cine-MRI via time-series segmentation and domain specific features. In: Pop, M., et al. (eds.) STACOM 2017. LNCS, vol. 10663, pp. 120–129. Springer, Cham (2018). https://doi.org/10.1007/978-3-319-75541-0_13

11. Khened, M., Alex, V., Krishnamurthi, G.: Densely connected fully convolutional network for short-axis cardiac cine MR image segmentation and heart diagnosis using random forest. In: Pop, M., et al. (eds.) STACOM 2017. LNCS, vol. 10663, pp. 140–151. Springer, Cham (2018). https://doi.org/10.1007/978-3-319-75541-0_15

12. Leng, Z., et al.: PolyLoss: a polynomial expansion perspective of classification loss functions. arXiv abs/2204.12511 (2022)
13. Li, C., et al.: APCP-NET: aggregated parallel cross-scale pyramid network for CMR segmentation. In: 2019 IEEE 16th International Symposium on Biomedical Imaging (ISBI 2019), pp. 784–788 (2019)
14. Lorch, B., Vaillant, G., Baumgartner, C.F., Bai, W., Rueckert, D., Maier, A.K.: Automated detection of motion artefacts in MR imaging using decision forests. J. Med. Eng. **2017** (2017)
15. Lyu, Q., et al.: Cine cardiac MRI motion artifact reduction using a recurrent neural network. IEEE Trans. Med. Imaging **40**, 2170–2181 (2021)
16. Ma, J.: Histogram matching augmentation for domain adaptation with application to multi-centre, multi-vendor and multi-disease cardiac image segmentation. In: Puyol Anton, E., et al. (eds.) STACOM 2020. LNCS, vol. 12592, pp. 177–186. Springer, Cham (2021). https://doi.org/10.1007/978-3-030-68107-4_18
17. Ma, J., et al.: Loss odyssey in medical image segmentation. Med. Image Anal. **71**, 102035 (2021)
18. Oksuz, I., et al.: Deep learning-based detection and correction of cardiac MR motion artefacts during reconstruction for high-quality segmentation. IEEE Trans. Med. Imaging **39**, 4001–4010 (2020)
19. Oksuz, I., et al.: Deep learning using K-space based data augmentation for automated cardiac MR motion artefact detection. In: Frangi, A.F., Schnabel, J.A., Davatzikos, C., Alberola-López, C., Fichtinger, G. (eds.) MICCAI 2018. LNCS, vol. 11070, pp. 250–258. Springer, Cham (2018). https://doi.org/10.1007/978-3-030-00928-1_29
20. World Health Organization: Cardiovascular diseases (CVDS) (2017). https://www.who.int/news-room/fact-sheets/detail/cardiovascular-diseases-(cvds)
21. Pérez-García, F., Sparks, R., Ourselin, S.: TorchIO: a python library for efficient loading, preprocessing, augmentation and patch-based sampling of medical images in deep learning. Comput. Methods Programs Biomed. **208**, 106236 (2021)
22. Wang, G., Li, W., Aertsen, M., Deprest, J.A., Ourselin, S., Vercauteren, T.K.M.: Aleatoric uncertainty estimation with test-time augmentation for medical image segmentation with convolutional neural networks. Neurocomputing **335**, 34–45 (2019)
23. Wang, S., et al.: The extreme cardiac MRI analysis challenge under respiratory motion (CMRxMotion) (2022)

Motion-Related Artefact Classification Using Patch-Based Ensemble and Transfer Learning in Cardiac MRI

Ruizhe Li[✉] and Xin Chen

Intelligent Modelling and Analysis Group, School of Computer Science,
University of Nottingham, Nottingham, UK
ruizhe.li@nottingham.ac.uk

Abstract. Cardiac Magnetic Resonance Imaging (MRI) plays an important role in the analysis of cardiac function. However, the acquisition is often accompanied by motion artefacts because of the difficulty of breath-hold, especially for acute symptoms patients. Therefore, it is essential to assess the quality of cardiac MRI for further analysis. Time-consuming manual-based classification is not conducive to the construction of an end-to-end computer aided diagnostic system. To overcome this problem, an automatic cardiac MRI quality estimation framework using ensemble and transfer learning is proposed in this work. Multiple pre-trained models were initialised and fine-tuned on 2-dimensional image patches sampled from the training data. In the model inference process, decisions from these models are aggregated to make a final prediction. The framework has been evaluated on CMRxMotion grand challenge (MICCAI 2022) dataset which is small, multi-class, and imbalanced. It achieved a classification accuracy of 78.8% and 70.0% on the training set (5-fold cross-validation) and a validation set, respectively. The final trained model was also evaluated on an independent test set by the CMRxMotion organisers, which achieved the classification accuracy of 72.5% and Cohen's Kappa of 0.6309 (ranked top 1 in this grand challenge). Our code is available on Github: https://github.com/ruizhe-l/CMRxMotion.

Keywords: Cardiac MRI · Motion artefacts · Ensemble learning · Patch-based classification

1 Introduction

Cardiac Magnetic Resonance Imaging (MRI) is widely used in clinical practice for cardiac function analysis and disease diagnosis. However, during cardiac MRI acquisition, respiratory motions introduce image artefacts, resulting in the inaccurate image-based analysis [1]. Therefore, it is essential to estimate the image quality of cardiac MRI. Manual annotation is time-consuming and laborious for human experts. Therefore, an automatic approach is needed to classify the image quality of cardiac MRI.

© The Author(s), under exclusive license to Springer Nature Switzerland AG 2022
O. Camara et al. (Eds.): STACOM 2022, LNCS 13593, pp. 429–438, 2022.
https://doi.org/10.1007/978-3-031-23443-9_40

Convolutional Neural Network (CNN)-based methods (e.g. ResNet [2] and Efficien-Net [3]) have shown excellent performance on natural image quality assessment tasks. Wu et al. extracted high-level features using a CNN-based model to learn a hierarchical feature degradation [4]. Varga et al. employed ResNet that achieved blind image quality assessment without any quality information for pristine reference images [5]. More recently, Vision Transformer (ViT) [6] also achieved remarkable performance on image quality assessment tasks [7].

Compared to natural images, the medical image is more challenging due to the lack of large datasets and the difficulty of annotation. Oksuz et al. proposed a k-space data augmentation strategy and a curriculum learning scheme to detect motion artefacts that achieved a good performance on a large cardiac MRI dataset [8]. Ettehadi et al. proposed a 3D residual Squeeze-and-Excitation CNN method for Diffusion MRI (dMRI) artefact detection [9]. Instead of binary-level classification, they achieved a remarkable result on multiclass classification for multiple artefacts. Besides, various other deep-learning-based approaches achieved acceptable performance on image quality detection and classification tasks in medical images [10–12].

However, the performance of the k-space based approaches highly depends on extensive effort and fine-grained design for parameter fitting, which is time costly and infeasible in practice [13]. In addition, most of the methods also rely on a large amount of balanced training data.

The CMRxMotion grand challenge (MICCAI 2022) dataset [14] consists of small, multi-class, and imbalanced data, which poses a huge challenge in training deep-learning based methods. To overcome these difficulties, we propose a patch-based ensemble learning method based on pre-trained models. Our design principle is three-fold, as described below. 1) To address the issue of small data with varying image sizes, each of the 3D volume is split into 2D slices and randomly sampled using image patches with a fixed size. Three different classifiers were trained for ensemble decision making, each of the classifier is fine-tuned on a pre-trained 2D classification model. This patch-based ensemble learning helps to avoid model overfitting. 2) To reduce the impact of large image intensity variations, the image gradient magnitude of the original image is used to combine with the original image for ensemble learning. 3) The majority voting is used to summarise the results from different models for decision making. We evaluated our method using 5-fold cross-validation. It was also tested on an independent validation set provided by the challenge organiser.

2 Methods

An overview of the proposed method is shown in Fig. 1. In the training phase, automatic segmentation of the heart region is firstly performed using an encoder-decoder deep CNN model (i.e. U-Net [15]). Each 2D slice in the input cardiac MRI is considered independently. A gradient magnitude map for each slice is calculated to reduce the appearance difference across images. After that, random sampling is applied to every 2D slice of both the original image and the gradient magnitude map to generate a set of image patches with a fixed size. These image patches are only around the heart region assisted by a heart segmentation model. The ensemble learning technique is then used to train multiple models with different architectures, each model is trained separately on the image patches of the original image and the gradient magnitude map.

In the test phase, image patches are randomly sampled from the original input image and its corresponding gradient magnitude map around the segmented heart region. For every sub-classifier, majority voting is used to make a slice-level decision based on the aggregation of predictive probability values from multiple image patches. Then a biased voting (described in Sect. 3.3) is applied to summarize the slice-level results to get the subject-level results. Finally, another majority voting is applied to produce the final subject-level decisions based on the outputs of all sub-classifiers.

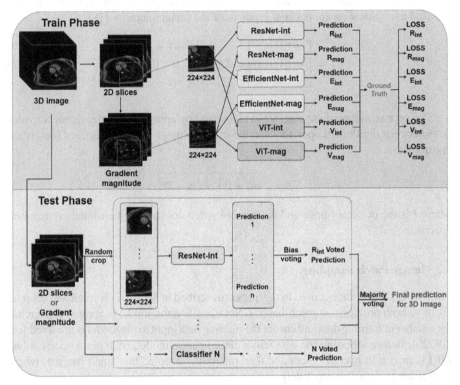

Fig. 1. Overview of the proposed framework. In the training phase, three networks (ResNet, EfficientNet and ViT) are trained on 2D image patches of original image and gradient magnitude map respectively, resulting in six classifiers. In the test phase, the input multi-slice image is sampled to multiple random patches. The final decision is voted by all classifiers on all patches.

2.1 Image Pre-processing Using Gradient Magnitude

The Cardiac MRI data acquired under different postures and breathing patterns (i.e. breath-hold, breath freely, etc.) have intensity and shape discrepancies, which pose challenges for both classification and segmentation tasks. In the context of motion artefact detection, machine learning models could be confused by motion artefacts and intensity changes. Image gradient magnitude is a commonly used method for feature representation [16, 17] to reduce intensity variations. The calculation of gradient magnitude on a 2D image is expressed in Eq. (1).

$$\nabla f(x, y) = \sqrt{\left(\frac{\partial f(x, y)}{\partial x}\right)^2 + \left(\frac{\partial f(x, y)}{\partial y}\right)^2} \tag{1}$$

where $\nabla f(x, y)$ indicates the gradient magnitude at pixel location (x, y). $\frac{\partial f(x,y)}{\partial x}$ and $\frac{\partial f(x,y)}{\partial x}$ are the gradients in the x and y directions respectively. Prewitt operators (Eq. (2)) are used to compute an approximal gradient of the image intensity [18].

$$h_x = \begin{bmatrix} +1 & 0 & -1 \\ +1 & 0 & -1 \\ +1 & 0 & -1 \end{bmatrix} \text{ and } h_y = \begin{bmatrix} +1 & +1 & +1 \\ 0 & 0 & 0 \\ -1 & -1 & -1 \end{bmatrix} \tag{2}$$

The approximal gradient map $\widetilde{\nabla} f$ is calculated by applying 2-dimensional convolutions between the Prewitt operators and the original image in both directions (described in Eq. (3)).

$$\widetilde{\nabla} f = \sqrt{(h_x * \mathrm{I})^2 + (h_y * \mathrm{I})^2} \tag{3}$$

where I is the original image and the symbol $*$ denotes the 2-dimensional convolution operation.

2.2 Image Patch Sampling

The CMRxMotion dataset used in this paper (described in Sect. 3.1) is small and diverse. To solve both problems, a patch-based CNN classification model is applied to increase the number of training data and ensure the training data input to the model is in a fixed size [19, 20]. Before inputting the data into a classification model, each multi-slice cardiac MRI scan is split into 2D slices and then randomly sampled into small patches, where each patch is classified separately.

We also assumed that the heart region is more important than other image regions for motion artefact detection in cardiac MRI. Hence, we propose to sample the image patches only around the heart region. To achieve this, an automatic cardiac tissue segmentation network is trained using an encoder-decoder network (U-Net [15]). The dataset from CMRxMotion task 2 (described in Sect. 3.1), is used to train this segmentation model. Then, according to the segmentation results, we randomly sample the image patches around the foreground region by ensuring that each image patch contains at least 80% of the foreground region in each image slice. In addition, the patch size is set to 224 × 224, which matches the input image size of pre-trained models (Sect. 2.3) and ensures good coverage of the heart region.

2.3 Transfer-Learning-Based Model Ensemble

Different deep learning models have their own advantages. An ensemble learning of combining different models may help to improve the accuracy and reliability of the predicted result. Additionally, the use of pre-trained models can help to avoid model overfitting to a small training dataset. In our method, three different networks are selected to construct an ensemble learning framework, including ResNet [2], EfficientNet [3] and ViT [6]. They have different filter sizes, model architectures and learning strategies (i.e. convolution vs. attention) that could potentially diversify the decision making process from different aspects. Rather than training from sketch, the official pre-trained models trained on ImageNet [21] were used to initialise the trainable weights. For each model, to match the input image dimension (1 for our dataset and 3 for the pre-trained model), we add a learnable convolutional layer with 3×3 kernel size at the beginning of each model, followed by a batch normalisation layer and a ReLU activation function.

During the training process, the pre-trained weights of all other layers except for the batch normalisation, the added convolutional layer and the decision-making layers (fully connected layers) are fixed. The training phase is to retrain the normalisation layers to refine it based on the mean and variance on the new domain specific dataset, the added convolutional layer for image feature conversion, and the fully connected layers to refine the decision making.

As illustrated in Fig. 1, the three pre-trained models are trained separately based on image patches of the original image and the gradient magnitude map, resulting in a total of six models in the ensemble framework. In the model inference stage, slice-level decisions are produced based on aggregating (i.e. average) the predictive probability values of each class using all image patches in each slice. A subject-level decision is made by a biased voting of slice-level predictions. Then the final decision is based on majority voting based on the six decisions from the sub-classifiers. In the case of an equal vote, the class that corresponds to a more severe motion artefact is selected.

3 Experiments and Results

3.1 Dataset

The CMRxMotion challenge Task 1 dataset is used to evaluate the proposed method. It consists of 320 Cardiac MRI scans from 40 volunteers. All the volunteers were asked to act in 4 situations, respectively: a) full breath-hold during acquisition; b) halve the breath-hold period; c) breathe freely; and d) breathe intensively. In each case, images at end-diastole (ED) and end-systole (ES) phases were acquired, resulting in 8 volumes per volunteer. The cardiac MRI scans were in a variety of widths and heights ranging from 400 to 600 pixels, as well as varying numbers of slices from 9 to 13.

The data was divided into 3 classes depending on the level of motion artefacts, 70, 69 and 21 scans for mild, intermediate and severe, respectively. To avoid the validation set and training set being too similar, the data were randomly divided into 5-fold on subject-level. Specifically, each volunteer's scans of 8 phases were within in the same fold for achieving fair cross-validation. In addition, due to the differences of respiratory motions for different volunteers, the data was not able to be split exactly evenly. By multiple random splits, each fold had almost balanced number of scans in mild and intermediate class and at least 3 scans in severe class. Each fold contained 32 scans.

The CMRxMotion challenge Task 2 dataset was used to train the segmentation model. The segmentation labels (left ventricle (LV), right ventricle (RV) and left ventricular myocardium (MYO)) are provided by the organiser. To simplify the segmentation task, all the 3 labels were combined into a single label (foreground) as a binary segmentation problem. We then trained a vanilla U-Net [15] on this simplified dataset. It achieved an acceptable result with dice coefficient of 0.899. This model was then applied in both the training and testing phases to detect an approximate heart region for image patch sampling.

3.2 Batch Balanced Training

In CMRxMotion dataset, the number of images in different classes is very unevenly distributed. Therefore, a batch-based data balancing strategy was designed to balance the data in each training batch. In each batch, the same number of 2D slices were acquired from each class to form a class-balanced batch for model training. Although the images from classes with fewer data appeared multiple times in the same epoch of training, the random patch-based approach was able to significantly reduce the repetition of data by sampling different image regions.

3.3 Bias Voting in Testing

Algorithm 1: Biased Voting for one subject

1: **Input:** number of slices predicted as mild, intermediate and severe (N_1, N_2, N_3)
2: bias ratio r_1, r_2
3: **IF** $N_2 + N_3 > r1 \times (N_1 + N_2 + N_3)$ **then**
4: **IF** $N_3 > r2 \times (N_2 + N_3)$
5: **return** severe
6: **else**
7: **return** intermediate
8: **end**
9: **else**
10: **return** mild
11: **end**

By analysing the data, we discovered that for each 3D volume, a small number of 2D slices with intermediate or severe motion artefacts, leads to an annotation of intermediate or severe for the subject. This shows that majority voting is not suitable at the slice-level. To solve this problem, we use a slice-based bias voting strategy. As shown in Algorithm 1, for each 3D volume, after the patch-based slice-level prediction, the number of slices predicted as mild, intermediate and severe are counted as N_1, N_2 and N_3 respectively. Then, we set N_2 and N_3 as a group, when the proportion of this group $(N_2 + N_3)$ in total number $(N_1 + N_2 + N_3)$ is greater than a ratio r_1, the subject-level precision will be chosen between intermediate and severe. Similarly, inside the group, when N_3 is greater than a ratio r_2, the final prediction will be set as severe. Through cross-validation on the training set, r_1 and r_2 are set to 0.4 and 0.25 respectively.

3.4 Other Parameter Setting

For the training of all classifiers, a fixed learning rate of 0.0001 was used. All models were trained with 10 epochs with Adam as optimiser. Dropout was added for every fully connected layer with a ratio of 0.5 to avoid overfitting. For both training and testing, the batch size was set to 30, where each batch included 10 patches in each class. In the testing phase, the number of randomly sampled patches was 20 for majority voting. The results were predicted from the best checkpoint based on a 5-fold cross-validation.

3.5 Experiments and Results

Table 1. The results for 6 different classifiers and the ensembled model on the training set over 5-fold cross-validation. Includes overall accuracy and accuracy for each level of motion artefacts (Mild, Intermediate (Inter) and Severe). Cohen's Kappa (Kappa) is also reported to measure the agreement between the ground truth and the prediction [22].

Classifier	Accuracy (%)				Kappa
	Overall	Mild	Inter	Severe	
ResNet-int	71.9	**87.1**	55.1	76.2	0.5492
ResNet-mag	70.6	82.9	52.2	**90.5**	0.5333
EfficientNet-int	73.8	84.3	63.8	71.4	0.5689
EfficientNet-mag	76.9	81.4	71.0	81.0	0.6227
ViT-int	73.1	85.7	58.0	81.0	0.5705
ViT-mag	73.8	71.4	**75.4**	76.2	0.5720
Ensembled	**78.8**	85.7	69.6	85.7	**0.6564**

Three different networks (Sect. 2.3) combined with two types of input images (original image and gradient magnitude map) were trained and evaluated separately before model ensemble, which result in six different classifiers named as ResNet-int, ResNet-mag, EfficientNet-int, EfficientNet-mag, ViT-int, ViT-mag. A 5-fold cross-validation was performed on each classifier. A total of 30 classifiers (6 models × 5 fold) were trained on CMRxMotion dataset. The results are reported in Table 1. According to Table 1, all models achieved 70% to 80% overall accuracies. Their Cohen's Kappa values were all larger than 0.53. The results from EfficientNet trained on gradient magnitude map achieved the highest Kappa value of 0.6227, which was a substantial agreement between the ground truth and the predicted results. However, the performance of different models in different classes varies significantly. ResNet-int and ResNet-mag achieved the best prediction results in mild and severe classes respectively. ViT-mag achieved the best result in intermediate class. This demonstrates the inherent instability of these sub-classifiers due to the insufficiency and imbalance of data.

Therefore, the above findings confirmed the necessity of using an ensembled framework. We combined the results of all models by majority voting (Sect. 2.3). As shown in Table 1. The Ensemble method yielded the best Cohen's Kappa, which achieved 0.6564 as a substantial agreement between the ground truth and the prediction. It also achieved the best overall accuracy (78.8%) and remarkable results (85.7%, 69.6% and 85.7%) for each class. The ensemble method not only achieved higher prediction accuracy, but also offered better stability and robustness compared to a single classifier.

The ensemble method achieved a competitive result on the validation dataset provided by the organiser through an online validation system, which achieved a classification accuracy of 70% and Cohen's Kappa of 0.5588. Subsequently, we submitted our code and trained model to the CMRxMotion organisers. Our final model was then independently evaluated on a test set by the CMRxMotion organisers, which achieved the classification accuracy of 72.5% and Cohen's Kappa of 0.6309 (**ranked top 1 in this grand challenge**).

4   Conclusions and Discussion

In this paper, we proposed a patch-based classification method based on ensemble and transfer learning for predicting three levels of motion artefacts in cardiac MRI. It was evaluated on CMRxMotion (MICCAI 2022) grant challenge dataset that consists of small, multi-class, and imbalanced data. Through the assistance of a heart segmentation model trained on the segmentation labels from task 2, the image patches were sampled around the heart region where the motion artefacts are more likely to be detected. For model training, the ensemble technique is used to train three state-of-the-art classification networks (ResNet, EfficientNet and ViT) on both the original intensity images and the gradient magnitude maps, resulting in six classification models. The final prediction is formed by majority voting on the outputs of all classifiers. The evaluation results of 5-fold cross-validation on the training set show the ensembled method outperforms all single models. The final model won first place in this grand challenge.

Furthermore, we also try to identify the shortcomings of our method. We noticed that the quality of the slices from the apical and basal regions of cardiac MRI may not contribute significantly to the quality score rated by humans as shown in Fig. 2. The

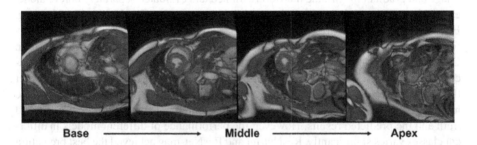

Base ⟶ Middle ⟶ Apex

Fig. 2. Examples of cardiac MRI slices which motion artefacts level is marked as mild at subject level based on the two middle slices. However, the Basal and Apex slices have more severe motion artefacts.

subject in Fig. 2 was rated as mild by humans, even the basal and apical slices contain more severe motion artefacts. For the methods like our framework, which is trained on 2D slices, this noisy labelling will interfere with the model's performance. Although patch-based training and ensemble learning can mitigate this effect, the contradictory labels may still confuse the model, leading to inaccurate prediction in certain cases. Therefore, in future work, we may design a method to remove these noisy labels and retrain a model with more representative examples.

References

1. McVeigh, E.R., Henkelman, R.M., Bronskill, M.J.: Noise and filtration in magnetic resonance imaging. Med Phys **12**, 586–591 (1985)
2. He, K., Zhang, X., Ren, S., Sun, J.: Deep residual learning for image recognition. In: Proceedings of the IEEE conference on computer vision and pattern recognition, pp. 770–778 (2016)
3. Tan, M., Le, Q.: Efficientnet: Rethinking model scaling for convolutional neural networks. In: International conference on machine learning, pp. 6105–6114 (2019)
4. Wu, J., Zeng, J., Liu, Y., et al.: Hierarchical feature degradation based blind image quality assessment. In: Proceedings of the IEEE International Conference on Computer Vision Workshops, pp. 510–517 (2017)
5. Varga, D., Saupe, D., Szirányi, T.: DeepRN: A content preserving deep architecture for blind image quality assessment. In: 2018 IEEE International Conference on Multimedia and Expo (ICME), pp 1–6 (2018)
6. Dosovitskiy, A., Beyer, L., Kolesnikov, A., et al.: An image is worth 16 x 16 words: Transformers for image recognition at scale (2020)
7. You, J., Korhonen, J.: Transformer for image quality assessment. In: 2021 IEEE International Conference on Image Processing (ICIP), pp 1389–1393 (2021)
8. Oksuz, I., Ruijsink, B., Puyol-Antón, E., et al.: Automatic CNN-based detection of cardiac MR motion artefacts using k-space data augmentation and curriculum learning. Med Image Anal **55**, 136–147 (2019)
9. Ettehadi, N., Kashyap, P., Zhang, X., et al: Automated multiclass artifact detection in diffusion MRI volumes via 3D residual squeeze-and-excitation convolutional neural networks. Front. Hum. Neurosci. 16 (2022)
10. Kelly, C., Pietsch, M., Counsell, S., Tournier, J.-D.: Transfer learning and convolutional neural net fusion for motion artefact detection. In: Proceedings of the Annual Meeting of the International Society for Magnetic Resonance in Medicine, Honolulu, Hawaii (2017)
11. Fantini, I., Rittner, L., Yasuda, C., Lotufo, R.: Automatic detection of motion artifacts on MRI using Deep CNN. In: 2018 International Workshop on Pattern Recognition in Neuroimaging (PRNI), pp. 1–4 (2018)
12. Ettehadi, N., Zhang, X., Wang, Y., et al: Automatic volumetric quality assessment of diffusion MR images via convolutional neural network classifiers. In: 2021 43rd Annual International Conference of the IEEE Engineering in Medicine & Biology Society (EMBC), pp. 2756–2760 (2021)
13. Zhang Y, Zhang W, Zhang Q, et al (2019) CMR motion artifact correction using generative adversarial nets. arXiv preprint arXiv:190211121
14. Wang, S., Qin, C., Wang, C., Wang, K., Wang, H., Chen, C., et al.: The extreme cardiac MRI analysis challenge under respiratory motion (CMRxMotion) (2022)

15. Ronneberger, O., Fischer, P., Brox, T.: U-net: Convolutional networks for biomedical image segmentation. In: International Conference on Medical image computing and computer-assisted intervention, pp 234–241 (2015).

16. Dalal, N., Triggs, B.: Histograms of oriented gradients for human detection. In: 2005 IEEE computer society conference on computer vision and pattern recognition (CVPR'05), pp 886–893 (2005)

17. Xue, W., Zhang, L., Mou, X., Bovik, A.C.: Gradient magnitude similarity deviation: A highly efficient perceptual image quality index. IEEE Trans. Image Process. **23**, 684–695 (2013)

18. Prewitt, J.M.S., et al.: Object enhancement and extraction. Picture processing and Psychopictorics **10**, 15–19 (1970)

19. Hou, L., Samaras, D., Kurc, T.M., et al.: Patch-based convolutional neural network for whole slide tissue image classification. In: Proceedings of the IEEE conference on computer vision and pattern recognition, pp. 2424–2433 (2016).

20. Feng, Z., Yang, J., Yao, L.: Patch-based fully convolutional neural network with skip connections for retinal blood vessel segmentation. In: 2017 IEEE International Conference on Image Processing (ICIP), pp 1742–1746 (2017).

21. Deng, J., Dong, W., Socher, R., et al.: ImageNet: A large-scale hierarchical image database (2009)

22. Landis, J.R., Koch, G.G.: The measurement of observer agreement for categorical data. Biometrics 159–174 (1977)

Automatic Image Quality Assessment and Cardiac Segmentation Based on CMR Images

Haixia Li, Shanshan Jiang, Song Tian, Xiuzheng Yue$^{(\boxtimes)}$, Weibo Chen, and Yihui Fan

Philips Healthcare, Hong Kong, China
xiuzheng.yue@philips.com

Abstract. This paper describes our methods for two tasks: automatic image quality assessment and cardiac segmentation based on cardiovascular magnetic resonance (CMR) images with respiration motion artifacts. For the quality assessment task, we developed a method fusing deep learning model results and radiomics model results. We trained an Efficientnet-b0 to capture image quality information from the global view. We trained multiple radiomics models to capture the segmented left ventricle quality information from the local view. Then we fused the global view results and local view results. We achieved an accuracy of 0.725 and kappa of 0.545 for the online validation set and got 2nd rank for the online test set with an accuracy of 0.7083 and kappa of 0.5493. For the segmentation task identifying the left ventricle blood pool (LV), myocardium (MYO), and right ventricle blood pool (RV), we used nnUNet as the backbone network and trained two cascaded models to predict the final three structures. The first model was trained by taking the three structures as one class, and the second was trained to segment each structure based on the first model's prediction. We also used the trained model to predict the data that have not been labeled in the training set due to low image quality and get their pseudo labels. Then we finally trained a new model with all available data, including unlabeled data with pseudo labels. Our online validation results for the cardiac segmentation task achieved top-1 rank in dice score of LV and top-10 rank in dice score of MYO, and RV blood pools in the challenge validation leaderboard. We achieved 5th rank on the online test set.

Keywords: Respiration motion artifact · CMR · Segmentation · Quality assessment

1 Introduction

Cardiac MRI (CMR) has emerged as a useful image for cardiovascular disease diagnosis. Moreover, the assessment of cardiac tissues with CMR is more accurate than other modalities [1]. However, the respiratory motion-caused artifacts can impair CMR image quality and therefore affect the diagnosis accuracy [2]. Respiratory motion artifacts will cause a mismapping of the signal and image blurring [2]. The CMRxMotion proposed two challenging tasks related to CMR with respiratory motion: I) image quality assessment, II) left and right ventricle blood pools (LV and RV), and myocardium

© The Author(s), under exclusive license to Springer Nature Switzerland AG 2022
O. Camara et al. (Eds.): STACOM 2022, LNCS 13593, pp. 439–446, 2022.
https://doi.org/10.1007/978-3-031-23443-9_41

(MYO) segmentation. In the CMRxMotion challenge, there are four breathing modes: full breath-hold, half breath-hold, free breath, and intensive breath. Each breathing mode provides images at the end-diastole (ED) and end-systole (ES) frames.

For task I, the quality control (QC) and artifact detection of CMR images are difficult and often require manual visual inspection. Thus, automatic assessment of CMR may accelerate the QC workflow. [3] describes some quality assessment methods such as intensity spatial variation, gradient entropy, temporal total variation, and other metrics. Recently, deep learning has become another mainstream approach that extracts image features without any human intervention. In [4], the CMR image quality is ranked from 0 to 4, 0 representing nondiagnostic and 4 representing excellent diagnostic value. The paper utilizes a network called IQ-DCNN, a 3D deep convolutional neural network, to assess the quality differences, which can reproduce expert decisions. [2] shows that the LV quality is an important factor in assessing the CMR image quality.

Task II is the segmentation of LV, RV and MYO. Cardiac segmentation based on deep learning methods occupies a large proportion in medical imaging in recent years. [5] collected cine MR scans from multicenter in multivendor. A standard UNet is used to automatically quantify the left ventricular function. From the dice coefficient, both endocardium and epicardium can be well segmented. Other related cardiac segmentation methods can be found in [6].

This paper aims to investigate automatic segmentation of LV, RV, and MYO and classification of the CMR image quality.

2 Methods

For CMR image quality, we trained both deep learning and radiomics models and ensemble them to assess the CMR image quality. For cardiac segmentation, we trained a cascaded nnUNet [7] model for the end-to-end cardiac segmentation that contains a whole cardiac area segmentation model and a separate structure segmentation model.

2.1 Task I: CMR Image Quality Classification

2.1.1 Global View Model Training

Cardiac MRI scans and annotations are not easy to acquire. Due to the limitation of dataset volume, we managed to use a lightweight CNN. In this challenge, the backbone was Efficientnet-b0 [8], a faster and tinier network compared to other CNNs, since a lightweight network may somehow inhibit overfitting when no external data is included. Furthermore, we adopted a self-supervised pretraining approach to this task with the training data, which may help the network better learn the representations and avoid overfitting. Then end-to-end fine-tuning was applied to the training set. In this step, the model first loaded the self-pretrained weights and then was trained with fewer iterations than learning from scratch.

In the first stage, self-supervised training, we followed the Simsiam [9] pipeline, which did not require a large training batch size. For data augmentation, random scale and crop, horizontal flipping, and intensity shift were applied. The optimizer was SGD with a momentum of 0.9 and a weight decay of 1e-4. The learning rate began at 0.4

and decayed to 0 in 200 epochs with a cosine scheduler. The loss function was cosine similarity in the range -1 to 1. The batch size was 32 and all the CMR slices were cropped or padded to 512x512.

The second stage was end-to-end classification fine-tuning. The model first loaded the pretrained weights. We set a small initial learning rate to 2e-3, with cosine scheduler and finally decay to 0 in 120 epochs. A learning rate warm-up stage was activated for the first 20 epochs, which increased linearly from 0. Two data augmentation methods, random horizontal flipping, and random scale crop with a ratio 0.8–1.0 were applied to all slices. The loss function was cross-entropy with label smoothing 0.1. Other losses we tried such as focal loss and seesaw were not very stable for this task. The batch size was set to 64 and image resolution was 512x512, same with the pretraining. The GPU we use was a single V100 with 32GB memory. Data transformation and augmentation were implemented by MONAI (version 0.6). Efficientnet was from mmcls (version 0.22). The deep learning pipeline was developed based on pytorch (version 1.12).

2.1.2 Local View Radiomics

Figure 1 shows an example of the heart including the left ventricle, right ventricle, and myocardium from a good quality slice and blurred slice. As the LV quality decrease can reflect the CMR image quality decrease, we used the LV results from the segmentation task, extracted their radiomics features and trained different machine learning models using the PyCaret [10] package. The input was normalized and resampled with a spacing setting of [1.0,1.0,1.0]. The wavelet, log with sigma setting of 1, 2, and 3, gradient, original, first order, glrlm, ngtdm, gldm, glcm, glszm and shape features were extracted with a bin count of 32. Feature selection was implemented, and 50 features were selected with default PyCaret feature selection method. Four machine learning models were selected based on internal validation results: extra trees, random forest, decision tree and gradient boosting classifier.

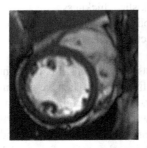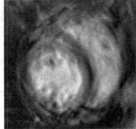

Fig. 1. Example of good quality (left) and blurred (right) CMR image slices.

2.1.3 Ensemble Efficientnet and Radiomics

We finally created a model called ER-vote to fuse the results of the global view deep learning model and the local view radiomics models. Figure 2 shows the workflow of the ensemble. The whole image volume was sent into the deep learning model (Efficientnet) to get the global view result. The LV area generated from the segmentation models in task 2 was used to extract the radiomics features and sent into four radiomics models to get four results. Finally, the output was decided by five votes with the same weights. Since there is one deep learning model and four radiomics models, the weights for deep learning and radiomics are 0.2 and 0.8, respectively.

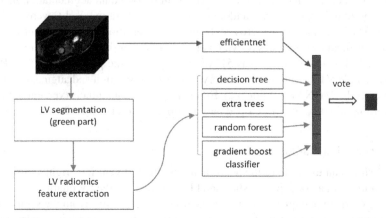

Fig. 2. Ensemble of the deep learning and radiomics models. Deep learning model: efficientnet. Radiomics models: decision tree, extra trees, random forest and gradient boost classifier.

2.2   Task II: CMR Image Segmentation

2.2.1   Model Training

In this task, there are three cardiac structures: LV, MYO, and RV. We used 2D nnUNet as the backbone architecture due to the thick slice thickness.

To delineate an object in a smaller receptive field is easier than in a larger one. This phenomenon is more obvious when the object is small, so we cropped the cardiac region first and then segmented LV, RV and MYO.

More specifically, the procedure consists of two steps. First, we merged the mask of three structures into one, and trained the first model. Then we generated new images by setting the pixel outside the mask to 0 and used the original three masks to train the second model. The input of the first model was the original image, and the output was a mask that could segment the whole cardiac area. The input of the second model was the result of multiplying the original image by the mask from the first model, and the output was the final three masks corresponding to the three structures. These two models were cascaded in a sequence to complete the whole segmentation task. We called this strategy two-step segmentation.

Due to limited training data and to leverage the extra information of the unlabeled data, in addition to the original training data, the model was further trained with pseudo masks generated from unlabeled data, which performed better in the validation set (Fig. 3).

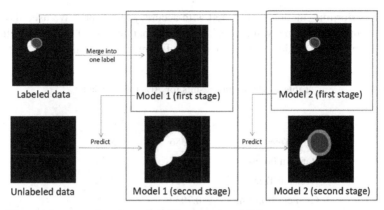

Fig. 3. The training procedure for task 2. Three masks of different structures were merged into one big cardiac mask, and this mask was used to train model 1. Then new images were generated by setting the pixel outside the cardiac mask to 0 and used the original three masks to train the model 2. These two models were cascaded in a sequence to complete the whole segmentation task.

2.2.2 Image Normalization and Augmentation

We adopted spacing normalization and intensity normalization for the data. We uniformed the spacing using the medium spacing, and scaled the pixel value to the range [0,1]. During the experiment, we kept the original image resolution and trained the model by patch. We applied two types of augmentation. (1) Spatial transformations: rotation, flipping, scaling, or deformation of the original images; (2) intensity transformations: adding noise, changing brightness.

3 Results

3.1 Task I: CMR Image Quality Classification

We evaluated the performance of the 2d Efficientnet. The training set was stratified randomly split into 70% and 30% as internal training and validation set. Table 1 reports the results of internal training and validation.

2d Efficientnet results for 70% training data and 30% internal validation data.

Set	Accuracy	Kappa
Training	0.92	0.87
Internal Validation	0.86	0.78

We evaluated various machine learning models. The training data was split into 70% internal training data and 30% internal validation data. Note that this split was implemented by PyCaret by default, and thus this 30% internal validation data are different from the deep learning part. The 70% internal data was trained with 10-fold cross-validation. Table 2 summarizes the average 10-fold cross-validation results. The top four best-performing models were listed with accuracy and kappa. We used these four models for the 30% internal validation. Table 3 summarizes the 30% internal validation results. Then all four models were finally trained using 100% training data.

Table 2. Radiomics results for 70% training data.

Model	Accuracy	Kappa
Extra Trees	0.7386	0.5475
Random Forest	0.7295	0.5371
Decision Tree	0.7121	0.5131
Gradient Boosting	0.7030	0.4999

Table 3. Radiomics results for 30% internal validation data.

Model	Accuracy	Kappa
Extra Trees	0.7551	0.5794
Random Forest	0.7551	0.5836
Decision Tree	0.6327	0.3806
Gradient Boosting	0.6939	0.4739

The official challenge validation results are listed in Table 4. The fusing results of the global view 2d Efficientnet and the local view radiomics models achieved better results than single views. The fusing method was to vote from all five models including Efficientnet and radiomics models (ER vote) as is shown in Fig. 2. This method achieved an accuracy of 0.725 and kappa of 0.5454.

Table 4. Results for online validation data.

Model	Accuracy	Kappa
ER vote	0.725	0.5454
Radiomics	0.675	0.4676
2d Efficientnet	0.500	0.1064

According to the official challenge test set results, our method finally achieved an accuracy of 0.7083 and kappa of 0.5493 and got 2nd rank. The metric for ranking is Cohen's kappa statistics [11]. Task 1's LV segmentation model (trained from task 2) was updated at the docker submission stage for testing, so the results of using the docker to predict the official online validation set could be slightly different from those reported in this paper obtained by validation leaderboard closing date.

3.2 Task II: CMR Image Segmentation

The dice score (DSC) of our model is listed in Table 5. The DSC of the training set in the segmentation of LV, MYO, and RV are 0.997, 1.000, and 1.000, respectively. The DSC of the online validation set are 0.926, 0.828, and 0.905, respectively. Figure 4 shows one example of the segmentation result in the validation set.

Table 5. Dice score for the training and online validation set.

Set	Dice_LV	Dice_MYO	Dice_RV
Training	0.997	1.000	1.000
Validation	0.926	0.828	0.905

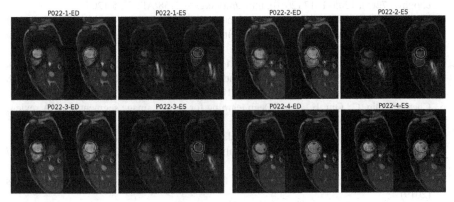

Fig. 4. One example of the segmentation results. The solid lines are the borders of the segmentation results.

4 Discussions and Conclusions

4.1 Task I: Classification

The quality classification validation results show that the deep learning method with few training samples is easy to overfit even with the self-supervised method. The radiomics methods focusing on the LV area perform similarly on the training, internal validation, and online validation dataset, but the accuracy and kappa are still not high. Fusing the two methods together helps to overcome the deep learning overfitting and radiomics underfitting issues.

4.2 Task II: Segmentation

The results show that we get satisfactory results from the segmentation of LV, MYO, and RV. As the background may affect the segmentation of the detail structures, we believe that the two-step segmentation is better than applying segmentation for three structures directly. Predicting the unlabeled data of the training set and then applying these data for training is the idea of semi-supervised training. Our experiment shows that semi-supervised learning can improve the segmentation performance.

References

1. Rajiah, P.S., et al.: Myocardial strain evaluation with cardiovascular MRI: physics, principles, and clinical applications. Radiographics **42**(4), 968–990 (2022). https://doi.org/10.1148/rg. 210174
2. Klinke, V., et al.: Quality assessment of cardiovascular magnetic resonance in the setting of the European CMR registry: description and validation of standardized criteria. J. Cardiovasc. Magn. Reson. **15**(1), 1–13 (2013)
3. Krupinski, E.A.: Current perspectives in medical image perception. Atten Percept Psychophys. **72**(5), 1205–1217 (2010). https://doi.org/10.3758/APP.72.5.1205
4. Piccini, D., et al.: Deep learning to automate reference-free image quality assessment of whole-heart MR images. Radiol. Artif. Intell. **2**(3), e190123 (2020). https://doi.org/10.1148/ryai.2020190123
5. Tao, Q., et al.: Deep learning-based method for fully automatic quantification of left ventricle function from cine MR images: A multivendor, multicenter study. Radiology **290**(1), 81–88 (2019). https://doi.org/10.1148/radiol.2018180513
6. Chen, C., Qin, C., Qiu, H., et al.: Deep learning for cardiac image segmentation: A review. Front. Cardiovasc. Med. **7**, 25 (2020)
7. Isensee, F., Jaeger, P.F., Kohl, S.A.A., et al.: nnU-Net: a self-configuring method for deep learning-based biomedical image segmentation. Nat. Methods **18**, 203–211 (2021). https://doi.org/10.1038/s41592-020-01008-z
8. Tan, M., Le, Q.V.: EfficientNet: Rethinking model scaling for convolutional neural networks (2019)
9. Chen, X., He, K.: Exploring simple siamese representation learning (2020)
10. Moez, A.: PyCaret: An open source, low-code machine learning library in Python. Version 3.0.0.rc3 PyCaret—pycaret 3.0.0 documentation (2020)
11. Wang, S., Qin, C., Wang, C., Wang, K., Wang, H., Chen, C., et al.: The extreme cardiac MRI analysis challenge under respiratory motion (CMRxMotion) (2022)

Detecting Respiratory Motion Artefacts for Cardiovascular MRIs to Ensure High-Quality Segmentation

Amin Ranem[1]([✉])[ID], John Kalkhof[1], Caner Özer[2], Anirban Mukhopadhyay[1], and Ilkay Oksuz[2]

[1] Darmstadt University of Technology, Karolinenpl. 5, 64289 Darmstadt, Germany
amin.ranem@gris.informatik.tu-darmstadt.de
[2] Computer Engineering, Istanbul Technical University, Istanbul, Turkey

Abstract. While machine learning approaches perform well on their training domain, they generally tend to fail in a real-world application. In cardiovascular magnetic resonance imaging (CMR), respiratory motion represents a major challenge in terms of acquisition quality and therefore subsequent analysis and final diagnosis. We present a workflow which predicts a severity score for respiratory motion in CMR for the CMRxMotion challenge 2022. This is an important tool for technicians to immediately provide feedback on the CMR quality during acquisition, as poor-quality images can directly be re-acquired while the patient is still available in the vicinity. Thus, our method ensures that the acquired CMR holds up to a specific quality standard before it is used for further diagnosis. Therefore, it enables an efficient base for proper diagnosis without having time and cost-intensive re-acquisitions in cases of severe motion artefacts. Combined with our segmentation model, this can help cardiologists and technicians in their daily routine by providing a complete pipeline to guarantee proper quality assessment and genuine segmentations for cardiovascular scans. The code base is available at https://github.com/MECLabTUDA/QA_med_data/tree/dev_QA_CMRxMotion.

Keywords: Cardiovascular MRI · Respiratory motion artefact detection · Semantic segmentation · Image quality assessment

1 Introduction

Respiratory motion artefacts are a common problem when performing image segmentation on cardiovascular magnetic resonance images (CMR). These motion artefacts can be more or less severe depending on the patient's ability to hold their breath. Over the past decades, there have been plenty of approaches trying to detect [4,9,14,15] and reduce [10,12,13,16,18,20,22] the artefacts produced by the patient's breathing patterns. These procedures are necessary since

---

A. Ranem, J. Kalkhof and C. Özer—Contributed equally to this work.

O. Camara et al. (Eds.): STACOM 2022, LNCS 13593, pp. 447–456, 2022.
https://doi.org/10.1007/978-3-031-23443-9_42

the breathing artefacts make anatomical boundaries unclear [22], and therefore, segmentation of them is more challenging or even impossible (see Fig. 1). Furthermore, these high-quality segmentations play a critical role, especially in treatments like image-guided radiation [22], as otherwise healthy tissue could be damaged [18].

Recent advances in deep learning help to perform high-quality cardiovascular segmentation on MRI images that contain mild to intermediate respiratory artefacts. However, it becomes much more difficult when the severity turns too strong. To ensure that the model always achieves sufficient performance, a classification model can be used as a first step in whether a cardiovascular MRI image has been acquired sufficiently well to perform segmentation.

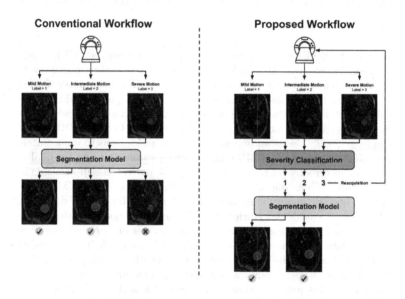

Fig. 1. Segmentation models can generally deal with different levels of motion artefacts, although they break down at a certain point. Therefore, detecting when this breakdown occurs due to severe motion artefacts is essential to prevent the model from making incorrect predictions.

We propose combining a segmentation model and a respiratory motion classifier as part of the 'Extreme Cardiac MRI Analysis challenge under Respiratory Motion (CMRxMotion)' registered in MICCAI 2022. The first step is to use the classifier to predict whether the given MRI image has sufficiently mild respiratory artefacts for our segmentation model to produce high-quality segmentations. Then, after filtering out low-quality scans, segmentation is performed on the remaining samples.

Our proposed classification model achieves 67.5% accuracy in classifying the severity of cardiovascular MRI images. The segmentation model then reaches an average Dice accuracy of 86.18% on the remaining images containing mild to intermediate motion artefacts. Our contributions are two-fold and can be summarised as follows:

- We introduce an ordinal regression-based pipeline that reaches 67.5% accuracy in predicting the severity of respiratory motion in MRI images.
- We produce a segmentation model achieving 86.18% Dice accuracy robust to images with mild to intermediate respiratory motion artefacts.

2   Method

We describe the components of our proposed workflow defined in Fig. 1. First, the respiratory motion classifier (in Sect. 2.1) predicts the severity of CMRs. After the images with severe motion artefacts are omitted, the remainder is segmented with our segmentation model (in Sect. 2.2).

2.1   Image Quality Assessment of Respiratory Motion Artefacts

We propose a Cardiac MRI motion artefact identification system to distinguish the scans with mild, intermediate and severe motion artefacts. Although it is possible to use a generic classification framework with this objective, which uses a Softmax classifier at the end, we would also like to involve the relative label information between the artefact levels so that we can apply more supervision while training our models. In this regard, we adopt the work of Cao et al. (CORAL) [2] and Shi et al. (CORN) [19] to predict the artefact level of a medical scan while considering the rank consistency among predictions.

Rank Consistent Neural Networks: For both CORAL and CORN, suppose we have a deep neural network composed of a feature extractor and a classifier designed to process 2D image slices. In addition, let us suppose that we have a dataset $\mathcal{D} = \{x^{(i)}, y^{(i)}\}$ where $x^{(i)}$ is the ith training sample, and $y^{(i)}$ corresponds to its label. Ordinal regression aims to minimise the cost function $L(r)$, where the mapping $r : \mathcal{X} \rightarrow \mathcal{Y}$ is called a ranking rule, such that each label has a level. The overall framework is shown in Fig. 2.

CORAL: In order to train a CORAL model, we start with extending the class label of a sample. For instance, if there are 3 classes in total, the binarised vector representation will be $[0, 0]$ for class 1, $[1, 0]$ for class 2, and $[1, 1]$ for class 3, different than the one-hot encoding scheme. Supposing that there exist K ordinal classes, we can model them by using $K - 1$ binary classifiers where we can predict the rank index of the sample as follows:

$$q = 1 + \sum_{k=1}^{K-1} f_k(\mathbf{z}_i). \tag{1}$$

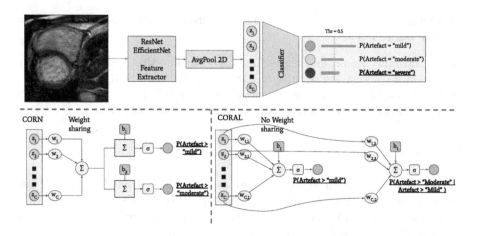

Fig. 2. CORAL and CORN approaches for satisfying rank consistency.

In Eq. (1), $f_k(.)$ is the thresholded output of the kth binary classifier, and $\mathbf{z}_i$ is the vector output of the feature extractor of the ith sample, or in other words, $\mathbf{z}_i = g(x_i; \mathbf{W})$ where $\mathbf{W}$ denotes the parameters of the feature extractor. This equation has a rank monotonicity assumption that, in order to predict the rank index of the image, all of the binary classifier outputs until some level should output 1 whereas after that particular level, the output should be 0. What is special about CORAL is that, the rank consistency is maintained by sharing the weight parameters of a classifier which are denoted by $\mathbf{w}$. However, each of the binary classifiers has different bias units, and each of them are denoted as b_k ($\mathbf{b}$ in the vector form) to differentiate each of the predicted labels. As a consequence, we can define the probability value of a classifier k as $\hat{P}(y_k^{(i)} = 1) = \sigma(\mathbf{w}\mathbf{z}_i + b_k)$ where σ is the Sigmoid activation function.

This network is trained by minimising the weighted cross-entropy loss function in Eq. (2) where $\lambda^{(k)}$ denotes the importance of each of the binary classifiers.

$$L(\mathbf{w}, \mathbf{W}, \mathbf{b}) = -\sum_{i=1}^{N} \sum_{k=1}^{K-1} \lambda^{(k)} [log(\hat{P}(y_k^{(i)} = 1)) y_k^{(i)} + log(1 - \hat{P}(y_k^{(i)} = 1))(1 - y_k^{(i)})]$$

(2)

After training the network and obtaining the probability values, the final decision of the classifier k is obtained by

$$f_k(\mathbf{z}_i) = \mathbb{1}(\hat{P}(y_k^{(i)} = 1) > 0.5),$$

(3)

where $\mathbb{1}(\cdot)$ is the indicator function. The overall decision for a sample is called rank index, which is defined in Eq. (1).

CORN: In contrast to CORAL, CORN argues that each of the binary classifier outputs is conditioned on the previous one. Supposing that the binary classifier's output is expressed as $f_k(\mathbf{z}_i) = \hat{P}(y^{(i)} > y_k | y^{(i)} > y_{k-1})$, the marginal probability of output random variable being greater than the level of y_k is

$$\hat{P}(y^{(i)} > y_k) = \prod_{j=1}^{k} f_j(\mathbf{z}_i). \tag{4}$$

Finally, to train CORN, we generate conditional training subsets, S_k, and optimise Eq. (5). Prediction is performed as in Eq. (3).

$$L(\mathbf{w}, \mathbf{W}, \mathbf{b}) = -\frac{1}{\sum_{j=1}^{K-1} |S_j|} \sum_{j=1}^{K-1} \sum_{i=1}^{|S_j|} [log(f_j(x^{(i)}))\mathbb{1}(y^{(i)} > y_k)$$
$$+ log(1 - f_j(x^{(i)}))\mathbb{1}(y^{(i)} \leq y_k)] \tag{5}$$

2.2 CMR Image Segmentation with Realistic Respiratory Motion

Creating high-quality segmentation for three-label CMR scans requires the network to be robust against respiratory motion artefacts. We tackle the segmentation problem of the left (LV) and right ventricle (RV) blood pools along with the left ventricular myocardium (MYO) with the nnU-Net framework [7]. nnU-Net is used in a two-dimensional (2D) and three-dimensional (3D) setup. Its dynamic framework design performs all relevant pre- and postprocessing steps. Thus, the CMRxMotion dataset is not resampled, cropped or normalised for the segmentation task. Based on the dataset specifics, the adaptive framework configures a corresponding U-Net making the framework state-of-the-art for several medical segmentation challenges [8]. The recently published ViT U-Net V2 (2D) [17] from the Lifelong nnU-Net Framework [5] is a nnU-Net based architecture with a base Vision Transformer (ViT) [3] backbone and achieves state-of-the-art segmentation performances for medical segmentation while introducing the successful mechanism of self-attention. The ViT U-Net network is used with the assumption that the attention mechanism of transformers can be leveraged to concentrate the self-attention on the motion artefact/heart area leading to a more robust segmentation network in severe artefact cases.

Three different architectures were analysed to create a robust segmentation network for CMR scans. We utilise the same experimental setup for all three architectures. The networks are trained on a random split for 250 epochs while using the pre-processing steps of the nnU-Net framework. The networks are trained on a system with 256 GB DDR4 SDRAM, 2 Intel Xeon Silver 4210 CPUs and 8 NVIDIA Tesla T4 (16 GB) GPUs.

3 Results

We showcase the experimental results of the classifier (Sect. 3.1) and the segmentation (Sect. 3.2) from Synapse platform on the validation set in this section.

3.1   Image Quality Assessment of Respiratory Motion Artefacts

We evaluate the performance using classification accuracy and Cohen's Kappa as suggested by CMRxMotion challenge. In Table 1, the official validation results for CMRxMotion challenge for 4 trained models are provided. The first model utilises a ResNet-152 backbone, trained with SoftMax loss, while the rest of the models utilise an EfficientNet-B5 backbone [21] which are trained by using focal loss [11], CORAL [2] and CORN [19]. We use Optuna [1] to tune the learning rate and find the best-performing optimiser by holding 4 of the patients as the unofficial validation data. We re-ran the training 50 times and pruned it whenever the performance seemed insufficient over the course of iterations.

Table 1. Performance of the three EfficientNet-B5 versions trained on the CMRxMotion data, evaluated on the official validation dataset. Bold values indicate the best performance among the three networks.

Model name	Focal loss	CORAL	CORN	Acc ↑	Cohen's Kappa ↑
ResNet-152	-	-	-	0.55	0.307
EfficientNet-B5	+	-	-	0.650	0.416
	-	+	-	**0.675**	**0.451**
	-	-	+	0.650	0.421

Considering the numerical results, we see a positive impact of using the EfficientNet-B5 over the ResNet-152 model. Furthermore, if we compare the Efficient-Net-B5 models, we see a slight difference between the methods even though the figures are in favour of the CORAL model.

3.2   CMR Image Segmentation with Realistic Respiratory Motion

The analysis of the segmentation networks is split into two consecutive parts; quantitative and qualitative analysis. All results are extracted based on the provided CMRxMotion validation dataset and the Synapse leaderboard, where ground truth labels are unknown to the participants.

Quantitative Analysis: We evaluate the robustness of the segmentation models with the Sørensen-Dice (Dice) and Hausdorff (HD95) metrics. Table 2 summarises the results for the three trained networks using the validation dataset from CMRxMotion[1].

Analysing the table, one can deduce that the ViT U-Net achieves the best scores for every segmentation label leading to a more robust version than the plain nnU-Net. These results confirm the assumption that transformers' self-attention can be leveraged to create a more robust network for CMR segmentation tasks.

---

[1] The scores are extracted from the Synapse leaderboard of the challenge.

Table 2. Performance of the three nnU-Net versions trained on the CMRxMotion data, evaluated on the corresponding validation dataset. Bold values indicate the best performance among the three networks.

Architecture	Dice ↑			HD95 ↓		
	LV	MYO	RV	LV	MYO	RV
nnU-Net$_{2D}$	0.8927	0.8038	0.8427	9.9391	5.3801	7.2047
ViT U-Net$_{2D}$	**0.9003**	**0.8134**	**0.8718**	**9.8360**	**4.8026**	**6.4533**
nnU-Net$_{3D}$	0.8589	0.7811	0.8545	12.549	6.2128	8.6875

Qualitative Analysis: To further analyse the segmentation performance of each model, a comparison between the predictions for a random scan is illustrated in Fig. 3. As the random scan is from the validation set, we cannot provide a ground truth mask.

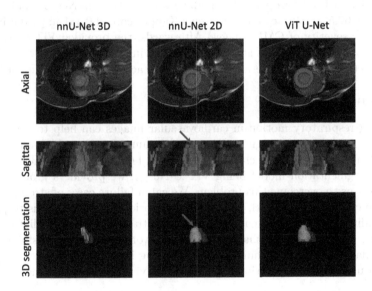

Fig. 3. Comparison of predictions between 3D/2D nnU-Net and ViT U-Net using a random CMRxMotion image from the validation set. The first row shows a random axial slice, the second row is a random sagittal slice and the last row is the 3D segmentation mask. Red arrows indicate the incomplete segmentations for nnuU-net, where ViT U-Net performs better. (Color figure online)

Closely analysing Fig. 3, one can easily see that there are differences in terms of robust segmentation masks among the three different architectures. The axial view clearly shows the lack of the RV (red) segmentation from the 3D nnU-Net, whereas no discernible difference can be seen between the 2D nnU-Net and ViT U-Net. The 3D segmentation masks show a significant difference between 3D nnU-Net and 2D nnU-Net/ViT U-Net regarding segmentation and robustness. When focusing on the top area of the sagittal view images (see red arrows), it is visible that the ViT U-Net segmentation for MYO and LV (green and red) is more complete than the one from the 2D nnU-Net, which is not that easy to spot in the 3D segmentation masks. This confirms the assumption made during the quantitative analysis that transformers lead to a more robust network.

4   Discussion

Detecting respiratory motion artefacts in cardiovascular images is not an easy task. Especially the inaccurately defined severity groups make it difficult to train a model that learns the different levels of severity. Our experiments have shown that well-known classification models like ResNet-152 [6] do not perform very well in predicting the severity of these images. Therefore, we had to implement a more sophisticated pipeline, using CORN and CORAL to put the different labels in a stricter relation.

Finally, with our proposed methodology using EfficientNet-B5 and CORN we achieve 67.5% accuracy and a Cohen's Kappa score of 0.451 in predicting the respiratory severity of CMR images. Afterwards, the provided ViT U-Net$_{2D}$ is used on samples that do not get classified as having strong respiratory artefacts. ViT U-Net$_{2D}$ reaches an average of 86.18% Dice accuracy on the different labels.

5   Conclusion

Detecting respiratory motion in cardiovascular images can help to catch problematic cases where the segmentation model cannot produce high-quality segmentation masks. With the classification model being EffcientNet-B5 + CORN and the segmentation model being ViT U-Net$_{2D}$, we provide a group of tools that allow us to overcome this problem. We catch failure cases of strong severity with the classification model, and the segmentation model allows us to predict high-quality segmentation masks on images with mild to intermediate motion artefacts. We have shown that our proposed workflow can successfully alleviate the daily routine of technicians by providing immediate feedback on image quality during CMR acquisition and therefore saving the hospital time and costs.

Reproducibility

Any data split and trained networks will be provided upon acceptance along with instructions on how to run and reproduce all experiments. The code will be made public under https://github.com/MECLabTUDA/QA_med_data/tree/dev_QA_CMRxMotion.

Acknowledgements. This paper has been produced benefiting from the 2232 International Fellowship for Outstanding Researchers Program of TUBITAK (Project No: 118C353). However, the entire responsibility of the publication/paper belongs to the owner of the paper. The financial support received from TUBITAK does not mean that the content of the publication is approved in a scientific sense by TUBITAK.

References

1. Akiba, T., Sano, S., Yanase, T., Ohta, T., Koyama, M.: Optuna: a next-generation hyperparameter optimization framework. In: Teredesai, A., Kumar, V., Li, Y., Rosales, R., Terzi, E., Karypis, G. (eds.) Proceedings of the 25th ACM SIGKDD International Conference on Knowledge Discovery & Data Mining, KDD 2019, Anchorage, AK, USA, 4–8 August 2019, pp. 2623–2631. ACM (2019). https://doi.org/10.1145/3292500.3330701
2. Cao, W., Mirjalili, V., Raschka, S.: Rank consistent ordinal regression for neural networks with application to age estimation. Pattern Recognit. Lett. **140**, 325–331 (2020). https://doi.org/10.1016/j.patrec.2020.11.008. https://www.sciencedirect.com/science/article/pii/S016786552030413X
3. Dosovitskiy, A., et al.: An image is worth 16x16 words: transformers for image recognition at scale. In: ICLR (2021)
4. Ferreira, P.F., Gatehouse, P.D., Mohiaddin, R.H., Firmin, D.N.: Cardiovascular magnetic resonance artefacts. J. Cardiovasc. Magn. Reson. **15**(1), 41 (2013)
5. González, C., Ranem, A., dos Santos, D.P., Othman, A., Mukhopadhyay, A.: Lifelong nnUNet: a framework for standardized medical continual learning (2022)
6. He, K., Zhang, X., Ren, S., Sun, J.: Deep residual learning for image recognition. In: Proceedings of the IEEE Conference on Computer Vision and Pattern Recognition, pp. 770–778 (2016)
7. Isensee, F., et al.: nnU-Net: self-adapting framework for u-net-based medical image segmentation. arXiv preprint arXiv:1809.10486 (2018)
8. Isensee, F., Petersen, J., Kohl, S.A., Jäger, P.F., Maier-Hein, K.H.: nnU-Net: breaking the spell on successful medical image segmentation. arXiv preprint arXiv:1904.08128, vol. 1, pp. 1–8 (2019)
9. Karakamış, K., Özer, C., Öksüz, İ.: Artifact detection in cardiac MRI data by deep learning methods. In: 2021 29th Signal Processing and Communications Applications Conference (SIU), pp. 1–4 (2021). https://doi.org/10.1109/SIU53274.2021.9477844
10. King, A.P., Buerger, C., Tsoumpas, C., Marsden, P.K., Schaeffter, T.: Thoracic respiratory motion estimation from MRI using a statistical model and a 2-D image navigator. Med. Image Anal. **16**(1), 252–264 (2012)
11. Lin, T., Goyal, P., Girshick, R.B., He, K., Dollár, P.: Focal loss for dense object detection. IEEE Trans. Pattern Anal. Mach. Intell. **42**(2), 318–327 (2020). https://doi.org/10.1109/TPAMI.2018.2858826
12. Oksuz, I., et al.: Detection and correction of cardiac MRI motion artefacts during reconstruction from k-space. In: Shen, D., et al. (eds.) MICCAI 2019. LNCS, vol. 11767, pp. 695–703. Springer, Cham (2019). https://doi.org/10.1007/978-3-030-32251-9_76
13. Oksuz, I., et al.: Deep learning-based detection and correction of cardiac MR motion artefacts during reconstruction for high-quality segmentation. IEEE Trans. Med. Imaging **39**(12), 4001–4010 (2020)
14. Oksuz, I., et al.: Deep learning using K-space based data augmentation for automated cardiac MR motion artefact detection. In: Frangi, A.F., Schnabel, J.A., Davatzikos, C., Alberola-López, C., Fichtinger, G. (eds.) MICCAI 2018. LNCS, vol. 11070, pp. 250–258. Springer, Cham (2018). https://doi.org/10.1007/978-3-030-00928-1_29
15. Oksuz, I., et al.: Automatic CNN-based detection of cardiac MR motion artefacts using k-space data augmentation and curriculum learning. Med. Image Anal. **55**, 136–147 (2019)

16. Özer, C., Öksüz, İ: Cross-domain artefact correction of cardiac MRI. In: Puyol Antón, E., et al. (eds.) STACOM 2021. LNCS, vol. 13131, pp. 199–207. Springer, Cham (2022). https://doi.org/10.1007/978-3-030-93722-5_22

17. Ranem, A., González, C., Mukhopadhyay, A.: Continual hippocampus segmentation with transformers. In: Proceedings of the IEEE/CVF Conference on Computer Vision and Pattern Recognition, pp. 3711–3720 (2022)

18. Seppenwoolde, Y., et al.: Precise and real-time measurement of 3D tumor motion in lung due to breathing and heartbeat, measured during radiotherapy. Int. J. Radiat. Oncol.* Biol.* Phys. 53(4), 822–834 (2002)

19. Shi, X., Cao, W., Raschka, S.: Deep neural networks for rank-consistent ordinal regression based on conditional probabilities (2021)

20. Sinclair, M., Bai, W., Puyol-Antón, E., Oktay, O., Rueckert, D., King, A.P.: Fully automated segmentation-based respiratory motion correction of multiplanar cardiac magnetic resonance images for large-scale datasets. In: Descoteaux, M., Maier-Hein, L., Franz, A., Jannin, P., Collins, D.L., Duchesne, S. (eds.) MICCAI 2017. LNCS, vol. 10434, pp. 332–340. Springer, Cham (2017). https://doi.org/10.1007/978-3-319-66185-8_38

21. Tan, M., Le, Q.V.: Efficientnet: rethinking model scaling for convolutional neural networks. In: Chaudhuri, K., Salakhutdinov, R. (eds.) Proceedings of the 36th International Conference on Machine Learning, ICML 2019, 9–15 June 2019, Long Beach, California, USA. Proceedings of Machine Learning Research, vol. 97, pp. 6105–6114. PMLR (2019). https://proceedings.mlr.press/v97/tan19a.html

22. Zhang, Q., Hu, Y.C., Liu, F., Goodman, K., Rosenzweig, K.E., Mageras, G.S.: Correction of motion artifacts in cone-beam CT using a patient-specific respiratory motion model. Med. Phys. 37(6Part1), 2901–2909 (2010)

3D MRI Cardiac Segmentation Under Respiratory Motion Artifacts

Yongqing Kou, Rongjun Ge[✉], and Daoqiang Zhang

School of Computer Science and Technology, Nanjing University of Aeronautics and
Astronautics, Nanjing, China
rongjun.ge@nuaa.edu.cn

Abstract. In cardiac magnetic resonance (CMR) imaging, 3D segmentation of
the heart is important for detailed description of its anatomy and thus estimation of clinical parameters. However, images acquired in the clinical routine often
contain artifacts caused by respiratory motion due to acquisition time and respira-
tory/cardiac motion limitations. Segmentation of these low-quality images using
conventional methods often does not yield accurate results. Here, we propose a
DenseBiasNet combined with Variational Auto-Encoder (VAE) network framework, a segmentation model robust to respiratory motion artifacts, for automatic
CMR image segmentation from extreme images. Given an image with respiratory motion artifacts as input, DenseBiasNet is utilized as the primary branch
for segmentation, and VAE network is utilized as the secondary branch to map
the low-resolution image of the DenseBiasNet encoded portion as input to the
low-dimensional space for reconstruction.

Keywords: Cardiac MR · Respiratory motion artifacts · Segmentation

1 Introduction

Cardiac MRI is the "gold standard [1]" for determining intracardiac structure and function and is clinically important in estimating clinical parameters such as ejection fraction,
ventricular volume, stroke volume and myocardial mass. Fine cardiac image segmentation is a major step in the estimation of clinical parameters, disease diagnosis, prognosis
estimation and surgical planning [2]. However, because its imaging requires prolonged
breath-holding, which is not feasible for patients with severe heart disease [12], cardiac
image segmentation is susceptible to artifacts generated by respiratory motion. Perva-
sive cardiac segmentation algorithms are able to achieve basic region segmentation,
but 3D CMR image segmentation with respiratory motion artifacts remains difficult to
perform in the clinical setting. Therefore, to address this problem, we propose a segmentation model that is robust to respiratory motion artifacts for automatic CMR image
segmentation from extreme images.

O. Camara et al. (Eds.): STACOM 2022, LNCS 13593, pp. 457–465, 2022.
https://doi.org/10.1007/978-3-031-23443-9_43

1.1 Related Works

In recent years, a large amount of work on 3D cardiac MRI segmentation with respiratory motion artifacts has been proposed to improve the efficiency of clinical parameter estimation, including traditional segmentation-based approaches and reconstruction-based approach, among others.

Traditional Segmentation-based Approach. Since the first application of FCN networks for image segmentation of functional regions of the cardiac on short-axis cardiac magnetic resonance (MR) images, many deep learning-based methods have been proposed with the aim of further improving the segmentation effect [2]. However, the cardiac is a constantly functioning organ and its shape changes during motion, so MRI images of the heart are susceptible to respiratory motion. Thus, deep learning-based algorithms are able to segment cardiac MRI images in real time with good accuracy. However, these algorithms still fail on those image acquisitions with poor image quality or significant artifacts.

Reconstruction-based Approach. There are also a large number of solutions for those images with poor image quality or significant artifacts. Check image quality before subsequent studies [3, 4], or predict segmentation quality to detect failure [5]. But this is time consuming for clinical studies, so it is important to have a deep learning-based approach whose ability to reduce effectively improves the efficiency of low quality image segmentation. Several researchers have used deep learning-based methods to improve image resolution [6], motion correction [7, 8, 13], artifact reduction [9], shadow detection [10], noise reduction [11], and latent space [12–14] after image acquisition.

Contributions. Inspired by He et al .[15] and Myronenko et al. [18], a combined framework of DenseBiasNet and Variational Auto-Encoder (VAE) networks is proposed as a segmentation model robust to breathing motion artifacts for automatic CMR image segmentation from extreme images.

2 Methods

Figure 1 illustrates our framework for cardiac MRI segmentation with respiratory motion artifacts. It has two components: 1) DenseBiasNet. A network based on dense bias connections has powerful representation capabilities for data at different scales due to its ability to fuse multi-resolution feature maps of data as well as multi-sensory fields. 2) Variational Auto-Encoder (VAE). An additional branch is used to reconstruct the original image. The input to this branch comes from the output of the DenseBiasNet encoder side and is intended to better cluster/group the features of the encoder endpoints and add additional guidance and normalization to the encoder part. VAE branches are used to provide a constraint on split branches, resulting in a better split.

2.1 DenseBiasNet Part

The most important feature of DenseBiasNet is its ability to fuse multi-sensory fields and multi-resolution feature maps, thus providing a powerful representation of multi-scale

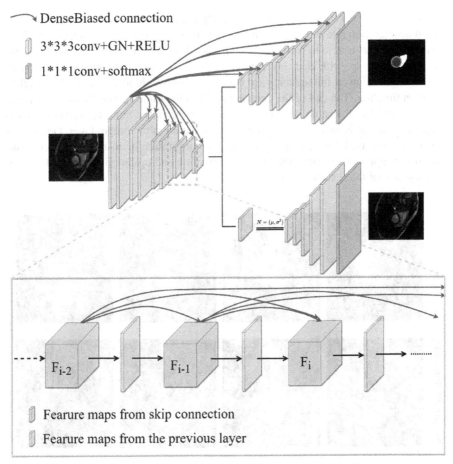

Fig. 1. A visual schematic of the network structure. The input is a four-channel 3D MRI image, followed by an initial 3x3x3 3D convolution with 32 filters. Each green block is a ResNet-like block with GroupNorm normalization. The output of the split decoder has a single channel (with the same spatial size as the input) followed by a sigmoid of the split map for the three split regions (LV, RV, MYO). The VAE branch reconstructs the input image as itself and is used to normalize the shared encoder only during training. The segmentation branch uses a biased join method and the reconstruction branch uses a variational autoencoder.

feature maps, which is achieved by transferring the feature maps of each layer to the forward layer through a dense bias connection. DenseBiasNet uses the encoding part and decoding part of the standard 3D U-Net [16]. As shown in Fig. 1, it has four resolution levels.

The encoding part is continuously downsampled to efficiently save GPU memory [17], and each layer contains two $3 \times 3 \times 3$ convolutions with GN+ReLU activation function followed by $2 \times 2 \times 2$ max pooling with stride of 2, after the convolution layer. The last layer of the encoding section yields the lowest resolution feature map, which is spatially smaller than the input image by a factor of 8. We decided not to reduce the size further to preserve more spatial content. By experimenting with a spatial dropout with an acceleration rate of 0.2 after the last layer of the coding part of DenseBiasNet. The decoding part is able to recover the low-resolution feature map to the original resolution feature map, with each layer having a 2×2×2 up-convolution operation with a stride of 2, followed immediately by two 3×3×3 convolutions and the GN+ReLU activation function. The end of the decoded part has the same spatial dimensions as the original image and the same number of features as the initial input feature size. A 1× 1×1 convolutional layer is used, followed by a softmax as the output layer. Dense bias connections are used throughout the network, incorporating multi-receptive fields and multi-resolution feature maps, and maximum pooling and upsampling are used to vary the resolution of the feature maps.

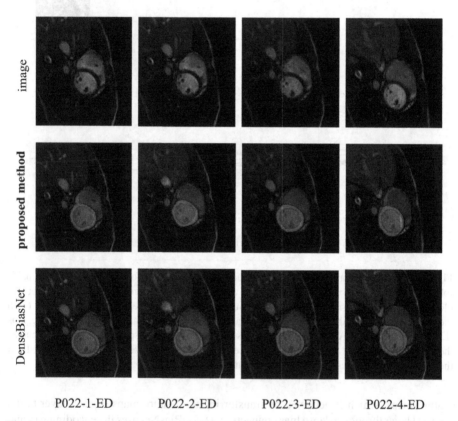

P022-1-ED   P022-2-ED   P022-3-ED   P022-4-ED

Fig. 2. A typical segmentation example with predicted labels overlaid on MRI axial, sagittal and coronal slices. The segmentation results include all visible labels (a combination of red, green and blue labels), where the LV class is shown as a red label, the MYO class is shown as green and the RV class is shown as blue. The segmentation results were also compared with those of DenseBiasNet alone. (Color figure online)

2.2 VAE Part

Using the output of the DenseBiasNet encoding part as the input to the VAE part, we first reduce the high-dimensional input to a low-dimensional space of 256 (128 for the mean and 128 for the standard deviation). Then, Then, a sample is taken from the Gaussian distribution [18], and given a mean and std for it. Finally, it is progressively upsampled and reconstructed as the input image dimension. We do not use interstage jumps of the encoder here.

2.3 Loss

Our loss function consists of 4 components:

$$L = L_{dice} + 0.1 \times L_{L2} + 0.1 \times L_{KL} \tag{1}$$

L_{dice} is a soft dice loss [19], applied to DenseBiasNet encoding part P_{pred} to match the segmentation mask P_{ture}:

$$L_{dice} = \frac{2 \times \sum P_{pred} \times P_{ture}}{\sum P_{pred}^2 + \sum P_{true}^2 + \epsilon} \tag{2}$$

where ϵ is a small constant to avoid zero division. Since the output of the DenseBiasNet decoding part has 3 channels (predictions for each cardiac partition), we simply add the three dice loss functions together.

L_{L2} is a reconstruction loss on the VAE branch output I_{pred} to match the input image I_{input}:

$$L_2 = \left\| I_{input} - I_{pred} \right\|_2^2 \tag{3}$$

L_{KL} is the cross-entropy of the model in a sample [20], L_{KL} is defined as follows:

$$L_{KL} = \frac{1}{N} \sum \mu^2 + \sigma^2 - \log\sigma^2 - 1 \tag{4}$$

where N is total number of image voxels.

2.4 Optimization

We use Adam optimizer with initial learning rate of $lr = 1e - 4$, weight_decay $= 1e - 5$ and decreases gradually according to:

$$\alpha = lr \times (1 - \frac{e}{N})^{0.9}$$

where e is an epoch counter, and N is a total number of epochs (200 in our case). We use batch size of 1, and put the data into the network in random order (making sure each data is used only once in each epoch).

2.5 Data Pre-processing

We normalize all data preprocessed to have zero mean and unit std.

3 Experiments and Results

Dataset. We used the CMRxMotion2022 training dataset (total 20 patients with 4 respiratory states each) for training, of which only 139 data with Ground turth were used for training without any additional internal dataset. The CMRxMotion2022 validation dataset (total of 5 patients with 4 respiratory states each) was used for validation, where all data without Ground turth were used. Since the test phase was submitted to the organizer for testing, no test set related presentation is available [21].

Table 1. Results of the validation dataset using the DenseBiasNet + VAE method and the Dense-BiasNet method alone in CMRxMotion 2022. Mean Dice and Hausdorff measurements for the proposed segmentation method. LV-Left ventricle, MYO-Left ventricle myocardium, RV-Right ventricle

	Network	Dice			Hausdorff (mm)		
Validation dataset	DenseBiasNet + VAE	LV	MYO	RV	LV	MYO	RV
		0.86	0.71	0.72	11.95	7.79	14.45
	DenseBiasNet	0.83	0.67	0.74	22.99	28.22	13.83

Implementation Details. We implemented our network in pytorch with the NVIDIA GeForce RTX 3060 Ti with 8GB GPU memory using the Adam optimizer ($\beta_1 = 0.9$, $\beta_2 = 0.999$) with the initial learning rate $= 1e-4$. There is no cropping of the raw data. The input to the network is 3D cardiac MRI images with one channel, and the output is a cardiac partition with three correlations.

Results: We report the results of our approach on CMRxMotion2022 validation (20 patients in total, four respiratory states each). We uploaded our segmentation results to the Synapse platform for evaluation of per class dice and Hausdorff distances. Figure 2 shows the segmentation results of our network on the CMRxMotion2022 validation dataset in the four breathing states. Table 1 shows the results of our model and DenseBiasNet on the CMRxMotion 2022 validation dataset. From the results, it can be seen that with the increasing severity of respiratory motion artifacts, there is a greater predominance of segmentation in both the left ventricular and left ventricular myocardial regions. And Table 2 shows the results of our model and DenseBiasNet on the CMRxMotion 2022 test dataset, this result is officially given by CMRxMotion 2022.

Table 2. Results of the tset dataset using the DenseBiasNet+VAE method in CMRxMotion 2022. Mean Dice (dsc) and Hausdorff_95 (95% HD) measurements for the proposed segmentation method. LV-Left ventricle, MYO-Left ventricle myocardium, RV-Right ventricle

	LV				MYO				RV			
	dsc_mean	dsc_std	Hd95_mean	Hd95_std	dsc_mean	dsc_std	Hd95_mean	Hd95_std	dsc_mean	dsc_std	Hd95_mean	Hd95_std
Test dataset	0.82	0.18	15.21	21.47	0.68	0.16	17.532	28.91	0.64	0.17	21.16	50.96

4 Discussions and Conclusion

In this work, we describe a segmentation model robust to respiratory motion artifacts for automatic CMR image segmentation from extreme images. While experimenting with the network architecture, we try several alternative approaches that allow our network to cope with different degrees of image degradation due to respiratory motion in simulated clinical practice. For instance, using three NVIDIA GeForce RTX 3060 Ti with 8GB GPU memory we can get faster training speed and verification speed than one NVIDIA GeForce RTX 3060 Ti with 8GB GPU memory. In addition, the extra Variational Auto-Encoder (VAE) branch not only improves performance but also helps to consistently achieve good training accuracy in any random initialization case.

References

1. van der Geest, R.J., Reiber, J.: Quantification in cardiac MRI. J. Magn. Reson. Imaging **10** (1999)
2. Khened, M., Varghese, A., Krishnamurthi, G.: Fully convolutional multi-scale residual DenseNets for cardiac segmentation and automated cardiac diagnosis using ensemble of classifiers. Med. Image Anal. **51**, 21–45 (2019)
3. Ruijsink, B., et al.: Fully automated, quality-controlled cardiac analysis from CMR: validation and large-scale application to characterize cardiac function. Cardiovasc. Imaging **13**, 684–695 (2019)
4. Tarroni, G., et al.: Learning-based quality control for cardiac MR images. IEEE Trans. Med. Imaging **38**, 1127–1138 (2019)
5. Zhou, B., Liu, C., Duncan, J.S.: Anatomy-Constrained Contrastive Learning for Synthetic Segmentation without Ground-truth. MICCAI (2021)
6. Wu, Q., et al.: IREM: High-Resolution Magnetic Resonance (MR) Image Reconstruction via Implicit Neural Representation. MICCAI (2021)
7. Dangi, S., Linte, C.A., Yaniv, Z.R.: Cine Cardiac MRI Slice Misalignment Correction Towards Full 3D Left Ventricle Segmentation. Medical Imaging (2018)
8. Tarroni, G., et al.: A Comprehensive Approach for Learning-Based Fully-Automated Inter-slice Motion Correction for Short-Axis Cine Cardiac MR Image Stacks. MICCAI (2018)
9. Öksüz, I., et al.: High-quality Segmentation of Low Quality Cardiac MR Images Using k-Space Artefact Correction. MIDL (2019)
10. Meng, Q., et al.: Weakly supervised estimation of shadow confidence maps in fetal ultrasound imaging. IEEE Trans. Med. Imaging **38**, 2755–2767 (2019)
11. Wolterink, J.M., Leiner, T., Viergever, M.A., Išgum, I.: Generative adversarial networks for noise reduction in low-dose CT. IEEE Trans. Med. Imaging **36**, 2536–2545 (2017)
12. Wang, S., et al.: Deep Generative Model-based Quality Control for Cardiac MRI Segmentation. MICCAI (2020)
13. Wang, S., et al.: Joint Motion Correction and Super Resolution for Cardiac Segmentation via Latent Optimisation. ArXiv, abs/2107.03887 (2021)
14. Chen, C., Hammernik, K., Ouyang, C., Qin, C., Bai, W., Rueckert, D.: Cooperative Training and Latent Space Data Augmentation for Robust Medical Image Segmentation. MICCAI (2021)
15. He, Y., et al.: Dense Biased Networks with Deep Priori Anatomy and Hard Region Adaptation: Semi-supervised Learning for Fine Renal Artery Segmentation. Medical Image Analysis (2020)

16. Çiçek, Ö., Abdulkadir, A., Lienkamp, S.S., Brox, T., Ronneberger, O.: 3d u-net: learning dense volumetric segmentation from sparse annotation. In: International Conference on Medical Image Computing and Computer-Assisted Intervention, pp. 424–432. Springer (2016)
17. Feng, C., Yan, Y., Fu, H., Chen, L., Xu, Y.: Task Transformer Network for Joint MRI Reconstruction and Super-Resolution. ArXiv, abs/2106.06742 (2021)
18. Myronenko, A.: 3D MRI brain tumor segmentation using autoencoder regularization. MICCAI (2018)
19. Milletari, F., Navab, N., Ahmadi, S.: V-Net: Fully convolutional neural networks for volumetric medical image segmentation. In: 2016 Fourth International Conference on 3D Vision (3DV), pp. 565–571 (2016)
20. Kingma, D.P., Welling, M.: Auto-Encoding Variational Bayes. CoRR, abs/1312.6114 (2014)
21. Wang, S., et al.: The Extreme Cardiac MRI Analysis Challenge under Respiratory Motion (CMRxMotion). arXiv preprint arXIv: 2210.06385 (2022)

Cardiac MR Image Segmentation and Quality Control in the Presence of Respiratory Motion Artifacts Using Simulated Data

Sina Amirrajab[✉], Yasmina Al Khalil, Josien Pluim, Marcel Breeuwer, and Cian M. Scannell

Department of Biomedical Engineering, Eindhoven University of Technology, Eindhoven, The Netherlands
s.amirrajab@tue.nl

Abstract. In this work, we propose solutions for the two tasks of the CMRxMotion challenge; 1) quality control and 2) image segmentation in the presence of respiratory motion artifacts. We develop a k-space based motion simulation approach to generate cardiac MR images with respiratory motion artifacts on open-source artifact-free data to handle data scarcity. For task 1, a motion-denoising auto-encoder is trained to reconstruct motion-free images from the pairs of images with and without simulated motion. The encoder part of the auto-encoder is used as a feature extractor for a fully-connected classifier. For task 2, an ensemble of modified 2D nn-Unet models is proposed to tackle different aspects of variations in the data with the purpose of improving the robustness of the model to images hampered by respiratory motion artifacts. All proposed models in this paper are trained using the images with simulated motion artifacts. The proposed quality control model achieves a classification accuracy of 0.75 with the Cohen's kappa coefficient of 0.64 and the ensemble model obtains the mean Dice scores of 0.922, 0.829, and 0.910 respectively for the left ventricle, myocardium, and right ventricle segmentation on the validation set of the CMRxMotion challenge.

Keywords: Respiratory motion artifacts · Cardiac image segmentation · Quality control · Motion artifact simulation

1 Introduction

Object motion during the acquisition of Magnetic Resonance (MR) images can negatively impact the image quality. For the application of cardiac MR imaging, the k-space data is acquired in multiple restricted segments distributed over multiple heart beats in order to reduce the beating motion artifact. Respiratory motion introduces inconsistency in the k-space data between different segments and the severity of artifact depend on the phase-encoding order and timing of

---

S. Amirrajab, Y. Al Khalil and C. M. Scannell—Contributed equally.

O. Camara et al. (Eds.): STACOM 2022, LNCS 13593, pp. 466–475, 2022.
https://doi.org/10.1007/978-3-031-23443-9_44

the motion [3]. The edges of the moving organ will be blurred if the motion occurs while collecting the high-frequency information at the edges of the k-space. Ghosting artifact, on the other hand, happens when the motion affects the central region of the k-space. Such artifacts represent a significant challenge in the clinical deployment of automated image analysis algorithms. Therefore, there is a need for further research to develop algorithms that can identify these artifacts in images and for the analysis algorithms to be robust in their presence. The Extreme Cardiac MRI Analysis Challenge under Respiratory Motion (CMRxMotion) [9] dataset was acquired with deliberate patient motion, of varying degrees, to allow the study of these problems.

In this paper, we propose solutions for both tasks of the CMRxMotion challenge [9] (http://cmr.miccai.cloud/). To augment the training data and tackle data scarcity, we develop a k-space based approach to simulate motion artifact on artifact-free images from previous publicly available cardiac MR databases of M&Ms-1 [2] and M&Ms-2 challenges (https://www.ub.edu/mnms-2/). We simulate images with different levels of respiratory motions and use these motion corrupted images for training all proposed deep-learning models, as explained in the following. Using publicly available database is allowed in the CMRxMotion challenge.

For Task 1, image quality assessment, a classifier is trained to directly predict the image quality score, on a slice-by-slice basis. The slice-based image quality scores are then aggregated in order to create a patient-level prediction. Since there are limited training data available for the CMRxMotion challenge, the feature extractor of the classification model is pre-trained using simulated motion artifacts on open-source data. This feature extractor is the encoder part of a motion-denoising auto-encoder that is trained to reconstruct motion artifact-free images from images that have simulated motion artifacts. The encoded image features are input to the classification model. This is proposed to be more effective than training the classifier from scratch due to the limited training data.

For Task 2, image segmentation, we propose an ensemble of models aimed at improving the performance robustness in the presence of severe motion artifacts, as well as variations in shape and intensity. This is done through data augmentation with simulated motion corrupted images, non-homogeneous batch sampling of basal and apical slices, and region-based training for segmentation of specific levels of the heart. We deploy the 2D nnU-Net [4] as the backbone of our segmentation models, with substantial modifications for improving the model robustness to motion artifacts.

2 Methods

2.1 Simulation of Respiratory Motion Artifact

Inspired by prior works on k-space based artifact simulation for brain motion [8] and respiratory motion [5,7], we model motion artifacts by applying translation to artifact-free images before transforming them to the Fourier domain. As depicted in Fig. 1a), the breathing motion is modeled as a simple sinusoidal translation of the image in one direction, as the first approximation. We assume that the k-space data is acquired in multiple blocks of segments at different respiration points corresponding to different amounts of translation. The combined k-space is composed of different sections (indicated with different colors) from the k-space data for each translated image. The final motion corrupted image is generated by transforming the combined k-space to the image domain via inverse Fourier transformation. We can change the severity of the motion artifact by tuning two parameters; the period of sinusoidal function corresponding to the number of breathing cycles during the acquisition time window and the amplitude corresponding to the maximum translation of organs during acquisition, i.e. breathing intensity. Note that for simplicity, we assumed one-dimensional homogeneous translation of all organs for modeling the breathing motion.

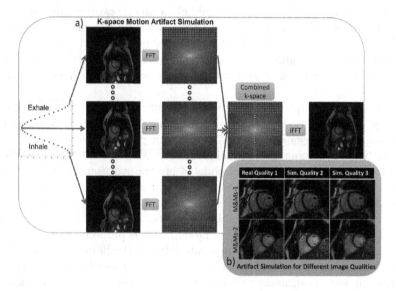

Fig. 1. a) K-space based method for simulating motion artifact on artifact-free cardiac MR images, and b) Motion artifact simulation to create two image quality levels on real cardiac MR images of M&Ms-1 and M&Ms-2 real data.

2.2   Image Quality Assessment

We train an auto-encoder to take an input image with simulated motion artifacts and to reconstruct the original image without artifacts. The reconstruction residual could be used for classification, however, recent work by Meissen et al. discussed the pitfalls of this [6]. Therefore, fully-connected layers are added to the encoder to directly predict the image quality score, and this is trained with training data and image quality labels from the CMRxMotion challenge. In other words, we are pre-training the feature extractor of a classification model to learn features relevant to the motion artifacts, and then combining this with the classification layers and re-training to directly predict the image quality score. The image quality predictions are trained on a slice-by-slice basis and the slice-wise predictions are combined to a single prediction for each image stack (one each for the end-diastolic and end-systolic images, as provided for the challenge). The pipeline is visualized in Fig. 2.

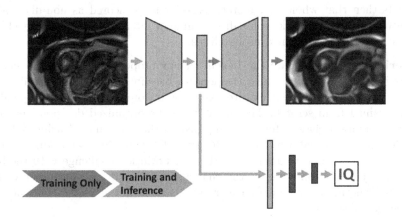

Fig. 2. A schematic representation of the image quality scoring pipeline developed in this work. Firstly, the auto-encoder is trained to reconstruct motion-denoised images. Then the encoder of this model, is used as input to fully-connected layers to classify images based on their image quality (IQ) score.

The auto-encoder is a fully convolutional 2D U-Net-like model (without the skip connections) and consists of six downsampling blocks followed by max-pooling with each downsampling block being made of up two convolution blocks (a convolutional layer, batch normalization, and ReLU activation). There is a corresponding number of upsampling blocks with transposed convolutions. The number of convolutional filters in the first block is 32 and this doubles for each downsampling block and halves for each upsampling block.

To ensure a pixel-wise correspondence for the L2 reconstruction loss function of the auto-encoder, this model is trained using simulated motion artifacts as then the ground-truth artifact-free image is available. Motion is simulated on the images with no motion artifacts from the CMRxMotion training data and the data from the M&Ms-1 challenge, as later described, for this purpose.

To train the classification model, the output of the encoder is flattened and passed to two further fully-connected layers using leaky ReLU activations (coefficient: 0.03) and dropout (probability: 0.2) with a final softmax output to predict one of three classes (mild motion, intermediate motion, or severe (non-diagnostsic) motion). The encoder with the added fully-connected layers are then re-trained to minimize the categorical cross-entropy loss using only the full CMRxMotion training data. Since images with severe motion artifacts are less common in the training data these are (5x) oversampled to balance the classes.

The choice of how to combine the predictions for each slice into a single prediction for the image stack is optimized on the validation set. Firstly, the probabilities predicted for the three classes are summed over all slices in an image. If the summed probability of the images having severe motion artifacts is > 0.5, corresponding to one positive prediction, the image stack is classified as having severe motion. The intuition is that if at least one slice has severe motion artifacts then that whole image stack is likely to be scored as non-diagnostic. Otherwise, the image stack is classified as mild or intermediate motion by choosing the larger of the summed probabilities.

Experiments: To evaluate the benefit of our proposed approach, its performance is compared against the corresponding baseline models. The classification model with the pre-trained feature extractor is compared to a corresponding model trained from scratch. The model with the optimized decision threshold for severe motion class is further compared to the same model using only the largest summed probability for classification. These model evaluations are performed on the validation data set of the CMRxMotion challenge with the best model chosen for submission to the challenge testing phase. Additionally, it is evaluated whether the reconstruction residuals of the auto-encoder could be used for the classification.

2.3   Cardiac Image Segmentation

Respiratory motion artifacts, commonly appearing in CMR imaging, often cause the appearance of fuzziness in the imaged tissue, as well as ghosting effects. Unclear cavity borders are a typical consequence of this effect, negatively impacting the performance of automated segmentation models, which are additionally required to adapt to variation in acquisition parameters, hardware and varying patient characteristics. In this work, we propose an ensemble of approaches aimed at developing a robust and generalizable segmentation model, particularly tailored to handle the appearance of respiratory motion artifacts in CMR images. To this end, we use the 2D nnU-Net [4] segmentation model as our baseline and adapt it to the task as follows:

1. To tackle the limitations in the number of training images, we utilize both external, open-source CMR data (M&Ms-1 and M&Ms-2 data-sets), **augmented with simulated, motion corrupted versions of this data** (see Sect. 2.1), in addition to training images from the CMRxMotion challenge;
2. To address the difficulties in segmenting the basal and apical regions of the heart, particularly in the presence of motion artifacts, we utilize **non-homogeneous batch sampling to over-sample basal and apical slices** seen during training by modifying the nnU-Net's data-loader. Thus, for each mini-batch during training, slices from the apex and the base of CMRxMotion images are selected within a probability of 2σ of a mean in a normal Gaussian distribution, while the basal and apical slices of other data-sets used for training are sampled within a probability of 1σ.

3. While over-sampling of basal and apical slices shows an improvement in segmentation across all regions, visual observation suggests cases of under-segmentation in both basal and apical regions. Thus, we employ a **region-based training approach**, where we train three separate nnU-Net models aimed at segmenting basal, mid-ventricular and apical slices, respectively. Both training and testing images are roughly split into different slices and merged back into an original 3D volume at inference time.

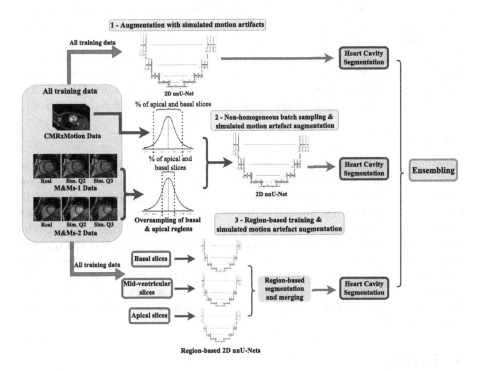

Fig. 3. An ensemble of proposed segmentation approaches for improved robustness to respiratory motion artifacts in CMR images.

In total, we obtain three sets of heart tissue segmentation maps, as seen in Fig. 3, consisting of predictions acquired from (i) a model trained utilizing only augmentation with simulated motion artifacts; (ii) a model trained using simulated motion artifact augmentation and non-homogeneous batch sampling; and (iii) a combination of three region-focused segmentation models, all trained using simulated motion augmentation. While for most models we utilize the same pre-processing and training set-up, determined by nnU-Net's fingerprinting method, as well as a 5-fold cross-validation approach, we train the two region-based models trained for apical and basal region segmentation using a Focal-Tversky loss [1], instead of a typical combination of cross-entropy and Dice losses. This choice is motivated by Focal-Tversky loss exhibiting a better trade-off between precision and recall when training on small structures compared to other losses, which we also confirm throughout our experiments. Our final submission consists of an ensemble of predicted softmax probabilities attained from the three models.

Experiments: To assess the robustness of the proposed segmentation pipeline, we compare it against a baseline model, consisting of a regular nnU-Net framework trained with a combination of M&Ms-1, M&Ms-2 and CMRxMotion images. We additionally study the influence of different segmentation modules included in the final ensemble, whereby we separately evaluate the performance of the model augmented with simulated motion-corrupted images, the model augmented with the same approach but trained using non-homogeneous batch sampling and a model trained using region-based segmentation with simulated motion-corrupted augmentation. All models are evaluated on the provided CMRxMotion validation set, using Dice and Hausdorff Distance (HD) scores across all three cardiac tissues (left ventricle, myocardium and right ventricle).

2.4   Data

We utilize publicly available databases of the M&Ms-1 [2] and M&Ms-2 challenge data to simulate motion artifacts. Corrupted versions of these databases are created in such a way to represent two versions of image qualities as depicted in Fig. 1b). The data with simulated artifacts is used in pre-training the proposed quality control model and data augmentation for the segmentation model. The CMRxMotion challenge database [9] includes 45 healthy volunteers (25 training and validation) with four levels of respiratory motions; 1) adhere to the breath-hold instructions; 2) halve the breath-hold period; 3) breathe freely; and 4) breathe intensively. Based on the image quality scores decided by radiologists, the images are divided into three classes: IQ1 (mild motion), IQ2 (intermediate motion), and IQ3 (severe motion).

3   Results

3.1   Image Quality Assessment

The normalized mean squared reconstruction residual between the IQ3 images and both IQ1 and IQ2 was significantly different (both $p < 0.01$), but there is substantial overlap between the classes and there is no significant difference between IQ1 and IQ2 ($p = 0.29$). It is, thus, clear that the reconstruction cannot be used directly for classification. The pre-trained classification model with optimized decision threshold achieved a classification accuracy of 0.75, and a Cohen's kappa coefficient of 0.64 on the validation data of the challenge. This improved over the baseline models using the model trained from scratch and the model without optimized decision threshold which gave accuracy and Cohen's kappa coefficient of 0.58 and 0.32, and 0.68 and 0.42, respectively. Our final submitted method achieved the test ($n = 120$ images) accuracy of 0.625 and Cohen's Kappa of 0.473, ranking third in the task 1.

3.2 Cardiac Image Segmentation

The differences in the mean Dice and HD scores between the baseline segmentation model and the proposed approaches are shown in Table 1. The obtained results suggest that augmentation with data containing simulated motion artifacts significantly improves the performance and adaptation of the segmentation model in the presence of motion artifacts, particularly across the myocardium and right-ventricle. Additionally, it aids with outlier reduction, confirmed by the reduction in HD scores. However, focusing the training on the basal and apical regions of the heart by utilizing non-homogeneous batch sampling and region-based segmentation shows major improvements compared to both the baseline and model using only augmentation. This confirms our assumption that slices around the apex and the base of the heart are the major source of segmentation errors. This is also reflected by stronger effects of motion artifacts in these regions in the form of blurring and ambiguous cavity borders. By ensembling the proposed models, we additionally reduce the under-segmentation in the base and the apex of the heart and improve the prediction around the borders throughout all heart regions. Our final submitted method achieved third ranking on the test set containing 102 unseen images for the task 2 in the challenge.

Table 1. Differences in mean Dice and 95th percentile HD scores between the baseline and the proposed approaches for CMR image segmentation undergoing respiratory motion artifacts. All models are evaluated across the CMRxMotion validation set (n = 40) for all cardiac tissues, using the provided submission and evaluation platform.

Approach	LV		MYO		RV	
	Dice	HD	Dice	HD	Dice	HD
Baseline	0.891	10.62	0.800	5.59	0.852	8.34
Augmentation	0.903	9.64	0.817	4.98	0.874	5.72
Batch sampling	0.919	8.08	0.825	4.17	0.883	5.45
Region-based train	0.920	8.31	0.828	**4.13**	0.908	4.28
Ensemble	**0.922**	**7.97**	**0.829**	4.14	**0.910**	**4.17**
Submitted (test results n = 102)	**0.924**	**3.98**	**0.863**	**3.16**	**0.918**	**3.53**

4 Discussion and Conclusion

This paper presented our solutions for both tasks of image quality assessment and segmentation of cardiac MR images in the presence of respiratory motion artifacts, as a participating team in the CMRxMotion challenge. We demonstrated that our simple k-space based motion simulation approach was effective in handling data scarcity by simulating different levels of motion artifacts on artifact-free publicly available cardiac MR images. While we modeled the organ motions due to breathing as a simple sinusoidal translation, the heart and organ motion during breathing is more complex, involving rotation and deformation.

Classification of images by the level of motion artifacts was found to be challenging and modest accuracy and Cohen's kappa coefficient were reported. The limited amount of training data contributed to the challenge of this task. It also exacerbated the class imbalance problem leading to very few images with severe motion artifacts, although, this was improved through the use of simulated data. We noticed difficulties due to the fact that not all slices in an image classified as IQ2 or IQ3 were affected by the artifacts, and future work could, thus, use weakly-supervised learning to improve performance.

The addition of simulated motion corrupted images to the training helped the segmentation network adapt to artifacts appearing due to respiratory motion during the acquisition by reducing typical segmentation errors caused by blurring, occluded and ambiguous tissue boundaries, and degraded image quality. In addition, we identified that motion artifacts tended to affect the slices around the base and apex of the heart more severely, consequently becoming the largest source of segmentation errors. To tackle this, we employed non-homogeneous sampling, forcing such slices to appear more frequently during training. Finally, we introduced region-based segmentation by training separate networks, each focused on segmenting a specific heart region.

Acknowledgments. This research is a part of the openGTN project, supported by the European Union in the Marie Curie Innovative Training Networks (ITN) fellowship program under project No. 764465.

References

1. Abraham, N., Khan, N.M.: A novel focal Tversky loss function with improved attention U-Net for lesion segmentation. In: 2019 IEEE 16th International Symposium on Biomedical Imaging (ISBI 2019), pp. 683–687. IEEE (2019)
2. Campello, V.M., et al.: Multi-centre, multi-vendor and multi-disease cardiac segmentation: the M&Ms challenge. IEEE Trans. Med. Imaging **40**(12), 3543–3554 (2021)
3. Ferreira, P.F., Gatehouse, P.D., Mohiaddin, R.H., Firmin, D.N.: Cardiovascular magnetic resonance artefacts. J. Cardiovasc. Magn. Reson. **15**(1), 1–39 (2013)
4. Isensee, F., Jaeger, P.F., Kohl, S.A., Petersen, J., Maier-Hein, K.H.: NNU-Net: a self-configuring method for deep learning-based biomedical image segmentation. Nat. Methods **18**(2), 203–211 (2021)
5. Lorch, B., Vaillant, G., Baumgartner, C., Bai, W., Rueckert, D., Maier, A.: Automated detection of motion artefacts in MR imaging using decision forests. J. Med. Eng. **2017** (2017)
6. Meissen, F., Wiestler, B., Kaissis, G., Rueckert, D.: On the pitfalls of using the residual error as anomaly score (2022). https://doi.org/10.48550/ARXIV.2202.03826. https://arxiv.org/abs/2202.03826
7. Oksuz, I., et al.: Automatic CNN-based detection of cardiac MR motion artefacts using K-space data augmentation and curriculum learning. Med. Image Anal. **55**, 136–147 (2019)

8. Shaw, R., Sudre, C.H., Varsavsky, T., Ourselin, S., Cardoso, M.J.: A K-space model of movement artefacts: application to segmentation augmentation and artefact removal. IEEE Trans. Med. Imaging **39**(9), 2881–2892 (2020)
9. Wang, S., et al.: The extreme cardiac MRI analysis challenge under respiratory motion (CMRxMotion) (2022). https://arxiv.org/abs/2210.06385

Combination Special Data Augmentation and Sampling Inspection Network for Cardiac Magnetic Resonance Imaging Quality Classification

Xiaowu Sun[✉], Li-Hsin Cheng, and Rob J. van der Geest

Division of Image Processing, Department of Radiology, Leiden University Medical Center, Leiden, The Netherlands
x.sun@lumc.nl

Abstract. Cardiac magnetic resonance imaging (MRI) may suffer from motion-related artifacts resulting in non-diagnostic quality images. Therefore, image quality assessment (IQA) is essential for the cardiac MRI analysis. The CMRxMotion challenge aims to develop automatic methods for IQA. In this paper, given the limited amount of training data, we designed three special data augmentation techniques to enlarge the dataset and to balance the class ratio. The generated dataset was used to pre-train the model. We then randomly selected two multi-channel 2D images from one 3D volume to mimic sample inspection and introduced ResNet as the backbone to extract features from those two 2D images. Meanwhile, a channel-based attention module was used to fuse the features for the classification. Our method achieved a mean accuracy of 0.75 and 0.725 in 4-fold cross validation and the held-out validation dataset, respectively. The code can be found here (https://github.com/xsunn/CMRxMotion).

Keywords: Image quality assessment · Cardiac MRI · Data augmentation

1 Introduction

Cardiac magnetic resonance imaging (MRI) is considered as the standard reference for the evaluation of cardiac function due to its excellent image resolution and soft-tissue contrast. However, the MR scanner's hardware itself or the interaction of patient with hardware can result in artifacts in MRI, yielding a low quality imaging, which is often detrimental to the analysis of cardiac function especially in the large-scale imaging studies [1]. Although the artifacts can be minimized by carefully designed image protocols, they still cannot be fully eliminated [2]. Visual inspection of imaging quality is time-consuming and high-cost labor, and also relies on experienced radiologists. Therefore, an automatic method is needed to classify the MR image quality.

In the field of natural images, the approaches to image quality assessment (IQA) can be divided into two categories: full-reference and no-reference, depending on the availability of the original reference image. Meanwhile, recent Convolutional Neural Network

O. Camara et al. (Eds.): STACOM 2022, LNCS 13593, pp. 476–484, 2022.
https://doi.org/10.1007/978-3-031-23443-9_45

(CNN) based methods, such as ResNet [3] and VGG [4], demonstrate promising performance in the automatic image classification task. Bosse et al. [5] employed a Siamese network to extract the features from the distorted and reference patch respectively and fused the difference of those features for IQA. Su et al. [6] proposed a self-adaptive hyper network to blindly assess image quality in the wild without any reference.

However, unlike IQA in natural images, in medical imaging it is particularly challenging for several reasons. There is no large-scale publicly available medical image dataset for IQA. In addition, the distinction between the diagnostic and non-diagnostic imaging is not always evident. Therefore, the labels annotated by radiologists are often subjective [7]. Previously, Fu tried to integrate the information from different colorspaces at feature-level and prediction-level to assess retinal image quality [8]. Oksuz et al. proposed a CNN model to automatically detect and correct motion-related artifacts in cardiac MRI using the K-space lines [9]. Lyu et al. used a recurrent generative adversarial network to reduce motion artifacts in cardiac MRI [10].

The CMRxMotion challenge aims to encourage the participants to develop an IQA model and a segmentation method for the extreme cardiac MRI dataset. In this paper we focuses only on the task of image quality assessment. Our contributions are as follows: (1) We designed specific data augmentation methods to enlarge the given limited data. (2) We proposed a two-branch network and combined a channel-based attention mechanism to fuse features from two random samples of the 3D volume, improving the IQA performance.

2   Materials and Methods

2.1   Dataset

The challenge provides short-axis cardiac MR images of 45 healthy volunteers (20 for training, 5 for validation and 20 for testing), obtained through the same 3T MR system (Siemens MAGNETOM Vida) under four different levels of respiratory motion, including full breath-hold, half breath-hold, free breath and intensive breath. Only the images of the end-diastolic (ED) and end-systolic (ES) phase are available. Therefore, there are 160 (20 volunteers × 4 scans × 2 phases), 40 and 160 3D volumes for training, validation and testing. The number of slices in one phase ranges from 9 to 13. The image resolution varies from $0.66 \times 0.66 \times 9.6 \text{ mm}^3$ to $0.76 \times 0.76 \times 10 \text{ mm}^3$, and the range of field of view (FOV) varies from $400 \times 512 \text{ mm}^2$ to $512 \times 512 \text{ mm}^2$.

Independent from motion levels, all images were reviewed and scored by multiple radiologists using a standard 5-point Likert scale. More details about the annotation standard can be found here (https://www.synapse.org/#!Synapse:syn32407769/wiki/618241). For better reproducibility, the organizer divided those images into three classes based on the 5-point scores: mild motion, intermediate motion, and severe motion.

During data preprocessing, we excluded slices outside of the heart region and selected the 9 slices in the center to make each processed case having the same number of slices. Afterwards, all the cases were cropped or zero-padded into a uniform matrix size of 192 × 192 × 9 and the image intensity was normalized to [0,1] using the min-max method.

2.2 Data Augmentation

In this section, we describe the specially designed data augmentation method for IAQ in detail. The first two strategies are based on weighted interpolation of images from the same subject, while the third strategy employs histogram matching plus interpolation to generate new images. All of the data augmentation methods are based on the 3D volume.

Generating Transition Phases Between ED and ES. The ED and ES phases capture the two extreme scenarios in a cardiac cycle. The transition phases between ED and ES in the same cardiac cycle have almost identical intensity distribution [12]. Therefore, given the available ED and ES phases, we first generate new transition phases between ED and ES using weighted interpolation defined as formula (1).

$$wp = wI_1 + (1 - w)I_2, \ n\_label \approx wL_1 + (1 - w)L_2 \tag{1}$$

where wp is the generated volume and n_label is its corresponding label, I_1, I_2, and L_1, L_2 are the 3D volume and labels of ED and ES phases, and w is the weight. In this work, we used three values for w, namely 0.2, 0.5 and 0.8, to generate transition phases. Figure 1 shows an example of the generated images using this approach.

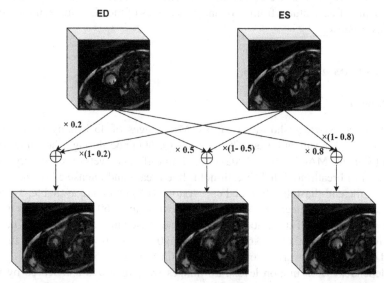

Fig. 1. An example of generated new phases by weighting ED and ES phases from the same case. The first row implies the 3D volume selected from ED and ES phase. The second row presents 3D volume selected from the generated phases using the weights of 0.2, 0.5 and 0.8, respectively.

Generating Intermediate Images from Different Levels of Respiratory Motion. Interestingly, each volunteer was scanned four times under different levels of respiratory motion. Those paired cases from the same volunteer have the same anatomy structures but with different image qualities. Therefore, we used the paired images at the same phase but from different respiratory motion to generate new images. As illustrated in

Fig. 2, two paired images (P004-1-ED, P004-4-ED for example) both from the ED phase of the same volunteer, but possibly with different image qualities, are selected randomly as the source images. After an intensity-based registration, the method described in formula (1) is used to generate the new image and its corresponding label. Similar as the previous augmentation strategy, the new images are generated using weights of 0.2, 0.5 and 0.8.

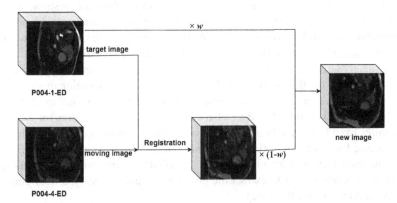

Fig. 2. An illustration of using two ED phases under different respiratory motion levels of the same volunteer to generate a new image.

Generating Degraded Images with Histogram Matching and Linear Interpolation. Within the 160 training cases, the numbers of cases with mild, intermediate, and severe motion artifacts are 70, 69, 21, respectively. To enlarge the subset with severe motion artifacts, the cases with intermediate artifacts were degraded into a lower-quality ones using the linear interpolation approach. As shown in Fig. 3, a 3D volume with severe artifacts is randomly selected as the reference and another one with intermediate artifacts is considered as the source image. The pixel intensity distribution of the source image is matched to that of the reference image. We then randomly choose 5% of the pixels from the matched image, and apply the linear interpolation approach on those selected pixels to expand into a new image. The label of the generated image is assigned as severe.

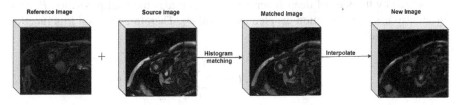

Fig. 3. The procedure of image degradation includes histogram matching and linear interpolation. The generated result is considered as a new image with lower quality.

2.3 Sampling Inspection Network Architecture

To mimic the sampling inspection, the quality of a 3D volume is determined by estimating the quality of random samples drawn from the volume. The advantage of the random selection strategy is that it can generate more data from a single volume to train a model. However, because the selected sample occasionally missed certain critical slices, we introduced another sample with different combinations of 2D slices, mimicking ensembling two times of sampling inspection. The model architecture is shown in Fig. 4.

The two samples were regarded as 2D multi-channel images, and was each processed by a convolution layer. In addition, according to our intuition, the slices from the apical, middle and basal regions contribute differently for IQA. The slices in the middle section, with a relatively larger size of the left ventricle than those in apex and base, have a significant impact on the quality assessment. Therefore, the channel attention module (CAM) proposed in [11] was introduced to exploit the intra-channel relationship of the input. After that, ResNet was introduced as a backbone to extract the features for each input. The features from those two branches were concatenated along the channel dimension, and a Feature Fusion Module (FFM) was introduced to fuse those features. The FFM block contains one channel attention module to exploit the inter-channel relationship and one averaged pooling layer to extract the global information. Lastly, a fully connected layer was used to predict the result.

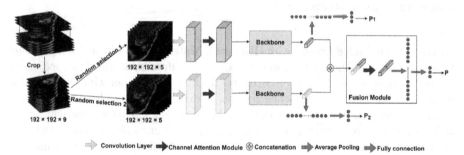

Fig. 4. Sampling inspection network architecture. The input of each branch is a multi-channel 2D image.

Inspired by the idea of deep supervision [13], besides the final prediction P, each branch also has one prediction denoted as P_1 and P_2. Therefore the total loss function can be expressed as:

$$Loss = CE(P, L) + \sum_{i=1}^{2} CE(P_i, L) \tag{2}$$

where CE is cross-entropy loss. Only the prediction P was used for the validation and testing. During the validation and testing, the sampling inspection was repeated 50 times for one 3D volume, the averaged result was regarded as the final prediction.

3   Experiments and Results

The training data was divided into 4-fold for cross validation. The metrics, including accuracy, precision, recall, F1-Score and Cohen's Kappa were used to evaluate the performance. All the results are reported as the mean value of four folds. All the experiments were implemented with Pytorch trained on a machine with a NVIDIA Quadro RTX 6000 GPU with 24 GB memory. Adam was employed as the optimizer with 0.00001 as the learning rate.

The ResNet was employed as the baseline, and it took a multi-channel 2D image with a size of $192 \times 192 \times 9$ as the input. We first compared our method with the baseline. Due to the small and imbalanced dataset, the baseline failed to predict the severe class, yielding a relatively poor result with accuracy being 0.41, as reported in Table 1. The performance indicated that a larger and balanced dataset is needed.

We also evaluated the performance of the proposed network and data augmentation (DA) techniques. The new data generated from the offline data augmentation approach was used to pre-train the model and the pre-trained model was fine-tuned using the original training data. Table 1 also reports the classification results derived from the proposed methods. It shows that the overall accuracy increased from 0.68 to 0.75 after using DA. Although on the class of mild, the accuracy using DA is a little lower than that without DA, the accuracy for the other two classes is better. The method using DA achieved the best performance on the metric of F1-Score in all classes. The confusion matrix in Fig. 5 further reveals that the number of false negatives for the severe class reduced after introducing DA. Therefore, the performance confirmed that the carefully designed DA works well for the IQA task.

Table 1. The 4-fold cross validation performance. Over-Acc: the overall accuracy based on all classes. DA: data augmentation. P: Precision. R: Recall. F: F1-Score

Model	DA	Over-Acc	Cohen's Kappa	Severity Level	Acc	P	R	F
ResNet	No	0.41	−0.04	Mild	0.76	0.42	0.76	0.54
				Intermediate	0.19	0.38	0.19	0.25
				Severe	0.00	0.00	0.00	0.00
Ours	Yes	**0.75**	**0.58**	Mild	0.77	**0.83**	0.77	**0.80**
				Intermediate	0.81	0.69	0.81	0.75
				Severe	0.48	0.71	0.48	0.57
	No	0.68	0.45	Mild	**0.86**	0.74	**0.86**	0.79
				Intermediate	0.68	0.64	0.68	0.66
				Severe	0.10	0.40	0.10	0.15

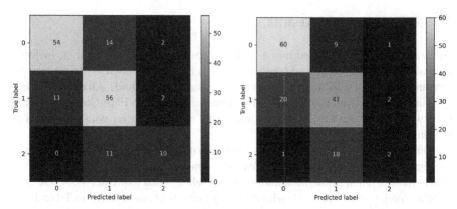

Fig. 5. Confusion matrix derived from the proposed network. 0, 1, 2 represent the classes of mild, intermediate and severe. The left one is the result using data augmentation and the right one is the result without data augmentation.

For the validation part, the labels were hidden by the organizer, we submitted our predicted results and evaluated the performance online. Our method achieved a competitive results, yielding accuracy of 0.725 and Cohen's Kappa 0.645. The best model in the validation data was chosen as the final model, and we submitted it to the organizer and evaluated the performance in the testing dataset with 120 image volumes [14], achieving accuracy of 0.6417 and Cohen's Kappa 0.456.

Ablation. In the proposed classification network, a module named FFM was used to fuse the features from two branches. To reveal the effectiveness of FFM, we evaluated the accuracy and confusion matrix derived from the three predictions P, P_1, P_2 as Reported in Table 2, Fig. 5 and Fig. 6. P_1, P_2 were derived from two individual braches, while P was generated using the FFM. Compared with the other two predictions, P achieved the best performance on all those classes and the overall.

Table 2. Comparison of the accuracy for each class derived from different branches

	P	P_1	P_2
Mild	**0.77**	0.71	0.73
Intermediate	**0.81**	0.79	0.80
Severe	**0.48**	0.29	0.29
Overall	**0.75**	0.69	0.71

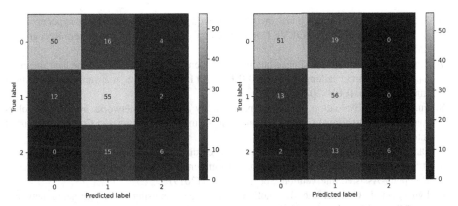

Fig. 6. Confusion Matrix of two predictions P_1, P_2. The left one is from the result P_1, and the right one is derived from P_2.

4 Conclusion

In this paper, we designed three data augmentation methods to enlarge the dataset and balance the classes for the cardiac MR image quality assessment task. Inspired by the idea of sample inspection, to enlarge the training data and to extract sufficient features, we randomly selected different combinations of 2D slices as the input of each branch of the network. The proposed method was trained and evaluated using four-fold cross validation. The results of the classification accuracy, precision, recall and F1-Score demonstrate that our method performed better than the baseline, and the results on the validation dataset shows a competitive performance against the other participants' methods.

Declaration. The authors of this paper declare that they did not use any additional medical image datasets other than those provided by the organizers. They also would like to acknowledge the organizer of the CMRxMotion challenge for collecting and sharing the dataset.

References

1. Krupa, K., Bekiesińska-Figatowska, M.: Artifacts in magnetic resonance imaging. Pol. J. Radiol. **80**, 93 (2015)
2. Zhang, L., et al.: Automated quality assessment of cardiac MR images using convolutional neural networks. In: International Workshop on Simulation and Synthesis in Medical Imaging. Springer, Cham (2016)
3. He, K., et al.: Deep residual learning for image recognition. In: Proceedings of the IEEE Conference on Computer Vision and Pattern Recognition (2016)
4. Simonyan, K., Zisserman, A.: Very deep convolutional networks for large-scale image recognition. arXiv preprint arXiv:1409.1556 (2014)
5. Bosse, S., et al.: Deep neural networks for no-reference and full-reference image quality assessment. IEEE Trans. Image Process. **27**(1), 206–219 (2017)

6. Su, S., et al.: Blindly assess image quality in the wild guided by a self-adaptive hyper network. In: Proceedings of the IEEE/CVF Conference on Computer Vision and Pattern Recognition (2020)

7. Ma, J.J., et al.: Diagnostic image quality assessment and classification in medical imaging: Opportunities and challenges. In: 2020 IEEE 17th International Symposium on Biomedical Imaging (ISBI). IEEE (2020)

8. Fu, H., et al.: Evaluation of retinal image quality assessment networks in different colorspaces. In: International Conference on Medical Image Computing and Computer-Assisted Intervention. Springer, Cham (2019)

9. Oksuz, I., et al.: Detection and correction of cardiac MRI motion artefacts during reconstruction from k-space. In: International Conference on Medical Image Computing and Computer-Assisted Intervention. Springer, Cham (2019)

10. Lyu, Q., et al.: Cine cardiac MRI motion artifact reduction using a recurrent neural network. IEEE Trans. Med. Imaging **40**(8), 2170–2181 (2021)

11. Wang, Q., et al.: ECA-Net: Efficient channel attention for deep convolutional neural networks. In: Proceedings of the 2020 IEEE/CVF Conference on Computer Vision and Pattern Recognition. IEEE, Seattle, WA, USA (2020)

12. Zhang, Y., et al.: Semi-supervised cardiac image segmentation via label propagation and style transfer. In: International Workshop on Statistical Atlases and Computational Models of the Heart. Springer, Cham (2020)

13. Wang, L., Lee, C.Y., Tu, Z., Lazebnik, S.: Training deeper convolutional networks with deep supervision. arXiv preprint arXiv:1505.02496 (2015)

14. Wang, S., et al.: The Extreme Cardiac MRI Analysis Challenge under Respiratory Motion (CMRxMotion). arXiv preprint arXIv: 2210.06385 (2022)

Automatic Cardiac Magnetic Resonance Respiratory Motions Assessment and Segmentation

Abdul Qayyum[1,3]($\boxtimes$), Moona Mazher[2], Steven Niederer[3], Fabrice Meriaudeau[4], and Imran Razzak[5]

[1] ENIB, UMR CNRS 6285 LabSTICC, 29238 Brest, France
engr.qayyum@gmail.com
[2] Department of Computer Engineering and Mathematics, University Rovira I Virgili, Tarragona, Spain
[3] Department of Biomedical Engineering, King's College London, London, UK
[4] ImViA Laboratory, University of Bourgogne Franche-Comté, Dijon, France
[5] University of New South Wales, Sydney, NSW, Australia

Abstract. Cardiac magnetic resonance imaging (CMR) is a powerful non-invasive tool for diagnosing a variety of cardiovascular diseases. However, the quality of CMR imaging is susceptible to respiratory motion artifacts. Recently, an extreme cardiac MRI analysis challenge was organized to assess the effects of respiratory motion on CMR imaging quality and develop a robust segmentation framework under different levels of respiratory motion. In this paper, we have presented two different deep learning frameworks for CMR imaging quality assessment and automatic segmentation. First, we have developed 3D-DenseNet to assess the image quality, followed by 3D-deep supervision UNet with the residual module using pseudo labelling for automatic segmentation task. Experiments on the Challenge dataset showed that 3D ResNet with deep supervision using Pseudo Labeling with nnUNet achieved significantly better performance (8.747 LV, 3.787 MYO, and 5.942 RV) HD95 score than 3D-UNet.

Keywords: CMR imaging quality · 3D Segmentation · Pseudo Labeling · Deep-supervision 3D UNet · 3D DenseNet

1 Introduction

Cardiovascular diseases profoundly affect indisposition and mortality globally, taking an estimated 17.9 million lives each year. Identifying patients at the highest risk of cardiac disease and ensuring they receive appropriate treatment can prevent premature deaths. The cardiac index (CI) is an important parameter that reflects both right and left heart function. Besides, it is also related to both diastolic and systolic ventricular function. Predicting the CI from cardiac magnetic resonance (MR) images is assist in diagnosis and identification of cardiac diseases. In particular, accurate quantification

© The Author(s), under exclusive license to Springer Nature Switzerland AG 2022
O. Camara et al. (Eds.): STACOM 2022, LNCS 13593, pp. 485–493, 2022.
https://doi.org/10.1007/978-3-031-23443-9_46

and identification of cardiac diseases from left ventricle (LV) cardiac imaging is a very imperative and recurrently demanding task [1].

During the clinical practice, LV segmentation methods manually generate the myocardium borders or manually measured the myocardium contouring borders [2]. The process requires accurate and reliable quantification of the myocardium. Besides, contouring myocardium border manually is a tedious, time consuming, subjective to the high intraobserver unevenness, and typically insufficient to the end-diastolic (ED) and end-systolic (ES) frames. Such factors make the segmentation and analysis inadequate for dynamic function analysis. Due to the lack of edge information and variability in shape, LV segmentation still requires significant efforts that involves efficient techniques. Cardiac MR imaging is the key modality to evaluate the structure and cardiac functionality.

Deep learning and machine learning-based approaches have been used in CMR challenges [3–6] and various others medical applications [7–11], However, the performance of existing methods is mainly impacted by inconsistent exist in population shifts (pathological vs normal cases) imaging environments (e.g., vendors and protocols), as well as unexpected behaviors (e.g., movement of body). It may be useful to explore the prospective failure modes by examining the trained framework to extreme cases in a 'stress test'.

Recent work on model generalizability mainly focused on vendor variability and anatomical structural variation and behavioral implications are comparatively less explored. For acquisition of CMR, respiration motion is one of the main problems. Acute symptoms patients can not adhere to the breath-hold instructions, resulting in degradation of quality of image and inaccurate analysis. We have tried various 2D and 3D segmentation models for the automatic segmentation of CMR images; however, we did not perform better on this dataset using basic 3D UNet. In the first submission, our model performed badly on the segmentation task. We have used deep supervision and some interpolation technique to improve the results. However, for a quality assessment task, our proposed approach works well.

The key contributions of this paper are as follows:

i.   proposed 3D Dens Net using 3D input volume for CMR quality assessment. In the Dense layer, the feature maps of all preceding layers are used as inputs, and their feature maps are used as inputs into all subsequent layers.
ii. simple and lightweight 3D convolutional layers' blocks before the proposed residual block have been presented for CMR motion segmentation. Proposed 3DUnet with deep supervision approach using the residual module at encoder side of proposed model and ensemble the prediction with various cross-validation data.
iii. 3DUnet with deep supervision is used to generate pseudo labels on validation dataset and further nnUNet used pseudo labels with training data to get the final prediction.

2   Respiratory Motions Assessment and Segmentation

The challenge dataset is designed two in tasks: 1) quality assessment and 2) segmentation. Quality assessment task focus on the development of efficient model for quality assessment of an image whereas segmentation task focuses on the development of segmentation techniques that are robust to artifacts in respiratory motion. In this work, we have proposed two different deep learning-based models for each task. For Task1, we have developed a 3D deep learning model to assess image quality. The model has been trained using the input 3D volume along with labels. The block diagram is shown in Fig. 1. In the dense layer, the feature maps of all preceding layers are used as inputs, and their feature maps are used as inputs into all subsequent layers. Four dense blocks have been used in our proposed model. In contrast, each dense block is composed of 6 layers, as shown in Fig. 1. The dense module alleviates the vanishing-gradient problem, strengthening feature propagation, encouraging feature reusability, and substantially reducing the number of parameters.

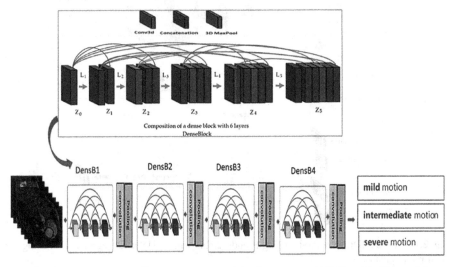

Fig. 1. CMR image quality assessment model using proposed 3D DensNet201 model

The proposed framework includes only a small fraction of feature to the collective knowledge while remaining feature maps remained unchanged. Finally, the last layer decides based on all feature maps which are reused in the model. Our proposed framework uses fewer number of parameters as well as improves the flow of information and gradients, making it easier to train. Notice that each layer in the framework has direct access to the gradients from the loss function and the input, leading that enforces implicit deep supervision which helps to train the deeper network. Further, we also noticed that dense connections also have a regularizing effect, that helps to reduce the overfitting with even small training data. In this work, we have trained the network from scratch using the CMRxMotion dataset.

In task2, our proposed model consisted of two stages. In the first stage, we proposed 3DResUNet with a deep supervision technique. The proposed model was trained on the training dataset, and the validation dataset was used to predict the labels. These labels are called pseudo labels. The nnUNet [11] model was trained in the second stage using pseudo and training datasets. The pseudo labels with validation cases were used in the second stage with the original training dataset to train the nnUNet. A detailed description is shown in Fig. 2.

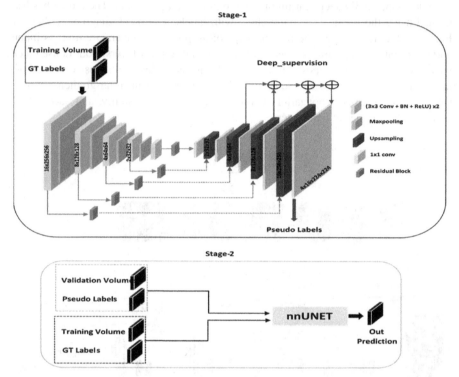

Fig. 2. Proposed 3DresNet model using Pseudo labels with nnUNet for segmentation using the heart motion dataset.

3D-ResUnet with Deep Supervision: A framework of the proposed model is presented as an encoder, a decoder, and a baseline module. The 1×1 convolutional layer with softmax function has been used at the end of the proposed model. The 3D strides convolutional layer has been used to reduce the input image spatial size. The convolutional block consists of convolutional layers with batch normalization and ReLU activation function to extract the different feature maps from each block on the encoder side.

In the encoder block, the spatial input size has been reduced with an increasing number of feature maps and on the decoder side, the input image spatial size will increase using a 3D Conv-Transpose layer. The input features' maps are obtained from every encoder block are concatenated with every decoder block feature map to reconstruct the semantic information. The convolutional ($3 \times 3 \times 3$conv-BN-ReLu) layer used the input feature maps extracted from every convolutional block on the encoder side and further passed these feature maps into the proposed residual module. The spatial size doubled at every decoder block and feature maps are halved at each decoder stage of the proposed model. The residual Bloch has been inserted at each encoder block with skip connection. The feature concatenation has been done at every encoder and decoder block except the last 1×1 convolutional layer. The three-level deep-supervision technique is applied to get the aggregate loss between ground truth and prediction. We have used nnUNet with one-fold cross-validation, we have modified training and optimization parameters as compared to the original nnUNet. The batch size in nnUNet was $96 \times 160 \times 160$ using 500 epochs.

3 Dataset and Evaluation Measures

CMRxMotion challenge dataset is a realistic cardiac MRI dataset. It consists of extreme cases having different levels of respiratory motion. The dataset has been collected from volunteers to obtain the extreme-cases. Clinical CMR scans have been performed using the same MRI scanner. Volunteers are trained to act in 4 manners, respectively: (a) adhere to the breath-hold instructions; (b) halve the breath-hold period; (c) breathe freely, and (d) breathe intensively. Therefore, for a single volunteer, a set of paired CMR images under 4 levels of respiratory motion is collected. First, image quality is assessed by radiologists and identified bad quality images. For diagnostic quality images, radiologists further segment the left and right ventricle and left ventricle myocardium. The dataset consists of images, quality scores, and ground-truth segmentation of 25 volunteers for training and validation. A detailed description can be found [13].

4 Implementation Details

In this section, we briefly describe the implementation setup.

4.1 Pre-processing

We have used the following preprocessing steps for data cleaning:

- Cropping strategy: None
- Resampling Method for anisotropic data:

  The nearest neighbor interpolation method has been applied for resampling.

- Intensity Normalization method:

  The dataset has been normalized using a z-score method based on mean and standard deviation.

4.2 Post-processing

In post-processing, we did not use any post-processing step except to interpolate the prediction mask to the original input size of each input volume. The bilinear interpolation method is used to make the original size.

4.3 Environments and Requirements

The proposed deep learning model is implemented in PyTorch and other libraries based on python are used for preprocessing and analysis of the datasets. The SimpleITK is used for reading and writing the nifty data volume. The ITK-SNAP is used for data visualization. The environments and requirements of the proposed method are shown in Table 1.

Table 1. Environments and requirements.

GPU	Nvidia V100
RAM	16 × 2 GB
CPU	Intel(R) Core (TM) i9-7900X CPU@3.30 GHz
CUDA version	11.3
DL framework	Pytorch (Torch 1.7.0, torchvision 0.2.2)
Dependencies	SimpleITK, Numpy, Skimage, Scipy, Nibabel, ITK-SNAP

4.4 Training Protocols

We have set the learning rate to 0.0004. We have used Adam optimizer and binary cross-entropy function. Batch-size of 2 with 200 epochs has been used with 20 early stopping steps. The best model weights have been saved for prediction in the validation phase. The $256 \times 256 \times 16$ input image size was used for training and prediction resample with the original input size at prediction using the nearest-neighbor interpolation method. The Pytorch library is used for model development, training, optimization, and testing. We have used V100 tesla NVidia-GPU machine. The data augmentation methods mentioned in Table 2 are used to improve the results further. As the dataset cases have different intensity ranges, thus, we have normalized between 0 and 1 using the max–min intensity normalization. The training protocol is described in Table 2.

Table 2. Training protocols.

Data augmentation	VerticalFlip (p = 0.5), RandomGamma (p = 0.8), HorizontalFlip (p = 0.5)
Batch size	2
Patch size	256 × 256 × 16
Number of epochs	200
Optimizer	Adam
Initial learning rate	0.0001
Training time	hours

4.5   Testing Protocols

The same preprocessing was applied at testing time. The training size of each image is fixed ($256 \times 256 \times 16$), and used linear interpolation method to resample the prediction mask to the original shape for each validation volume. The prediction mask produced by our proposed model has been resampled such that it has the same size and spacing as the original image and copies all of the meta-data, i.e., origin, direction, orientation, etc.

5   Results

For Task1, challenge organizers used accuracy and Cohen's kappa between the submission and the ground-truth labels. For Task2, the Dice and HD95 scores are used to for rankings. For each case, we computed the three Dice scores and three HD95 measures between the ground truth and the segmentation of LV, MYO, and RV, respectively.

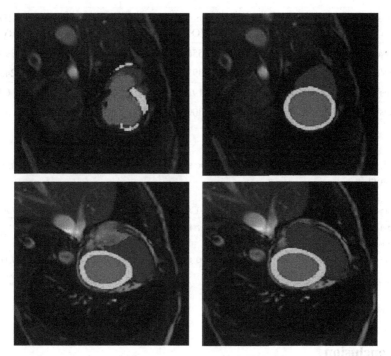

Fig. 3. Qualitative analysis of prediction using validation case (P22-1-ES).

Figure 3 shows the prediction produced by the proposed model and base 3D UNet using P22-1-ED and ES. In some cases, the model produced a good segmentation mask, and base 3DUNet failed to produce the prediction for myocardium border and RV.

The scores for task1 and task2 are shown in Table 3 and Table 4. These scores are computed based on the validation dataset and shown on the challenge website. The performance of proposed model on unseen test dataset for task1 is shown in Table 4 and for task2 is shown in Table 5. Our proposed model improved the performance using useen test dataset in task2.

Table 3. Performance assessment for CMR quality assessment using proposed models

Proposed models	Accuracy	Cohens Kappa
3D-DensNet121	0.525	0.2619
3D-DensNet201	0.575	0.3436
Performance on test unseen dataset		
3D-DensNet201	0.567	0.382

Table 4. Performance assessment for automatic segmentation using proposed models

Proposed models	Dice LV	Dice MYO	Dice RV	HD95 LV	HD95 MYO	HD95 RV
3D-UNet-Monai	0.4220	0.1556	0.3163	14.98	42.9	39.8
3D-UNet-base	0.6023	0.4392	0.5417	60.159	35.32	24.15
3D-ResNet + DS	0.9043	0.8197	0.8714	9.662	4.432	6.819
3D-ResNet + DS + nnUNet	0.9179	0.8300	0.8898	8.7472	3.7870	5.942

Table 5. Performance assessment for automatic segmentation using proposed models on unseen test dataset

Proposed models	LV_dsc_mean	LV_dsc_std	MYO_dsc_mean	MYO_dsc_std	RV_dsc_mean	RV_dsc_std
3D-ResNet+DS+nnUNet	92.82075	4.274334	86.12846	3.540047	90.66745	5.566093
HD95 on unseen dataset						
	LV_hd95_mean	**LV_hd95_std**	**MYO_hd95_mean**	**MYO_hd95_std**	**RV_hd95_mean**	**RV_hd95_std**
3D-ResNet+DS+nnUNet	3.894681	3.834829	2.613385	2.385594	4.865491	3.896979

6 Conclusion

In this paper, two-stage method has been proposed for segmentation and 3DDenseUNet is used for motion classification assessment. In the first stage, the proposed 3D-ResNet with deep supervision was used and the second stage used nnUNet to get the final prediction. Our proposed approach produced an optimal performance as compared to existing methods. In Future, we will use the 3D transformer-based method to further enhance the performance.

References

1. Karamitsos, T.D., Francis, J.M., Myerson, S., Selvanayagam, J.B., Neubauer, S.: The role of cardiovascular magnetic resonance imaging in heart failure. J. Am. College Cardiol. 54(15), 1407–1424 (2009)
2. Peng, P., Lekadir, K, Gooya, A., Shao, L., Petersen, S.E., Frangi, A.F.: A review of heart chamber segmentation for structural and functional analysis using cardiac magnetic resonance imaging. Magn. Reson. Mater. Phys., Biol. Med. 29(2), 155–195 (2016)
3. Bernard, O., et al.: Deep learning techniques for automatic MRI cardiac multi-structures segmentation and diagnosis: is the problem solved?. IEEE Trans. Med. Imaging 37(11), 2514–2525 (2018)
4. Lalande, A., et al.: Deep learning methods for automatic evaluation of delayed enhancement-MRI. The results of the EMIDEC challenge. Med. Image Anal. 79, 102428 (2022)
5. Mazher, M., Qayyum, A., Benzinou, A., Abdel-Nasser, M., Puig, D.: Multi-disease, Multi-view and Multi-center Right Ventricular Segmentation in Cardiac MRI Using Efficient Late-Ensemble Deep Learning Approach. In: International Workshop on Statistical Atlases and Computational Models of the Heart, pp. 335–343. Springer, Cham, 2021, September
6. Chen, Z., et al.: Automatic deep learning-based myocardial infarction segmentation from delayed enhancement MRI. Comput. Med. Imaging Graph. 95, 102014 (2022)
7. Qayyum, A., Mazhar, M., Razzak, I., Bouadjenek, M.R.: Multilevel depth-wise context attention network with atrous mechanism for segmentation of covid19 affected regions. Neural Computing and Applications, pp. 1–13 (2021)
8. Ma, J., et al.: Fast and Low-GPU-memory abdomen CT organ segmentation: The FLARE challenge. Med. Image Anal., 102616 (2022)
9. Payette, K., et al.: Fetal brain tissue annotation and segmentation challenge results. arXiv preprint arXiv:2204.09573 (2022)
10. Kausar, A., Razzak, I., Shapiai, I., Alshammari, R.: An improved dense V-network for fast and precise segmentation of left atrium. In: 2021 International Joint Conference on Neural Networks (IJCNN), pp. 1–8. IEEE (July 2021)
11. Noreen, N., Palaniappan, S., Qayyum, A., Ahmad, I., Alassafi, M.O.: Brain tumor classification based on fine-tuned models and the ensemble method. Comput., Mater. Contin. 67(3), 3967–3982 (2021)
12. Isensee, F., Jaeger, P.F., Kohl, S.A., Petersen, J., Maier-Hein, K.H.: nnU-Net: a selfconfiguring method for deep learning-based biomedical image segmentation. Nat. Methods 18(2), 203–211 (2021)
13. Wang, S., et al.: The extreme cardiac MRI analysis challenge under respiratory motion (CMRxMotion). arXiv preprint arXIv: 2210.06385 (2022)

Robust Cardiac MRI Segmentation with Data-Centric Models to Improve Performance via Intensive Pre-training and Augmentation

Shizhan Gong[1]([✉]), Weitao Lu[2], Jize Xie[2], Xiaofan Zhang[2,3], Shaoting Zhang[2,3], and Qi Dou[1]

[1] Department of Computer Science and Engineering, The Chinese University of Hong Kong, Hong Kong, China
szgong22@cse.cuhk.edu.hk
[2] School of Electronic Information and Electrical Engineering, Shanghai Jiao Tong University, Shanghai, China
[3] Shanghai Artificial Intelligence Laboratory, Shanghai, China

Abstract. Segmentation of anatomical structures from Cardiac Magnetic Resonance (CMR) is central to the non-invasive quantitative assessment of cardiac function and structure, and deep-learning-based automatic segmentation models prove to have satisfying performance. However, patients' respiratory motion during the scanning process can greatly degenerate the quality of CMR images, resulting in a serious performance drop for deep learning algorithms. Building a robust cardiac MRI segmentation model is one of the keys to facilitating the use of deep learning in practical clinic scenarios. To this end, we experiment with several network architectures and compare their segmentation accuracy and robustness to respiratory motion. We further pre-train our network on large publicly available CMR datasets and augment our training set with adversarial augmentation, both methods bring significant improvement. We evaluate our methods on the cine MRI dataset of the CMRxMotion challenge and obtain promising performance for the segmentation of the left ventricle, left ventricular myocardium, and right ventricle.

Keywords: Cardiac segmentation · Network pre-train · Robust learning

1 Introduction

Cardiac Magnetic Resonance (CMR) imaging is the golden standard sequence for non-invasive evaluation of cardiac anatomical structures and functionalities [1]. Anatomical segmentation allows for analysis of what is of interest to the clinicians. By discarding irrelevant information, the images can be smaller in size, which can reduce the post-processing time and computing power for the downstream analysis. It is also a crucial pre-requisite for the calculation of several image-based biomarkers [2–4] with diagnostic value. Recently, with the development of artificial

O. Camara et al. (Eds.): STACOM 2022, LNCS 13593, pp. 494–504, 2022.
https://doi.org/10.1007/978-3-031-23443-9_47

intelligence, fully-automatic segmentation algorithms based on deep learning start to surpass manual segmentation with faster speed, less subjective bias, and comparable or even higher accuracy. However, in clinical practice, the model performance highly relies on image qualities. For CMR acquisition, respiration motion is one of the major causes of degenerated image qualities, as it may be difficult for certain patients with acute symptoms to follow the instructions and hold their breath for a long time during the scan. Images contaminated by respiratory motion have seen a significant drop in model performance and result in obvious failure cases.

Currently, most automatic cardiac segmentation models are based on deep learning methods, which learn a function to map the input images to the segmentation masks. This method highly relies on the quantity and quality of the training data, and is based on the assumption that the hold-out data has alike distribution to the training data. In practical clinic scenarios, the number of available training data is limited, and the quality of most images is high to guarantee diagnostic value. With such training data as supervision, the model is trained to only perform well on clean images and can easily fail when encountering images with low quality.

Pre-training and data augmentation are two important data-driven methods which have proven effect in improving model robustness. These two methods focus on enlarging the training data and better utilizing the existing data respectively. Pre-training exposes the model to a large dataset other than the training set, so as to broaden the model's horizon. It is recognized to yield better results than training from scratch [21,22]. Although some researchers argue that pre-training offers little benefit for certain tasks or light weight architectures [23,24], it is still undeniable that pre-training can enhance the model robustness and improve its performance on hold-out individuals [25,26]. Data augmentation is another standard trial to build robust segmentation models. It exposes the network to higher variability through perturbations of the training data. This includes native approaches such as cropping, rotation, and flipping. As respiration motion will cause spatial transformation of the anatomic structure, deformation-based augmentation [6] is also rewarding for training. A recently proposed adversarial data augmentation method [5] can generate plausible and realistic signal corruptions that are difficult for the models to analyze and therefore increase the model's adversarial robustness.

In this work, we explore these two data-driven approaches in the context of the CMRxMotion challenge [32]. The main goal of this challenge is to build a model for segmentation of the left ventricle (LV), left ventricular myocardium (MYO), and right ventricle (RV), based on limited training data. This model should be robust under different levels of respiration motion as the test data composes of images with diverse quality levels. We pre-train our model on large publicly available datasets with the same tasks, and increase the data variability through both random augmentation and adversarial augmentation. We find that, through intensive pre-training and strong data augmentation, even without novel DCNN architectures, the model can still reveal high robustness towards the qualities of the images, regardless of reasons for low-quality images, such as respiration motion.

2   Methods

We first give a brief introduction to the dataset of the CMRxMotion challenge and then explain our proposed approaches in detail.

2.1   Dataset

The CMRxMotion dataset [32] consists of short-axis (SA) cine MRI acquisitions of 45 healthy volunteers. Each volunteer is trained to act in 4 manners during the scanning process, namely a) stick to the breath-hold instructions, b) halve the breath-hold period, c) breathe freely, and d) conduct intensive breathing. The pixel sizes vary from ~0.66 to ~0.76 mm, the image resolution ranges from 400 to 512 pixels, the number of slices is between 9 and 13, and the slice thickness ranges from ~9.6 mm to 10 mm. For images with diagnostic qualities, LV, MYO, and RV at the end of systole (ES) and end of diastole (ED) are manually segmented by radiologists. Exams of 20 volunteers with both images and ground-truth segmentation as well as 5 volunteers with only images are released for training and validation respectively. The remaining 20 volunteers are withheld for testing.

2.2   Network Architecture

We try three variants of U-Net: nnU-Net [14], Swin-UNETR [13], Swin-UNet [8].

nnU-Net [14] is a pure CNN-based method, which is good at catching the local pattern of the image. One modification we make is to replace convolution operation at the bottleneck layer with deformable convolution [10], which has shown increases in performance since it allows for a flexible receptive field [20]. The sampling offset learned by deformable convolutions is expected to counteract some of the shifts caused by respiratory motion.

On the contrary, Swin-UNet [8] is a pure transformer-based network, which is better at capturing global features through self-attention and shifted windows [11]. However, the segmentation result of raw Swin-UNet usually contains zigzag margins since Swin-UNet uses patch rather than pixel as the smallest unit to operate on. To overcome this limitation, we add two convolutional layers with layer normalization and leaky-relu at the end of the network, which helps produce smooth segmentation results.

Swin-UNETR [13] is a combination of transformer-based encoder and CNN-based decoder. which is expected to utilize the global information of the images and generate a refined segmentation map.

The input volumes only contain around 10 slices, which is not enough to be divided into multiple patches and windows. Therefore, for nnU-Net, we try both 2D and 3D variants, while for Swin-UNETR and Swin-UNet, only the 2D version is used.

2.3   Pre-training

Our training data is only composed of 20 healthy volunteers, which is too few for the model to even exhibit strong robustness towards the inherent variations of anatomical structures among different people, let alone unstable image quality or perturbations. Pre-training has proven to be effective in improving model performance as well as enhancing its robustness. To increase the model robustness towards unseen data with different appearances and qualities, we collect five public datasets of cine MRI with the same segmentation tasks listed in Table 1 for pre-training. We fuse these datasets and use labeled images from both the training phase and testing phase for pre-training. As transformer-based methods can benefit more from pre-training, the Swin-Unet is additionally pre-trained with ImageNet [19]. The encoder of Swin-UNETR is pre-trained with extra public CMR datasets without labels [28–31], following multi-task self-supervised learning manners [27]. Although no deliberate respiration motion is conducted during the acquisition of these images, they exhibit significant variations in many other aspects including but not limited to scanner type, acquisition center, protocols, and health condition of the subjects. We believe these variations allow the model to ignore pixel-wise noisy signals and turn to catch the essential and high-level features useful for cardiac segmentation.

Table 1. Public CMR dataset used for pre-training

Dataset	Number of cases	Number of labeled SA images per case	Note
ACDC [15]	100	2	
M&Ms [16]	344	2	
M&Ms-2 [16]	360	2	Use only SA images
MyoPS [17,18]	25	1	Use only bSSFP
MS-CMRSeg [17,18]	45	1	Use only bSSFP

2.4   Data Augmentation

We use both random data augmentation and adversarial data augmentation.

For random data augmentation, we follow the same schema as the default mode of nnU-Net [14], which contains rotation, scaling, Gaussian noise, Gaussian blur, brightness, contrast, simulation of low resolution, gamma correction, and mirroring.

As the inherent property of our data is that it is contaminated by respiratory motion, which can result in diffeomorphic deformation or spatial transformation of the raw image. Adversarial data augmentation proves to be more effective than random data augmentation in terms of improving model robustness towards a certain type of perturbations [7]. In adversarial data augmentation, given a model, the optimal perturbation of certain types is learned which will impair the model performance to the utmost extent. The model is then trained to resist this perturbation, which in our case, can be deformation caused by respiration motion. In this work, we apply AdvChain [5] to improve our model's robustness towards diffeomorphic deformation and spatial transformation. To be more specific, we first train the network with random data augmentation, then fine-tune it with AdvChain. In each iteration of the fine-tuning phase, we turn off Gaussian noise, rotation, and scaling from the random data augmentation while keeping the rest. A perturbation consists of a series of Gaussian noise, spatial transformation, and diffeomorphic deformation is randomly generated, whose parameters are trainable. We freeze the network parameters and optimize the perturbation parameters to increase the consistency loss (MSE and contour loss [9] in our case) between two predicted segmentation maps before and after the perturbation. Finally, we fix this perturbation and update the network parameters by propagating supervised loss together with consistency loss. The above steps are conducted for multiple iterations until convergence.

2.5   Training Protocol

Pre-processing methods follow a similar framework to the default mode of nnU-Net [14], which consists of adjusting the pixel size of all images to ~0.66 with third-order spline interpolation, cropping or padding the resulting images to the resolution of 512×512 pixels, normalization, and intensity clipping (0.5 and 99.5 percentiles). For transformer-based methods, we crop the images to the same resolution of 224×224 pixels. We posit the heart in the middle of the images using ground-truth segmentation labels (training set) or pseudo labels predicted by nnU-Net (validation and test set) as reference. We also apply min-max normalization to each image. For inference, we apply test time augmentation by mirroring along all axes. In post-processing steps, we select only the largest connected component of each structure in the predicted segmentation mask. We conduct this operation twice, in both a slice-wise manner and a volume-wise manner. And then we remove the RV from slices predicted to have no LV nor MYO. We train our model on an NVIDIA A100 GPU with 80 GB memory. All networks are implemented using the PyTorch framework. The optimization of nnU-Net follows its default setting. The Swin-UNETR is trained using the AdamW optimizer with initial learning rate of 0.0004 and weight decay of 0.00005. The Swin-UNet is trained using the SGD optimizer with initial learning rate of 0.05, weight decay of 0.0001 and momentum of 0.9. A weighted sum of cross-entropy loss and dice loss is used for back-propagation to optimize our model.

3   Experiments

We conduct a series of comparison and ablation studies so as to select the best network architecture (Sect. 3.1) and evaluate the effectiveness of different data-driven approaches (Sect. 3.2). We randomly split the 20 volunteers in the training set into five non-overlapping folders. For quick comparisons of proposed methods, we use four folders to train networks and use the rest folder to determine the convergence of the training process. The Dice coefficients and 95% Hausdorff distance for three anatomical structures of the online validation set are reported.

3.1 Architectural Variants

We first compare the performance of 2D nnU-Net, 3D nnU-Net, Swin-UNETR, and Swin-UNet. For a fair comparison, all models are pre-trained with public datasets. The results on the hold-out dataset are depicted in Table 2.

From the results, we find that 2D nnU-Net performs better than 3D nnU-Net, which agrees with existing literature and may be due to the large between-slice distance of the SA sequence [12]. We also find that transformer-based methods have comparable (slightly lower) results to CNN-based methods on LV segmentation and RV segmentation, but are inferior to CNN on MYO segmentation. Through assessing the performance on individual cases and slices, we further observe that the segmentation map generated by nnU-Net is very smooth and regular, while the results of the transformer have twisted margins, especially when the edge in the raw image is vague. We infer that the patch division and window partition process of the swin-transformer is the reason for rough segmentation maps. As the raw input is discretized into multiple windows, it can perform well in segmenting large and convex-shaped structures such as LV and RV but is hard to segment exquisite structures such as MYO.

Moreover, when comparing results of different respiratory motion intensities, we find that nnU-Net is more robust. The respiratory motion can result in an unclear margin between RV and its adjacent tissue, to which transformer-based methods are more vulnerable. The transformer is prone to include nearby objects with similar pixel intensities in the predicted RV segmentation map. In addition, when the margin of MYO is ambiguous, nnU-Net seems to follow a population-based prior to give a smooth and rounded segmentation map, while the results of transformation and be more distorted and angular, which are more faithful to the raw pixel intensities (Fig. 1).

Table 2. Quantitative comparison of proposed architectures on the online validation set of 5 volunteers, in terms of Dice and Hausdorff distance

Methods	Dice score			95% Hausdorff distance		
	LV	MYO	RV	LV	MYO	RV
2D nnU-Net [14]	**0.9184**	0.8284	**0.8994**	**8.4581**	4.3903	**5.0750**
3D nnU-Net [14]	0.9164	**0.8295**	0.8879	9.1120	**3.9550**	6.9939
Swin-UNETR [13]	0.9145	0.8133	0.8925	9.1525	4.9728	5.2112
Swin-UNet [8]	0.9167	0.8206	0.8962	8.5567	4.3373	5.4207

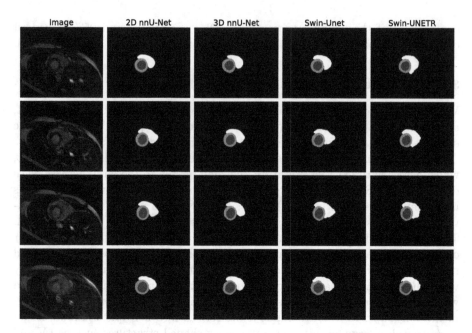

Fig. 1. Qualitative segmentation results of different network architectures. From top to bottom are images from a single volunteer with breath-holding, half breath-holding, regular breathe and intensive breathe. Cases are selected from online validation set.

3.2 Data-Driven Methods with Pre-training and Augmentation

To evaluate the effectiveness of pre-training and adversarial data augmentations, we choose the vanilla 2D nnU-Net as our baseline. We then compare it to both a nnU-Net pre-trained on the public dataset (Sect. 2.3) and a nnU-Net with AdvChain (Sect. 2.4). The results are reported in Table 3.

From the results, we can conclude that both pre-training and adversarial augmentations improve performance compared to the baseline. Pre-training can significantly improve the segmentation accuracy of LV and MYO, while AdvChain promises considerable gain in RV segmentation. Composition of both methods can further enhance the segmentation accuracy.

3.3 Ensemble

Based on the above comparison and analysis, we combine the network architecture and data-driven methods proven to bring out improvements over their baseline in an ensemble manner as our final submission to the CMRxMotion challenge. To this end, we train four networks mentioned in Sect. 3.1 with pre-training and adversarial augmentations in a 5-fold cross-validation setting on the training dataset, and average the outputs of each network to obtain the ensemble prediction. Finally, in the online validation set, our model achieves dice scores of

Table 3. Segmentation performance of baseline, pre-training, and AdvChain, in terms of Dice and Hausdorff distance.

Methods	Dice score			95% Hausdorff distance		
	LV	MYO	RV	LV	MYO	RV
Baseline	0.9141	0.8264	0.8915	8.4561	4.2720	5.2156
+Pre-training	0.9184	0.8284	0.8994	8.4781	4.3903	5.0750
+AdvChain [5]	0.9160	0.8296	0.9019	8.6097	4.0094	4.8959
+Both	**0.9201**	**0.8306**	**0.9062**	**8.0947**	**4.0088**	**4.6297**

0.9220, 0.8352, and 0.9069 for segmentation of LV, MYO, and RV respectively, and 95% Hausdorff distance of 8.07, 3.70, and 4.69. And in the test phase, our model achieves dice scores of 0.9372, 0.8738, and 0.9239 for segmentation of LV, MYO, and RV respectively, and 95% Hausdorff distance of 3.13, 2.57, and 3.59, which is ranked the first place among all participants.

4   Discussion and Conclusion

In this work, we propose a data-centric model for the tasks of cardiac segmentation which can achieve high performance even when the image quality is degenerated by intensive repository motion. We compare multiple networks or methods and find that pre-training and adversarial augmentation are two effective data-driven approaches that can significantly improve the model performance. Deep learning is essentially a data-driven methodology, whose performance highly relies on data quantity and quality. Collecting sufficient data and making utmost use of the data is second to nothing in terms of improving model performance and robustness. We believe this philosophy can guide the real application of deep learning. Instead of designing novel DCNN architectures or fancy training schema, rethinking how to utilize the data can be far more important.

Intensive respiratory motion impairs the images by making its margin indistinct so that it becomes hard for the model to decide which structure each pixel belongs to. During the experiments, we find even if the boundary between LV and MYO or between MYO and RV is ambiguous, the model can always predict it well. However, it is hard for the model to delineate the boundary between MYO, RV, and their surrounding tissues, especially for RV, which shows great irregularity and variability in terms of shape and pixel intensity. Currently, we are mainly using some common data augmentations such as rotation or flip. In the future, we may design some exclusive perturbations that can better reflect this fuzzy-boundary property.

Another future direction entails the use of inter-slice information. Although the performance of 3D nnU-Net is inferior to that of 2D nnU-Net due to discontinuity between slices, we find that the ensemble of both models results in a better performance. Therefore, we believe the inter-slice information is beneficial for the cardiac segmentation task. Furthermore, the influence of respiratory motion on slices may differ from each other. A well-designed architecture may utilize the information of clean slices to help do segmentation on dirty slices.

References

1. Schulz-Menger, J., et al.: Standardized image interpretation and post-processing in cardiovascular magnetic resonance - 2020 update: Society for Cardiovascular Magnetic Resonance (SCMR): Board of Trustees Task Force on Standardized Post-Processing. J. Cardiovasc. Magn. Reson. **22**(1), 19 (2022). https://doi.org/10.1186/s12968-020-00610-6
2. Alfakih, K., Plein, S., Thiele, H., Jones, T., Ridgway, J.P., Sivananthan, M.U.: Normal human left and right ventricular dimensions for MRI as assessed by turbo gradient echo and steady-state free precession imaging sequences. J. Magn. Reson. Imaging **17**(3), 323–329 (2003). https://doi.org/10.1002/jmri.10262
3. Bai, W., et al.: A bi-ventricular cardiac atlas built from 1000+ high resolution MR images of healthy subjects and an analysis of shape and motion. Med. Image Anal. **26**(1), 133–45 (2015). https://doi.org/10.1016/j.media.2015.08.009
4. Bai, W., et al.: Biventricular surface reconstruction from cine MRI contours using point completion networks. In: 2021 IEEE 18th International Symposium on Biomedical Imaging (ISBI), pp. 105–109 (2021). https://doi.org/10.1109/ISBI48211.2021.9434040
5. Chen, C., et al.: Enhancing MR image segmentation with realistic adversarial data augmentation. arXiv preprint (2022). https://arxiv.org/abs/2108.03429
6. Corral Acero, J., et al.: SMOD - data augmentation based on statistical models of deformation to enhance segmentation in 2D cine cardiac MRI. In: Coudière, Y., Ozenne, V., Vigmond, E., Zemzemi, N. (eds.) FIMH 2019. LNCS, vol. 11504, pp. 361–369. Springer, Cham (2019). https://doi.org/10.1007/978-3-030-21949-9_39
7. Madry, A., Makelov, A., Schmidt, L., Tsipras, D., Vladu, A.: Towards deep learning models resistant to adversarial attacks. In: International Conference on Learning Representations, pp. 1–23 (2017). https://arxiv.org/abs/1706.06083
8. Cao, H., et al.: Swin-Unet: Unet-like pure transformer for medical image segmentation (2021). https://arxiv.org/abs/2105.05537
9. Chen, C., et al.: Realistic adversarial data augmentation for MR image segmentation. In: Martel, A.L., et al. (eds.) MICCAI 2020, Part I. LNCS, vol. 12261, pp. 667–677. Springer, Cham (2020). https://doi.org/10.1007/978-3-030-59710-8_65
10. Zhu, X., Hu, H., Lin, S., Dai, J.: Deformable ConvNets V2: more deformable, better results, In: IEEE/CVF Conference on Computer Vision and Pattern Recognition (CVPR), pp. 9300–9308 (2019). https://doi.org/10.1109/CVPR.2019.00953
11. Liu, Z., et al.: Swin Transformer: hierarchical vision transformer using shifted windows. In: Proceedings of the IEEE/CVF International Conference on Computer Vision (ICCV), pp. 3464–3473 (2021). https://doi.org/10.48550/arXiv.2103.14030
12. Isensee, F., Jaeger, P.F., Full, P.M., Wolf, I., Engelhardt, S., Maier-Hein, K.H.: Automatic cardiac disease assessment on cine-MRI via time-series segmentation and domain specific features. In: Pop, M., et al. (eds.) STACOM 2017. LNCS, vol. 10663, pp. 120–129. Springer, Cham (2018). https://doi.org/10.1007/978-3-319-75541-0_13

13. Hatamizadeh, A., Nath, V., Tang, Y., Yang, D., Roth, H.R., Xu, D.: Swin UNETR: swin transformers for semantic segmentation of brain tumors in MRI images. In: Crimi, A., Bakas, S. (eds.) Brainlesion: Glioma, Multiple Sclerosis, Stroke and Traumatic Brain Injuries. BrainLes 2021. LNCS, vol. 12962. Springer, Cham (2022). https://doi.org/10.1007/978-3-031-08999-2_22

14. Isensee, F., Jaeger, P.F., Kohl, S.A.A., et al.: nnU-Net: a self-configuring method for deep learning-based biomedical image segmentation. Nat. Methods **18**, 203–211 (2021). https://doi.org/10.1038/s41592-020-01008-z

15. Bernard, O., Lalande, A., Zotti, C., Cervenansky, F., et al.: Deep learning techniques for automatic MRI cardiac multi-structures segmentation and diagnosis: Is the Problem Solved? IEEE Trans. Med. Imaging **37**(11), 2514–2525 (2018). https://doi.org/10.1109/TMI.2018.2837502

16. Campello, V.M., et al.: Multi-centre, multi-vendor and multi-disease cardiac segmentation: the M&Ms Challenge. IEEE Trans. Med. Imaging **40**(12), 3543–3554 (2021). https://doi.org/10.1109/TMI.2021.3090082

17. Zhuang, X.: Multivariate mixture model for myocardial segmentation combining multi-source images. IEEE Trans. Pattern Anal. Mach. Intell. **41**(12), 2933–2946 (2019). https://doi.org/10.1109/TPAMI.2018.2869576

18. Zhuang, X.: Multivariate mixture model for cardiac segmentation from multi-sequence MRI. In: International Conference on Medical Image Computing and Computer-Assisted Intervention, pp. 581–588 (2016). https://doi.org/10.1007/978-3-319-46723-8_67

19. Deng J., Dong, W., Socher, R., Li, L.J., Li, K., Li, F.: ImageNet: a large-scale hierarchical image database. In: IEEE Conference on Computer Vision and Pattern Recognition, pp. 248–255 (2009). https://doi.org/10.1109/CVPR.2009.5206848

20. Fulton, M.J., Heckman, C.R., Rentschler, M.E.: Deformable Bayesian convolutional networks for disease-robust cardiac MRI segmentation. In: Puyol Antón, E., et al. (eds.) STACOM 2021. LNCS, vol. 13131, pp. 296–305. Springer, Cham (2022). https://doi.org/10.1007/978-3-030-93722-5_32

21. Yosinski, J., Clune, J., Bengio, Y., Lipson, H.: How transferable are features in deep neural networks? In: Advances in Neural Information Processing Systems, pp. 3320–3328 (2014). https://doi.org/10.48550/arXiv.1411.1792

22. Zhuang, F., et al.: A comprehensive survey on transfer learning. arXiv preprint (2019). https://arxiv.org/abs/1911.02685

23. He, K., Girshick, R., Dollar, P.: Rethinking imagenet pre-training. arXiv preprint (2018). https://arxiv.org/abs/1811.08883

24. Raghu, M., Zhang, C., Kleinberg, J., Bengio, S.: Transfusion: understanding transfer learning for medical imaging. In: Advances in Neural Information Processing Systems, pp. 3347–3357 (2019). https://doi.org/10.48550/arXiv.1902.07208

25. Hendrycks, D., Lee, K., Mazeika, M.: Using pre-training can improve model robustness and uncertainty. In: Proceedings of the International Conference on Machine Learning (2019). https://doi.org/10.48550/arXiv.1901.09960

26. Mathis, A., et al.: Pretraining boosts out-of-domain robustness for pose estimation. In: Proceedings of the IEEE/CVF Winter Conference on Applications of Computer Vision, pp. 1859–1868 (2019). https://doi.org/10.48550/arXiv.1909.11229

27. Tang, Y., et al.: Self-supervised pre-training of swin transformers for 3D medical image analysis. In: IEEE/CVF Conference on Computer Vision and Pattern Recognition (CVPR), pp. 20730–20740 (2022). https://doi.org/10.48550/arXiv.2111.14791

28. Radau, P., Lu, Y., Connelly, K., Paul, G., Dick, A.J., Wright, G.A.: Evaluation framework for algorithms segmenting short axis cardiac MRI. MIDAS J. Cardiac MR Left Ventricle Segmentation Challenge (2009). https://hdl.handle.net/10380/3070

29. Petitjean, C., Zuluaga, M.A., Bai, W., et al.: Right ventricle segmentation from cardiac MRI: a collation study. Med. Image Anal. **19**(1), 187–202 (2015). https://doi.org/10.1016/j.media.2014.10.004

30. Tobon-Gomez, C., Geers, A.J., Peters, J., et al.: Benchmark for algorithms segmenting the left atrium from 3D CT and MRI datasets. IEEE Trans. Med. Imaging **34**(7), 1460–1473 (2015). https://doi.org/10.1109/TMI.2015.2398818

31. Second annual data science bowl: transforming how we diagnose heart disease (2015). https://www.kaggle.com/competitions/second-annual-data-science-bowl/overview

32. Wang, S., Qin, C., Wang, C., Wang, K., Wang, H., Chen, C., et al.: The extreme cardiac MRI analysis challenge under respiratory motion (CMRxMotion). arXiv preprint (2022). https://doi.org/10.48550/arXiv.2210.06385

A Deep Learning-Based Fully Automatic Framework for Motion-Existing Cine Image Quality Control and Quantitative Analysis

Huili Yang[1,2]([⊠]), Lexiaozi Fan[1,2]([⊠]), Nikolay Iakovlev[1], and Daniel Kim[1,2]

[1] Department of Radiology, Northwestern University, Chicago, IL, USA
{huiliyang2023,lexiaozifan2019}@u.northwestern.edu
[2] Department of Biomedical Engineering, Northwestern University, Evanston, IL, USA

Abstract. Cardiac cine magnetic resonance imaging (MRI) is the current standard for the assessment of cardiac structure and function. In patients with dyspnea, however, the inability to perform breath-holding may cause image artifacts due to respiratory motion and degrade the image quality, which may result in incorrect disease diagnosis and downstream analysis. Therefore, quality control is an essential component of the clinical workflow. The accuracy of quantitative metrics such as left ventricular ejection fraction and volumes depends on the segmentation of the left ventricle (LV), myocardium (MYO), and right ventricle (RV). The current clinical practice involves manual segmentation, which is both time-consuming and subjective. Therefore, the development of a pipeline that incorporates efficient and automatic image quality control and segmentation is desirable. In this work, we developed a deep learning-based fully automated framework to first assess the image quality of acquired data, produce real-time feedback to determine whether a new acquisition is necessary or not when the patient is still on the table, and segment the LV, MYO, and RV. Specifically, we leverage a 2D CNN, incorporating some basic techniques to achieve both top performance and memory efficiency (within 3 GB) for the quality control task and nnU-Net framework for top performance for the segmentation task. We evaluated our method in the CMRxMotion challenge, ranking first place for the quality control task on the validation set and second place for the segmentation on the testing set among all the competing teams.

Keywords: Cardiac cine MR images · Motion artifacts · Quality control · Segmentation · Deep learning

1 Introduction

Cardiovascular disease is a leading cause of death and sickness in the world. Cardiac cine magnetic resonance imaging (MRI) is the current standard for the assessment of cardiac structure and function and is an essential component of all cardiovascular MRI examinations [1].

However, the cardiac cine imaging quality is susceptible to respiratory motion artifacts, which may degrade the image quality. Therefore, quality control is an essential

O. Camara et al. (Eds.): STACOM 2022, LNCS 13593, pp. 505–512, 2022.
https://doi.org/10.1007/978-3-031-23443-9_48

component of a clinical pipeline to avoid potentially incorrect disease diagnosis and downstream analysis. Ideally, this quality control processing should be performed while the patient is still on the table so that a repeat scan can be performed as needed [2]. Radiologists or cardiologists are usually not available during MRI scans to assess the image quality, whereas the technicians, especially inexperienced ones, may not be confident in determining the image quality. It is therefore necessary to have an automatic real-time quality control method.

The accuracy of downstream analysis such as the measurement of ventricular volumes, masses, and ejection fraction quantification depends upon accurate segmentation of the left ventricle (LV), myocardium (MYO), and right ventricle (RV). Manual delineations are still the gold standard for segmentation, which is tedious, time-consuming, and subjective. Thus, the development of an efficient and automatic method is desirable.

In the past decade, deep learning (DL) has been widely used in image quality control [3, 4] and segmentation [5, 6]. However, few studies [2] have integrated quality control into a segmentation framework, although there is a clear relationship between image quality and segmentation accuracy. The CMRxMotion challenge aims to first develop a model for image quality assessment under respiratory motion artifacts. For those images with diagnostic quality, even if they might have different levels of respiratory motion artifacts, to have a robust segmentation model at one's disposal. In this challenge, the cohort is comprised of 45 healthy volunteers (20 for training, 5 for validation, and 20 for testing), and each of them underwent clinical CMR scans using the same MRI scanner (Siemens 3 T MRI scanner MAGNETOM Vida) and was trained to act in 4 manners, respectively, to produce 4 levels of respiratory artifacts: (a) adhere to the breath-hold instructions; (b) halve the breath-hold period; (c) breathe freely; (d) breathe intensively. For reference, the radiologists first assessed the image quality and recognized the images with bad quality. For those images with diagnostic quality, radiologists further segmented the LV, MYO, and RV.

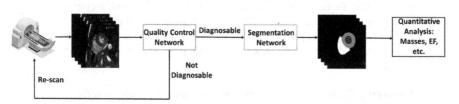

Fig. 1. Pipeline of the proposed method. The acquired cine images will be sent to a quality control network to assess the image quality. If diagnosable, the images will propagate to the segmentation network for segmentation to prepare for the quantitative analysis. If not, there will be real-time feedback to require a new scan.

In this paper, we proposed to incorporate the automatic quality control network into the segmentation framework so that there will be real-time feedback to determine whether the image quality of the acquired data is readable or not. If not, we can acquire new cine data when the patients are still in the scanner (see Fig. 1 for the pipeline). For the quality control challenge, we leveraged the 2D CNN by incorporating some basic techniques to achieve both top performance and memory efficiency (within 3 GB).

For the segmentation challenge, we applied the nnU-Net framework, as it has shown state-of-the-art performance for segmentation by configuring itself automatically [5].

2   Methods

2.1   Datasets

This CMRxMotion challenge [7] prospectively enrolled 45 healthy volunteers to acquire cine imaging using the clinical 'TrueFISP' sequence. The MRI scan was performed based on the recommendations reported in a previously published paper [8]. Typical image parameters were used, including matrix size ranging from 400 * 448 to 512 * 512, spatial resolution 2.0 mm * 2.0 mm, slice thickness 8.0 mm, slice distance 4.0 mm, and 9 to 13 slices based on cardiac size. Each subject was trained to act in 4 manners, respectively, to produce 4 levels of respiratory artifacts: a) adhere to the breath-hold instructions; b) halve the breath-hold period; c) breathe freely; d) breathe intensively. Only the end-diastole (ED) and end-systole (ES) frames are provided for this challenge. The dataset was split into three cohorts: 160 training cases (20 volunteers*4 scans*2 frames), 40 validation cases (5 volunteers*4 scans*2 frames), and 160 test cases (20 volunteers*4 scans*2 frames).

2.2   Quality Control

The datasets for each volunteer acquired with 4 different levels of respiratory motion were first scored by a radiologist using the standard 5-point Likert scale (5, excellent diagnostic quality; 4, more than adequate for diagnosis; 3, adequate for diagnosis; 2, questionable for diagnosis; 1, non-diagnostic). To enhance the reproducibility, 3 levels of motion artifacts were defined based on the original 5-point scores. Images with quality scores 4–5 were labeled as having mild motion artifacts (label 1), images with quality score 3 were labeled as having intermediate motion artifacts (label 2), and images with quality score 1–2 were labeled as having severe motion artifacts (label 3).

We employed a 3-class 2D CNN as the basic network. Based on that, we extended the image size to 512*512 by padding instead of cropping the image size to 400*400 to maintain all features in the images because the respiratory motion artifacts are spread across the whole image as shown in Fig. 2. Data augmentation was performed during training, including vertical flip, horizontal flip, and image rotation, sequentially. The probability of flip was 50% and the degree range of image rotation was ±180°. Batch size of 2 was used as a previous study showed that it was beneficial to the performance [5]. Test time augmentation (TTA) proved to improve the network performance when there are limited datasets [9]. We passed three transformed versions of the same images to the model, whose results were then averaged to predict the final motion levels. Figure 3 shows the architecture of the network. The Cross-entropy was used as the loss function. For training, we used the Adam optimizer. The initial learning rate was set to 1e-4 and decayed by 5% after every epoch.

Cohen's Kappa and accuracy were calculated between the network prediction and ground-truth labels provided by the radiologist to evaluate the performance of our network.

2.3 Segmentation

The images with diagnostic quality (label 1 and label 2 from abovementioned quality control task) were segmented by an experienced radiologist, including contours of LV MYO, and RV.

We first combined a bi-directional convolutional LSTM (Bi-ConvLSTM) and a 2D U-Net based on the network architecture proposed by Bai et al. [10] for the segmentation task. A previous study [11] showed that recurrent neural network (RNN) image was able to capture features in consecutive frames of a cardiac cycle for improved performance. Instead of capturing the temporal continuity across the time frames, we hoped to leverage RNN to capture the spatial continuity in the slice dimension in order to improve the performance. Figure 4 shows the architecture of the network. The loss function consists of cross-entropy loss, DICE loss, and Hausdorff distance loss. We used the Adam optimizer. The learning rate was 1e-4.

Recently, nnU-Net network [6] has shown state-of-the-art performance for segmentation [12]. Based on the original paper, the network automatically configures itself, including preprocessing, network architecture, training, and post-processing for any new task in the biomedical domain. A previous Multi-Centre, Multi-Vendor and Multi-Disease Cardiac Segmentation (M&M) challenge results review paper [13] showed that the top3 teams all employed nnU-Net for LV, MYO and RV segmentation. To fully exploit the limited data, we applied the "insane" data augmentation instead of the default data augmentation in nnU-Net. In comparison, the "insane" version uses enhanced Gaussian noise, Gaussian blur, brightness, contrast, and gamma correction. We used an ensemble of a 2D and a 3D version of the nnU-Net.

To evaluate the performance of the segmentation network, Dice score and 95% Hausdorff distance were calculate between the network outputs and ground-truth LV, MYO, RV contours segmented by radiologist.

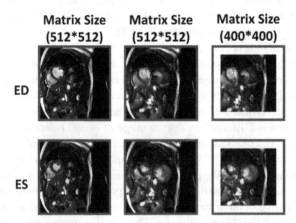

Fig. 2. Cine images from one representative case. From left to right, each column shows mild motion cine images with matrix of 512*512, severe motion cine images with matrix of 512*512, severe motion cine images with matrix of 400*400. We can tell that the respiratory artifacts spread across the whole image (column1 VS. column2) and cropping the image will result in a partial loss of the features (column2 VS. column3).

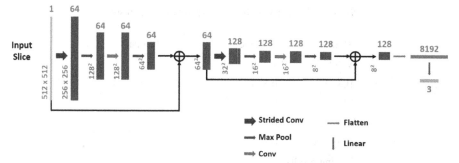

Fig. 3. Architecture of the 2D CNN network used for quality control task.

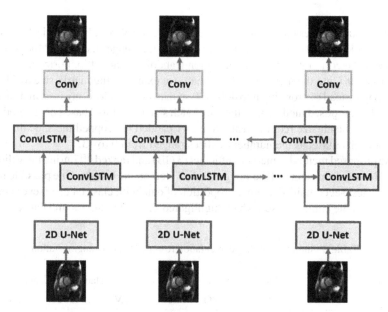

Fig. 4. Network architecture. A bi-directional convolutional long short-term memory network (Bi-ConvLSTM) was combined with a 2D U-Net which extracts features from the time frames. Conv: 2D convolutional layer.

3 Results

The validation and testing results are shown in Table 1 (Quality Control) and Table 2 (Segmentation). Figure 5 shows segmentation results from one representative volunteer from 2 different motion artifact levels: mild motion and intermediate motion. The performance of the nnU-Net is robust to images with different levels of respiratory motion artifacts as it shows similar Dice scores and 95% Hausdorff distances for mild motion images and intermediate motion images (upper two rows VS. lower two rows).

Table 1. Summary of the image quality control results.

Method	Accuracy	Cohen's Kappa
Validation results		
Baseline	0.575	0.262
Method1	0.675	0.421
Method2	0.750	0.569
Proposed	0.800	0.658
Testing results		
Proposed	0.650	0.447

Table 1 summarizes the results of image quality control task. Baseline is the basic 2D CNN network. To unify the matrix size, we cropped the image size to 400*400. Based on baseline, we incorporated data augmentation (random vertical flip, horizontal flip, and image rotation) into the method1. In method2, we extended the matrix size to 512*512 by padding instead of cropping it to 400*400 so that we don't lose any information in the images. The proposed method modified the batch size from 1 to 2 based on method2, and to satisfy the batch size requirement, we either padded or cropped the slice dimension to 12 for each dataset as the number of slices ranged from 9 to 13.

The proposed method is memory efficient (3 GB) and ranked 1st during the validation phase. However, this method did not generalize well on the test set. One possible reason is that the test set contained a larger proportion of data with class label 3 (severe motion artifact) than the validation set, which highlighted the issue of class imbalance.

Table 2 Summary of the segmentation results. DA: data augmentation

Method	Dice score (%)			95% Hausdorff distance		
	LV	MYO	RV	LV	MYO	RV
Validation results						
RNN	87.1	73.9	81.0	10.91	10.05	12.95
nnU-Net	91.8	82.7	89.8	8.30	3.99	5.16
nnU-Net+ "insane" DA	92.2	83.1	89.9	8.21	3.85	5.23
Testing results						
nnU-Net+ "insane" DA	93.4	87.1	92.3	3.42	2.37	3.68

Table 2 summarizes the results of the segmentation task. The nnU-Net outperforms RNN in this task. The "insane" data augmentation further boosted the performance of nnU-Net. The nnU-Net (ensemble of 2D and 3D models) combined with the "insane" data augmentation (enhanced Gaussian noise, Gaussian blur, brightness, contrast, and gamma correction) won 2nd place in the testing phase.

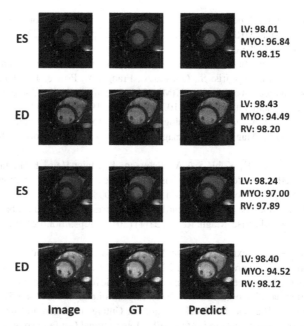

Fig. 5. Segmentation results from one representative volunteer. From left to right, each column shows cine images, ground truth, and prediction of nnU-Net. From upper to lower, the first two rows show results from motion level1 (mild motion) and the last two rows show results from motion level2 (moderate motion). The network performance is robust to different levels of respiratory motion artifacts.

4 Conclusions

In this paper, we proposed a framework that incorporates efficient and automatic image quality control and segmentation based on DL methods. It has the potential to be embedded into clinical routine for improved disease diagnosis and quantitative analysis (masses, EF, etc.). We chose to use the 2D CNN as the basic network for quality control task in order to enable memory efficiency (within 3 GB) so that this model could be easily loaded to any workstation without the necessary of hardware upgrade. The proposed method for quality control ranked first among all the competitive teams in the validation phase. For the segmentation task, the nnU-Net with "insane" data augmentation achieved best performance for robust LV, MYO, and RV segmentation across images with different motion artifacts levels. This method won 2nd place in the segmentation task of the CMRxMotion Challenge.

References

1. Tao, Q., et al.: Deep learning-based method for fully automatic quantification of left ventricle function from cine MR images: A multivendor, multicenter study. Radiology **290**(1) (2019). https://doi.org/10.1148/radiol.2018180513

2. Machado, I.P., et al.: A deep learning-based integrated framework for quality-aware under-sampled cine cardiac MRI reconstruction and analysis, pp. 1–10, 2022, [Online]. Available: http://arxiv.org/abs/2205.01673

3. Zhang, L., et al.: Automated quality assessment of cardiac MR images using convolutional neural networks. In: Tsaftaris, S., Gooya, A., Frangi, A., Prince, J. (eds) Simulation and Synthesis in Medical Imaging. SASHIMI 2016. Lecture Notes in Computer Science, vol. 9968. Springer, Cham. https://doi.org/10.1007/978-3-319-46630-9_14

4. Küstner, T., et al.: A machine-learning framework for automatic reference-free quality assessment in MRI. Magn. Reson. Imaging **53** (2018). https://doi.org/10.1016/j.mri.2018.07.003

5. Isensee, F., Jaeger, P.F., Kohl, S.A.A., Petersen, J., Maier-Hein, K.H.: nnU-Net: a self-configuring method for deep learning-based biomedical image segmentation. Nat. Methods **18**(2) (2021). https://doi.org/10.1038/s41592-020-01008-z

6. Leiner et al., T.: Machine learning in cardiovascular magnetic resonance: basic concepts and applications. J. Cardiovasc. Magn. Reson. **21**(1) (2019). https://doi.org/10.1186/s12968-019-0575-y

7. Wang, S., et al.: The extreme cardiac MRI analysis challenge under respiratory motion (CMRxMotion). arXiv preprint arXIv: 2210.06385 (2022)

8. Wang, C., et al.: Recommendation for cardiac magnetic resonance imaging-based phenotypic study: imaging part. Phenomics **1**(4) (2021). https://doi.org/10.1007/s43657-021-00018-x

9. Shanmugam, D., Blalock, D., Balakrishnan, G., Guttag, J.: When and why test-time augmentation works. In: Proceedings of the IEEE International Conference on Computer Vision (2021)

10. Bai, W., et al.: Recurrent neural networks for aortic image sequence segmentation with sparse annotations. In: Frangi, A., Schnabel, J., Davatzikos, C., Alberola-López, C., Fichtinger, G. (eds) Medical Image Computing and Computer Assisted Intervention – MICCAI 2018. MICCAI 2018. Lecture Notes in Computer Science, vol. 11073. Springer, Cham (2018). https://doi.org/10.1007/978-3-030-00937-3_67

11. Zhang, D., et al.: A multi-level convolutional LSTM model for the segmentation of left ventricle myocardium in infarcted porcine cine MR images. In: Proceedings—International Symposium on Biomedical Imaging, vol. 2018, April 2018. https://doi.org/10.1109/ISBI.2018.8363618

12. Mariscal Harana, J., et al.: Large-scale, multi-vendor, multi-protocol, quality-controlled analysis of clinical cine CMR using artificial intelligence. Eur. Hear. J. Cardiovasc. Imaging **22**(Supplement_2) (2021). https://doi.org/10.1093/ehjci/jeab090.046

13. Campello, V.M., et al.: Multi-centre, multi-vendor and multi-disease cardiac segmentation: the MMs challenge. IEEE Trans. Med. Imaging **40**(12), 3543–3554 (2021). https://doi.org/10.1109/TMI.2021.3090082

Author Index

Printed in the United States
by Baker & Taylor Publisher Services